Modeling of Casting and Welding Processes

Modeling of Casting and Welding Processes

Proceedings of a symposium sponsored by the Solidification Committee of The Metallurgical Society of AIME, the Process Modeling Activity of American Society for Metals, and the Engineering Foundation, held at Rindge, New Hampshire, August 3-8, 1980.

Edited by

HAROLD D. BRODY
University of Pittsburgh
Pittsburgh, Pennsylvania

and

DIRAN APELIAN
Drexel University
Philadelphia, Pennsylvania

A Publication of The Metallurgical Society of AIME

A Publication of The Metallurgical Society of AIME
P.O. Box 430
420 Commonwealth Drive
Warrendale, Pa. 15086
(412) 776-9000

Printed in the United States of America.
Library of Congress Card Catalogue Number 81-83753
ISBN Number 0-89520-380-4

Foreword

This conference is a result of the increasing activity in the modeling of solidification processes and the realization of the value of a sound understanding of manufacturing processes toward increasing productivity and quality. It became apparent that an opportunity was needed for those active in the modeling of casting and welding to establish the current level of capabilities in process modeling and the extent of application of process modeling in casting, ingot making, and welding operations. In addition to establishing the state of the art, the Organizing Committee's goal was to define areas of needed developments in modeling techniques, encourage modeling activity in critical areas, and define potential areas for the application of process models to process control and design and to computer-aided manufacturing.

We felt the goals of the conference were approached closely. We were fortunate to bring together for a week a group of people of diverse backgrounds, each of whom could contribute some expertise to the wide-ranging and lively discussions which crossed many disciplinary boundaries. The written papers presented herein have the advantage of longer consideration than the original oral presentations, but this volume does not capture the content and spirit of the discussion. A few participants summarized their evaluations and impressions during the discussion of the last day of the conference and prepared papers for inclusion in this volume.

The conference was sponsored by the Engineering Foundation, the Solidification Committee of The Metallurgical Society of AIME, and the Process Modeling Activity of American Society for Metals. We are grateful to each organization for its support.

The need for a conference on process modeling was reinforced by the formation within ASM of the Process Modeling Activity under the initial chairmanship of Harris Burte. The interest that evolved from general sessions on process modeling at technical meetings of the Process Modeling Activity indicated the desirability of having a meeting on casting and welding. However, because the backgrounds of individuals that could contribute to the field are broader than would attend the meetings of a single society, we desired a format that would encourage broad participation. The Engineering Foundation Conferences filled that need. The administrative support of the Engineering Foundation, especially Harold Comerer, in organizing the conference and in making the arrangements allowed the Organizing Committee to concentrate on the technical organization of the conference. The need to involve the Solidification Committee of TMS-AIME in a conference of this type is an obvious necessity. The offer of The Metallurgical Society to publish the proceedings was accepted quickly as we considered it important to document the current state of the art in modeling of solidification processes for the benefit of workers in the field who could not attend the conference and to catalyze further participation in the field. The considerable help of John Ballance of TMS-AIME in preparing the proceedings for publication is greatly appreciated.

The participants at the conference expressed a strong feeling that activity in the field would merit a second conference in 2-3 years. They felt the format of this conference contributed to its success and that future conferences should be limited to a group of about 100 individuals active in the field. The participants should cover a broad range of disciplines. Formal and informal discussion should be encouraged by again holding the conference in a secluded location, requiring participants to stay for the full week, and limiting the time alloted for formal presentations to allow more time for discussion.

We have organized the conference and this volume such that the various aspects of modeling of casting and welding processes are addressed in a general and yet coherent manner. The chapter titled "Heat Flow in Castings" includes contributions on computer approaches to calculating and representing heat transfer in sand mold castings, permanent mold castings, and in die castings. Modeling of heat flow in conjunction with mass flow in remelting processes is included in Chapter IV. Similarly, in the chapter titled "Heat Flow in Welding Processes" we have included a tutorial paper focusing on the heat transfer analysis techniques, which is equally applicable to casting processes, as well as contributions on modeling of heat flow in specific processes, vertical electroslag welding, and pulsed GTA welding. Under the chapter titled "Stress and Deformation" are excellent tutorial and review papers which offer models of stresses and deformations in solidifying bodies, including mechanical, physical, and thermal data for modeling solidification processes in steels. In addition, a paper specifically addressing models of stresses and deformation due to welding was included.

The paramount importance of the fluid flow during solidification for the control of segregation and grain structure has been well documented. In the section titled "Fluid Flow in Castings and Ingots" we have collected papers which address the effect of natural and forced convection on macrosegregation including the modeling of vacuum consumable-arc-remelted and electroslag-remelted ingots.

Significant advances in modeling of growth kinetics and attainment of different solidification morphologies have been realized and coherently presented recently as the result of determined work over several years. Credit here goes to the pioneering work of the authors of the papers of this section. The chapter titled "Growth Kinetics and Morphology" contains papers on the simulation of crystal growth mechanisms and the modeling of eutectic and dendritic growth at the limit of stability.

Contributions on computer graphics for heat transfer simulations in castings as well as a status report on geometrical modeling are included in the section titled "Graphics and Geometric Modeling."

Lastly, we have collected the thoughts of several industrial and academic leaders concerning the future directions and needs of modeling of casting and welding processes in the section titled "Future Directions and Needs."

Credit for the success of the conference must go to the diligent and committed Organizing Committee. Many of the committee members also chaired the various sessions during the conference. We strongly felt that a conference on modeling of casting and welding processes was needed and timely, and asked the Organizing Committee to finalize the program in a very short period of time, seven months from start of planning to the conference date. It has been proven before and we have also proven once more that, with the right team, wonders can be achieved.

D. Apelian
Drexel University
Philadelphia, Pennsylvania

H. D. Brody
University of Pittsburgh
Pittsburgh, Pennsylvania

July, 1981

ORGANIZING COMMITTEE

Prof. Diran Apelian
Drexel University
Co-Chairman

Prof. John Berry
Georgia Institute of Technology

Dr. Owen Richmond
United States Steel

Prof. Martin Glicksman
Rensselaer Polytechnic Institute

Prof. Jack Keverian
Drexel University

Dr. Herschel Smartt
Department of Energy

Prof. Harold Brody
University of Pittsburgh
Co-Chairman

Lt. William Baeslack
Wright-Patterson AFB

Dr. Robert Mehrabian
National Bureau of Standards

Dr. William Erickson
Los Alamos Scientific Laboratory

Prof. Merton Flemings
Massachusetts Institute of Tech.

Table of Contents

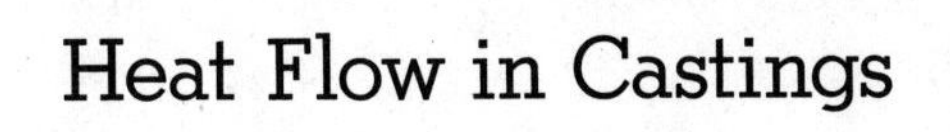

Heat Flow in Castings

SIMULATIONS IN THE DESIGN OF SAND CASTINGS

Robert A. Stoehr

Department of Metallurgy and Materials Engineering
848 Benedum Hall
University of Pittsburgh
Pittsburgh, Pennsylvania 15260

Programs using the finite difference and finite element methods are available to aid the castings engineer in the design of sand mold castings, but further development will undoubtedly extend their usefulness. The desirable attributes of such a program, including simplicity, realism, flexibility, accuracy, stability, and ease of interpretation are often attainable only at the expense of each other or at the expense of the economic justifiability of the simulation. Some compromises to reduce these conflicts are suggested and some instances where currently available programs are being used in the design of production castings are reported.

Introduction

Computer modeling of sand castings can make an important contribution to the economical production of new designs. The objective is to assist the castings engineer in translating the drawings and specifications for a finished part into the proper design of the mold and the proper casting practice, including, as shown in Fig. 1:

1. Placement of gates and risers.
2. Selection of sands and other mold materials.
3. Selection and locations of chills.
4. Allowances for shrinkage, machining tolerances, and drafts.
5. Choice of pouring temperatures, hot-topping procedures, etc.

The program should enable the castings engineer to determine the location of liquidus and solidus, the temperature at any point within the casting, the local cooling rate, the local temperature gradient, and the rate of liquidus and solidus advance, all at appropriate time intervals within the casting process.

With this knowledge, the castings engineer can draw conclusions concerning the likelihood of the formation of voids and porosity, cracks, chill zones, microsegregation, and certain microstructures.

Background

Development of digital computer simulation of the casting process has been going on for some years and application of various forms of the finite difference method to solidification has been described by a number of authors (1-13). In early work, Schniewind (1) developed a method for calculating the movement of the solidification front between nodal points in a pure metal.

Of particular note is the pioneering work of Henzel and Keverian (2) in using a large finite difference mesh (200 to 1000 nodes) to represent the irregular shapes found in real sand castings. They adapted a general purpose transient heat transfer program which solved the implicit form of the finite difference equations by the Gauss-Seidel technique, and they reported good agreement between their calculated and experimentally measured temperatures and cooling times.

Mizikar (3) modeled heat flow in a continuously cast steel slab using a one-dimensional finite difference technique. Lait, Brimacombe, and Weinberg (4) developed a one-dimensional finite difference heat transfer model for continuously cast steel which considered convective mixing and latent heat released throughout the solidification range.

Sciama (5-7) calculated temperature distributions in the solidification of various shapes such as cylindrical bars, plates, spheres, "L"s, "T"s and crosses. Narrone, Wilkes, and Pehlke (8) modeled the solidification of "L"-shape and "T"-shape low carbon steel castings utilizing an artificially raised heat capacitance to account for heat of fusion. Pehlke,Kirt, Narrone and Cook (9) simulated the solidification of silicon brass and aluminum castings with a two-dimensional explicit finite difference method. Kirt and Pehlke (10) used the correlation of computer simulations with experimental results to determine thermal properties of materials and heat transfer coefficients. Jevarajan and Pehlke (11) modeled a casting with a chill.

Brody and Stoehr (12-13) have reported on a system for calculating solidification of very large alloy steel roll castings using a two-dimensional, axisymmetric, finite difference program. This program accommodates up to 20 materials with temperature dependent properties, variable heat transfer coefficients that adjust for the formation of an increasing gap between the metal and mold as solidification progresses, and the ability to simulate the "double pouring" of rolls. This program also features simplified entry of complex two-dimensional meshes, built-in diagnostics that check the input data for internal consistency, and simplified methods for producing pictorial output on a line printer.

Although most of the work in the casting industry has been done by the finite difference method, some people have been looking at the finite element method. Das (14) has advocated the flexibility and versatility of the finite element method for application to the design of castings. Orivuori (15) and Brunch and Zyvoloski (16) reported the use of the finite element method to simulate heat conduction in bars and other shapes. Comini, Del Guidice, Lewis and Zienkiewicz (17) examined nonlinear heat conduction by the finite element method with special reference to a change of phase. They tested their work on the solidification of an infinite slab and a corner region. Morgan, Lewis, Zienkiewicz (18) developed a more numerically stable finite element solution. They incorporated the heat of fusion into the enthalpy, representing it by a smooth curve. The accuracy of using finite element solutions to the transient heat conduction equation has been verified by Donea (19) and Cella (20). Mathew and Brody (21-23) have used the finite element method to model heat flow and thermal stresses in the solidification of axisymmetric continuous castings.

Current Status and the Need for Future Development

As a result of this work, programs are now available, both commercially and from private sources, which can be used for the simulation of the casting process in sand molds. No attempt will be made to catalog them here because this will be done in other papers at this conference. Nevertheless, their use is not widespread in the foundry industry.

At the University of Pittsburgh we have developed several two- and three-dimensional programs for the simulation of sand mold castings using either the finite element (24, 13) or the implicit finite difference method (12, 13). Some designers are using these programs at the present time to test proposed sand mold designs. In order to make the programs more widely useful, we are continuing to develop them with the user's needs in mind.

In this paper we will attempt to outline the things which must be done to make computer simulation a universally accepted tool in sand mold casting design.

In developing a program for modeling the solidification of sand mold castings, one tries to attain certain desirable attributes, commonly including:

1. Simplicity from the user's viewpoint.
2. Realistic representation of the casting and mold design.
3. Flexibility to represent a wide variety of designs.
4. Accuracy of results as indicated by:
    a. Agreement with analytical solutions.
    b. Agreement with experimental results.
5. Stability.
6. Easily interpreted results.
7. Economically justifiable costs.

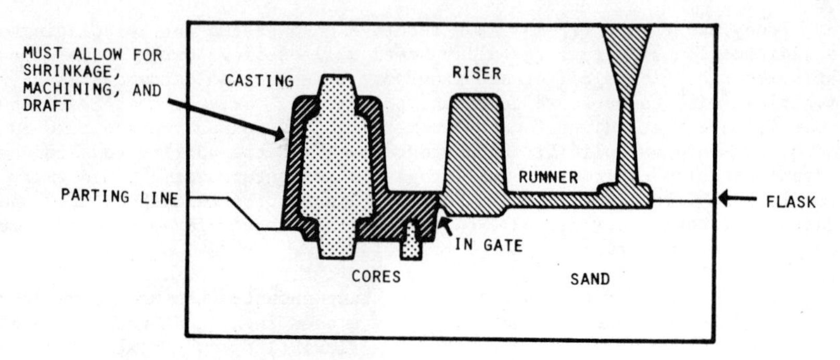

Fig. 1. The typical sand casting to be modeled is a complex assemblage of numerous materials, which dictates to a large degree the complexity of the mesh used to represent it.

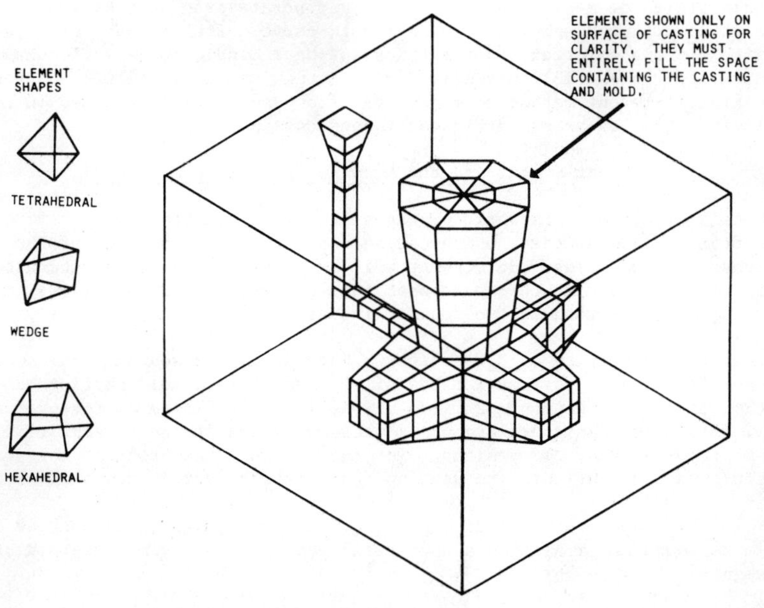

Fig. 2. The casting and rigging must be divided into a mesh consisting of tetrahedral, wedge, or hexahedral-shaped elements when using the finite element or finite difference methods. The mesh must follow the contours of the mold and casting in such a way that no element contains more than one material.

Unfortunately, attaining any one of these attributes is often in conflict with attaining the others.

Simplicity vs. Realistic Representation

In both the finite difference method (FDM) and the finite element method (FEM), geometrically complex castings are represented by dividing the space which contains the mold and casting into a number of subdivisions called elements. The elements themselves must be simple geometric shapes which completely fill the space. In two-dimensions they are usually triangles and/or quadrilaterals. In three-dimensions they are usually tetrahedrons, wedges, and/or hexahedrons. These elements are generally bounded by straight lines and planes (Fig. 2).

The number, exact shape, and location of these elements are generally chosen with two ideas in mind: (1) the boundaries between elements should follow the actual boundaries of the materials as closely as possible (dictated by the requirement that each element should contain only one material), and (2) their spacing determines the spatial resolution of the temperatures to be calculated.

Geometrically speaking, the major difference between the finite element method and the finite difference method is that in the FEM the temperatures are calculated at nodes which are located at the corners of the elements, while in the FDM the temperatures are calculated at nodes which are located at the centers of the elements.

Generating the mesh of elements and nodes is not a trivial chore. Typically, representation of a casting and mold design in two-dimensions will require 30 to 50 rows of elements in each direction for a total of 900 to 2500. Including the third dimension with the same resolution increases the number to 27,000 (for a 30 x 30 x 30 mesh) or to 125,000 (for a 50 x 50 x 50 mesh). Even the 27,000 elements stretch the capability of most computers.

One may ask whether such resolution is necessary. Our experience indicates it is. For example, on large steel cast rolls, a difference of 1-in. in the neck radius of a roll with an overall mold radius of 40-in. makes the difference between a sound casting and a casting with voids at a critical location (Fig. 3). The voids form, of course, when metal in the cope neck freezes before solidification in the body is complete.

The total number of elements may be kept down by making maximum use of any symmetry which exists in the design. Each plane of symmetry obviously can cut the mesh size in half. Even if the plane of symmetry is only approximate it may allow the problem to be solved as two smaller independent problems. If the shape is axisymmetric it reduces a three-dimensional problem to a two-dimensional problem at tremendous savings. The shape shown in Figure 1 contains only one plane of symmetry (parallel to the paper), whereas the shape shown in Figure 3 contains an axis of symmetry. Figure 3 also contains a horizontal plane of approximate symmetry in the middle of the body, permitting the top and bottom halves to be calculated separately.

Mesh generation usually must be done by a person who understands the requirements of both the casting design and the analytical method (FDM or FEM). This person must use intuition, experience, and rules of thumb to generate a mesh which is reasonably efficient and still an accurate representation of the actual mold and casting. There are few such people so far and this has inhibited the use of such systems. Even the commercially available mesh generation systems such as UNISTRUC do not remove the need for such a

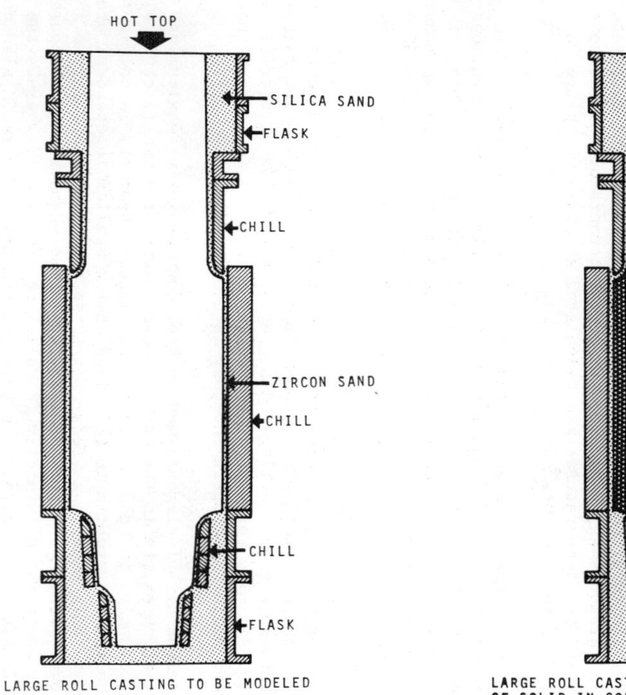

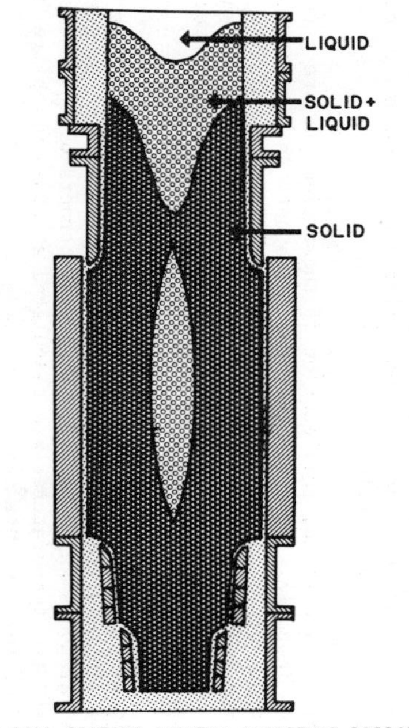

Fig. 3. Large steel roll castings such as shown in this figure have been modeled successfully to predict the occurrence of premature bridging of solid in the neck which results in voids below this region. Such premature bridging is shown in the drawing on the right. Further simulations showed the modifications needed in the design to prevent this type of defect.

person but simply serve as aids which speed the translation of his judgments into numerical data.

It is important for the ultimate success of modeling systems for sand castings that they include improved methods for translating proposed mold and casting designs into the mesh for the solidification calculations. Such systems should make maximum use of digitizers or similar devices to translate points on a drawing into numerical form. They should contain default parameters that will relieve the designer of the task of choosing the element size which is needed to give appropriate resolution.

A system designed to generate the mesh with the minimum input from the user is currently being developed at Battelle-Columbus Laboratories for the University of Pittsburgh and the Army Tank Command (25).

Realistic representation of a casting and mold design also involves the physical properties of the materials. Density, thermal conductivity, and specific heat are temperature-dependent properties. Latent heat of solidification for alloys is given off over a range of temperatures and the temperature dependence is not generally linear between the liquidus and solidus. How closely the simulation follows the physical reality can have a considerable influence on the computer time required. Approximations which retain the linear nature of the heat transfer equations are highly desirable.

Our programs (13) are designed to accept segmented linear representations of these properties (Fig. 4). The length of the segments can be controlled by the user in some versions. This means that, for example, these properties can be represented as constants over the entire solid range or they can be tabulated at certain temperatures throughout the range and the program will perform linear interpolations for temperatures between tabulated values.

Flexibility

Programs can be written to be very general, but often this results in more cumbersome programs with high costs. For example, programs which are written to handle two-dimensional mesh can be more efficient than using a three-dimensional program for the job. Axisymmetric geometry can be handled most efficiently in a two-dimensional program designed to handle this particular geometry. If average physical properties are to be used, it is best to eliminate those features in the program which are designed to accommodate their temperature dependence. Arrays should preferably be dimensioned quite close to the size that will actually be used. This means that if the simulation is written in a language like FORTRAN, it should be preceded by a utility program which operates at the monitor level to make the desired changes and selection of options before loading and compiling it. Here, too, the programs should be written so that it can be used by persons who are not experts at computer programming.

Accuracy

Accuracy of results is influenced by the computational method and by the accuracy of the representation of physical and geometric factors. Errors arising out of the computational method can frequently be evaluated by comparison with analytical solutions for simple shapes. Other errors must be evaluated by comparison with experiments.

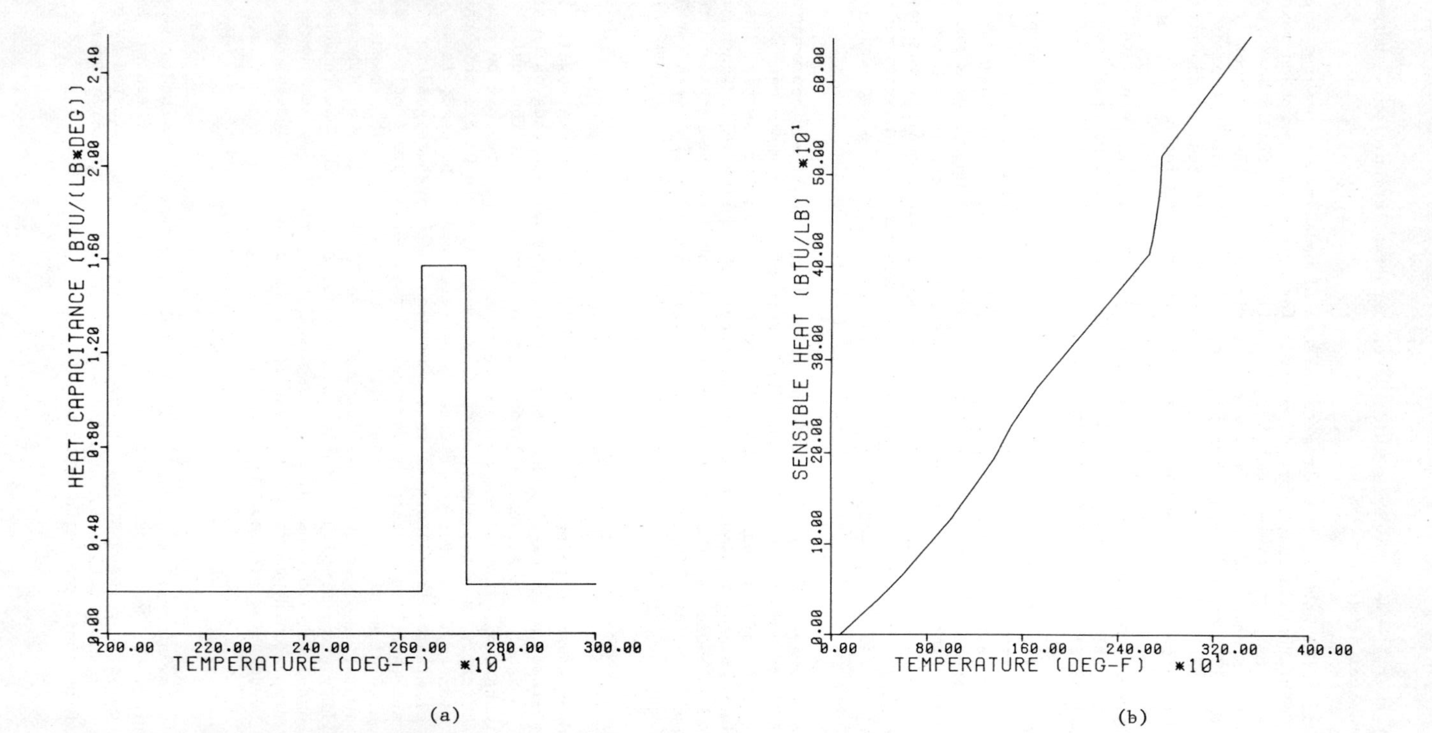

Fig. 4. Two ways of representing the heat capacity and latent heat of a solidifying metal are shown in this figure. (a) An artificially high value of the Cp in the solidification range is used to account for the latent heat of fusion. (b) A segmented approximation of the sensible heat includes the effect of a temperature - dependent specific heat and a latent heat of fusion distributed over the solidification range.

Figure 5 shows comparisons of results generated by our finite element program with results obtained by classic analytical techniques for the heating of a steel bar when one end is brought into contact with a heat source at constant temperature. It is essential that any simulation program pass these tests, but in themselves they are not sufficient.

Agreement with experimental results is a more demanding but more nebulous objective. Many details must be known. Proper representation often requires knowledge of such things as temperature-dependent physical properties, transformation temperatures, and heat-transfer coefficients with more accuracy than has been known before. The modeling program thus gives rise to the need for more experimental work.

The availability of a modeling program also provides an additional way to evaluate the physical parameters needed, so long as the effects of various variables can be separated. For example, the conductivity of molding sands could be evaluated under actual casting conditions if the other parameters could be held constant or could be accurately measured (10). This separation of effects is often not easy and great care should be taken to avoid the application of compensating errors which may force the solution to conform to the experimental results but cause other solutions to be farther from the truth.

Stability

Application of either the finite difference method or finite element method to transient heat transfer problems can introduce oscillations in temperature which are not present in nature and which are in fact prohibited by the laws of thermodynamics. To avoid these problems certain relationships must be observed between the geometric grid spacing (Δx) and the time step (Δt). In the explicit form of the finite difference equations the criterion for stability is $\Delta t \leq \frac{1}{2}\frac{(\Delta x)^2}{\alpha}$ for a one-dimensional mesh (or $\leq \frac{1}{4}\frac{(\Delta x)^2}{\alpha}$ for a two-dimensional mesh, or $\leq \frac{1}{6}\frac{(\Delta x)^2}{\alpha}$ for a three-dimensional mesh, where α is the thermal diffusivity). In the finite element method the criterion for stability is just the reverse and a minimum time step must be specified according to the criterion $\Delta t \geq \frac{1}{4}\frac{(\Delta x)^2}{\alpha}$. This is true in at least some commercially available finite element programs (26) as well as our own (24). It may arise from the use of an explicit finite difference step in the time domain along with the finite element representation in the geometric domain.

These criteria become particularly restrictive when the mesh has been drawn to represent a mold and casting design containing several materials with widely differing α's, and where the values of Δx have been dictated by their actual thickness in the metal.

The implicit form of the finite difference equations is inherently stable and so values of Δt and Δx may be chosen independently. A price is paid for this in two ways: (1) The heat transfer equations must be solved as a set of simultaneous equations instead of consecutively. (2) The model behaves as though it is completely damped and so it is more sluggish than the real system.

Solution of the heat transfer equations as a set of simultaneous equations is not as laborious as it may appear at first, since only the coefficients representing heat transfer between nearest neighbor elements are needed. Thus if the system contains 1000 elements, the 1000 equations needed may be represented by two 1000-element column arrays T and B (representing

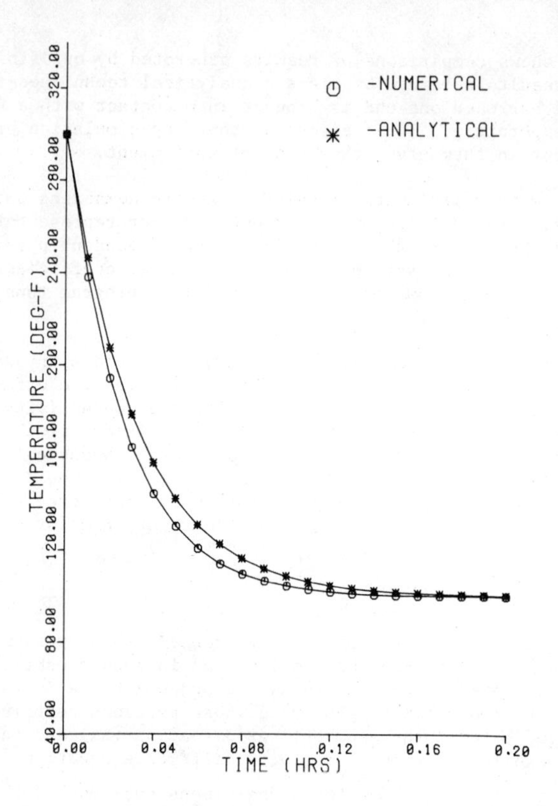

Fig. 5. A comparison of the analytical and finite element solutions for convective cooling of a fin are shown above. It is important that any system for modeling the solidification of castings should be verifiable in this way.

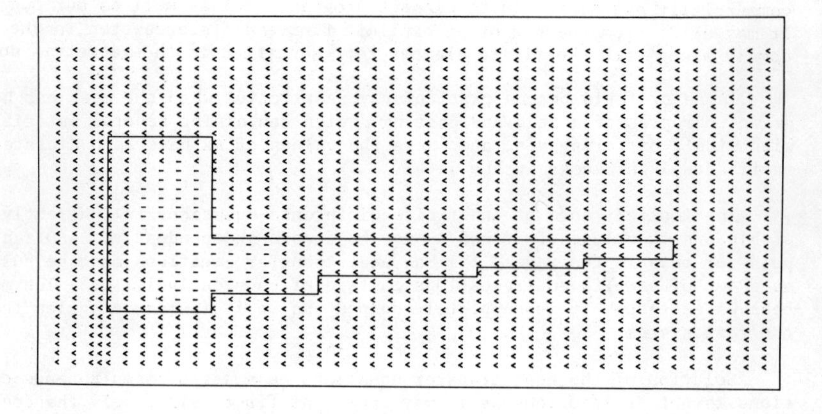

Fig. 6. Location of material between liquidus and solidus is shown on this plotten drawing by the dashes (-). Material below the solidus is shown by the left caret (<).

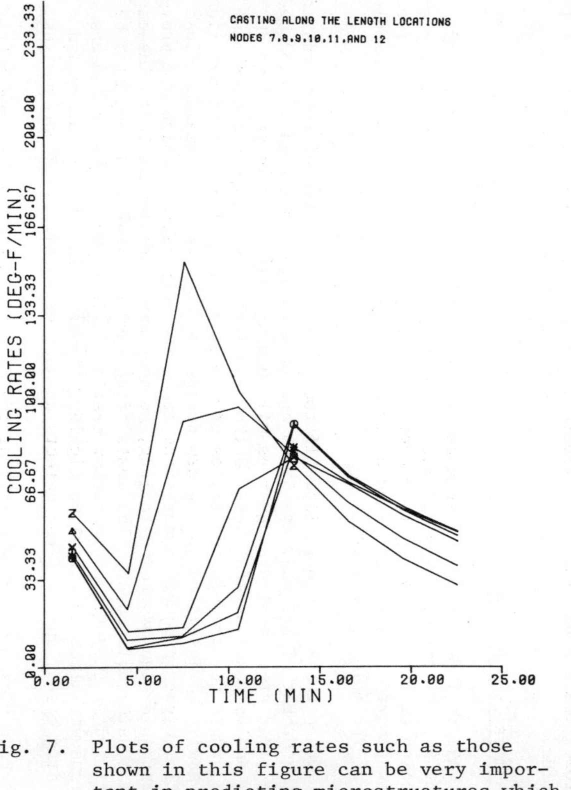

Fig. 7. Plots of cooling rates such as those shown in this figure can be very important in predicting microstructures which will form in castings.

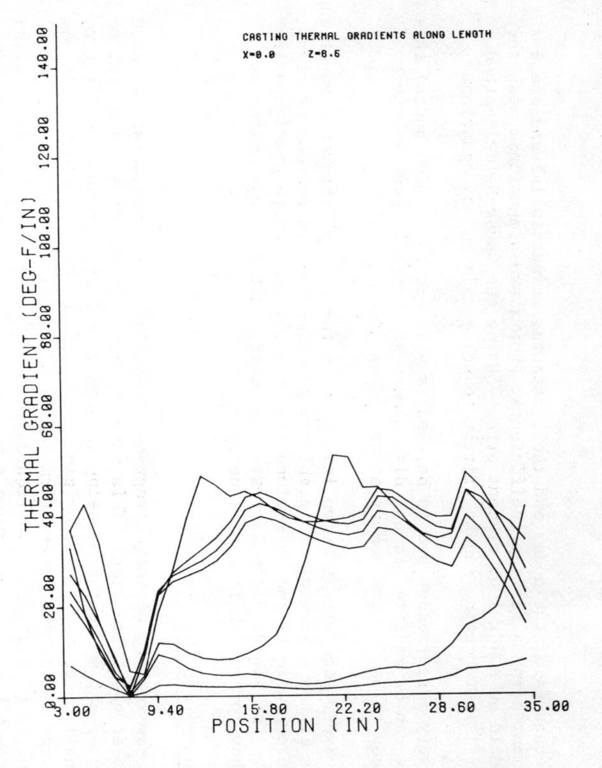

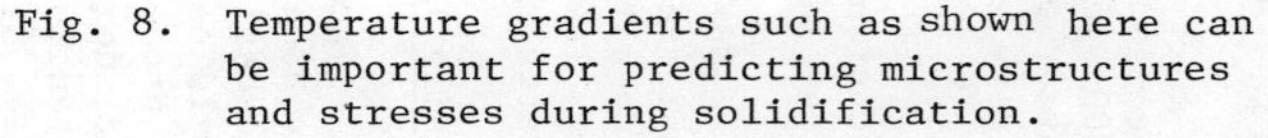

Fig. 8. Temperature gradients such as shown here can be important for predicting microstructures and stresses during solidification.

the temperatures of the nodes and the constants, respectively) and one 6 x 1000 element array A for the coefficients. For a two-dimensional mesh the latter would be a 4 x 1000 element array. Using the Gauss-Seidel method, a very efficient routine can be written to solve this set of equations.

If very large meshes are to be used (say in excess of 3000 nodes) this technique can be modified to use disc memory so that only part of these arrays need to be in core memory at any given time.

Our experience with the implicit form of the finite difference equations indicates it has absolutely no stability problems even in regions of high temperature gradients and in regions in the mushy zone where gradients are very low. The finite element program, however, must obey the criterion stated above or instability will develop.

Easily Interpreted Results

The raw results of these programs are in the form of temperatures at the nodes at each time step. This form of output is not directly useful to the designer and so post-processing is used to put it into more useful form. It is usually best to put the original output file onto tape so that it may be processed later to whatever form the user desires. Graphical outputs (24) include plots of isotherms and liquidus/solidus locations (Fig. 6), cooling rates through certain temperature ranges (Fig. 7), and temperature gradients (Fig. 8). The possibility of utilizing plots produced on a line printer may provide the lowest cost method at some installations (Fig. 9). The ratio of temperature gradient to rate of liquidus or solidus advance (the G/R ratio) can be calculated and displayed as an aid to predicting both macro- and microsegregation and the microstructure which will be found in parts of the casting.

Economically Justifiable Costs

All of the factors listed above contribute to the overall costs of using a casting simulation program. Computer costs may be a relatively small fraction of the overall costs. Overall costs of a simulation may run from a few hundred to a few thousand dollars, depending primarily on the complexity of the mold and casting.

These costs are justifiable if they reduce the number of trial castings or the number of bad castings which are made when a new design is produced. Some large steel castings, such as mill backup rolls, may weigh 40 to 80 tons and be worth several hundred thousand dollars each. A producer may take orders for them in quantities of one at a time. The cost of modeling may easily be justified when a new design is required. For smaller, less costly, higher production parts the lead time required to design and prove a certain design may be quite long, so that any reduction in this time can save considerable money. The costs of performing simulations may also be justified if they result in better casting designs, where "better" means designs with less conservative rigging, higher yields, lower energy and cleaning costs, and/or fewer defects and better metallurgical properties.

Foundries may find the cost of the modeling system to be more attractive if the same computer and graphics installation can be used for drafting, estimating costs, quality control, material tracking and record keeping, so that the total cost may be distributed over more functions.

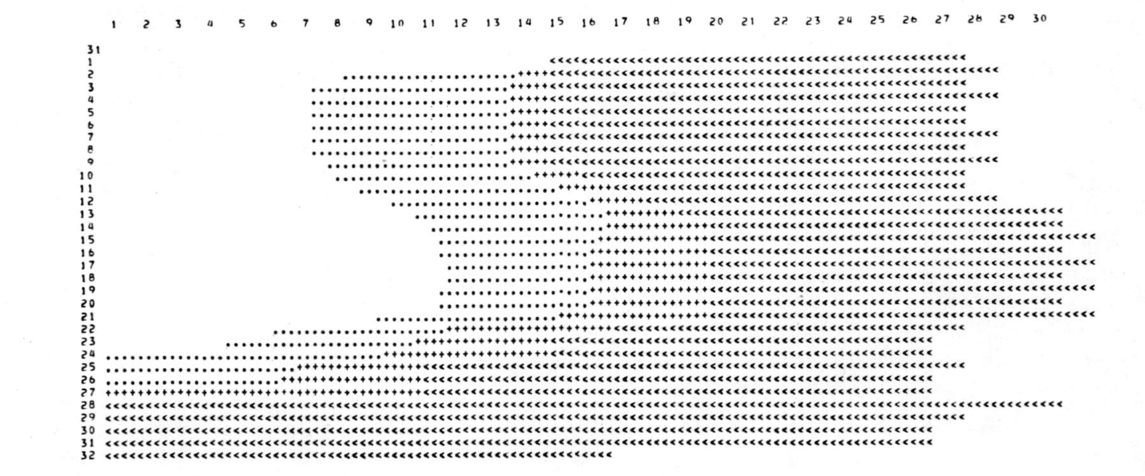

Fig. 9. A line printer or terminal can be used to produce graphical output as shown in this representation of a roll casting similar to those shown in Fig. 3. The center line of the casting is at the left. Vertical distances have been condensed much more than horizontal distances because horizontal resolution is much more important in predicting the soundness of these castings. Symbol key used in this output: liquid -- blank; mush -- period (.); solid -- plus (+); mold material -- left caret (<).

Conclusions

Currently available systems provide simulations of the solidification of sand castings which are useful to designers in their present form. The process of making them more acceptable to the user continues to go on along the lines discussed in this paper. These improvements involve both the interface with the user and the internal efficiency of the programs, both of which are necessary to expand the number of applications where the use of these models may be justified economically.

Acknowledgements

Work at the University of Pittsburgh described in this report has been performed primarily with financial support from the U. S. Army Tank and Automotive Research and Development Command, Contract No. DAAK30-78-C-107, and the Roll Manufacturers Institute, Pittsburgh, PA, Project Nos. 41, 41A and 41B.

The CalComp Plotter figures (Figs. 4 - 8) were produced by Walter Schwarz as part of his M.S. thesis at the University of Pittsburgh (24).

References

1. Schniewind, J., "Solution of the Solidification Problem of a One-dimensional Medium by a New Method", Journal of the Iron and Steel Institute, Vol. 201 (July, 1963), pp. 394- .

2. Henzel, J.G., Jr., and Keverian, J., "The Theory and Application of a Digital Computer in Predicting Solidification Patterns", Journal of Metals, Vol. 17 (May, 1965), pp. 561-568.

3. Mizikar, E.A., "Mathematical Heat Transfer Model for Solidification of Continuously Cast Steel Slabs", Trans. Met. Soc. of AIME, Vol. 239 (1967), pp. 1747-1753.

4. Lait, J.E., Brimacombe, J.K., and Weinberg, F., "Mathematical Modeling of Heat Flow in Continuous Casting of Steel", Ironmaking and Steelmaking, Vol. 1 (1974), pp. 90-97.

5. Sciama, G., "Computation of Cooling Time in Permanent Mold Cast Cylindrical Bars - Checking Tests", Cast Metals Research Journal, Vol. 4 (1968), pp. 62-68.

6. Sciama, G. and Jeancolas, M., "Temps de Solidification de Pieces Elementaires Coulees en Sable", Fonderie, Vol. 303 (July, 1971), pp. 239-250.

7. Sciama, G., "Etude de la Solidification de Profils Types a Points Chauds en Fonte", Fonderie, Vol. 306 (Nov., 1971), pp. 363-370.

8. Marrone, R. E., Wilkes, J.O., and Pehlke, R.D., "Numerical Simulation of Solidification", Parts I and II, Cast Metals Research Journal, Vol. 6 (Dec., 1970), pp. 184-192.

9. Pehlke, R.D., Kirt, M.J., Marrone, R.E., and Cook, C.J., "Numerical Simulation of Casting Solidification", Cast Metals Research Journal, Vol. 9 (June, 1973), pp. 49-55.

10. Kirt, M.J. and Pehlke, R.D., "Determination of Material Thermal Properties Using Computer Techniques", Transactions of American Foundrymen Society, Vol. 81 (1973), pp. 524-528.

11. Jevarajan, A. and Pehlke, R.D., "Computer Simulation of a Casting with a Chill", Transactions of American Foundrymen Society, Vol. 84 (1976), pp. 647-652.

12. Stoehr, R.A. and Brody, H.D., Solidification Contours of Selected Rolls, The Roll Manufacturers' Institute, Pittsburgh, Pa. (1976).

13. Brody, H.D. and Stoehr, R.A., "Computer Simulation of Heat Flow in Casting", Journal of Metals Vol. 32, No. 9 (Sept. 1980) pp. 20-27.

14. Das, P.K., "Finite Element Analysis Method in the Design of Castings", Transactions of American Foundrymen Society, Vol. 13 (1979), pp. 597-600.

15. Orivuori, S., "Efficient Method for Solution of Nonlinear Heat Conduction Problems", International Journal for Numerical Methods in Engineering, Vol. 14 (1979). pp. 1461-1476.

16. Brunch, J.C. and Zyvoloski, G., "Transient Two-dimensional Heat Conduction Problems Solved by the Finite Element Method", International Journal for Numerical Methods in Engineering, Vol. 8 (1974), pp. 481-494.

17. Comini, G., Del Guidice, S., Lewis, R.W., and Zienkiewicz, O.C., "Finite Element Solution on Non-linear Heat Conduction Problems with Special Reference to Phase Change", International Journal for Numerical Methods in Engineering, Vol. 8 (1974), pp. 613-624.

18. Morgan, K., Lewis, R.W., and Zienkiewicz, O.C., "An Improved Algorithm for Heat Conduction with Phase Change", International Journal for Numerical Methods in Engineering, Vol. 12 (1978). pp. 1191-1195.

19. Donea, J., "On the Accuracy of Finite Element Solutions to the Transient Heat Conducting Equation", International Journal for Numerical Methods in Engineering, Vol. 8 (1974), pp. 103-110.

20. Cella, A., "On the Accuracy and Stability of the Finite Element Approximation for Parabolic and Hyperbolic Operations", Conf. on Mathematics of Finite Elements and Applications, Brunel University, 1975 (New York: Academic Press, Inc., 1976), pp. 183-189.

21. Mathew, J. and Brody, H.D., "Analysis of Heat Transfer in Continuous Casting Using Finite Element Method", Computer Simulation for Materials Applications, Nuclear Metallurgy, R.J. Arsenault, et al. (eds.), Vol. 20, Part 2 (1976), pp. 1138-1150.

22. Mathew, J. and Brody, H.D., "Simulation of Thermal Stresses in Continuous Casting Using a Finite Element Method", Computer Simulation for Materials Applications, Nuclear Metallurgy, R.J. Arsenault, et al. (eds.), Vol. 20, Part 2 (1976), pp. 978-990.

23. Mathew, J. and Brody, H.D., "Simulation of Heat Flow and Thermal Stress in Axisymmetric Continuous Casting", Solidification and Casting of Metals, The Metals Society (London, 1979), pp. 244-249.

24. Schwarz, W., Simulation of Heat Flow During Solidification, M.S. Thesis, University of Pittsburgh (1980).

25. Badawy, A., Schreiber, K. and Akgerman, N., "CSTMS3, A Finite Element Mesh Generator for Castings", paper presented at this conference.

26. ANSYS User Information Manual, Control Data Corporation, 1978.

THERMAL MODELING OF A PERMANENT

MOLD CASTING CYCLE

John W. Grant

Engineering & Research Staff
Ford Motor Company
Dearborn, Michigan

Permanent molds incorporating active cooling systems have significant usage in the casting of aluminum automotive components. There is a considerable potential for improving the productivity of the process by thermal modeling if it can be shown that modeling accurately predicts the temperatures and heat flows in operating molds. The complete operating cycle of a permanent mold system used to cast experimental aluminum brake drums has been modeled. Mold temperatures and cooling system heat flow predictions of the model were verified by comparison to actual values measured on the operating mold. Locations of potential macroshrinkage in the casting as predicted by the model agree with those observed in actual castings. The observed agreement between the model and the actual mold system gives confidence in the use of these techniques as mold design and analysis tools.

INTRODUCTION

Permanent mold casting is generally characterized by the use of steel or iron molds into which the metal is introduced by the action of gravity or low pressure air. Its use in nonferrous casting production is differentiated from high pressure die casting by the relatively low gate velocities of the metal. Because of these low velocities it is often selected for the casting of components requiring high structural integrity or the use of sand cores for complex internal passages. The process is currently used in the casting of such aluminum automotive components as wheels and cylinder heads.

When viewed as a total production process, a permanent mold can be a rather complex system. The molds are operated on a cyclic basis and remain at elevated temperature from cycle to cycle. The lengths of the casting cycles for many commercial applications are typically in the 2-10 minute range. Active mold cooling systems are frequently employed with air or water used as the cooling media, to remove the heat of the solidifying casting. Specific cooling locations within the mold are usually selected to obtain directional solidification in the casting, thereby avoiding the formation of shrinkage voids. Gas or electric heaters are sometimes used to further affect the solidification direction and to prevent premature freezing in some areas of the mold.

As production molds are often complex and constructed of hardened tool steel (typically H-13), mold heat flow design by trial and error although commonly done, can be expensive in terms of lost time and mold modification costs. There is then considerable potential for thermal modeling as a tool in mold design. Prior to its acceptance as such a tool, however, modeling must be shown to be reasonably accurate in terms of predicting the time variant temperature and heat flow fields in a complete production type permanent mold system.

In order to evaluate the predictive capability of standard thermal modeling techniques as applied to the permanent mold process, an operational mold system was modeled and the models' predictions then verified by comparison to data obtained from the actual mold.

Mold System

The particular mold which was modeled is one which had been used to cast experimental aluminum brake drums such as those shown in figure (1). The drums were cast in a single cavity mold using the low pressure process. Figure (2) shows a cross section of the mold and casting system. The shaded area of the figure represents the aluminum in the casting cavity and the feed tube (G) from the low pressure furnace. The letters (A) through (E) denote circumferential air cooling passages in the mold. A "fountain" or spot type cooling passage (also air) is located in the sprue pin and is designated by the letter (F). A gas heater is applied continuously to the feed tube, in the area denoted by the letter (H), to maintain the aluminum in a liquid state below the ingate. The mold is constructed of H-13 steel.

At the start of the low pressure casting machine cycle, the mold is closed by a hydraulic ram and the furnace is pressurized (to 8-10 psia), forcing aluminum up the feed tube and into the casting cavity. Pressure is maintained on the metal during casting solidification to feed shrinkage. After casting solidification is complete, furnace pressure is released, the mold is opened and the casting ejected. While the mold is open, it is checked and cleaned of any debris. A typical cycle for this particular

Figure 1 - Brake Drum Castings

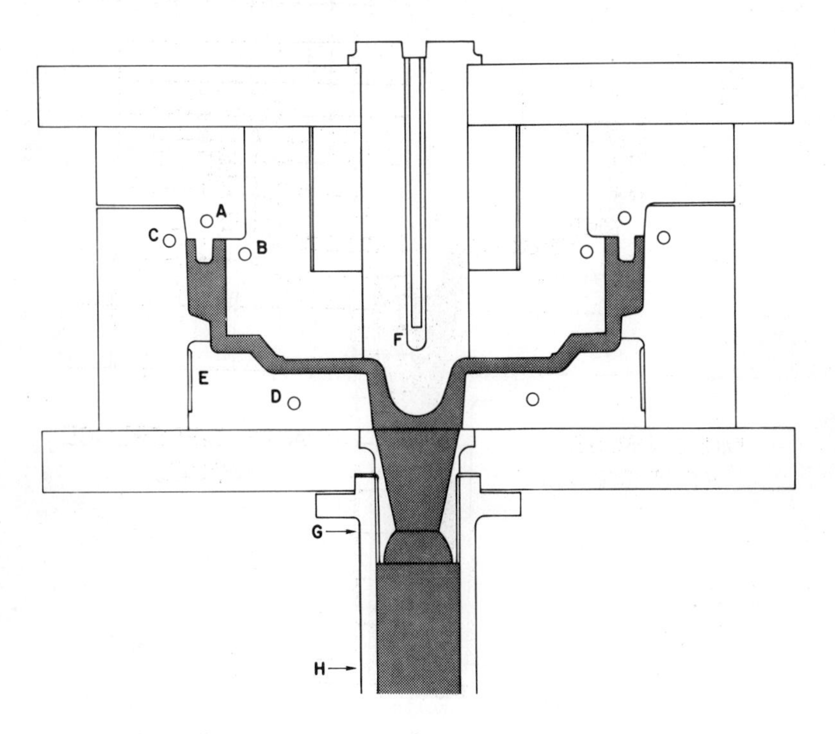

Figure 2 - Low Pressure Brake Drum Mold Cross Section

casting is of 135 seconds duration and consists of a 90 second closed mold period and a 45 second open mold period.

Mold/Casting Thermal Model

Finite Difference Model

To model the time-variant temperature and heat flow in the mold/casting system, the finite difference (or lumped parameter) method of heat transfer analysis was used. As this technique is well documented in the literature, (i.e., references (1), (5) and (6)) only a general comment on its methodology will be rendered here. In the application of this technique the physical system to be modeled is conceptually divided into a mesh of small volume elements or "nodes". Each node is then represented as a concentrated mass or thermal capacitance connected to neighboring nodes by finite thermal resistances. The temperature change of each node is then calculated over finite time intervals. Figure (3) shows the particular nodal mesh arrangement used in modeling the brake drum mold and casting. The system was treated primarily as a two dimensional axisymmetric segment. An exception to the two dimensional assumption was imposed in the area of the casting where the brake drum's peripheral cooling

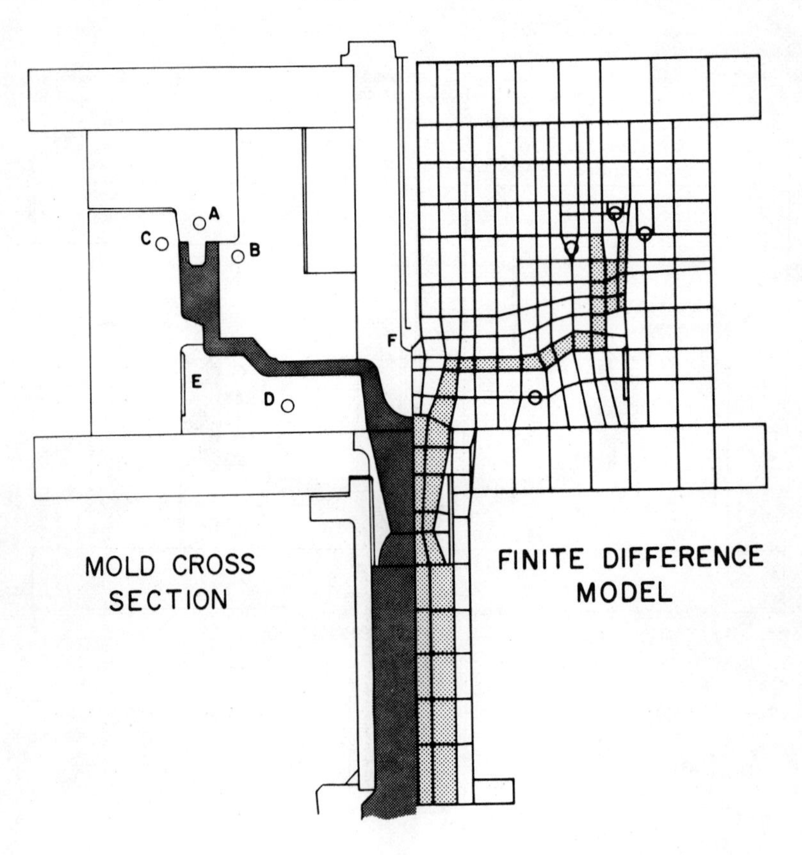

Figure 3 - Brake Drum Mold Finite Difference Model

fins are located (see figure (1)). In this area, circumferential heat transfer was permitted between the fins and their mold images. A total of 357 internal and surface nodes are used in the model.

The finite difference solution of the heat transfer in the system was done using a general purpose heat transfer program, TRUMP(1)*. This code was selected because of several features desirable in the modeling of general casting solidification. Among these are the abilities to simulate constant temperature latent heat rejection, temperature variant material properties, and three dimensional shapes.

The closed and open mold aspects of the complete casting cycle were simulated by adjusting the model composition to account for the characteristics peculiar to each of the two parts of the cycle. An instant fill of the aluminum nodes was assumed at the start of the closed mold period. During the open mold period, the aluminum nodes were removed from the model, and the mold cavity nodes were allowed to exchange heat with the environment via the appropriate boundary conditions.

Boundary and Initial Conditions

Boundary conditions imposed on the model to simulate the complete mold system are indicated in figure (4). Cooling air passage turbulent heat transfer coefficients were calculated based on appropriate Reynolds numbers using correlations from references (2) through (4). Individual passage Reynolds numbers were obtained from actual air mass flow rates in the operating mold system. Mean air temperature in each passage was used to specify air properties and boundary temperatures. Table I lists the heat transfer coefficients used in the model. Cooling in the fountain cooler (F) is characterized principally as impinging type heat transfer. That is, the heat transfer coefficient is a maximum where the air jet impinges on the surface and decays significantly with distance(3),(4). This decay was included in the model's boundary conditions, the maximum and minimum values being shown in Table I.

TABLE I - Cooling Passage Heat Transfer Parameters

Passage	Heat Transfer Coefficient BTU/Hr - ft^2 - F°	Mean Gas Temperature °F
A	68	175
B	63	170
C	69	185
D	63	190
E	55	255
F	200 - 90	140

The exterior surfaces of the mold reject heat to the environment by both natural convection and radiation. For the normal operating temperatures of the mold (500-600°F) both are of about equal significance. Natural convection coefficients for the mold surfaces were calculated using the correlations of reference (5). Greybody radiation to the environment was assumed for the mold surfaces with a metal emissivity of 0.85(6) and a geometric shape factor of unity.

* The computer code TRUMP was developed at Lawrence Livermore Laboratory, University of California. A copy of it was obtained from the Radiation Shielding Information Center at the Oak Ridge National Laboratory for this work.

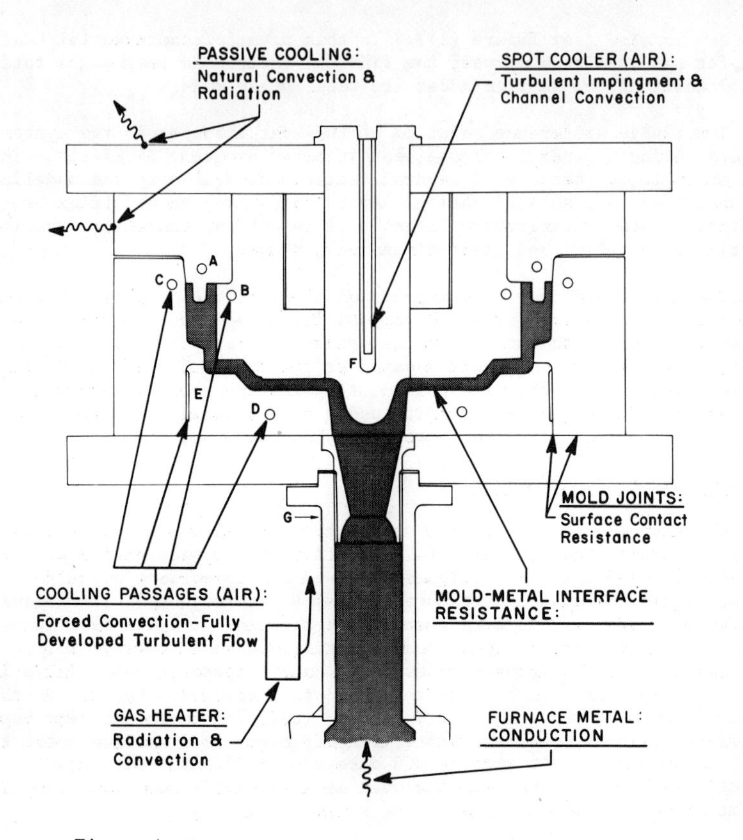

Figure 4 - Boundary Conditions: Brake Drum Mold

Although not boundary conditions in the strict mathematical sense, mold joint resistances and mold-metal interface resistance are grouped in figure (4) as they are modeled by heat transfer coefficients similar to boundary conditions. In formulating the nodal connections for use in TRUMP, the user is allowed to include discrete thermal resistances between the nodes. To model the effects of mold joints then, contact resistance coefficients(6) typical of metal-to-metal joints were included in the appropriate nodal connections in the model. The mold-casting interface coefficient was similarly included in the casting to mold nodal connections. A constant value of 500 BTU/hr-ft^2-°F was selected for the interface coefficient based on typical values reported in the literature(7),(8),(9) for similar applications.

With respect to the specification of an initial condition for the heat transfer model of a permanent mold casting process, one must consider the overall nature of the process. As indicated schematically in figure (5) it is characterized thermally by a starting transient period, followed by quasi-steady state operation. During the starting transient, cyclic average mold temperatures are raised over time to the quasi-steady state range. As the starting transient faces, the mold temperatures oscillate about their respective quasi-steady state values through each casting cycle. This cyclic variation during the steady operation corresponds to

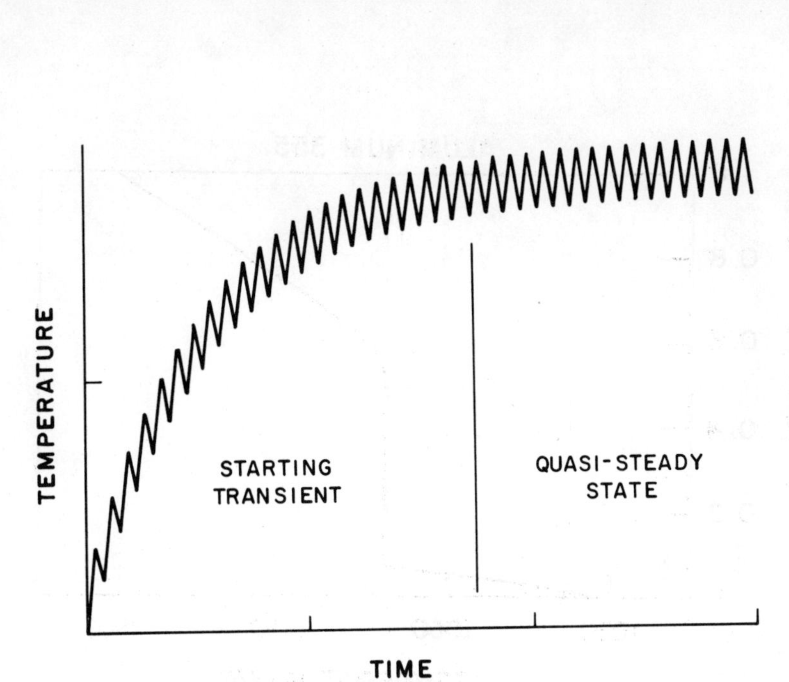

Figure 5 - Behavior of Mold Temperature in Permanent Mold Casting Operatings

typical production conditions. This then is generally the period that is of interest for modeling rather than the starting transient. However, in most cases, it is necessary to model the starting transient by repeatedly cycling the model in order to develop the correct steady temperature levels. For this particular mold, about ten cycles are required in order to bring average mold temperatures to within 15°F of their steady values. This is analogous to the start-up period in the actual casting operations in which the mold temperature is raised by heaters and/or repeated casting cycles.

Casting Alloy Properties

The particular aluminum alloy used for the brake drum castings was A355. Properties for this alloy, as used in the model, are shown in figures (6) through (8). As seen from figure (6), the total latent heat of the alloy (191BTU/lb) is released over a considerable temperature range(10). In the model the temperature variant latent heat release at either end of the curve was accounted for by using a variable specific heat spike (pseudo specific heat method). The constant temperature portion of the curve (1070°F) was calculated using TRUMP's capability to model eutectic type heat release as referred to above. The temperature variant nature of both the conductivity and actual specific heat as shown in figures (7) and (8) were included in the modeling (11).

The model was run on a DECsystem 10 computer and required about 90K words of core storage. Of this, about 15K is used for the program and required system routines, the remainder being used for variable storage. To simulate the complete 135 second casting cycle, about 25 minutes of central processor time (CPU) was required. The main factor accounting for the CPU time is a reduced time stepping imposed by the

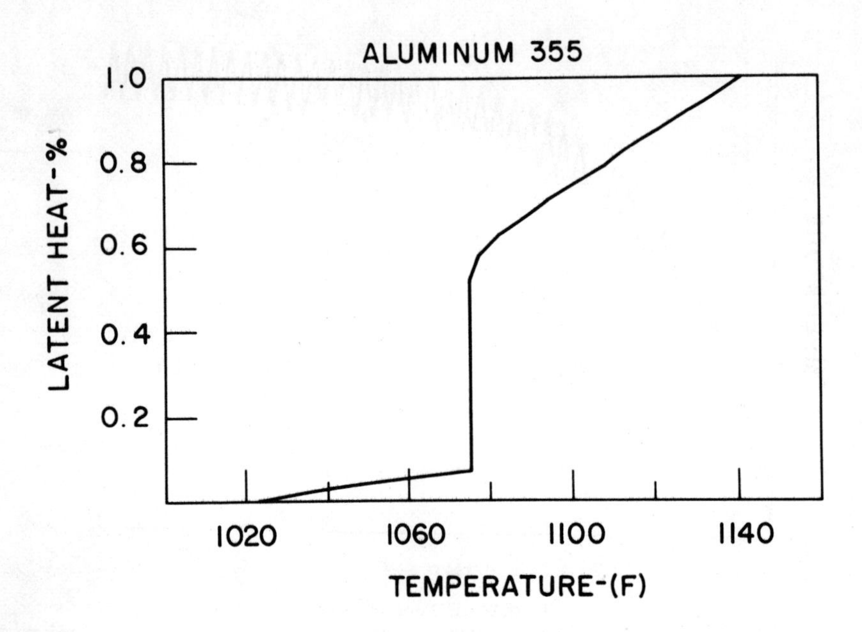

Figure 6 - Latent Heat Distribution: 355 Aluminum Alloy

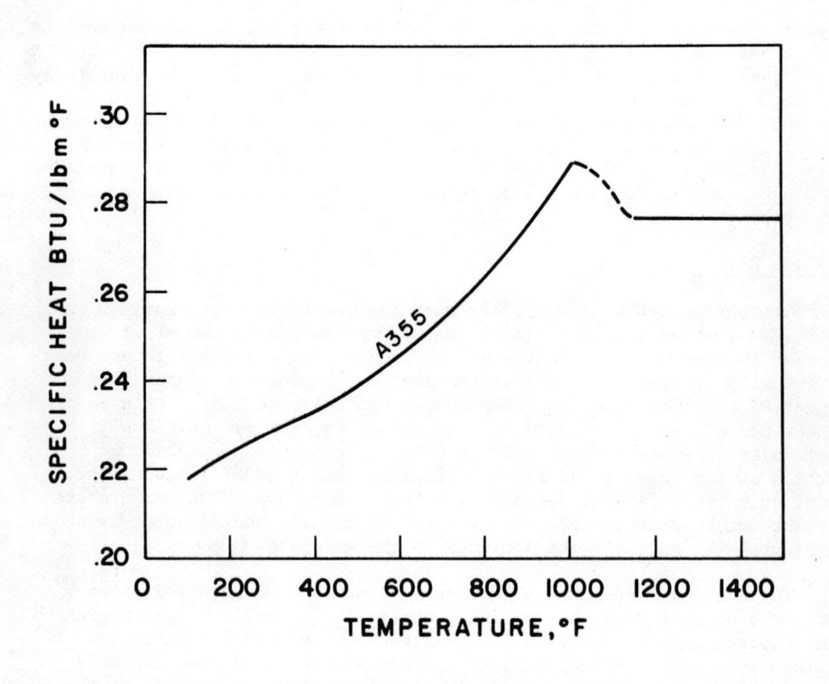

Figure 7 - 355 Aluminum Alloy Specific Heat

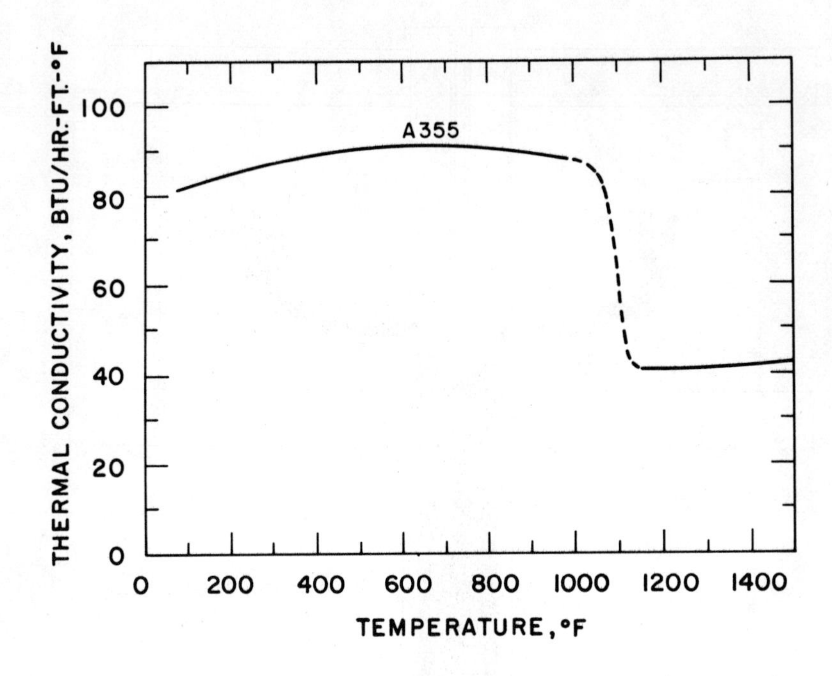

Figure 8 -355 Aluminum Alloy Thermal Conductivity

program when going through dramatic property changes such as those in the liquidus-solidus range. It was found that the CPU time can be reduced by a factor of ten by using constant properties.

Inasmuch as only about 50% of the A355 alloy's latent heat (see figure (6)) is rejected at constant temperature, the use of constant properties might be expected to affect the overall accuracy of the simulation. However, the magnitude of this effect has not been established.

Predicted and Experimental Results

Experimental Procedures

The thermal model as described predicts mold and casting temperature as a function of time during the casting cycle. In order to verify the model's simulation and increase confidence in the predictive techniques, temperatures in the actual system were measured and compared to predicted values. Figure (9) shows the locations of thermocouple installed in the brake drum mold for this purpose. Thermocouples were installed at seven different radii in the mold as indicated by the triangular symbols in the figure. The numbers shown adjacent to the symbols are for identification and will be referred to in later discussion. In order to check circumferential temperature variation, duplicate thermocouples at equal radial distances were installed at several locations around the mold. Type K (chromel-alumel) thermocouples encased in 1/16 inch, grounded, stainless steel sheathes were employed. Access holes to the mold interior for the thermocouples were cut into the mold by electric discharge machining (EDM).

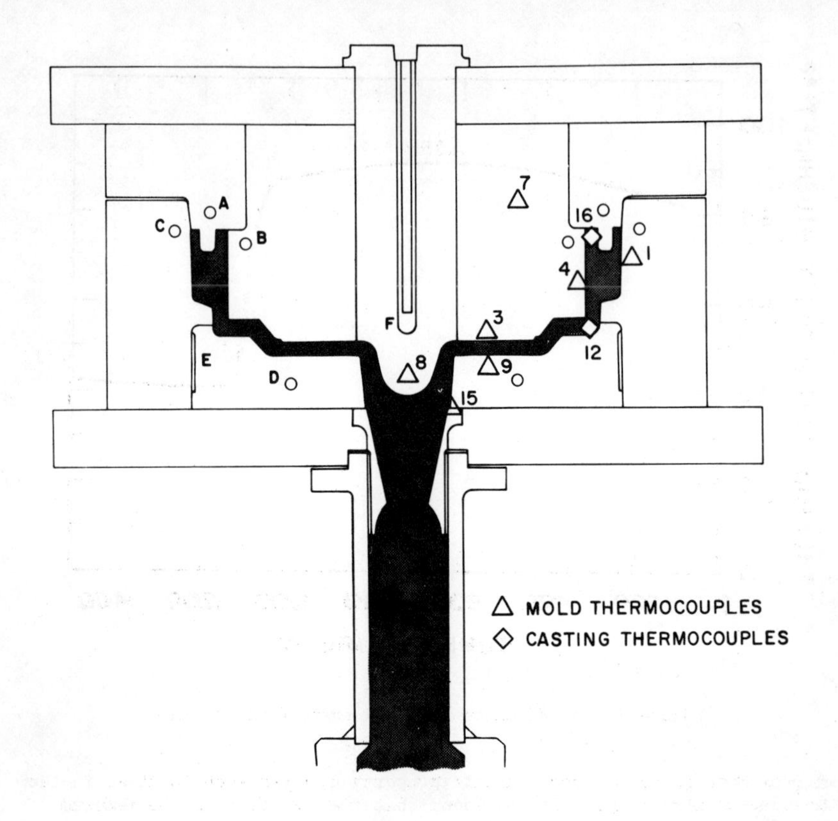

Figure 9 - Brake Drum Mold Internal Thermocouple Locations

As indicated by the diamond shaped symbols in figure (9), additional sheathed thermocouples were extended through the mold and into the casting cavity to measure casting temperature. These were positioned so as to prevent destruction during casting ejection and could be used over many casting cycles. Mold coating was applied to the sheathes except at the tip to protect them from attack by the molten aluminum and to aid in their release from the castings during ejection. Temperature data was recorded at two second intervals using a Fluke Data Logger system.

In addition to the instrumentation indicated on figure (9), data to establish the heat pickup by the individual air cooling passages was obtained during casting operations. Cooling air mass flow rates and temperature rise data for each passage was recorded for this purpose.

Results

Results of both the modeling and experimental efforts are shown in figures (10) through (17) in the form of temperature versus time plots during a typical casting cycle. Each plot corresponds to one of the thermocouple locations shown in figure (9). The particular data shown was taken after the mold had warmed up to operational or quasi-steady state temperature levels.

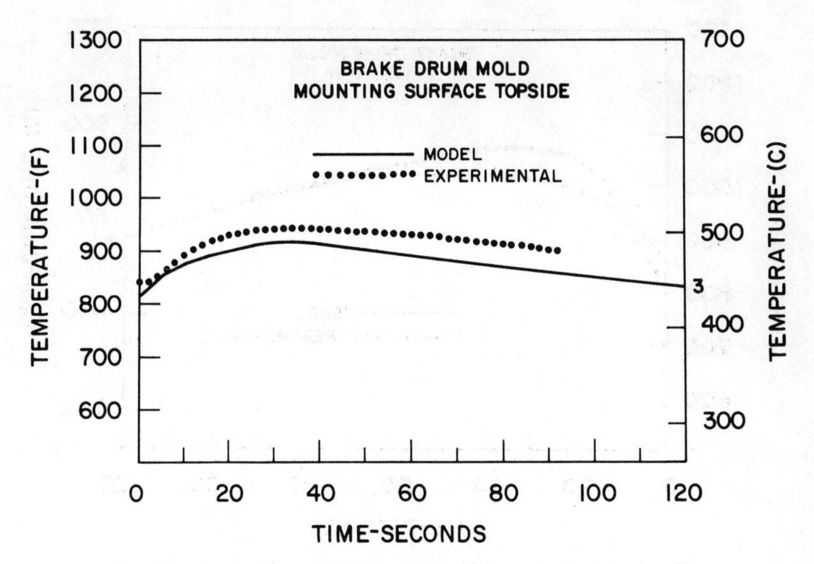

Figure 10 - Mold Cyclic Temperature: Thermocouple #3

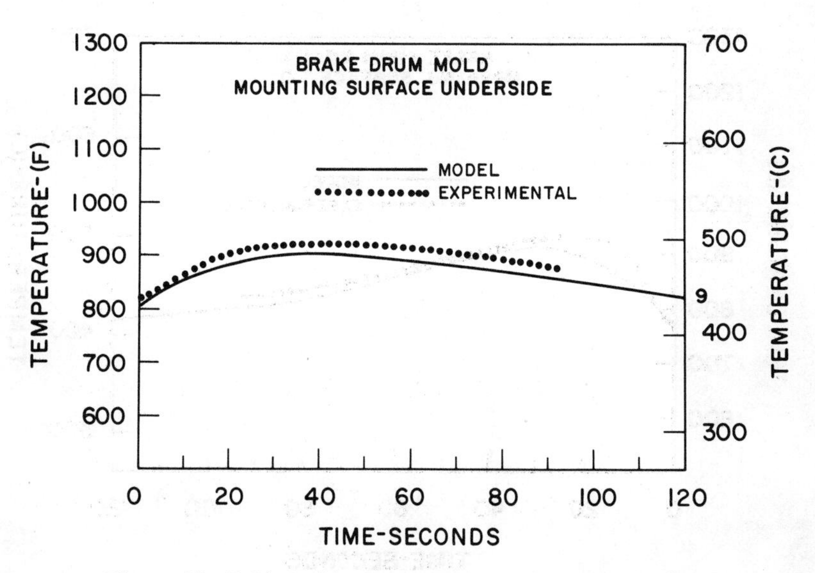

Figure 11 - Mold Cyclic Temperature: Thermocouple #9

Figures (10) and (11) present data for thermocouple locations #3 and #9. The model predictions and the experimental data agree fairly well, both in terms of overall temperature level and in terms of the cyclic variation shape and magnitudes. Figures (12) and (13) present similar data for thermocouple locations #8 and #4. In both these figures, although the model-experimental agreement is generally good, some disparity can be noted in the slopes. This may be due to a higher heat transfer coefficient on these surfaces than that used in the model (500 BTU/hr ft^2 - °F).

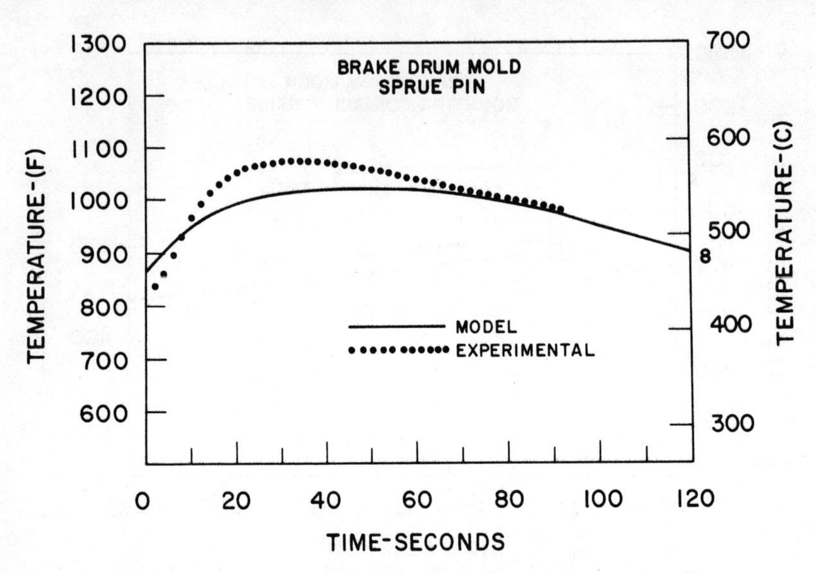

Figure 12 - Mold Cyclic Temperature: Thermocouple #8

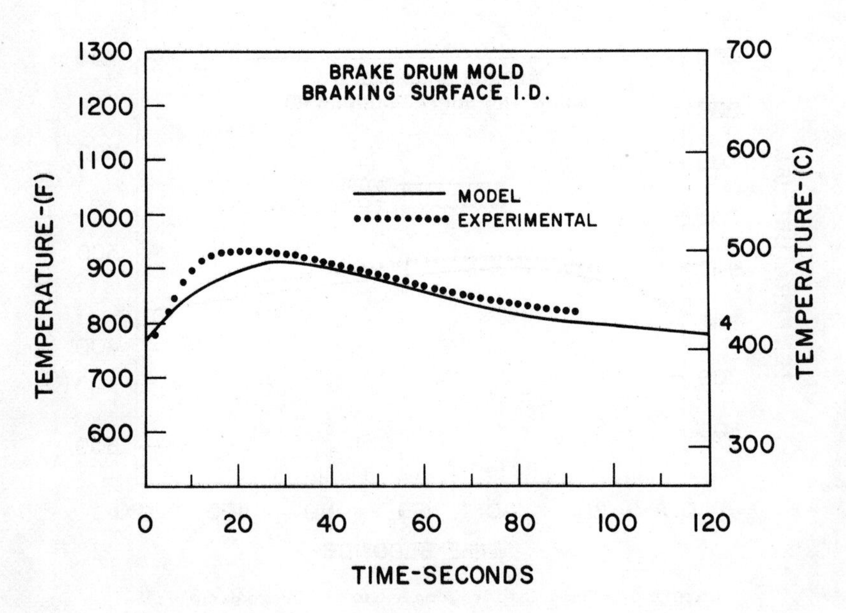

Figure 13 - Mold Cyclic Temperature: Thermocouple #4

It can be noted from figure (9) that both locations #8 and #4 are situated in areas that could be considered as internal radii in relation to the casting. During solidification, therefore, the casting shrinks in upon these surfaces. This could cause a local increase in the interface coefficient because of the increased contact pressure. It could also increase the coefficient because of the distress to the mold coating over repeated casting cycles as the casting is ejected from these surfaces.

This hypothesis is reinforced by the fact that no such disparity was noted in the data for thermocouples #3 and #9 which are located in lateral surfaces where casting contraction would not be expected to be significant. The opposite effect is noted in figure (14), the temperature time data for thermocouple #1. Although agreement is generally good, the actual initial slope is slightly less than the model's prediction. As thermocouple #1 is located in a surface which is an external radius to the casting, one would expect the casting to draw away from it. This would then cause less contact pressure, less resulting distress to the mold coating and, therefore, a lower heat transfer coefficient. Estimated mold-metal interface coefficients which provide after the fact matching of the temperature-time initial slopes are given in Table II.

Table II. Mold-Metal Interface Coefficients

Location	Interface Coefficient BTU/hr ft^2 - °F
Internal Surfaces	
TC#8	1500
TC#4	1200
Lateral Surfaces	
TC#3&9	500
External Surface	
TC#1	275

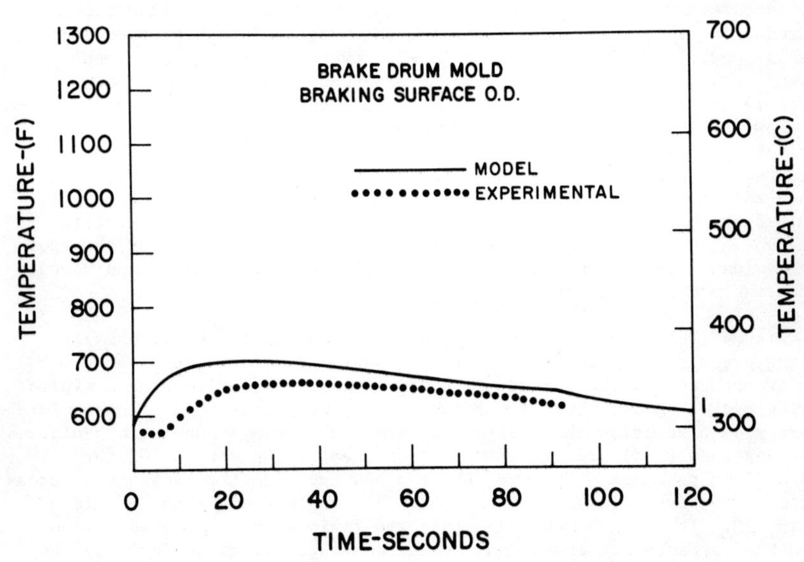

Figure 14 - Mold Cyclic Temperature: Thermocouple #1

In figure (15) temperature-time data as measured and predicted in the mold surface of the sprue area (thermocouple location #15) is shown. The sharp drop-off in the latter part of the cycle in both the model and the actual data corresponds to the opening of the mold.

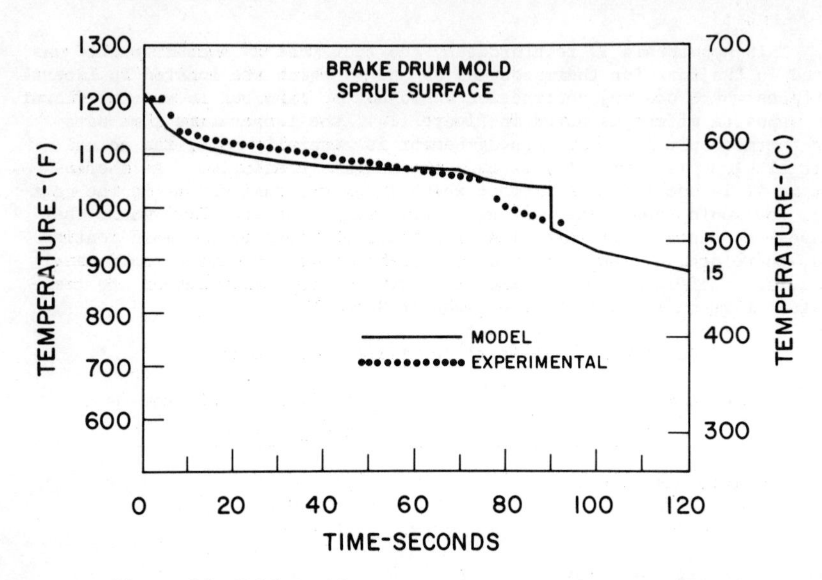

Figure 15 - Mold Cyclic Temperature: Thermocouple #15

Figures (16) and (17) show the temperature-time data in the casting itself at the two locations (#12 and #16) shown on figure (9). The relative temperature difference between the two locations is of importance in predicting the occurrence of casting shrinkage. It can be seen that this is predicted very well by the model. Overall absolute agreement between the model's prediction and the measured data is fairly good except in the thermal arrest area. It is noted that the measured data does not show the extended thermal arrest predicted by the model. This is attributed to two factors, the first being the probable existence of subcooling in the liquid. The second factor is the small size of the thermocouple tip in relation to the model node size. The nodal temperature is retained at constant temperature until all the eutectic latent heat is rejected. The effect of this is to give average conditions over the whole nodal volume. The thermocouple on the other hand gives local conditions at the thermocouple tip itself.

Figure (18) shows the results of the monitoring of individual air cooling passage heat pickup. The figure shows this in terms of actual heat transferred to the individual cooling passages (A-E) during a typical casting cycle. Model predictions for the passages are shown. In absolute accuracy of prediction the four circumferential passages having circular cross sections (A-D) are highest, the agreement being within 10% for each of four passages. Passage "E", the passage with the rectangular cross section (see figure (3)) is within 14%. The impingement tube "F", is within 17%. It would be expected that the impingement tube predictions would be less accurate than those of the conventional circumferential tubes as heat transfer coefficient correlations are not as well developed for this type of air flow. The total active cooling for the complete system as predicted by the model is within 5% of that measured. The model predicts that 65% of the heat rejected by the mold/casting system is absorbed by the cooling air, the remainder going to the environment by radiation and natural convection.

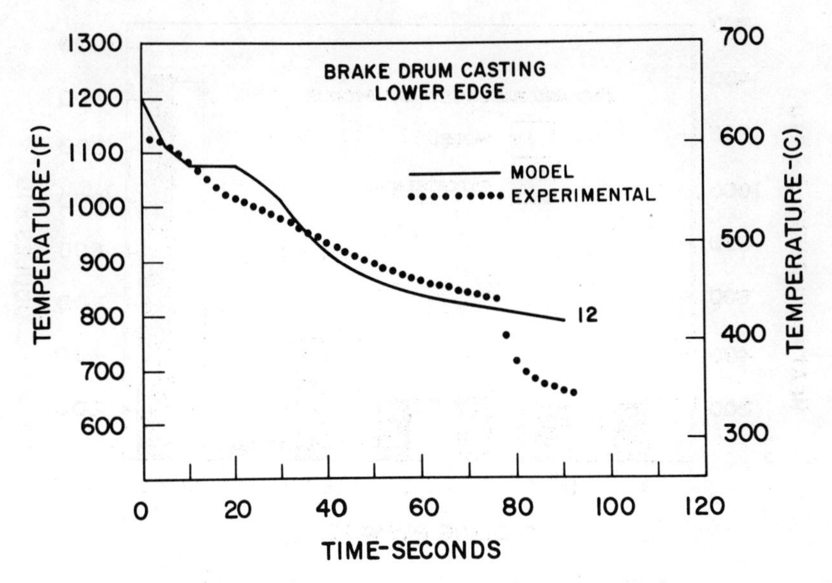

Figure 16 - Casting Temperature: Thermocouple #12

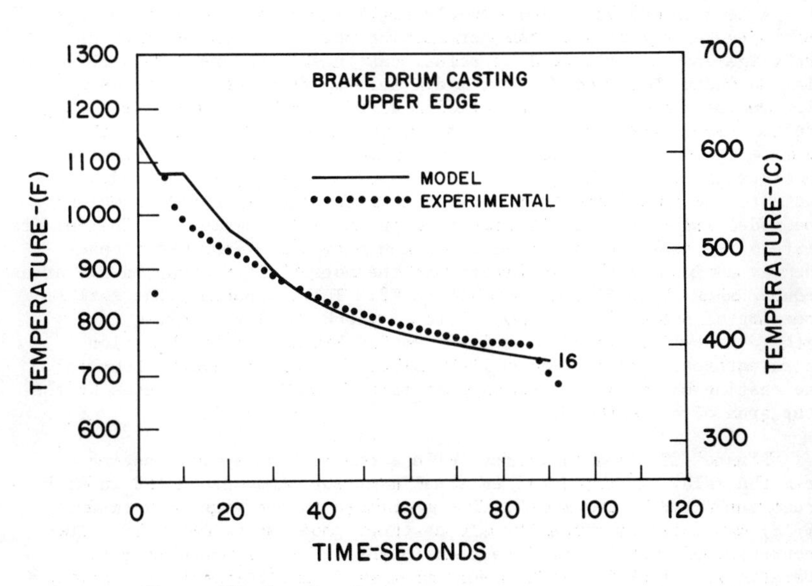

Figure 17 - Casting Temperature: Thermocouple #16

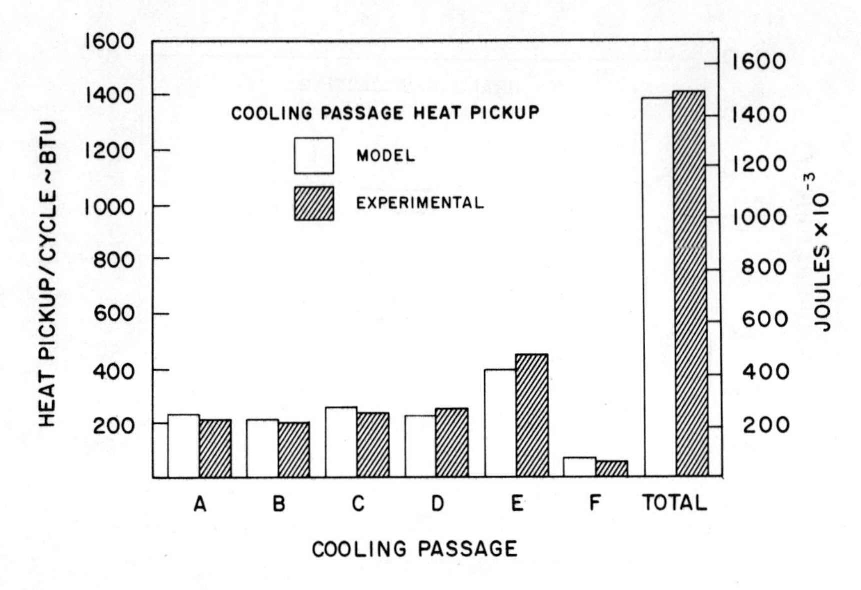

Figure 18 - Mold Cooling Passage Heat Pickup

A less quantitative but equally significant aspect of the modeling technique verification is the correlation between predicted casting shrinkage and that observed on actual castings. The prediction of potential shrinkage locations by the model follows from its ability to predict the sequence of casting solidification from the casting heat flow field. Figure (19) shows such a sequence for the brake drum casting model. The figure represents a section of the casting and shows the node structure used in the model. The varied shading in the nodes indicate the time frame from the start of the casting cycle in which the model indicates solidification of the node is completed. The numbers next to the nodes are for identification purposes. From the figure one may anticipate that shrinkage has the potential to occur in the areas around nodes 21 to 23 and nodes 16 to 18. This is because the earlier freezing of nodes 19 and nodes 13 to 15 block the liquid metal feed paths. Node 19 solidifies early primarily because of its location at an external corner. Nodes 13-15 solidify early primarily because the casting has a thinner section at their location than it does in the step area of nodes 16-18.

Figure (20) is a photograph of a section of an actual casting from the mold. Internal and external macroshrinkage are noted in both areas predicted by the model. The shrinkage in the step area (nodes 16-18) was noted in virtually all castings produced by the mold. The occurrence of that in the face area (nodes 21-23) is somewhat more sporadic but still frequent enough to justify confidence in the model.

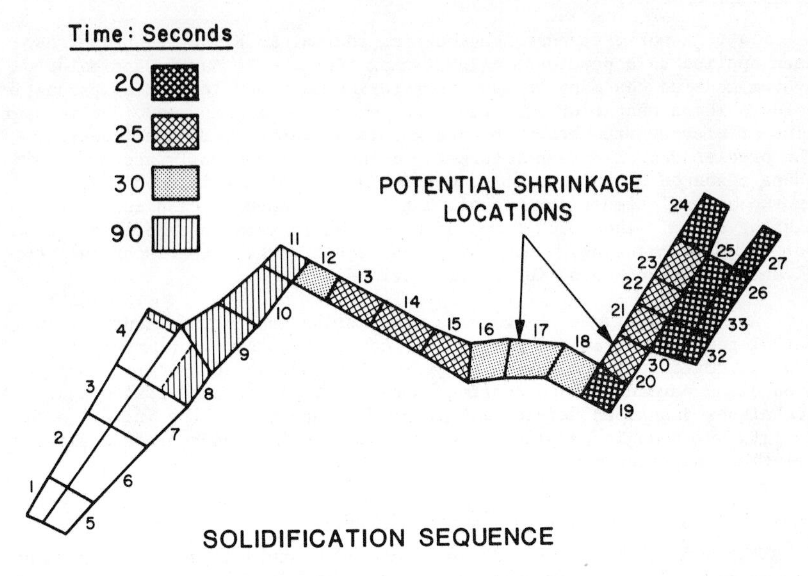

Figure 19 - Predicted Casting Solidification Sequence

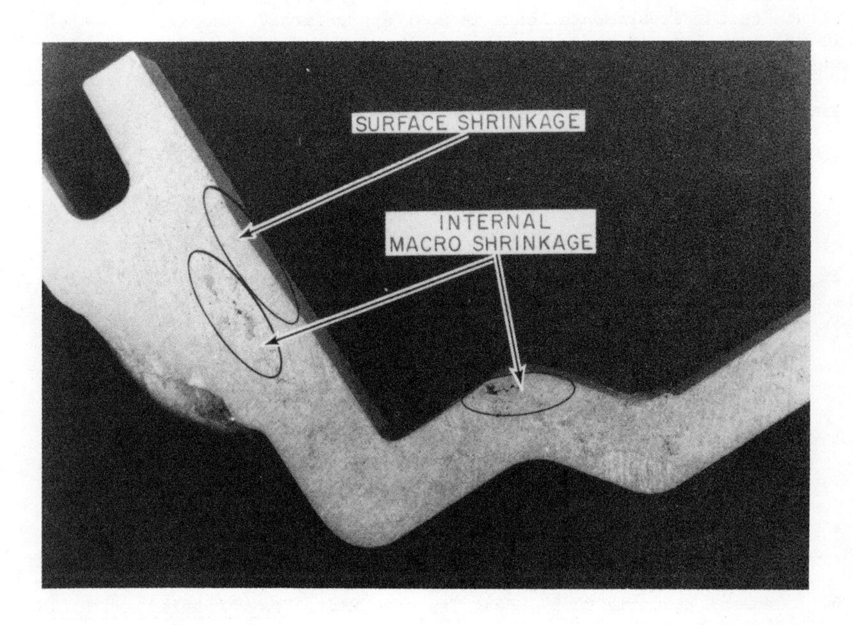

Figure 20 - Section of Casting Showing Macroshrinkage

Summary & Conclusion

In this work, standard engineering thermal modeling techniques have been applied to a permanent mold casting cycle. The particular mold system modeled has many of the characteristics found in typical production molds. These characteristics include active cooling systems, cyclic operation at elevated temperatures, and multiple section mold construction. The predicted cyclic temperatures from the model generally agree with those measured on the operating mold. This agreement is found not only in the overall temperature level, but also in the variation during the casting cycle. The predictions of the model in terms of the locations of potential macroshrinkage have also been verified by comparisons to significant shrinkages found in actual castings.

Agreement between the model's predictions and events occurring in actual casting operations lead to the conclusion that modeling of this type is sufficiently accurate to justify its use as an analytic tool in production permanent mold casting operations. It is anticipated that the techniques described herein will be useful not only in the initial mold designs but also in helping solve problems and improving productivity of existing operating molds.

ACKNOWLEDGEMENTS

The author gratefully acknowledges the contributions of the personnel in the Metal Melting and Processing Department of the Research Staff, Ford Motor Company for their assistance in preparing this work. In particular, the efforts of Mr. Richard Fekete and Mr. Charles W. Bagwell in obtaining the experimental data is most appreciated. The assistance of Dr. S. A. Weiner in reviewing this manuscript and offering valuable suggestions is also recognized.

References

1. Edwards, A. L., "TRUMP: A Computer Program for Transient and Steady State Temperature Distributions in Multidimensional Systems", UCRL Report 14754 Rev. II, Lawrence Radiation Lab., Livermore, California, Sept., 1972.

2. Kays, W. M., Convective Heat and Mass Transfer, McGraw-Hill Book Co., New York, N.Y., 1966.

3. Gardon, R. and J. Cobonpue, "Heat Transfer Between a Plate and Jets of Air Impinging on It", Proceedings of International Heat Transfer Conference, Part II, p. 454-460, (published by ASME), 1961.

4. Huang, G.C., "Investigations of Heat Transfer Coefficients for Air Flow Through Round Jets Impinging Normal to a Heat Transfer Surface", Trans. ASME, Journ. of Heat Transfer, Vol. 85, 1963, p. 237-243.

5. Kreith, F., Principles of Heat Transfer, International Textbook Co., Scranton, Pa., 1959.

6. General Electric Corp., Heat Transfer Div., Heat Transfer and Fluid Flow Data Book, July, 1974.

7. Prates, M., and H. Biloni, "Variables Affecting the Nature of the Chill Zone", Metallurgical Trans., Vol. 3, June, 1972, p. 1501-1510.

8. Sun, R. C., "Simulation and Study of Surface Conductance for Heat Flow in the Early Stages of Casting," AFS Case Metals Res. Jour., Vol. 6, No. 3, September, 1970.

9. Sully, L.J.D., "The Thermal Interface Between Castings and Chill Molds," Trans. AFS, Vol. 84 p. 735, 1976.

10. Jorstadt, J. L., "390 Alloy Technology - Foundry Characteristics", Research and Development, Reynolds Metals Co., Richmond, Va., June, 1975.

11. Ruhlandt, G. K., and H. C. Chin, "Development of Computer-Aided Thermal Analysis Techniques for Casting Tooling", Low Pressure Die Casting Seminar Transactions, (published by Soc. of Die Casting Engineers), Oct., 1979.

APPLICATION OF A SOLIDIFICATION MODEL

TO THE DIE CASTING PROCESS

Otto K. Riegger

Tecumseh Products Company
3869 Research Park Drive
Ann Arbor, Michigan

A basic heat transfer program has been adapted to model the solidification sequence for die castings. The model, which is capable of fully describing complex three-dimensional castings, has been used to simulate the thermal history of a connecting rod, a piston, and a small engine cylinder block including the cooling fins. The solidification sequence, total time to solidify, and die temperature were studied in detail for each example. In turn, the theoretical results were verified with experimental data from the die casting shop. Use of the model demonstrates the potential for eliminating macroporosity due to shrinkage. In addition, simulation of die temperatures can be applied to the problem of the heat checking of steel dies.

The computer model was further enhanced to simulate the complete operating thermal cycle of die casting machine. Die temperatures are determined by modeling heat transfer during solidification, die open time, and die lubrication. Die start-up conditions, regular machine cycles, and interrupted cycles have all been investigated. The model is used to identify process control schemes which will be helpful in establishing favorable casting conditions within the die.

Introduction

The die casting process has traditionally had a reputation of being a high speed, low cost casting process, but the quality of parts produced has been of some concern when compared to other methods such as permanent mold or sand casting. The purpose of the several studies summarized in this paper was to improve the reliability of the castings produced thru control of the process. The casting process is basically a thermal one and therefore, governed by the laws of heat transfer. An understanding of the heat transfer during the casting cycle was therefore considered fundamental to identifying process control schemes for improving the quality of castings. The heat transfer model developed to facilitate these studies is described in detail as it was applied to a small engine piston, a air-cooled engine crankcase, and later to the entire casting cycle for an engine connecting rod.

Early work in the application of thermal analysis to the solidification of castings can be traced to the study by Keverian and Henzel of large steel castings.[1] Wallace[2] investigated aspects of heat transfer in the die casting process and Caswell and Lorentzen[3] described a computer method specifically for die casting applications. Pehlke directed the research which resulted in an AFS monograph on the computer simulation of solidification.[4] In the monograph a computer program is described in detail and its application to some simple casting shapes is discussed.

Much of the recently-published work has centered on the computer programming necessary to handle the mathematics of heat transfer which became overpowering if exact solutions are sought for even very simple casting geometries having well-known boundary conditions. Hence, most studies have dealt only with one-or two-dimensional analysis. For practical purposes, where exact mathematical solutions are not required, a general computer program using finite-element analysis which computes approximate solutions for complex geometries has been developed by Gatecliff.[5]

The use of this program to simulate the solidification of the several castings in a steel die is described. Despite several simplifying assumptions in treating the mathematical and heat transfer aspects of the problem, the results and trends predicted by the model correlate well with experimental measurements from the die casting shop. Results of the simulation indicate that the method could be applied to the investigation of reducing shrink porosity in critical areas, increasing mechanical properties through regulating solidification rates, and increasing die life by control of die temperatures, as described in previous publications, 6,7,8.

Basis for the Computer Model

In applying the finite element analysis to a heat transfer problem, the first step is to subdivide the system into a number of small finite volumes called nodes and assign a reference number to each. Then assuming that each node is at temperature corresponding to its center, the physical system is replaced by a network of imaginary heat-conducting rods between the centers or nodal points as indicated in Fig. 1. If a heat flow corresponding to the conductance of the material between nodal points is assigned to each rod, the heat flow in the rod network will approximate the heat flow in the continuous system. In a three-dimensional cubic system an interior nodal point will have six neighbors and six imaginary conducting rods emanating from it. Modification of the finite-element method for use with a casting of complex shape is simple. Rather than using the cubic network, the size and shape of the nodes are chosen for their convenience in describing the geometry of the casting. This method entails the computation of individual heat transfer areas, lengths, volumes and masses for irregularly shaped nodes.

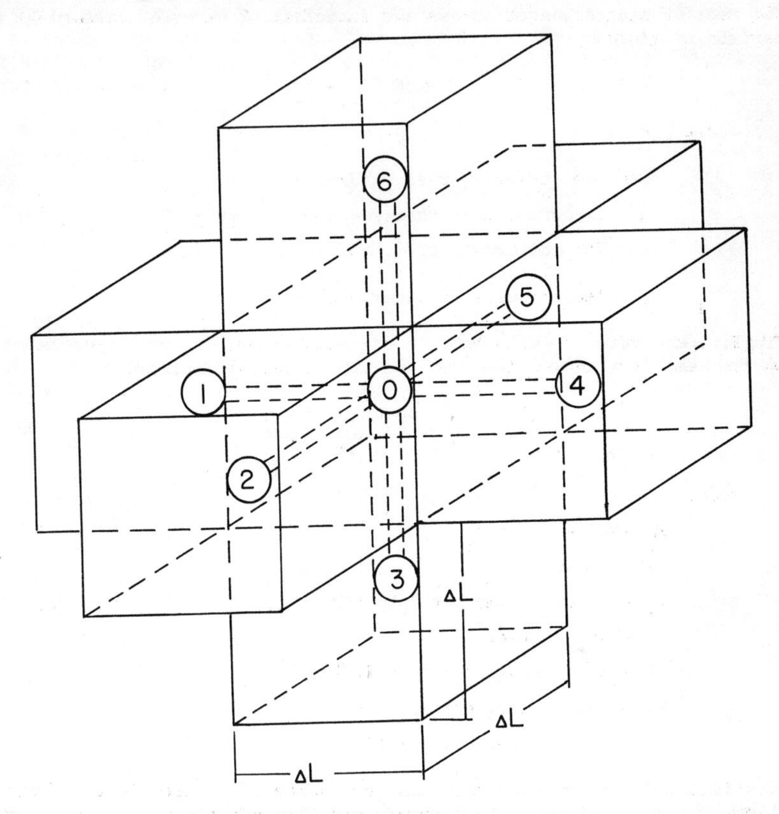

Fig. 1. Sketch showing a cubic center node with heat-conducting "pipes" to the six adjacent nodes. The length, ΔL, would be used to calculate face areas.

Once the nodal network is chosen, a heat balance is written for each node. The mathematical statement for the heat balance requires that the heat input to the die equals the sum of the latent heat of fusion of the casting, the superheat of the molten metal, and the heat released by the cooling of the solidified portion of the casting. The mathematical relationship that governs the rate of conductive heat flow between nodes of the same material is given by Fourier's law of conduction.

$$\dot{q} = -KA \frac{(T_1 - T_2)}{X} \qquad (1)$$

where:

T_1 and T_2 = Temperatures at two locations in a body, (^{o}F)

$\dot{q}$ = Rate of heat flow (BTU/hr.)

K = Conductive heat transfer coefficient (BTU/hr. ft.^{o}F)

X = Heat transfer length (ft.)

A = Surface area (ft.^{2})

The rate of heat transfer across the interface of nodes of dissimilar materials is given by

$$\dot{q} = hA\ (T_1 - T_2) \qquad (2)$$

where:

$\dot{q}$ = Heat transfer rate (BTU/hr.)

h Interface heat transfer coefficient (BTU/hr. ft.^{2} ^{o}F)

A = Surface area (ft.^{2}), and

T_1 & T_2 = Temperatures of materials in contact (^{o}F)

Finally, the rate at which heat is accumulated or lost by a node is given by the heat lost in cooling over the time interval considered

$$\dot{q} = \frac{V\rho\ (T_2 - T_1)\ Cp}{\Delta t} \qquad (3)$$

where:

Δt = Time interval

ρ = Density

C_p = Specific heat (BTU/lb.^{o}F)

V = Volume (in.^{3})

$\dot{q}$ = Heat transfer rate (BTU/hr.)

T_1 = Initial temperature (^{o}F)

T_2 = Final temperature (^{o}F)

Substitution of known constants and the choice of a suitable time interval allows the computation of temperature profiles and the prediction of the solidification sequence of the various nodes.

Determination of a Nodal Network for a Piston

The piston to be modeled is shown in Fig. 2 and is typical of ones used in small engines. The part is first divided into small volumes or nodes. The finite element method, which in its basic form uses a cubic nodal network, can be modified simply for use with a casting of complex shape. Rather than using the cubic network, the size and shape of the nodes are chosen for convenience in describing the casting geometry. The piston closely approximates a simple cylinder. Because of the summetry, only one quarter of the circumference had to be examined. It was assumed that there was no net heat flow across the symmetry axes. Referring to Fig. 3, the first step in choosing the network is to divide the piston into a number of layers of rings and then overlay a vertical grid. This yields a group of near-rectangular solids to describe the skirt. The head is first cut into pie-shaped pieces and then further divided using a parallel rings. Oddly-shaped areas (such as the piston pin boss) are divided so that they fit with the rest of the network. The die nodal network (Fig. 4) is determined by extending the same grid used to describe the casting beyond the surfaces of the casting. The network is chosen so that as many nodes or groups of nodes as possible will have the same dimensions and heat transfer characteristics. This situation becomes important when calculating the heat transfer input data. Every node is given a number to identify location and thermal properties.

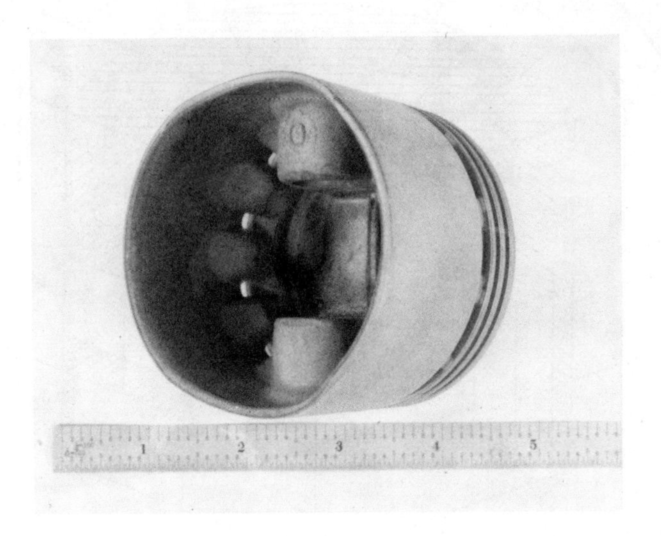

Fig. 2. Typical Small Engine Piston.

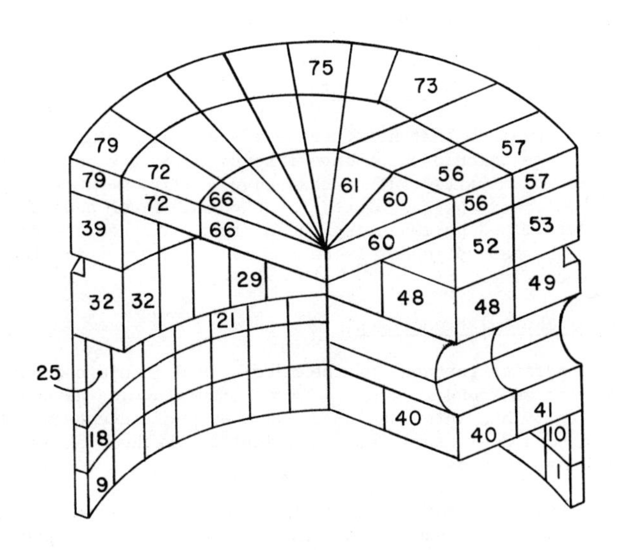

Fig. 3. Nodal network used to describe the piston casting.

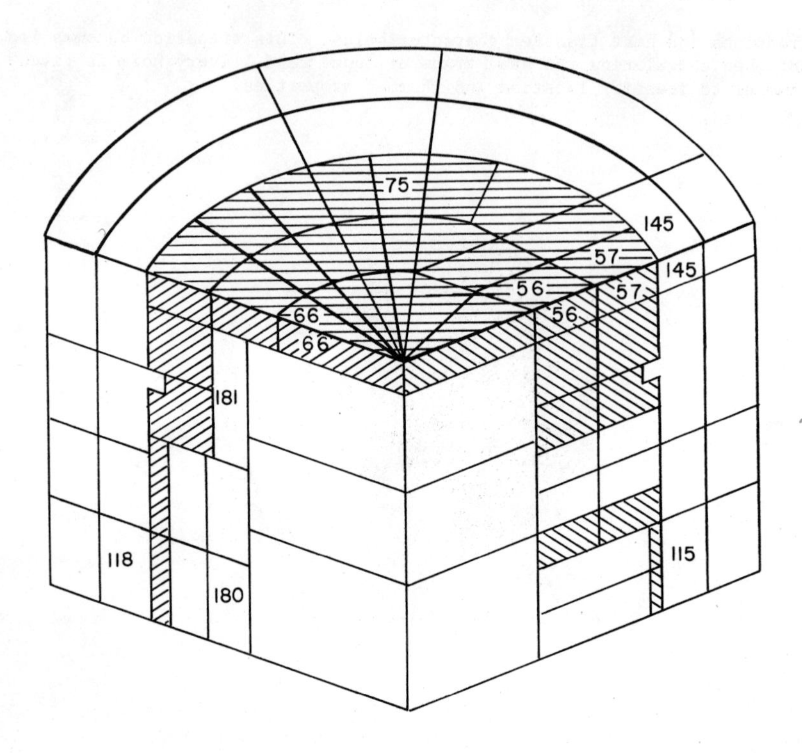

Fig. 4. Casting grid extended to form the grid used to divide the part of the die considered by the simulation.

Calculation of Heat Transfer Input Data

The program requires that heat transfer area, heat transfer length and mass of each node be calculated. Since irregularly-sized and -shaped volumes are chosen for complex parts (rather than equal, cubic volumes), the heat transfer input data must be calculated for the individual nodes. Heat transfer length is determined by measuring the distance between the centers of contacting nodes. The mass of nodes is determined by multiplying the volume of the node by its density.

Determination of Thermal Properties

Thermal properties for most of the metals used in die casting are fairly well known and can be found in the technical literature. The specific heat, thermal conductivity and interface heat transfer coefficient are needed for both the casting and die materials. In addition, the heat of fusion is required for the casting material.

Thermal coefficients for the casting and die were chosen for a casting of 380 aluminum-base alloy and a steel die. The temperature-independent constants used were:

- specific heat: Al = 0.23 BTU/F-lb
- thermal conductivity: K, Al = 62.5 BTU/hr-ft-F
- heat of fusion: Al - 167 BTU/lb
- specific heat: die = 0.14 BTU/F-lb
- thermal conductivity: K, die = 19.2 BTU/hr-ft-F

A laboratory experiment was used to determine the heat transfer coefficient of die to air. The temperature drop of a steel block was measured vs. time. The heat transfer coefficient of casting to die is an artificial coefficient for heat transfer between dissimilar materials. Choice was made after considering general handbook values and was then slightly modified to yield a good correlation between actual and predicted die temperatures. The temperature dependence of the thermal coefficient of heat transfer between dissimilar metals and the thermal coeffient of heat transfer between dissimilar metals and the phenomenon of gap formation between die and casting apparently combine to affect the heat flow in such a way that a constant coefficient will simulate the overall heat transfer in the temperature range examined. The values used were:

- heat transfer coefficient for die-open = 1.75 BTU/hr-ft^2-F
- heat transfer coefficient for casting to die = 2300 BTU/hr-ft^2-F

It was assumed that the solidification of the part (release of the heat of fusion) took place at a constant temperature of 1050F (566C). This is the middle of the solidification range (1000-1100F) (538-593C) for a 380 aluminum alloy. The program determined the amount of heat transferred from a solidifying node during the program calculation time step and subtracted that amount from the remaining heat of fusion until the heat of fusion was fully released. At this point, temperature reduction of the part was resumed.

Computer Run

The input data is tabulated in a specified format on computer cards, tape or other input devices. The format of typical input data is shown in Tables 1 and 2 which includes heat transfer area, heat transfer length, mass and the thermal properties - specific heat, conductivity between nodes and heat of fusion. When the term 0.0 is noted in the heat transfer coefficient column, transfer between dissimilar materials is indicated and the program chooses the proper coefficient from a subroutine. This method will allow the input of heat transfer data that is temperature-dependent, although in the present study a constant value was specified in the subroutine.

Table 1. Format of the Input Data to Describe the Geometric and Thermal Characteristics of the Heat Transfer Circuit

Node No.	Transfers Heat With Node No.	Heat Transfer Coefficient	Heat Transfer Area	Heat Transfer Length
1 (C)	2 (C)	62.	.029	.25
1 (C)	10 (C)	62.	.023	.312
1 (C)	105 (C)	0.0	.23	—
1 (C)	115 (D)	0.0	.078	—
1 (C)	170 (D)	0.0	.078	—

(C) node describing aluminum casting
(D) node describing steel die

Table 2. Format of the Input Data Describing the Thermal Properties of the Nodes

Node No.	Mass (lbs)	Specific Heat (BTU/F-lb)	Heat of Fusion (BTU/lb)
1	.00073	.23	167.
2	.00073	.23	167.

The time interval between each calculation (program time step) is specified as input data. In this instance time steps of 0.025 and 0.050 sec. were used. For each computer run the initial temperatures of the die and the melt temperature of the aluminum also must be chosen. These temperatures will depend on the nature of the investigation. The initial melt and die temperatures used in the piston study are shown in Table 3. The 1300F (704C) melt temperature approximates that used in practice. The die-cavity and die-core temperatures correspond to values that might be reached with different combination of water cooling, time of lubricant spray and the dwell time.

Table 3. Combinations of Melt and Die Temperatures Used as Initial Conditions for the Piston Simulation

	Temperatures, °F (°C)		
Condition	Aluminum Melt	Die Cavity	Die Core
1	1200(649)	500(260)	200(93)
2	1300(704)	200(93)	200(93)
3	1300(704)	500(260)	200(93)
4+	1300(704)	500(260)	500(260)
5	1300(704)	700(371)	200(260)
6	1300(704)	700(371)	700(371)

+initial die and melt temperatures simulating production conditions

Output from the program can be in any specified form. For this study the results were printed as the node number with its time to solidification. Temperatures of all casting and die nodes were also listed at specific time intervals. Typical output of the program is shown in Tables 4 and 5.

Table 4. Form of the Computer Output Showing Solidification Order

```
NODE 66 SOLIDIFIED AFTER 1.6500 SEC.
NODE 42 SOLIDIFIED AFTER 1.8500 SEC.
NODE 40 SOLIDIFIED AFTER 2.7520 SEC.
NODE 64 SOLIDIFIED AFTER 3.9750 SEC.
NODE 44 SOLIDIFIED AFTER 4.3750 SEC.
NODE 62 SOLIDIFIED AFTER 4.7000 SEC.
NODE 45 SOLIDIFIED AFTER 5.1750 SEC.
NODE 26 SOLIDIFIED AFTER 5.3500 SEC.
NODE 75 SOLIDIFIED AFTER 5.7500 SEC.
NODE 74 SOLIDIFIED AFTER 5.9000 SEC.
```

Table 5. Form of the Computer Output Indicating Node Temperatures

```
ELAPSED TIME ..... 2.5000 SEC.
T (   1) = 847.67 DEG. F
T (   9) = 854.55 DEG. F
T (  17) = 872.17 DEG. F
T (  25) = 958.84 DEG. F
T (  29) =1050.00 DEG. F
```

Results of the Simulation

To verify the accuracy of the model, the solidification of the piston was examined using actual temperatures from production dies as the initial values in the computations. Comparison of the calculated results with actual casting and solidification phenomena were made according to three criteria: 1) solidification pattern of the casting, 2) solidification time of the casting, and 3) temperature response of the die. The values used as initial conditions were 1300F (704C) melt temperature with 500F (260C) die-cavity and die-core temperature. These actual die-closing temperatures were determined experimentally by using an infrared pyrometer to survey the temperature on the cavity surface immediately before closing. This data was taken after the die had been running under production conditions for a long time and die temperatures had stabilized. Effects of changing the initial die temperatures on the solidification order, the solidification time and the die temperature response during solidification were also examined. Initial temperatures used are shown in Table 3.

Solidification Pattern

Figure 5 shows the computed solidification order of the piston. The areas indicated as last to solidify in the figure do so without feeding of liquid metal. Shrink porosity would be concentrated in these areas. Sample castings were sectioned and etched to reveal the location of porosity. An example piston casting with the nodal network superimposed is shown in Fig. 6. The casting has been severely etched with sodium hydroxide to accentuate the shrink porosity and to open up the smaller interdendritic porosity not ordinarily seen on polished or lightly etched surfaces. Comparison of the computed and actual locations for porosity shows that areas predicted to solidify without proper feed indeed correspond to areas of actual porosity. This accuracy is a good verification of the model.

Imposing other artificial die and melt temperature initial conditions to the piston caused only slight changes in solidification pattern. The changes were not major ones because the die temperatures are uniform temperatures and not temperature gradients. However, the results indicate that solidification order is somewhat dependent on the initial die temperature.

Solidification Time

Time to solidification resulting from production die temperatures as initial conditions for a number of nodes is shown in Table 6. The simulation predicts that all but one node is solidified by 7.0 sec. and that the last node solidifies at 7.9 sec. (condition 4). As a comparison, the dwell time of a piston casting cycle was reduced until the castings would not eject from the die because they had not solidified. This occurred occasionally between 6.5 and 7.5 sec. and consistently below a 6.0 sec. dwell time. While the correlation is not exact, the calculated values are within the limits of reasonable accuracy and will satisfy the second criterion.

Table 6 also shows the effect on solidification times due to other initial die temperature conditions. The total time to solidify is increased by increasing melt temperature as indicated by comparing conditions 1 and 3. Increasing the die cavity or core temperatures either singly or together tends to increase the total time to solidification. This data indicates that solidification time can be influenced by changes in die temperatures.

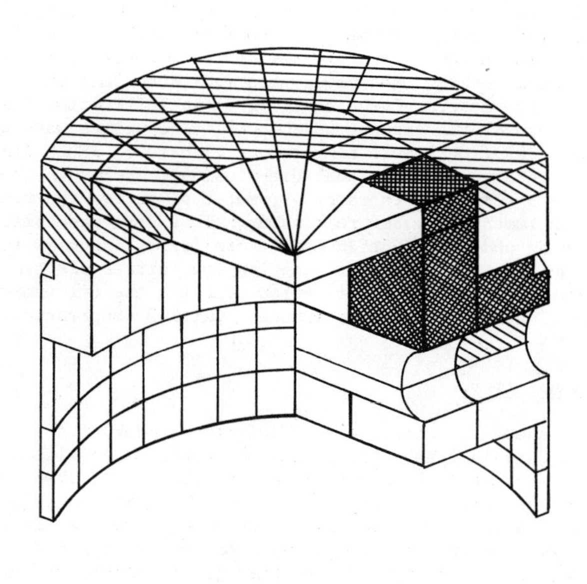

EARLY SOLIDIFICATION

MID-TIME SOLIDIFICATION

LAST TO SOLIDIFY

Fig. 5. Solidification pattern for the piston casting for initial conditions of a 500F cavity and core temperature and a 1300F melt temperature.

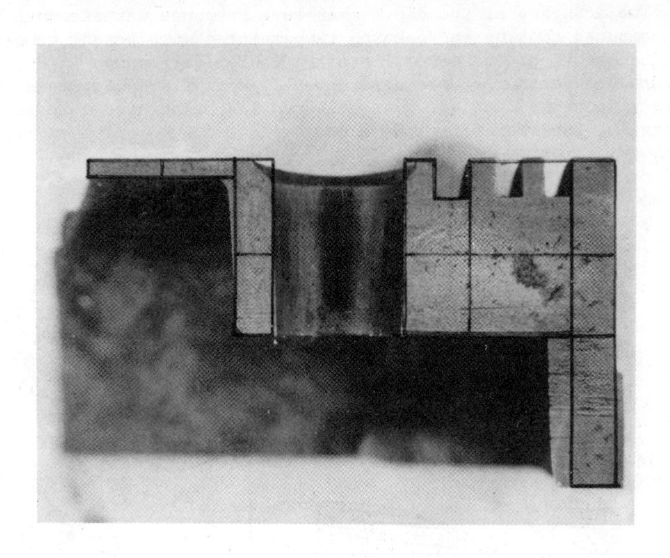

Fig. 6. Photograph of a piston casting that was sectioned and etched to reveal the porosity in the pin boss area. The nodal network has been superimposed onto the cut surface.

Table 6. Total Solidification Time of Five Nodes as a Function of Initial Temperatures

Condition No.	Solidification Time, seconds 1	2	3	4+	5	6
Melt Temp.	1200	1300	1300	1300	1300	1300F
Die Temp.	500	200	500	500	700	700F
Core Temp.	200	200	200	500	200	700F
Node No.*						
1	0.30	0.25	0.35	0.42	0.40	0.70
9	0.30	0.25	0.35	0.42	0.40	0.75
35	2.72	1.85	3.39	4.07	5.45	8.90
66	0.50	0.50	0.60	0.85	0.72	1.65
52 (last to solidify)	4.82	4.00	5.70	7.90	7.95	(did not solidify in 9 sec.)

*See Figure 3 for location of nodes

+Initial die and melt temperatures simulating production conditions

Die Temperature Response

The accuracy of the die temperature response was examined by comparing the computed closing and opening temperatures with actual temperature (Table 7). Locations of the nodes listed in Table 7 are shown in Fig. 7. The measured temperatures were determined by an infrared pyrometer survey immediately upon die opening. The measured and calculated temperatures correlate favorably, thus verifying the accuracy of the simulation and satisfying the third criterion.

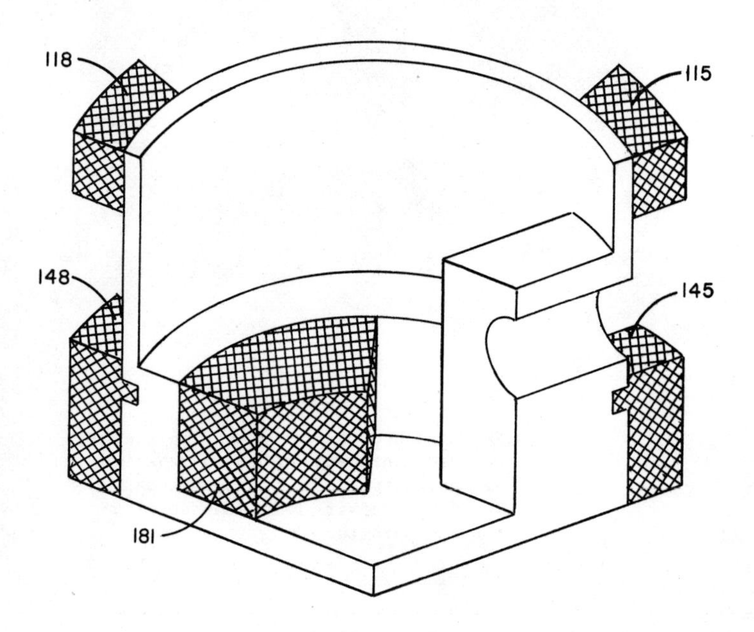

Fig. 7. Sketch showing the locations of nodes 118 and 148 used to verify the accuracy of the simulation.

Table 7. Comparison of Measured and Computed Die Temperature After 8.0 Seconds

	Computed Die Temperatures, °F		Measured Die Temperatures, °F	
Node No.	Close	Open	Close	Open
148	500	792	525	775
118	500	645	525	620
145	500	834	-	-
115	500	576	-	-
181	500	844	-	-

Melt Temperature 1300°F
*Computed temperatures based on condition 4 in Table 3

Table 8 shows the die temperatures for a number of nodes resulting from the imposed initial die temperature conditions. Increasing the cavity or core temperatures tends to increase the die temperatures. Die temperatures profiles for a civity surface node during solidification resulting from three initial die temperature conditions (2, 4 and 6 in Table 3) are compared in Fig. 8. The nodes with higher closing temperatures reach the highest maximum temperature and also have higher temperatures at a given time. These reults indicate that the temperature response of the die is affected by the initial die temperatures.

Table 8. Predicted Die Temperatures After 6 Seconds as a Function of Initial Temperatures

Condition No.	Temperature after 6 sec.					
	1	2	3	4+	5	6
Melt Temp.	1200	1300	1300	1300	1300	1300F
Die Temp.	500	200	500	500	700	700F
Core Temp.	200	200	200	500	200	700F
Node No.*						
148	772	586	797	839	961	971
118	553	382	561	647	679	818
145	804	628	834	863	966	972
115	566	413	576	671	681	837
181	814	662	844	903	1014	---

*see Figure 3 for location of nodes
+production temperatures

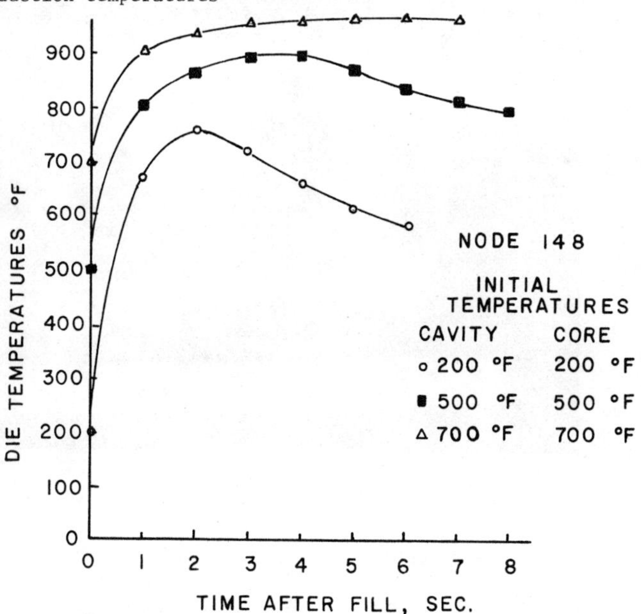

Fig 8. Temperature response of a die node on the cavity surface resulting from three initial temperature conditions.

Determination of a Nodal Network for an Engine Crankcase

The solidification model was also applied to a typical air-cooled engine crankcase shown in Fig. 9. The engine cylinder block could be adequately mapped with about 900 nodes. This large number of nodes presented a minor problem because the information required to describe the cylinder block was larger than could be handled by the available computer memory - a commercial 16k core. In the previously described analysis of the piston, the 16k core was adequate to handle the approximately 300 nodes required for that part. The core memory limitation, however, is not as difficult a problem as it might appear. In this instance, the cylinder block was separated into two distinct parts, the bore and the pan, which were analyzed individually. In both, considerations of symmetry allowed further simplification of the network. It was assumed that there would be no net heat flow across the axes of symmetry.

Referring to Fig. 10, the uniform spacing of the fins was convenient for slicing the bore section into layers as the first step in defining the network. In the cylinder walls, a pie-shaped grid extending from the center of the bore to the ends of the fins was imposed over each layer. The breather box was divided into rectangular solids. Odd-shaped nodes that would most closely describe the complex geometry were used in the valve area. Because of the regular geometry of the pan, rectangular solids would describe most areas, as shown in Fig. 11. The die nodal network was determined by extending beyond the casting surfaces the grid used for the nodes. Finally, each node is given a number to identify its location and thermal properties.

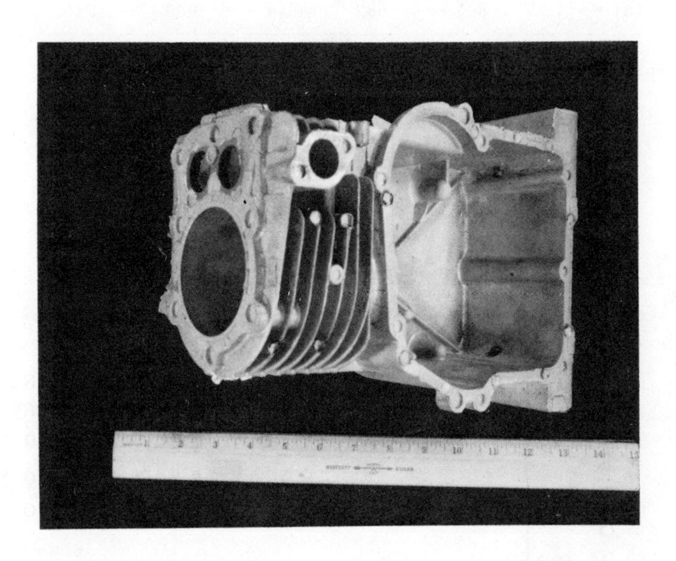

Fig. 9. Air Cooled Engine Crankcase

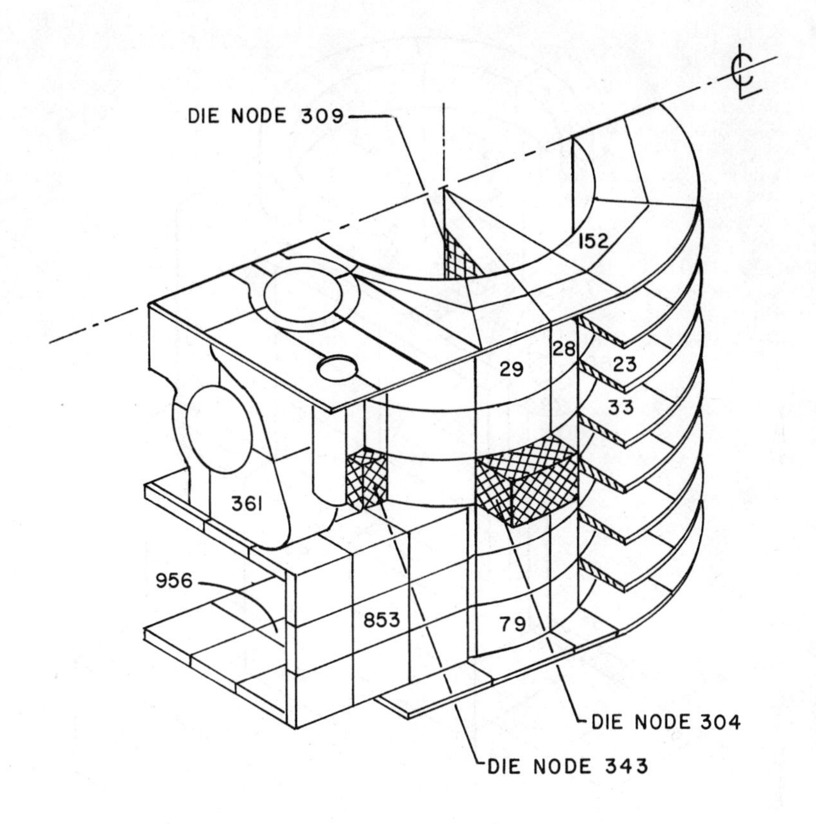

Fig 10. Nodal network used to describe the bore section of the cylinder block casting. The location of three die nodes are noted.

Calculation of Heat Transfer Input Data

The input data required is the same form as used with the piston. The network has been chosen so that as many nodes or groups of nodes as possible have the same dimensions and heat transfer characteristics in order to reduce the number of calculations to be made. The data for a typical node (No. 29) are shown in Table 9.

Table 9. Typical Input Data Calculated for Each Node

Node No.	Transfers Heat With Node Nos.	Heat Transfer Area, $in.^2$	Heat Transfer Length, in.	Mass of Node, lbs
29	26	0.25	0.75	0.01225
	28	0.125	1.0	
	30	0.125	1.0	
	32	0.25	0.375	
	659	0.50	-	
	429	0.50	-	

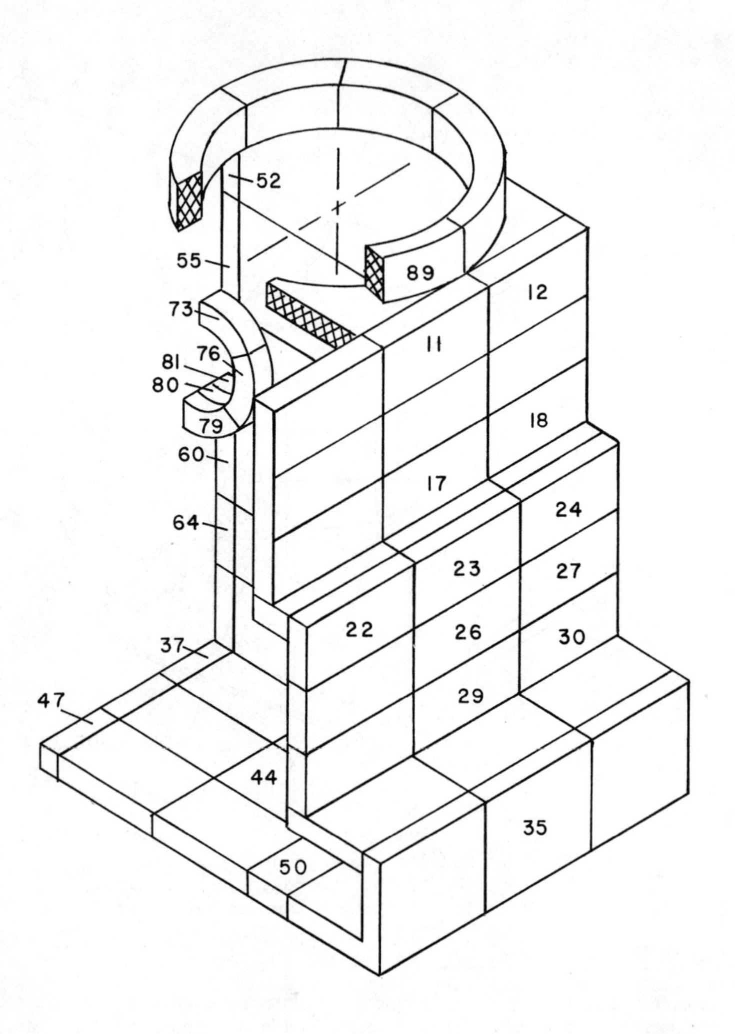

Fig. 11. Nodal network used to describe the pan section of the cylinder block casting.

Computer Run

For this study the print-out included the node number and its time to solidification. The temperatures of all casting and die nodes were also listed at specific time intervals. Typical output of the program is shown in Tables 10 and 11. For each computer run the initial temperatures of the die and the melt temperature of the aluminum must be chosen. These temperatures will depend on the nature of the investigation.

Results of the Simulation

To verify the model accuracy, solidification of the cylinder block was examined using actual temperatures from production dies as the initial values in the computations. Comparison of the calculated results with actual casting and solidification phenomena were made according to three criteria: 1) The solidification order of the bore and pan sections, 2) the solidification time of the casting, and 3) the temperature response of the die. The values used as initial conditions were 1250F melt temperature, with die temperatures ranging from 500 to 650°F depending on the node location. These die closing temperatures were determined experimentally by using an infrared pyrometer to survey the cavity surface temperature immediately before closing. The data were taken after the die had been running under production conditions for a long period and the die temperatures had become stabilized. Further studies were made with an artificial uniform 600°F die temperature.

Table 10. Form of the Computer Output Indicating Node Temperatures

```
NODE  589  SOLIDIFIED AFTER  2.0000 SEC.
NODE  556  SOLIDIFIED AFTER  2.0500 SEC.
NODE  560  SOLIDIFIED AFTER  2.2000 SEC.
NODE  455  SOLIDIFIED AFTER  2.3000 SEC.
NODE  362  SOLIDIFIED AFTER  2.5000 SEC.
NODE  456  SOLIDIFIED AFTER  2.6500 SEC.
NODE  960  SOLIDIFIED AFTER  2.8000 SEC.
NODE  558  SOLIDIFIED AFTER  3.0500 SEC.
NODE  457  SOLIDIFIED AFTER  3.9500 SEC.
```

Table 11. Form of the Computer Output Showing Solidification Order

```
ELAPSED TIME .... 1.9000 SEC.
  T (    1) = 816.93 DEG. F
  T (  154) = 896.31 DEG. F
  T (  159) = 933.82 DEG. F
  T (   12) = 804.45 DEG. F
```

Solidification Pattern

Sample castings were sectioned and etched to reveal the location of porosity. The crankshaft boss in the pan section is shown in Fig. 12 with the nodal network superimposed. The casting has been severely etched with sodium hydroxide to accentuate the shrink porosity and to open up the smaller interdentritic porosity not ordinarily seen on polished or lightly etched surfaces. Fig. 13 and 14 show the solidification sequence of the bore and pan sections. The areas indicated as last to solidify in the figures do so without feeding of liquid metal. Shrink porosity would be concentrated in these areas. Comparison of the computed and actual locations for porosity shows that areas predicted to solidify without proper feed indeed correspond to areas of actual porosity. This accuracy is a good verification of the model.

Imposing an artificial uniform 600°F initial die temperature distribution to the pan and bore sections caused changes in the solidification pattern. The changes were not major because the die temperature change was not extensive. However, the results indicate that the solidification sequence is dependent on the distribution of initial die temperature.

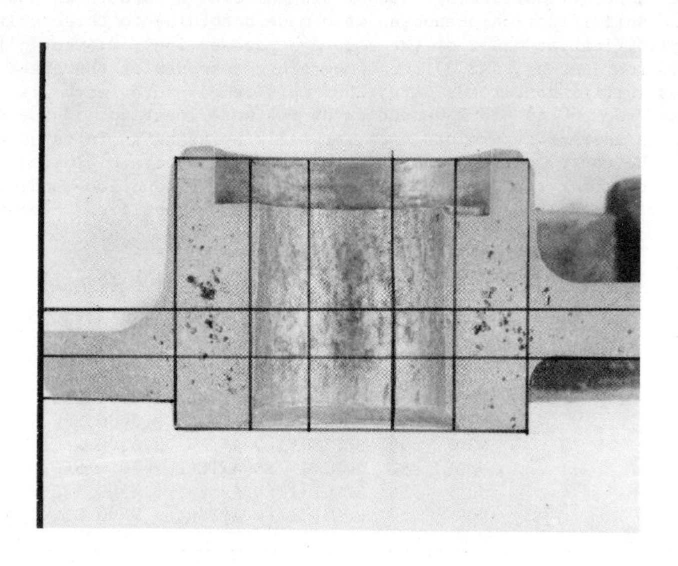

Fig. 12. Sectioned and etched casting with nodal network superimposed. Crankshaft boss in pan section.

Solidification Time

Time to solidification under the production die temperature distribution for a number of nodes in the pan and bore is shown in Table 12. Predicted solidification times are about 6 sec., as compared to actual production times which are near 9 sec. The discrepancy is probably due to the small die mass in the simulation as compared to the large mass of the casting. While the correlation is not exact, the calculated values are within limits of reasonable accuracy and will satisfy the second criterion.

Table 12 also shows the affect on solidification times due to an artificial 600°F initial die temperature. The higher overall closing temperatures result in longer calculated times to solidify. In the pan section, where closing die temperatures due to the production gradients can range from 500 to 650°F, the lower closing temperatures predicted a shorter solidification time. When the production die temperature was higher than 600°F the solidification time was not increased because of the temperature gradients imposed by the production conditions. Rapid heat flow to areas with lower temperature evidently causes faster solidification than from a uniform lower die temperature. These data indicate that the solidification time can be influenced by changes in die temperatures.

Die Temperature Response

The accuracy of the die temperature response was examined by comparing the computed closing and opening temperatures with actual temperatures in Table 13. The locations of the areas listed in Table 13 are shown in Fig. 14. The measured temperatures were determined by an infrared pyrometer survey immediately upon die opening. The measured and calculated temperatures correlate

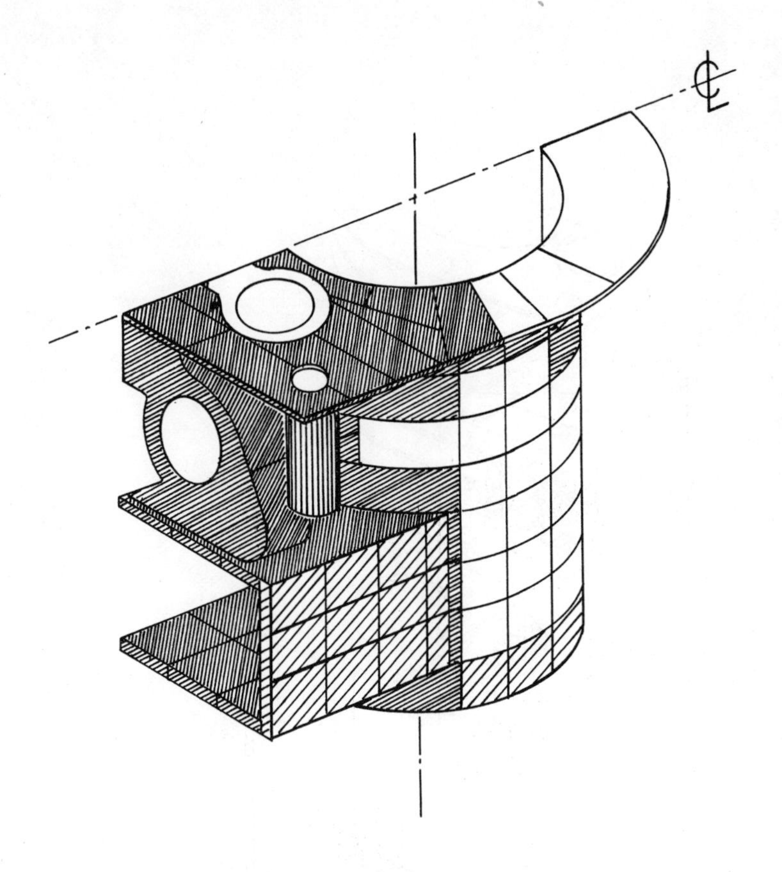

Fig. 13. Solidification pattern of the bore section.

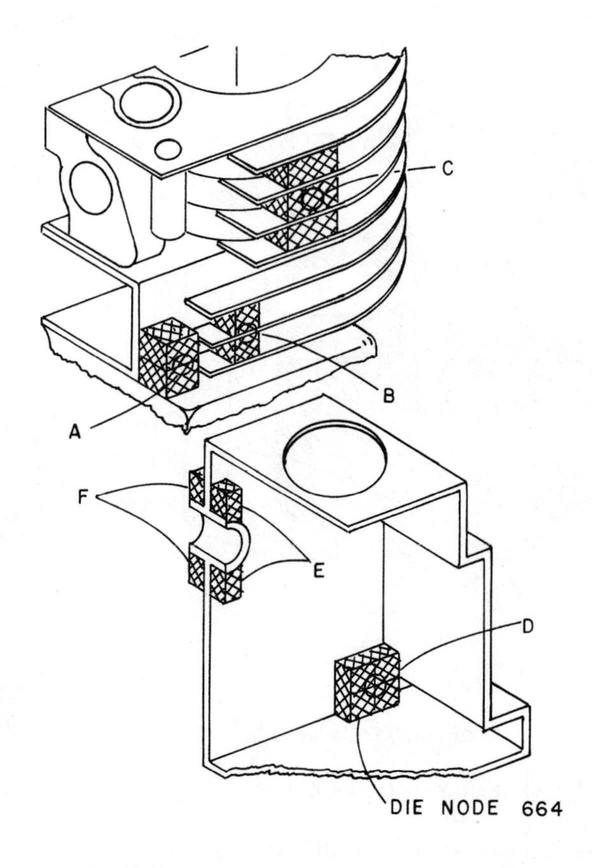

Fig. 14. Sketch showing the locations of die areas for which temperatures were measured in the die casting shop.

favorably, thus verifying the accuracy of the simulation and satisfying the third criterion.

Table 12. Comparison of Solidification Times Resulting from a Uniform 600°F Die Temperature and a Production Die Temperature Distribution

	Production Die Temperature		Uniform Die Temperature	
Node No.	Closing Temperature,*F	Time to Solidify, sec	Closing Temperature,*F	Time to Solidify, sec
Bore Section				
23	550	0.25	600	0.25
853	570	0.65	600	0.70
29	550	1.35	600	1.50
956 (last to solidify)	570	4.95	600	5.50
Pan Section				
26	600	1.70	600	1.65
80	650	3.40	600	3.50
89 (last to solidify)	550	5.60	600	6.00

*closing temperature of near-by die nodes

Table 13. Comparison of Measured and Computed Die Temperature

	Measured Die Temperatures, °F		Computed Die Temperatures, °F*	
Location**	Close	Open	Close	Open
Bore Section A	560	702/712	570	697
B	560	780	570	788
C	550	750	556	730
Pan Section D	560	725/740	540	766
E	650	750	650	750
F	540	730	524	688

**See Figure 14

*Computed temperatures are averages of several nodes in the area

Discussion of Solidification Simulation Results

The results of the solidification simulation indicate that the model chosen can adequately predict the solidification sequence of a real, three-dimensional casting, closely approximate the total time for solidification, as well as predicting the die temperature response.

The importance of being able to predict the solidification order of a casting is apparent if the directional solidification away from the critical areas is considered. The simulation indicates that solidification order can be predicted with reasonable accuracy and that initial die temperature conditions affect that order. The opportunity exists to study changes in die temperature gradients which would alter the solidification pattern so that critical areas would freeze with liquid metal feeding. This could result in castings with critical areas free from shrink porosity. Other changes affecting solidification order easily examined by the simulation could include use of insulating cores to regulate heat flow, relocation of gates and changes in section sizes to move hotspots due to differences in mass.

Accurate prediction of solidification times can lead directly to the question of solidification rates. For aluminum it is well known that mechanical properties are directly related to solidification rate and microstructure.[9] Faster solidification yields finer microstructural constituents and therefore improved mechanical properties. Since a change in solidification time is predicted by changing initial temperature conditions, increasing the rates in specific areas by die temperature control could yield localized higher strength.

Prediction of die temperature profiles during solidification can be applied to the investigation of die life. The temperature profiles define the maximum die temperature, maximum temperature difference and rate of temperature change. All these factors are important with respect to heat checking and thermal fatigue. Table 14 shows the maximum temperatures reached by the piston die core is dependent on initial conditions and indicates how die areas subject to thermal fatigue and heat checking could be analyzed. Under production conditions an attempt is made to water cool the die core of the piston to below 500F (260C) from an opening temperature of 700F (371C) to 800F (427C) to prevent soldering of aluminum to the core. Maximum temperatures of about 1029F (554C) are reached when the die core is cooled to 200F (93C) (conditions 2 and 3). Interestingly, if the die core is cooled to only 500F (260C) (condition 4), the core temperature reaches a maximum of 1042F (561C) indicating that the drastic cooling to 200F (93C) which causes thermal fatigue damage to the core only decreases the maximum temperature by 13F (23C).

Table 14. Predicted Maximum Temperature of the Piston Die Core for Node 181

Condition No.	Temperatures, F			Temperature, F
	Melt	Die	Core	
1	1200	500	200	1030
2	1300	200	200	1029
3	1200	500	200	1027
4	1300	500	500	1042
5	1300	700	200	1038
6	1300	700	700	1068

Some cooling is necessary, since if the die core is allowed to remain at 700F (371C) (condition 6), the core temperatures during solidification reach 1050F (655C) and above. Since 1050F (655C) is in the upper half of the solidification range of the alloy, this is a condition under which aluminum soldering will definitely occur. Thus it can be seen that the simulation is useful in studying methods of eliminating harmful thermal conditions in the die.

Computer Model of the Complete Casting Cycle

The computer simulation was also used to study in detail the die temperature during the complete operating cycle of a die casting machine. The die temperatures are determined by modeling heat transfer during casting solidification, die-open time and die lubrication. Die-open time includes such events as casting removal, spraying of other die members and unscheduled cycle interruptions. The simulation is used to investigate die start-up conditions, regular machine cycles and interrupted cycles. The use of auxiliary heaters as a means of controlling die temperatures during start-up and interrupted cycles is demonstrated. Establishment of favorable die temperature gradients through the use of heaters or waterlines is also discussed.

Computer Model

The same nodal analysis technique previously described has been applied to a connecting rod as shown in Fig. 15. The computer program is used to calculate the heat transferred during a given time interval to and from each node based on the initial temperature/thermal characteristics of the nodes. This computation results in a new set of temperatures. The newly-calculated temperatures become initial temperatures for the next calculation. The computations are repeated for the total period under consideration to model repetitive casting cycles.

Simulation of Machine Cycles

In one complete casting cycle the die cavity surface temperature generally respond to the heat of the molten aluminum by increasing rapidly during dwell and solidification and then decreasing during lubrication. A final stage in the cycle is the equalization of temperatures throughout the die. For the simulation the cycle has been broken into four stages:

- A 6-sec. dwell time during which the casting solidifies and cools.
- A 1-sec. interval with the cavity open to the air (corresponding to casting removal and time before spraying.)
- A 2-sec. interval during which lubricant is sprayed on the cavity surface.
- A 6-sec. period for lubrication of other cavities and the pause before closing.

When die-open times were simulated, the model was changed so that the die nodes that had transferred heat with the casting now only transferred heat with the air. To simulate die lubrication, the rate of heat transfer from the die was drastically increased because of the high velocity of lubricant spray and subsequent evaporation of the lubricant water base. When heaters were used in the simulation, nodes were constructed with thermal characteristics such that the BTU/hr. equivalent of 1000 watts of constant heat was received from each heater. In practice, the heaters could be either a circulating fluid or an in-die cartridge element, as opposed to radiant die surface

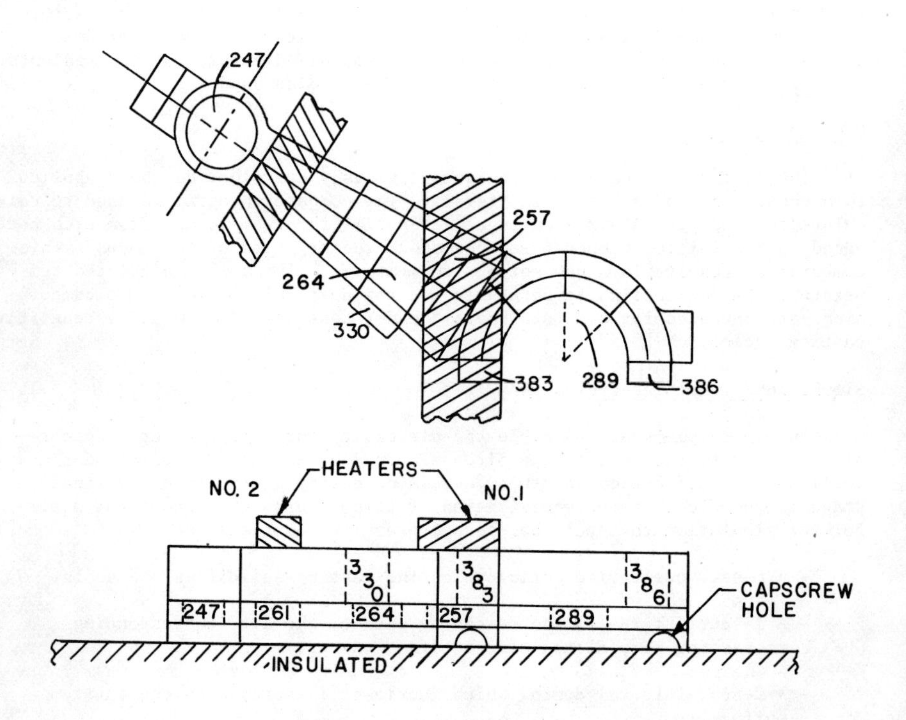

Fig. 15. Nodal network used to describe the connecting rod. The locations of the heaters and some die nodes are indicated.

heaters often used for start-up. The function is similar to the use of waterlines except that heat is added rather than removed.

The heat transfer coefficients of casting to die and for spraying are artificial coefficients for heat transfer between dissimilar materials. They were cohsen after considering general handbook values and were then slightly modified to yield good correlation between actual and predicted die temperatures. The values used were:

- Heat transfer coefficient for lubricant spraying = 1015 BTU/hr. ft.^{2} ^oF

- Heat transfer coefficient for Die-Open = 1.75 BTU/hr. ft.^{2} ^oF

- Heat transfer coefficient for casting to die = 2300BTU/hr. ft.^{2} ^oF.

Computed die cavity temperature profiles for the connecting rod are shown in Fig. 16. Die temperatures for nodes 247, 257, 264 and 289 as shown in Fig. 15 correlated favorably with temperature surveys taken during production. Actual die temperatures were measured on a production die using an infrared pyrometer to survey the cavity surface immediately after opening.

Normal Cycles Without Heaters

Simulated cycles which correspond to production were first calculated for the connecting rod die cavity without auxiliary heating and cooling. This situation would correspond to the normal cycles in the middle of a production run. The initial temperature of the die was chosen as a uniform value of 540°F (which turned out to be an equilibrium condition for casting and die). The temperature response of two types of nodes were examined: nodes on the cavity surface and nodes away from the cavity surface. The temperatures responses of the nodes (257, 261, 330, 383 and 386 as seen in Fig. 15), for successive cycles are shown in Fig. 17. The peak and minimum temperatures are indicated for cavity surface nodes. For nodes away from the cavity surface only the peak temperatures are shown.

The temperature responses for the cavity surface nodes (No. 261 and 257), show the expected rapid increase during solidification and dwell followed by a decrease during die-open and lubricating time. During the final die-open time the temperatures of the nodes equalize with those around them, in this instance showing a rise in temperature. The nodes away from the cavity surface (Nos. 330, 383, and 386) do not undergo the same wide temperature with successive cycles indicating that equilibrium has not been reached. The slight overall increase in temperature also holds true for cavity surface nodes.

The first cycle was run with a 2-sec. die-open interval followed by a 1-sec. spray to check the effect of lubrication time on temperature. The shorter spraying time decreased the temperature drop during the interval. The temperature after the 6-sec. open-to-air equalization period are about 50°F higher than with the longer spray. The result indicates that the die closing temperature can be influenced by the length of the spraying time.

Effect of Auxiliary Heaters on Cycles

The change in die temperature resulting from the use of auxiliary die heaters was examined by adding heater nodes to the simulation of the location indicated in Fig. 15. The general fluctuations in temperature response are the same as the changes for cycling with no heaters (Fig. 18). The maximum and minimum temperatures of the cycles are shown in Table 1 and 2. The effect of the heaters on die temperature depends on the location of the nodes with respect to the heater and cavity surface. Nodes either in contact with or near a heater tend to have higher temperatures. Nodes away from the heaters show no temperature change. In the first few cycles modeled, the die temperature gradients set up by solidification of the casting are not affected by the use of heaters.

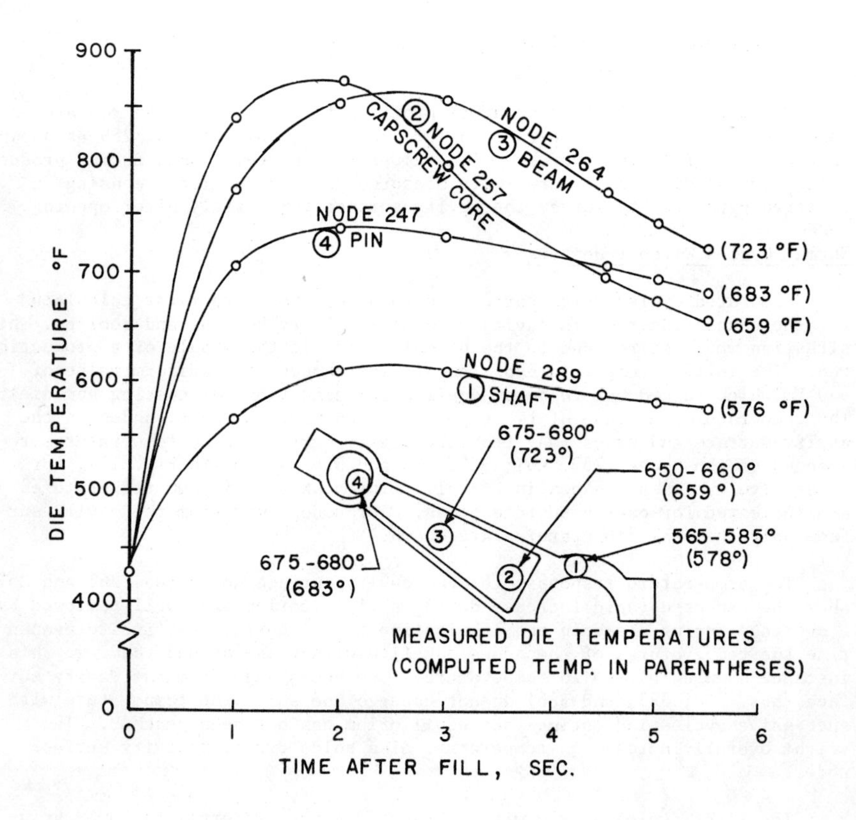

Fig. 16. Computed die temperature profiles for casting with 540°F uniform initial die temperature and 1300°F melt. Measured die temperatures in four areas are indicated for comparison.

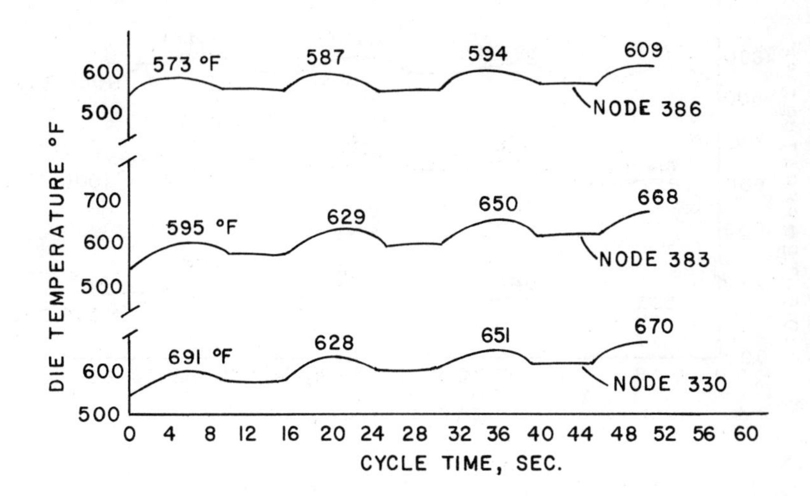

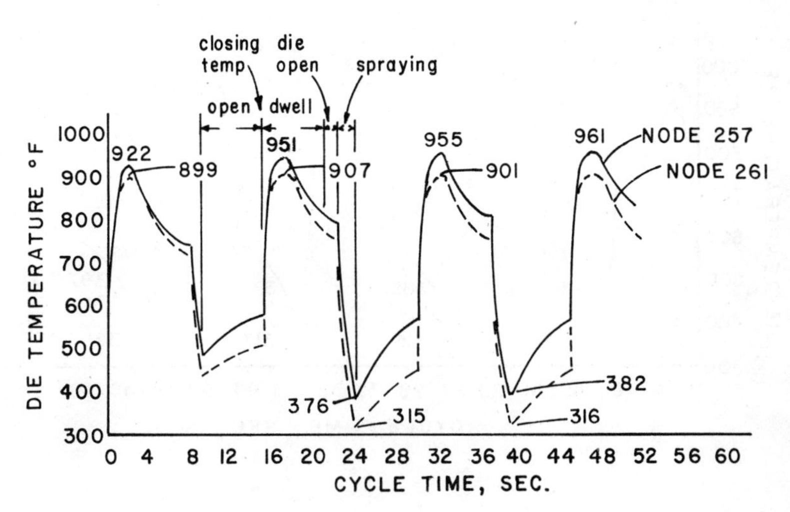

Fig 17. Computed temperature profiles for six die nodes during four normal cycles without die heaters.

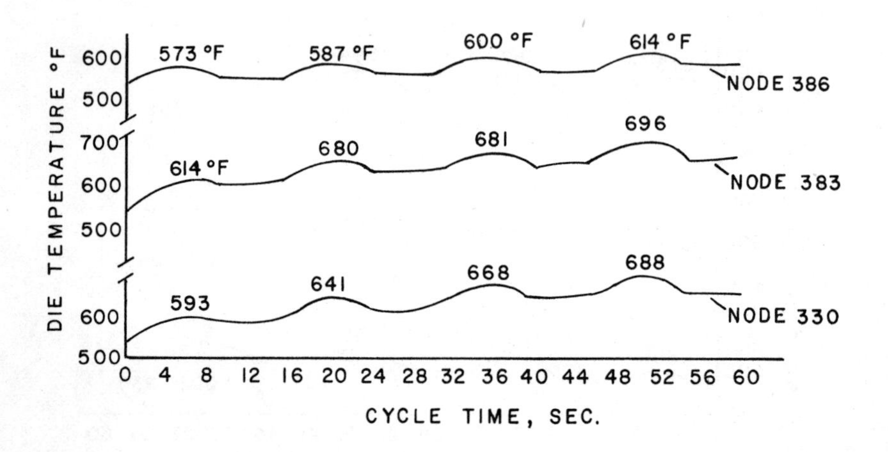

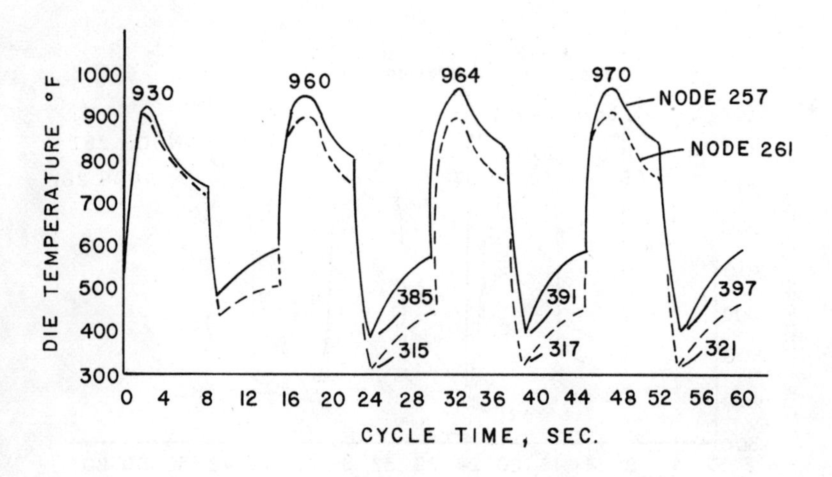

Fig 18. Computed temperature profiles for six die nodes during four cycles with two heaters in the die.

Effect of Heaters on Die Start-Up

To investigate the effect of heaters on die start-up, the results of two simulations were compared:

- Temperature response following several casting cycles in an initially cold (85°F), die.

- Temperature response produced by two heaters in an initially cold (85°F), die but without casting cycles.

The first simulation predicts die temperatures that would be encountered under normal casting procedures, while the second preducts those that would be expected if the die heaters were used.

Comparisons of the predicted temperature response for nodes 257 and 261 are shown in Fig. 19 and 20. With two heaters and no cycling, node 257 reaches 380°F and node 261 reaches 460°F in 60 sec. By superimposing the temperature response for cycling from a cold die for nodes 257 and 261 on the two-heater start-up response, it is shown that the closing die temperature for node 257 increased at about the same rate for both heating methods. On the other hand, the closing die temperature of node 261 when cycled from a cold die lags nearly 200°F behind the response due to the heaters (Fig. 20.)

Typical temperature distribution after 60 sec. as predicted by the simulations are shown in Fig. 21. The temperature due to cycling from a cold die ranges from 400°F at the large end of the rod to 250°F at the small end of the rod. The distribution resulting from warm-up by heaters ranges from 300°F at the large end to 450°F at the small end. An initial gradient, therefore, is imposed on the die that would be different from the one produced by the casting cycle. Thus, the possibility exists for changing the solidification pattern of the casting to promote directional solidification away from critical zones.

Table 14. Effect of a Heater on Peak Temperatures

No. of Heaters:	Peak Temperatures, °F											
	Cycle 1			Cycle 2			Cycle 3			Cycle 4		
	0	1	2	0	1	2	0	1	2	0	1	2
Node Nos.												
257	922	930	930	951	960	960	955	964	965	961	970	971
261	899	899	900	907	907	920	901	902	919	906	907	923
330	591	593	594	628	641	646	651	668	672	670	688	700
383	595	614	614	629	660	660	650	681	681	668	696	697
386	575	573	573	587	587	587	597	600	600	609	614	614

Effect of Heaters on Short Interruptions

Unscheduled interruptions in the casting cycle that are short duration (one minute or less), were examined. This condition was simulated with and without heaters. It is predicted that if the interruption occurs just before casting removal, the die will cool at about 15°F/min. when it is without heaters and open to the air. After spray, the drop is translated into about a 100°F difference in the closing temperature as compared to an uninterruped cycle. When heaters are placed in the die under the same conditions, the temperature actually increased during the interruption.

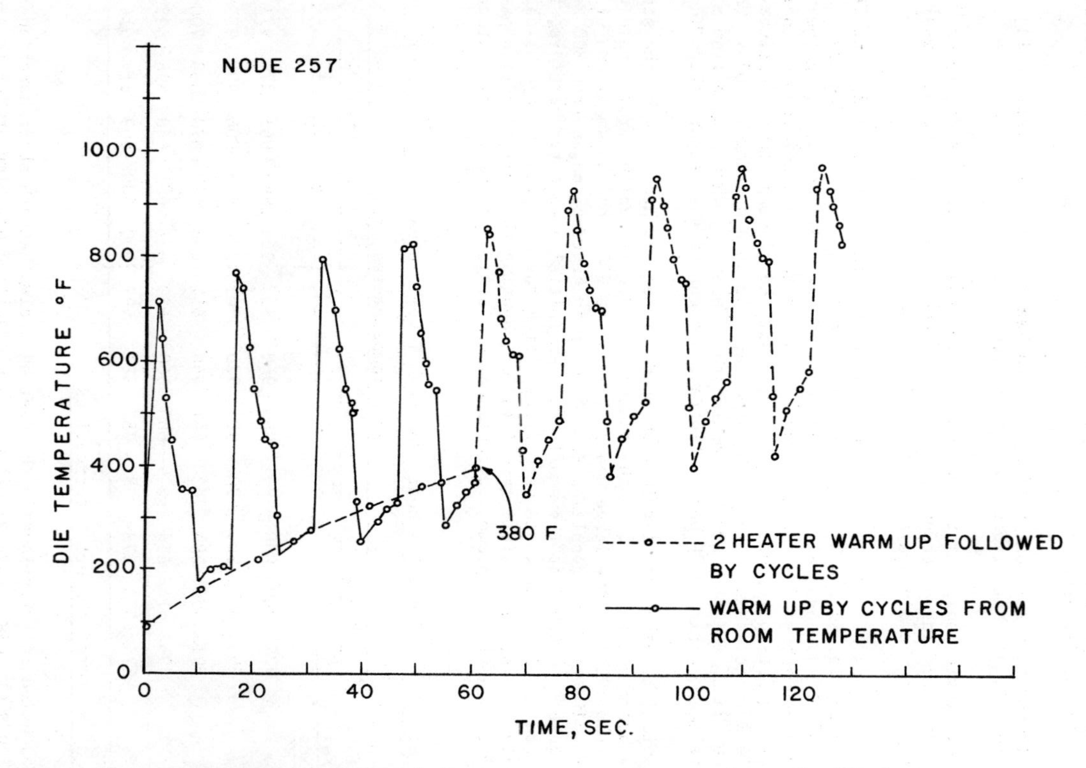

Fig. 19. Comparison of predicted die temperature response for node 257 during a warm-up by cycling and a warm-up using two heaters followed by cycling.

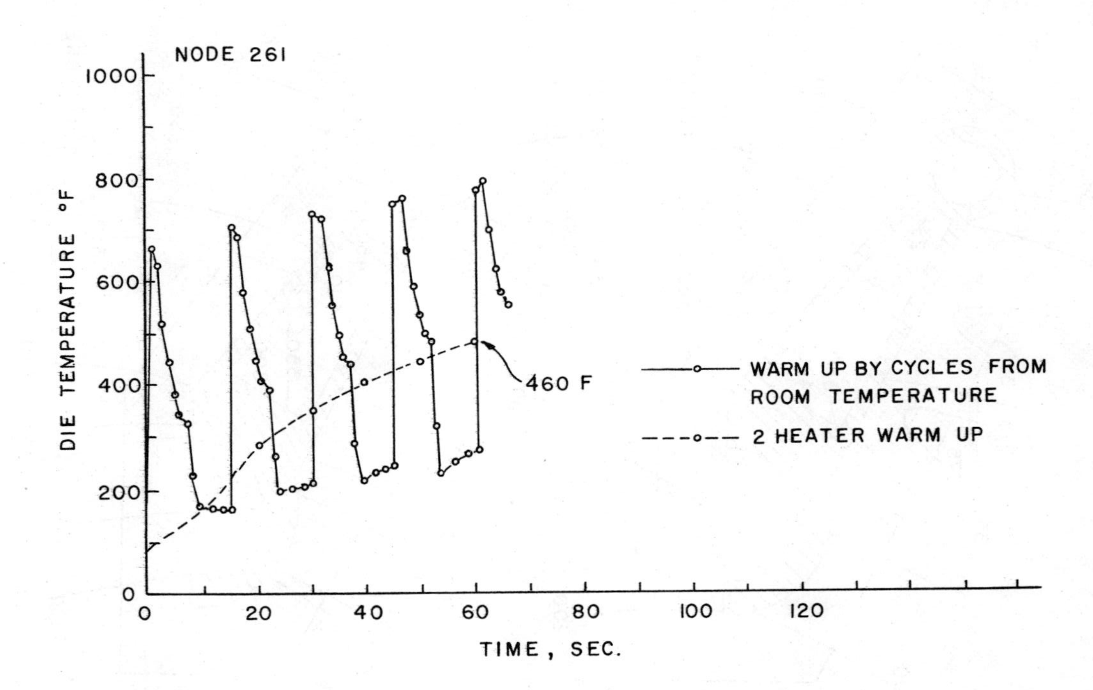

Fig. 20. Comparison of predicted die temperature response for node 261 during warm-up by cycling and a warm-up using two heaters.

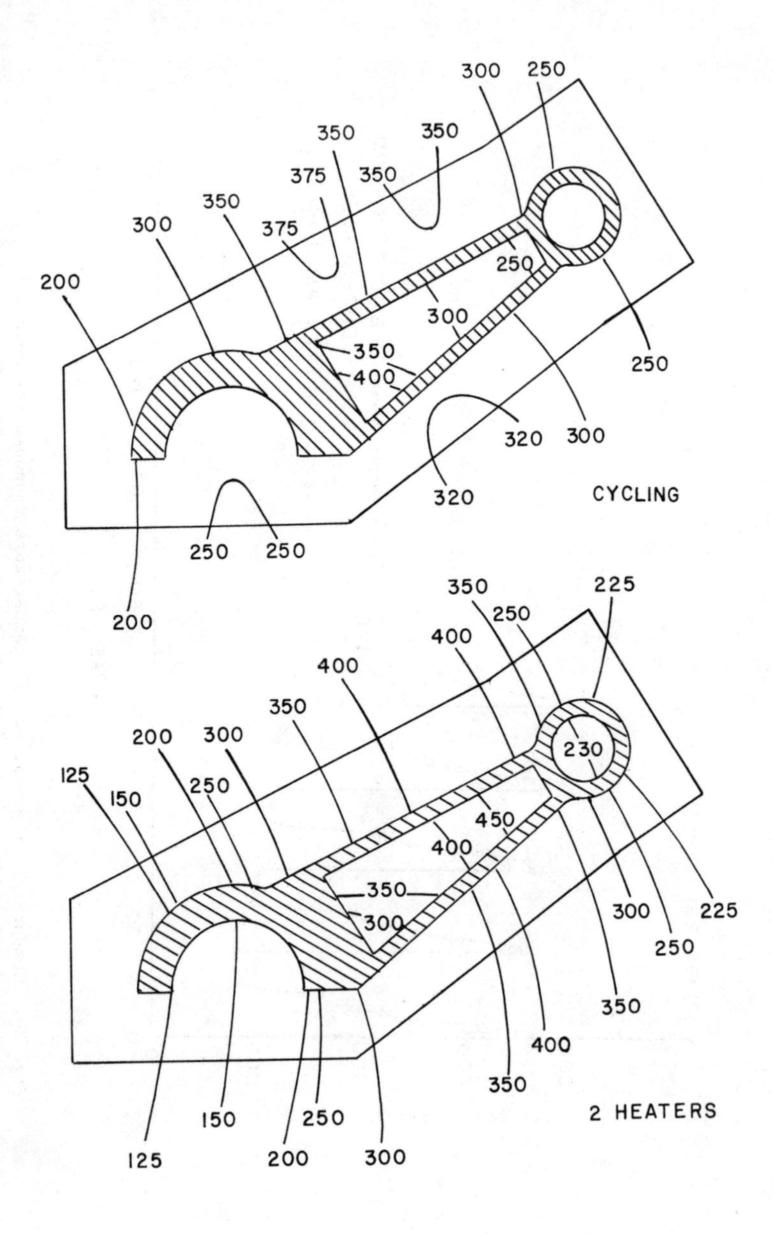

Fig. 21. Temperature profiles in the die resulting from normal cycling from a cold die (top) and from a two heater warm-up (bottom).

Table 15. Effect of a Heater on Minimum Temperatures

	Minimum Temperatures, °F											
No. of	Cycle 1			Cycle 2			Cycle 3			Cycle 4		
Heaters:	0	1	2	0	1	2	0	1	2	0	1	2
Node Nos.												
257	473	478	479	376	385	385	382	391	392	389	387	399
261	440	440	448	315	315	328	316	317	331	320	321	335
330	574	582	583	691	615	622	623	640	652	643	659	673
383	574	602	607	597	630	637	616	648	657	643	661	671

Conclusions

A relatively simple computer model has been applied to simulate the solidification process in a die casting. The versatility of the program has been demonstrated by its application to several real casting, including a complex cylinder block casting.

It has been shown that simulation will predict with reasonable accuracy the solidification pattern of the casting, the die temperature response, and the solidification time. The availability of such a program lays the groundwork for additional studies. For example:

1. By calculating the solidification pattern and noting which areas freeze without feeding of liquid metal, the location of shrink porosity was predicted. The possibility exists, therefore, of investigating ways to eliminate or move proosity. Changes that could be examined easily include application of temperature gradients, insulating cores, relocating gates and changes in section size.

2. Predicted solidification times could be used to examine solidification rates. It is known that faster solidification produces better mechanical properties. Hence, microstructural refinement through localized faster heat removal could improve mechanical properties in critical areas.

3. Investigations of die temperature profiles yield methods of increasing die life by identifying the location of hot spots. Further, it is shown that thermal shock can be reduced by appropriate die temperature control.

The simulation program was also used to study thermal conditions in the die during the total die casting cycle. The results have been verified by comparing predicted and actual die temperatures. The use of the model to study the effect of machine cycle stages and auxiliary in-die heaters on die temperature response has been described.

References

1. J.G. Henzel and J. Keverian, "The Theory and Application of a Digital Computer to Predicting Solidification Patterns, "Journal of Metals, Vol. 17, (May, 1965), pp. 561-568.

2. J.F. Wallace and W.J. Stuhrke, "Gating of Die Castings," Modern Castings, Vol. 44, (January, 1966), pp. 51-79.

3. B.F. Caswell and P. Lorentzen, Some Theoretical Aspects of Heat Transfer in the Die Casting Process," Die casting Engineer, (March-April, 1968), pp. 26-32.

4. R.D. Pehlke, R.E. Marrone and J.O. Wilkes, Computer Simulation of Solidification, AFS Monograph, 1976.

5. G.W. Gatecliff, "A Digital Simulation of a Reciprocating Hermetic Compressor Including Comparisons with Experiments," Ph.D. Thesis, University of Michigan, 1969.

6. R.B. Weatherwax, O.K. Riegger, "Theoretical and Experimental Studies of Die Casting Techniques for Small Engine Connecting Rods," Transactions, Society of Automotive Engineers, 1971.

7. R.B. Weatherwax and O.K. Riegger, "Computer-Aided Solidification Study of a Die Cast Aluminum Piston," Transactions, American Foundrymen's Society, Vol. 85, (1977) pp. 317-322.

8. R.B. Weatherwax and O.K. Riegger, "Solidification Studies of Die Cast Aluminum Small Engine Crankcase Castings," Transactions, Society of Die Casting Engineers, Vol. 9, (1977), Paper No. G-T77-016.

9. M.C. Flemings, S.Z. Uram and H.F. Taylor, "Solidification of Aluminum Castings," Modern Castings, Vol. 38, No. 6, (December, 1960) pp. 66-80.

COMPUTER PROGRAMS FOR HEAT TRANSFER IN METAL CASTINGS

John L. Jechura and James O. Wilkes
Department of Chemical Engineering

and

A. Jeyarajan and Robert D. Pehlke
Department of Materials and Metallurgical Engineering
The University of Michigan
Ann Arbor, Michigan 48109

Abstract

The merits of 25 different finite-difference and finite-element programs are considered for simulating the three-dimensional transient heat transfer occurring during the solidification of metal castings. A short list of eight programs is recommended for further study. For program testing, three representative castings have been selected: interconnected aluminum discs, a gun-metal flanged barrel, and a steel bearing cap. The number of arithmetic operations involved in representative finite-difference and finite-element procedure is also studied, and formulas are derived that will allow computing times to be extrapolated to new situations.

Introduction

Preliminary research toward the goal of recommending a computer software system capable of simulating the heat transfer in metal casting has been done in three areas:

1. Identification of computer programs for unsteady-state heat transfer.
2. Development of representative casting problems for checking and comparing the computer programs.
3. Comparison of the number of arithmetic operations for the two major classifications of numerical methods used: the finite-difference method (FDM) and the finite-element method (FEM).

The purpose of this paper is to report the progress made in each of the above areas, and to outline the work remaining to be done.

Identification of Computer Programs for Unsteady-State Heat Transfer

Regarding software identification, preference has been directed toward existing general-purpose programs, mainly because the development of new computer programs for each metal casting problem is a major task involving program development, debugging, testing, and verification.

In developing criteria for evaluating software, all acceptable programs should be able to model the following:

1. Transient heat conduction; this presupposes that the effects of the natural convection can be approximated using enhanced liquid thermal conductivity.

2. Various boundary conditions, including both convective and radiation boundary heat, and also a mold-metal interface resistance.

3. Problems in three spatial dimensions.

4. Variable thermal properties due to temperature and/or time effects.

5. Latent heat effects due to solidification, either directly (such as the method reported by Eyers, et al. (2)) or indirectly using a modified specific heat (3,4).

A group of 25 existing computer programs was examined as to their cababilities for modeling metal casting. The results of this survey are summarized in Tables 1, 2, and 3. Table 1 lists those programs that meet the criteria for performing the metal-casting thermal analysis. Table 2 summarizes other programs examined that do not have the needed computational capabilities. Finally, Table 3 provides a summary of various programs which may be applicable, but contact with their administering organizations for the purpose of confirming their capabilities has not yet been completed.

It should be cautioned that these results, shown in the summary tables, reflect the status of the programs determined as of Fall 1980. Programs that are currently unsuitable could be modified in the near future and become more applicable. A prime example of this continuous upgrading of computer software is MSC/NASTRAN. At the end of 1979, the program used a constant specific heat, making it unacceptable for use in this study. However, the recently released Version 60 has new capabilities for a user-supplied variable specific heat.

Classification of the software as either FDM (finite-difference method) or FEM (finite-element method) not only denotes the method used for the spatial dependency of the temperature field, but also serves to give a rough indication of the program's capabilities. In general, FDM programs, however, are generally geared toward structural-type analysis, with the thermal analysis being just one of many capabilities.

The eight existing programs listed in Table 1 have been identified as candidates for simulating the transient heat transfer during solidification; in each case User's Manuals have been secured for further study. During the continuation of this work we shall select the most promising programs, hoping to come to a short list of two or three. We plan to run sample castings (see next section) with each of the short-listed programs to obtain benchmarks regarding computing time, cost, accuracy, and ease of use. Then, we shall make final recommendations as to the program or programs that appear most suited to the problems of the foundry industry.

TABLE 1. Computer Software Applicable to Metal-Casting Simulation

Program	Organization and Contact	Comments
ANSYS	Swanson Analysis System, Inc. Houston, PA David Dietrich (412) 746-3304	FEM - Frontal solution method can deal with a wavefront of over 1000 degrees of freedom.[11]
HEATING-5	Oak Ridge National Laboratory Oak Ridge, TN Betty McGill (615) 574-6176	FDM - Geometry is approximated as a set of rectangular parallelipipeds whose faces are parallel to the coordinare planes.[12]
MARC	MARC Analysis Research Corp. Palo Alto, CA Michael B. Hsu (415) 326-7511	FEM - Originally developed for welding problems.
MSC/NASTRAN	MacNeal-Schwendler Corp. Los Angeles, CA Jerry A. Joseph (213) 254-3456	FEM - Recently issued Version 60 can deal with temperature-dependent specific heats.[9]
SINDA	COSMIC Library University of Georgia Athens, GA Stephen J. Horton (404) 542-3265	FDM - Solves the general diffusion-type equations.
SMART II	Institut für Statik und Dynamik der Luft- und Raumfahrtkonstruktionen University of Stuttgart Pfaffenwaldring 27 7000 Stuttgart 80 West Germany Professor J. H. Argyris 0711-7841	Large FEM package, available on CDC, IBM and UNIVAC equipment.
TRUMP	Oak Ridge National Laboratory Oak Ridge, TN Betty McGill (615) 574-6176	FDM - Treats the latent heat effect as a heat release at constant temperature, using a modified specific heat or using a combination of both.[13]
WECAN	Westinghouse Pittsburgh, PA S. E. Gabrielse (412) 256-5040	FEM - Originally developed and used in-house, now being marketed on the outside.[14]

TABLE 2. Computer Software Not Applicable to Metal-Casting Simulation

Program	Organization	Comments
ADINAT	K. Bathe Massachusetts Institute of Technology	FEM - Constant specific heat, no other provision for latent heat effects.[14,15]
AYER	Los Alamos Scientific Lab.	FEM - Only two-dimensional geometry.[1]
ATLAS	Boeing	FEM - No transient heat transfer.[10]
FETE	Foster Wheeler Corp.	FEM - Two dimensions without capabilities for latent heat effect.[17]
GT STRUDL	CDC Strudl	FEM - No heat transfer.[10]
MCAUTO STRUDL	McAuto	FEM - No heat transfer.[10]
SAPIV	University of Southern California at Berkeley	FEM - No heat transfer.[10]
STARDYNE	CDC	FEM - No heat transfer.[10]
SUPERB	SDRC Cincinnati, OH	FEM - No transient heat transfer.[16]
SAP6	University of Southern California	FEM - No heat transfer.[10]

TABLE 3. Computer Software with Pending Status

Program	Organization	Comments
ASAS	Atkins Research and Development	FEM
BETA	Boeing	FDM
CINDA-3G	Center for Information and Numerical Data Analysis and Synthesis Purdue University	FDM - Presumably superseded by MITAS and SINDA
FESS FINESSE	University of Wales at Swansea	FEM
MITAS	Martin-Marietta	FDM
NISA	Engineering Mechanics Research Corporation	FEM

Development of Representative Casting Problems for Checking and Comparing the Computer Programs

Existing computer programs are commonly checked using simple heat-transfer problems that can be solved analytically (7,8). Comparison of two general-purpose programs with a specific heat transfer problem is also reported.[17] The casting solidification is a complex problem in heat transfer due to factors such as liberation of latent heat, temperature-dependent thermal properties of casting and molding materials, and intricate geometry of commercial castings. To get an evaluation helpful to the foundrymen using these programs, the existing computer programs will be checked using actual casting conditions and geometry. For example, the latent-heat evolution largely determines the temperature distribution during solidification, and hence any test case must include metal solidification.

It is also desirable that the castings chosen should be commercial castings or laboratory castings with the complexities commonly encountered in commercial castings. This would facilitate examination of the ease with which casting geometry and other parameters can be input to the program.

An ideal way of verifying the results of the simulation is to check them against the experimental results obtained under identical casting conditions. In reality, the complexity of the casting process necessitates simplifying assumptions in simulation and introduces deviations from the desired conditions in experiments. For example, instantaneous filling of the mold cavity with liquid metal is usually assumed in simulation even though mold filling takes place over a short period. The difficulties encountered in accurately locating thermocouples in experiments could be an example of source of experimental error.

Casting A

The first casting proposed for study is in the form of interconnected discs attached to a shoe-shaped riser, as shown in Fig. 1. Experimentally monitored and computed temperature distributions are both reported[18] for a pure aluminum casting of this shape poured at 1350°F into a dry sand mold at 70°F. The mold, of diameter 5 in. and length 11 in., was hand-rammed to a nominal density of 100 lb/ft^3 using 4% western bentonite, 2.5% water and AFS 80 mesh New Jersey silica sand with a nominal green compressive strength of 5.0 at a nominal green hardness of 75. The mold was dried at 350°F for 24 hours. The thermal properties of this particular mold material have been measured and reported (19). The solidification of the casting was monitored using thermocouples of 30 gage butt-welded chromel-alumel wires protected by 1 mm I.D. fused silica tubes.

Casting B

The second casting is a flanged barrel, as shown in Fig. 2, formed from 85-5-5-5 leaded gun metal poured at 2050°F. The mold is 10 in. diameter and length of 20 in., and is initially at 70°F; it consists of core sand inside the barrel and dry sand surrounding the casting. Not shown in Fig. 1 are graphite chills placed on: (a) the end of the barrel, (b) the cylindrical surfaces of the barrel, and (c) the lower end of the flange. Even though the casting is axisymmetric, the heat transfer during solidification is not axisymmetric due to the placement of the chills. Study of the flanged barrel shape is important, since the "L" shaped joint encountered in the cross section occurs in many commercial castings and poses a complexity in design of castings using empirical methods. Also, the modifications in the temperature

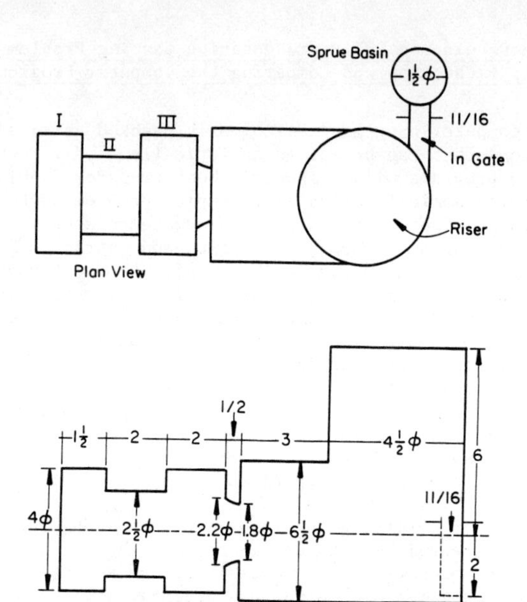

Fig. 1. Plan and cross section elevation views of Casting A.

profiles produced by the chills are of interest.

Casting C

The third casting is a bearing cap, shown in Fig. 3. This is a complex commercial casting for which empirical designs and their successes in actual practice are reported by Wlodawer (6). It is poured from steel at 3000°F into a dry sand mold at 70°F with graphite chills present. The mold dimensions are 400 cm x 250 cm x 200 cm. The complex shape of the casting offers an opportunity to study the ease of casting geometry input to the program.

A Comparison of the Number of Arithmetic Operations Needed for FDM and FEM Solutions of the Heat-Conduction Equation

This section gives additional perspective on the relative computing times of the FDM and FEM solutions by directly counting the number of arithmetic operations involved in each when advancing the solution over a single time step. In view of the number of simplifying approximations that are made, the comparison will have some obvious limitations. However, the results will be useful at least when making rough comparisons of the FDM and FEM, and also when extrapolating known execution times to situations involving a finer mesh and/or a higher number of space dimensions.

The following simplifying assumptions are made:

1. Density, ρ, and specific heat, C_p, do not change with time. However, the thermal conductivity, k, is a function of position and/or temperature. Thus, the thermal diffusivity, $\alpha = k/\rho C_p$, is a variable, and the heat-

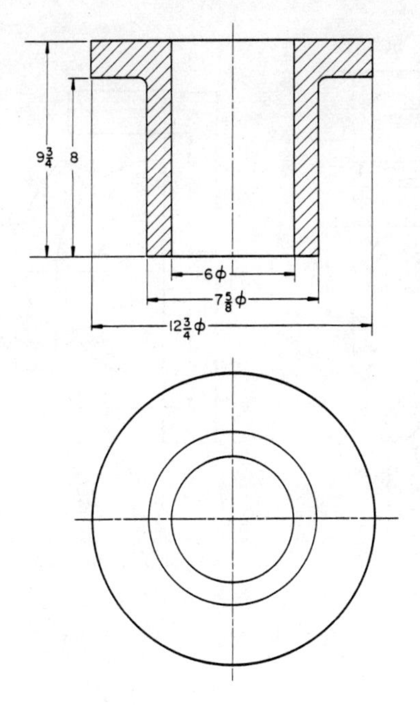

Fig. 2. Flanged barrel casting.

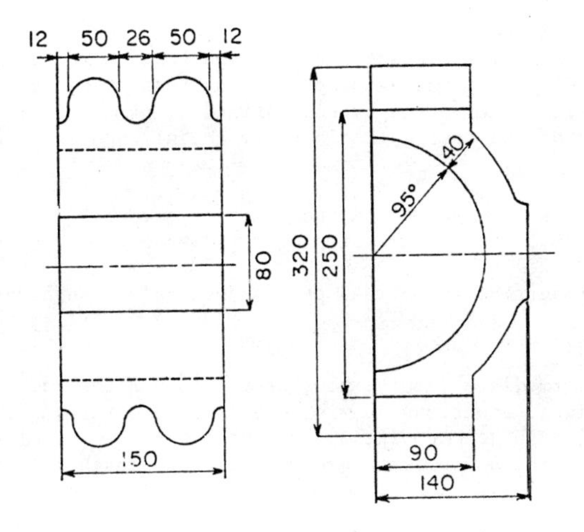

Fig. 3a. Bearing-cap casting.

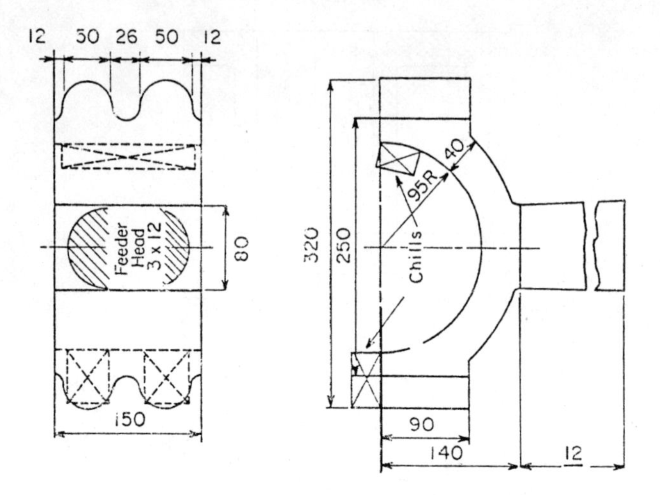

Fig. 3b. Riser and chill design for bearing-cap casting.

conduction equation (in one, two, or three space dimensions) is:

$$\nabla \cdot \alpha \nabla T = \frac{\partial T}{\partial t} .$$

That is, we have made some, but not complete, accommodation of varying physical properties.

2. Only simple geometries will be examined, namely: a line, a square, and a cube, in one, two, and three space dimensions respectively. Any special considerations at the boundaries will be ignored, the number of unknowns being either N, N^2, or N^3.

3. The four basic operations of addition, subtraction, multiplication, and division (A,S,M,D) are assumed to consume equal computing times. Although this is a relatively poor assumption, our results can still be expected to give useful order-of-magnitude estimates, particularly if the relative occurrences of the four operations are roughly the same for each method used.

4. All operations are counted on the basis of advancing the computed temperatures across a single time step of duration Δt.

5. Physical properties at various points need only be evaluated once per time step.

Useful intermediate results have been obtained for the solution of the n simultaneous linear equations $A\bar{v} = \bar{b}$, involved in both the FDM and FEM. If A is tridiagonal, a LU decomposition solution takes $8n - 7$ operations. And if A is banded symmetric, a Doolittle decomposition approach consumes

$$\frac{1}{2}[n(3B^2 + 13B + 2) - 2B(B+1)(B+3)]$$

operations, where the semi-bandwidth B is defined as the number of nonzero diagonals above the main diagonal.

When investigating the FDM solution of $\nabla \cdot \alpha \nabla T = \partial T / \partial t$, we have studied implicit methods that have second-order accuracy both in space and time. In one, two, and three space dimensions, these amount to: (i) the Crank-Nicolson method, (ii) the implicit alternating-direction (IAD) method, and (iii) a Crank-Nicolson modification of IAD, as outlined by Carnahan, Luther, and Wilkes (5). Each case involves the solution of tridiagonal systems, and the total number of arithmetic operations has been determined and is given in Table 4 below.

For compatibility, we have investigated FEM solutions that are also second-order accurate both in space and time. In one dimension, line segments are employed. In two dimensions, both linear triangular and linear quadrilateral elements are involved. And in three dimensions, both linear tetrahedral and trilinear cube elements are examined.

The number of arithmetic operations per time step for the FEM and FEM are summarized in Table 4. (For brevity, the results for the FEM are approximate, taking only the leading two terms of most significance.

TABLE 4. Number of Arithmetic Operations for A Single Time Step

Number of Space Dimensions	FDM		
1	$18N - 7$	$25N - 18$	
2	$2N(19N - 7)$	$\frac{3}{4}N^4 + \frac{27}{2}N^3$	(linear triangles)
		$\frac{3}{2}N^4 + \frac{41}{2}N^3$	(linear quadrilaterals)
3	$3N^2(34N - 7)$	$\frac{3}{2}N^7 - N^6$	(linear tetrahedra)
		$\frac{3}{2}N^7 + 2N^6$	(trilinear cube)

By inserting representative values for N (such as $N = 10$), it appears that the FEM takes substantially more operations than the FDM, particularly for two and three space dimensions. However, this apparent disadvantage of the FEM must be tempered by three important considerations not yet fully explored:

1. The possible use of the FEM of quadratic and isoparametric elements for reducing the total number of unknowns.

2. Use of equation solvers in the FEM that recognize sparseness within the bandwidth.

3. Difficulties encountered in the FDM of accurately representing curved boundaries and unusual geometric shapes.

When properly accounted, these considerations are expected to make the FEM much more competitive, and such matters will also be pursued during the continuation of our work.

References

1. W. C. Erickson, "Computer Simulation of Solidification," AFS International Cast Metals Journal, March 1980, pp. 30-41.

2. N. R. Eyers, D. R. Hartree, J. Ingham, R. Jackson, R. J. Sarjant, and J. R. Wagstaff, "The Calculation of Variable Heat Flow in Solids," Philosophical Transcations of the Royal Society of London, Series A, 240 (1948) pp. 1-57.

3. G. Comini, S. DelGuidici, R. W. Lewis, and O. C. Zienkiewicz, "Finite Element Solution of Nonlinear Heat Conduction Problems with Special References to Phase Change," International Journal of Numerical Methods in Engineering, 8 (1974) pp. 613-624.

4. R. D. Pehlke, R. E. Marone, and J. O. Wilkes, Computer Simulation of Solidification, American Foundrymen's Society, DesPlaines, IL, 1976.

5. B. Carnahan, H. A. Luther, J. O. Wilkes, Applied Numerical Methods, John Wiley & Sons, 1969.

6. R. W. Wlodawer, Directional Solidification of Steel Castings, pp. 12 and 61; English Translation, Pergamon Press, Oxford, 1966.

7. HEATING-5: Generalized Heat Conduction Code System, report from the RISC Computer Code Collection, March 1977.

8. TRUMP: A Program for Transient and Steady-State Temperature Distributions and Multidimensional Systems, report from the RISC Computer Code Collection.

9. Personal communication, Jerry Joseph, MacNeal-Schwindler Corporation.

10. Internal Digital Equipment Corporation report.

11. Personal communication, David Dietrich, Swanson Analysis Systems, Inc.

12. HEATING-5: Generalized Heat Conduction Code System, report from the RISC Computer Code Collection, March 1977.

13. TRUMP: A Program for Transient and Steady-State Temperature Distributions and Multidimensional Systems, report from the RISC Computer Code Collection.

14. Personal communication, Dr. Karl Bathe, Department of Mechanical Engineering, Massachusetts Institute of Technology.

15. ADINAT Manual available from Dr. Karl Bathe, Department of Mechanical Engineering, Massachusetts Institute of Technology.

16. Personal communication, Wayne Simon, SDRC.

17. C. T.Hsu, "Comparison of a Finite Element and a Finite Difference Computer Code in Heat Transfer Calculations," ASME paper 79-PVP-63.

18. A. Jeyarajan, Computer-Aided Design of Castings, Ph.D. Thesis, Department of Materials and Metallurgical Engineering, The University of Michigan.

19. M. J. Kirt and R. D. Pehlke, "Determination of Material Thermal Properties Using Computer Techniques," AFS Trans., 81 (1973) pp. 524-528.

THERMAL PROPERTIES OF MOLD MATERIALS*

J. G. Hartley and D. Babcock
Georgia Intitute of Technology
School of Mechanical Engineering
Atlanta, Georgia 30332

Reliable predictions of the cooling and solidification of metal castings depend significantly upon accurate thermal property data of both the cast metal and the mold medium. Whether the predictions account for the presence of the mold through a detailed thermal analysis or by incorporating only gross effects, the importance of accurate property data cannot be overemphasized.

The available thermal properties of sand and sand-based mold media are compiled and evaluated, and the influence of moisture content, binder content and ramming density on the thermal properties of sand-based mold materials is discussed. Recent experimental data are included which indicate that an optimum moisture content and an optimum binder content exist for these materials.

* This work is supported by the National Science Foundation under Grant No. DAR78-24301

Introduction

In the metal casting industries the principal heat-transfer medium involved in the solidification and cooling of the casting is, of course, the mold. Mold practice can conveniently be broken down into so-called "soft" and "hard" mold processes. The biggest portion of soft mold practice is comprised of green sand techniques, where essentially a particulate silica sand is used. The silica sand is bonded with bentonite clays and thus contains a considerable portion of associated water (up to 6%) plus other additives (wood, flour, sea coal, etc.). Also coming more into prominence are chemically bonded sands with either organic or inorganic type bond materials.

Thermal transport properties or chilling power of molding sands have commanded considerable attention of foundry researchers over the last thirty years or more. The methods utilized in their determination have been classified by Berry (1) along the following lines:

a. Steady state methods for measuring thermal conductivity and thermal diffusivity of the mold,

b. Unsteady state methods, for example those involving the making of castings, for determination of mean thermal conductivity, mean thermal diffusivity and mean heat diffusivity,

c. Methods of calculating thermal properties.

The steady state and the transient experimental methods each have advantages and disadvantages although the transient techniques seem to be preferred in more recent investigations. Atterton (2) and Whitmore and Ingerson (3) used a steady state method based on measuring the temperature disbtirubtion in a cylindrical sample heated from within to determine the thermal conductivity of sand specimens. A variety of transient measurement techniques have been developed for thermal property determination. These methods can be grouped into two general categories. The first consists of methods based on the hot wire or thermal probe technique used, for example, by Ninomiya, et. al. (4). The second category includes methods based on the actual thermal response of a mold material during pouring and solidification of a casting. Investigators using such methods include Ruddle and Mincher (5), Seshadri and Ramachandran (6), and Davies, Hansen and Clausen (7).

Of the materials of primary interest to the present study, significant thermal property research has been limited to bentonite-bonded silica, olivine and zircon sands in the past. For each material the mass fraction of bentonite has been maintained at approximately four percent. However, recent innovations in molding practices, for example:

a. Sodium - silicate or other inorganically bonded sand,

b. No bake or isoset sand mixtures,

c. Shell and other core sand developments,

are not adequately characterized in the literature in terms of their thermal properties.

Recently some very interesting research has been directed toward controlling the thermal conductivity of sand-based materials. Inorganic particulate fillers were added to sands to produce materials much like the bentonite-bonded molding sands, and the addition of binders (e.g. wax, oil, adhesives) produced materials similar to the resin-bonded sands.

Farouki (8) added binders or fillers such as kaolinite, calcium carbonate, fly ash and lime to a graded sand. He found that the maximum thermal conductivity of the air-dried mixtures occurred with binder contents of about eight percent by mass for mixtures prepared with less than four percent moisture by mass. At this binder content the thermal conductivity was found to increase by more than 100 percent over the values for sand without binders. He further concluded that the apparent thermal conductivity of the mixtures was relatively insensitive to the type of binder added to the sand. This is due in part, to the relatively low values of binder thermal conductivity encountered.

Jackson (9) conducted a very thorough analytical and experimental investigation of sand-based mixtures in an effort to develop high thermal conductivity backfill materials for underground electrical cable installations. Several of his findings are of interest to foundry applications. A "unit cell" model was developed which can predict the effective thermal conductivity of a granular medium composed of sand particles, thermal binders and air. The model has been verified experimentally with room temperature measurements, and the greatest increase in thermal conductivity occurred with volumetric binder contents of about 10 percent (corresponding to what Jackson calls the meniscus structure). Further addition of binder beyond this value resulted in only small increases in the thermal conductivity of the mixtures. This result should be directly applicable to resin-bonded sands.

Jackson also found that the thermal conductivity of silica sand-binder mixtures could be increased further by the addition of filler particles such as novaculite, kaolin and graphite. He found that adding filler particles increased the apparent thermal conductivity until an optimum mass fraction of filler particles was reached. Further addition of filler beyond the optimum value caused the apparent thermal conductivity to decrease. The optimum mass fraction of filler was between about eight percent and 16 percent depending on the binder used (e.g. water, wax, latex adhesive, etc.).

As a first step to understanding the complex behavior of heat transfer from castings, the properties of mold materials must be accurately determined. Without a knowledge of properties, no analysis of the heat transfer rates and no estimate of casting-sand temperatures as a function of time can be considered to be complete. Therefore, the thermal properties of sands are of considerable importance to the foundry industry and the overall objectives of the present work.

Experimental Determination of Thermal Conductivity

The current thermal conductivity measurements are based on the thermal probe method introduced by Hooper and Lepper (10). The thermal probe is an experimental approximation of a line heat source in an infinite medium, the analytical solution for which is given by Carslaw and Jaeger (11) as

$$T(t) = T_i - \frac{q'}{4\pi k}\left[\ln\left(\frac{4\alpha t}{r_p^2}\right) - \text{Const.}\right] \quad (1)$$

where r_p is the radius of the probe, T is the temperature of the probe, q' is the power dissipated per unit length of the probe, and α and k are the thermal diffusivity and thermal conductivity, respectively, of the medium. This equation is termed the long time solution since it is valid only for large values of $\alpha t/r^2_p$. The thermal conductivity is inversely proportional to the slope of the straight line and can be calculated from experimental measurements of the temperature-time response of a specimen

$$k = \frac{q' \ln(t_2/t_1)}{4\pi \ (T_2/T_1)} \tag{2}$$

Materials and Sample Preparation

The sands under investigation in the present study are zircon, chromite and olivine sands and a well-graded silica sand (ASTM C-109 Ottawa sand). Two different types of bonding materials are used: bentonite, representive of the filler/particle bonding, and a no bake (alkyd bond) resin binder representative of adhesive bonding.

Samples bonded with bentonite are compacted moist by hand after milling in a small mill to the desired dry density. The samples are compacted in 1-cm layers and the depth of each layer is accurately controlled by a series of spacer rings each 1-cm in height. Samples with resin bonds are poured into compaction molds, and densities are determined by weighing the sample after curing is complete. Bentonite bonded samples are furnace dried prior to testing to remove the moisture from the samples.

Results and Discussion

The first phase of the present study has been focused on room temperature thermal conductivity measurements of bonded and unbonded zircon, chromite, olivine, and silica sands. In addition, the influence of water content, binder content and ramming density on thermal conductivity has been investigated. This section presents a summary of data available in the literature as well as that drawn from initial experimental results associated with the present investigation.

The initial moisture content of bentonite bonded sands has a significant influence on the thermal conductivity of the dry mixture. Jackson's (9) results are useful in interpreting this phenomenon. At low moisture contents moisture is absorbed by the sand and bentonite, and a thin film of water will be present on the sand particles. Water will not be available in sufficient quantity to achieve a complete meniscus structure around the sand particles. As a result, only a fraction of the bentonite particles will coalesce at the sand particle contact points. While this increases the thermal contact between particles thereby increasing the dry thermal conductivity of the mixture, the maximum increase in thermal contact cannot be achieved.

As the initial moisture content is increased further while the dry density is maintained constant, a gradual increase in thermal contact results until an optimum moisture content is reached. At this optimum moisture content, the meniscus structure is nearly complete, and most of the bentonite particles coalesce at the sand particle contact points. The thermal conductivity of the dry mixture which results from mixing at the optimum moisture content cannot be increased substantially by mixing at higher moisture contents. This phenomena is illustrated in Figure 1. For the particular silica sand used in this study, the optimum moisture content was found to be about eight percent of the dry mass of the mixture. At high binder contents, the optimum moisture content would likely increase. However, the binder content also exhibits an optimum value.

The use of bentonite to bond sands not only improves mechanical properties of mold materials but enhances mold thermal properties as well. This has been explained in terms of the improved contact area available for conduction heat transfer. However, it has been observed experimentally (8,9) that for silica sands there exists an optimum binder content beyond which the further addition of binder causes the thermal conductivity of the bonded sand to decrease.

As bentonite is added to the sand while the dry density is maintained constant, higher thermal conductivity sand particles are replaced with lower thermal conductivity binder (as in the case of bentonite). This alone would decrease the thermal conductivity of the mixture; however, as explained previously, thermal contact is increased and the net result is an increase in conductivity. The improvement in conductivity continues as the binder content is increased until further improvements in contact area diminish. Beyond this point (the optimum binder content) the net effect of further addition of binder is to replace high thermal conductivity sand with low thermal conductivity clay. Therefore, the value of the optimum binder content would depend on the relative magnitudes of the thermal conductivities of the sand particles and the binder particles.

The relationship between dry thermal conductivity and binder content for a silica sand is shown in Figure 2. Each mixture was compacted to fixed dry density after being mixed at or above the optimum moisture content. The optimum binder content for the silica sand, Figure 2, is between 10 and 20 percent by dry mass. For this range of binder contents the thermal conductivity of the mixture is about three times the value for the unbonded sand. Similar results were obtained for zircon, olivine and chromite sands. The optimum binder content for

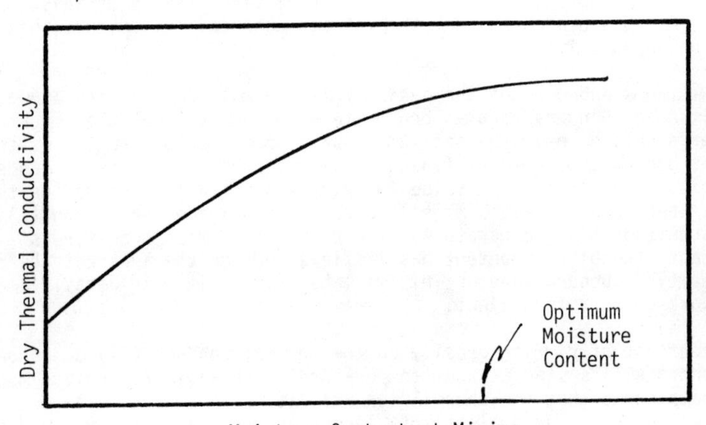

Figure 1. Typical Curve of Dry Thermal Conductivity of a Bonded Sand as a Function of Moisture content at Mixing.

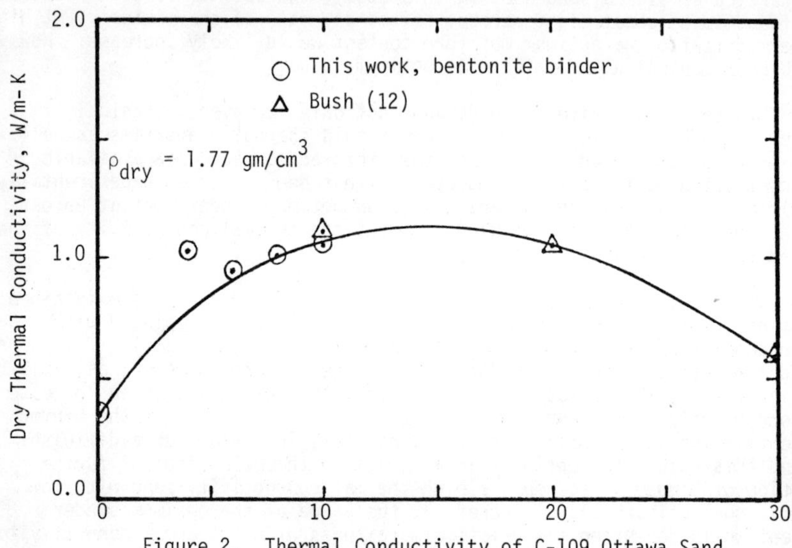

Figure 2. Thermal Conductivity of C-109 Ottawa Sand Versus Binder Content. Mixed at 8% Moisture, Furnace Dried.

zircon sand appears to be much more distinct and occurs at approximately 6 percent binder by dry weight. At this point the thermal conductivity of the mixture is more than 3.5 times the value for the unbonded zircon. The results for olivine and chromite sands are, as yet, incomplete, but it appears that the optimum binder content for olivine sand is at least 8 percent. At this binder content the thermal conductivity is more than twice that of unbonded olivine.

From the standpoint of increasing mold thermal conductivity alone, the addition of binders such as bentonite is effective, but the amount of binder should be near the optimum value. Common mold practice is to add 4 percent bentonite to medium fineness silica, zircon and olivine sands. While this is a near-optimum value for zircon sand, higher binder contents could be used to advantage with silica sand and olivine. Atterton (2) suggests that at high temperatures, where radiation heat transfer is significant, the binder content has little affect on the apparent thermal conductivity of bonded sands. This will be studied in more detail in future work under this project.

The effect of ramming density on the thermal conductivity of bonded and unbonded silica sand at room temperature is shown in Figure 3. Much of the data shows that the thermal conductivity of the sands with constant binder content has a gradual, almost linear increase as the ramming density increases. The data from Atterton, however, shows a much steeper rate of increase of thermal conductivity with increasing density.

The specific heat of bonded sands increases uniformly with temperature. The largest increase in specific heat occurs from room temperature to about

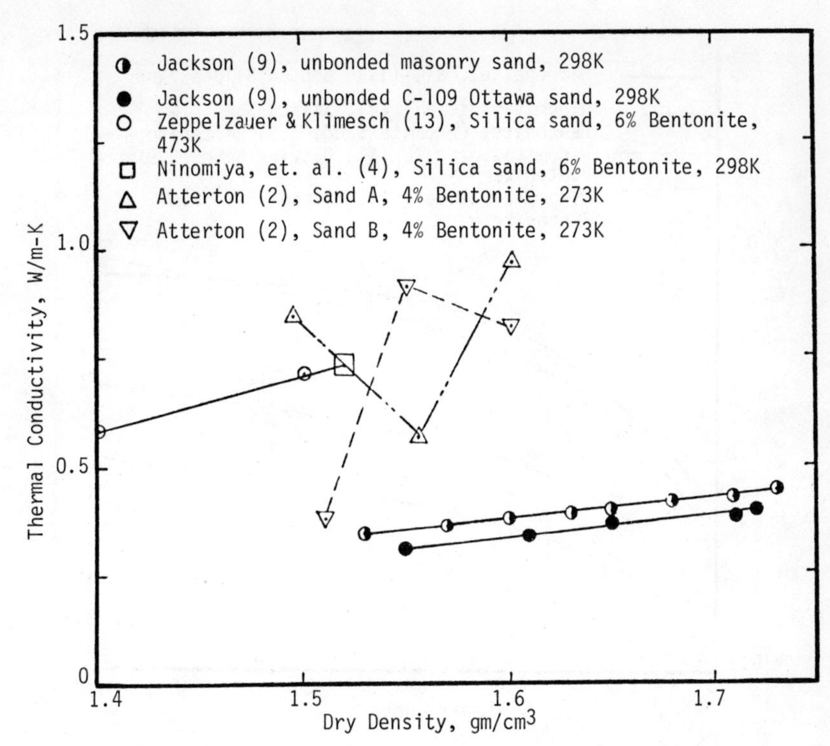

Figure 3. Thermal Conductivity of Bonded and Unbonded Silica Sands Versus Dry Density, Low Temperature.

400 or 500°C. The specific heat as a function of temperature for several sands is shown in Figure 4. The data from Marrone, et. al. (14) represents a compilation of data on silica sand from various sources. The remaining data is from Ninomiya, et. al. (4) who used a carlorimetric method to determine specific heat. The silica sand data agree very well, and the specific heat of the bonded chromite sand is very nearly the same as that for bonded silica sand at all temperatures. Bonded zircon sand has the lowest specific heat of the four sands shown in Figure 4.

The specific heat of a bonded sand can be determined, with reasonable accuracy, from the specific heats of the individual components. Therefore, measurements of the specific heats of the pure sands and of pure bentonite would be of considerable value. Measurements of these values will be obtained during subsequent work on the present project.

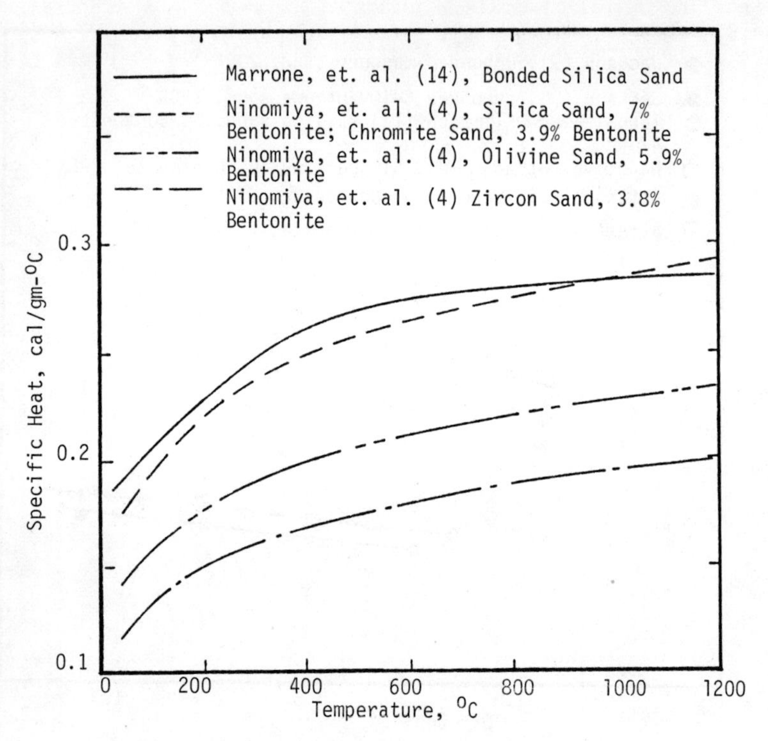

Figure 4. Specific Heats of Bonded Sands Versus Temperature

Conclusions

Analysis of thermal property data from the literature and from experimental results from this project has led to the following conclusions:

1. The amount of moisture present during mixing of bonded sands, for example those bonded with bentonite, has a significant influence on the thermal conductivity of the dry mixture. An optimum moisture content exists which is associated with the development of full meniscus structure. It is this structure which results in the most significant increase in thermal contact area between sand grains for a given binder content. The optimum moisture content for silica sand with up to 10 percent binder by dry mass was found to be about 8 percent water by dry mass.

2. An optimum binder content also exists for bonded sands. The thermal conductivity increases as clay binders are added until the optimum binder content is reached. Further additions of binder cause the thermal conductivity to decrease.

3. The thermal conductivity of bonded and unbonded sands increases as the dry density is increased. The increase is nearly linear and shows the same general behavior even at elevated temperatures.

4. The specific heat of bentonite-bonded sands increases uniformly with temperature. Data on specific heats from various sources in the literature show much better agreement than do the thermal conductivity data.

References

1. J. T. Berry, V. Kondic and G. Martin, "Solidification Times of Simply Shaped Sand Castings", Transactions, AFS, Vol. 67, (1959), pp. 449-476.

2. D. V. Atterton, "The Apparent Thermal Conductivities of Moulding Materials at High Temperatures", J. Iron Steel Inst., Vol. 174, (1953), pp. 201-211.

3. D. H. Whitmore and Q. F. Ingerson, "Apparent Thermal Conductivity of Molding Sand at Elevated Temperatures", Transactions, AFS, Vol. 68, (1960), pp. 49-57.

4. M. Ninomiya, Y. Nozaki, Y. Sakaguchi, R. Kurokawa, and M. Sato, "Heat Transfer Properties of Several Kinds of Sand Molds", Reports of the Government Industrial Research Institute, Nagoya, Vol. XXVII, No. 8, (1978), pp. 262-268.

5. R. W. Ruddle and A. L. Mincher, "The Thermal Properties and Chilling Power of Some Non-Metallic Mould Materials", J. Inst. Metals, Vol. 76, (1949), pp. 43-90.

6. M. R. Seshadri and A. Ramachandran, "A Transient Heat Flow Method of Determining Thermal Properties of Mould Materials", The British Foundryman, Vol. 58, (1962), pp. 385-392.

7. V. de L. Davies, P. N. Hansen and B. Clausen, "Bestemmelse af termiske data for formsand", Report of Thermal Materials Processing Group, AMT, Danish Technical University, 1977.

8. O. T. Farouki, "Physical Properties of Granular Materials with Reference to Thermal Resistivity", Highway Res. Record, Vol. 128, (1966), pp. 26-44.

9. K. W. Jackson, "Enhancement of Thermal Energy Transport Through Granular Media", Ph.D. Thesis, Georgia Institute of Technology, Atlanta, Georgia, June, 1980.

10. F. C. Hooper and F. R. Lepper, "Transient Heat Flow Apparatus for the Determination of Thermal Conductivities", Heating, Piping and Air Conditioning, August, 1950, pp. 129-134.

11. H. S. Carslaw and J. C. Jaeger, Conduction of Heat in Solids, Oxford Press, Second Edition, 1959, pp. 261-262.

12. R. Bush, Georgia Power Company, Research Lab, Forest Park, Georgia, private communication, 1979.

13. K. Zeppelzauer and B. Klimesch, "Beitrag zur Kenntnis Thermischer Grundgrossen von Formstoffen", 35th International Foundry Congress, Kyoto, October 1968.

14. R. E. Marrone, J. O. Wilkes and R. D. Pehlke, "Numerical Simulation of Solidification, Part 1: Low Carbon Steel Casting - "T" Shape", Cast Metals Res. J., Vol. 6, No. 4, (1970), pp. 184-188.

THE USE OF EMPIRICAL, ANALYTICAL AND NUMERICAL MODELS TO DESCRIBE SOLIDIFICATION OF STEEL DURING CONTINUOUS CASTING

T.W. Clyne, A. Garcia, P. Ackermann and W. Kurz
Department of Materials
Swiss Federal Institute of Technology (EPFL)
Lausanne - Switzerland

A brief outline is given of the different approaches that can be used to model solidification during the continuous casting of steel by means of unidimensional heat flow analysis. Mention is made of the effect of a finite thermal resistance across the metal/mould interface, leading to errors in the classical analytical treatments, and the limitations of empirical expressions in describing the resulting behaviour are described. The flexibility of the recent analytical solution of Garcia, Clyne and Prates, in which a finite interfacial conductance is permitted, is then demonstrated by comparison with experimental data.

The application of numerical techniques to the unidirectional solidification problem is then illustrated by the use of an explicit finite difference model. Among important aspects of the process examined in this way are the handling of latent heat evolution, particularly when this occurs over a finite freezing range. The effect in this context of the nature of the solute redistribution within the mushy zone is outlined. Finally, simulation of the thermal effects of melt superheat and liquid convection is described. Some conclusions are drawn from these studies concerning both the significance of experimental pool profile data and the practical effects of changes in operating conditions.

Introduction

Heat flow during the continuous casting of steel is normally such that treatment is possible with a number of simplifying boundary conditions. Among the most important of these is to neglect axial heat conduction in the billet, which often makes only a small contribution to the total heat transport in this direction. The shape of the thermal field in a transverse plane depends on the sectional geometry, which is often such that heat flow is predominantly unidirectional (i.e. corner or contour effects are small in many cases). It follows that 1-D models (based on heat flow in a descending volume element) can give useful descriptions of the process. The configuration is illustrated in Fig. 1 and it is clear that depth below the liquid level corresponds to a dwell time co-ordinate (related via the casting speed).

While heat transfer coefficients are, strictly speaking, operative at both coolant/mould and mould/billet interfaces, it is common to introduce a global value h_g describing coolant/billet exchange, as its value can be estimated experimentally by monitoring the cooling water. (This implies that the mould is perfectly chilled to a constant temperature T_0). In practice the values of h_i and h_g are very close for a thin, efficiently cooled mould wall of high conductivity material. The most important period of growth is in general that occurring within the mould and the following treatments and discussions refer primarily to this regime.

Simple Models and Empirical Equations

A number of analytical solutions are available for the limit $h \to \infty$ (zero interfacial resistance). The most general is that due to Schwarz and this reduces to the Stefan solution for a perfectly chilled mould. These models, which are included in the review of Jones [1], all predict parabolic solidification behaviour. In practice, the interfacial resistance is normally significant so that the observed growth varies between parabolic and linear limits. Attempts have been made to describe this deviation by fitting empirical equations of the form $S = Bt^b$ where $0.5 < b < 1$. These often give poor agreement with experiment and modification, such as permitting B and b to vary with dwell time, have been introduced. It is clear, however, that their use is severely limited for extrapolative purposes, in that neither interfacial transfer nor thermal conduction is rigorously treated.

Interfacial Resistance Models

The recently developed exact analytical solution of Garcia, Clyne and Prates [2] is of interest in that it permits finite interfacial conductance. This model has been termed the Virtual Adjunct Method (VAM), as it is based on the principle of representing the thermal resistance of the interface by imaginary extra thicknesses of mould and billet material. It has been demonstrated [3] that the analysis may be used to account for solidification behaviour over a range of conditions. The equations required to apply the model are presented in Appendix A.

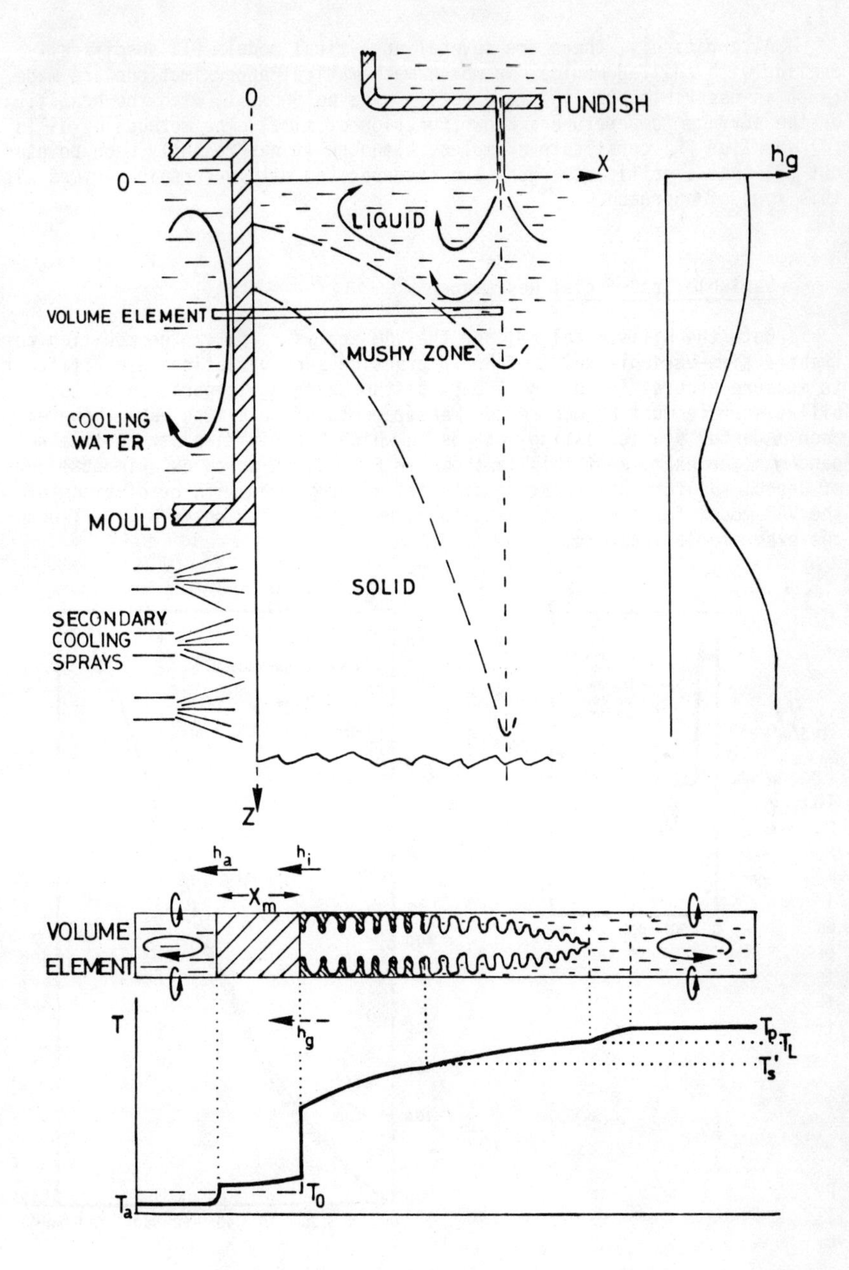

Fig. 1 Schematic illustration of the sectional geometry of continuous casting in the vicinity of the mould.

Alternatively, there are several analytical models [1] specifically dealing with chilled moulds, in which mathematical approximations are made (such as describing the thermal profile by a polynomial, with the heat flux or the surface temperature a known function of time). The methods of Hills [4] and Tien [5] constitute examples, although it has recently been pointed out [6] that significant errors are incorporated within certain regimes with this type of approach.

Variable Interfacial Resistance

Both the Hills model [7] and the VAM analysis [8] can be modified to treat a time-variable value of h. In practice such variations are difficult to measure accurately during CC because the changing temperature of the billet surface must be monitored. Measurements of this type have, however, been made for static casting set-ups in which h exhibits a strong time dependence. An example of this is shown in Fig. 2, in which experimental data of Jacobi [9] for chill casting of steel are compared with predictions of the VAM model for the growth behaviour, using a variation in h modelled on the experimental results.

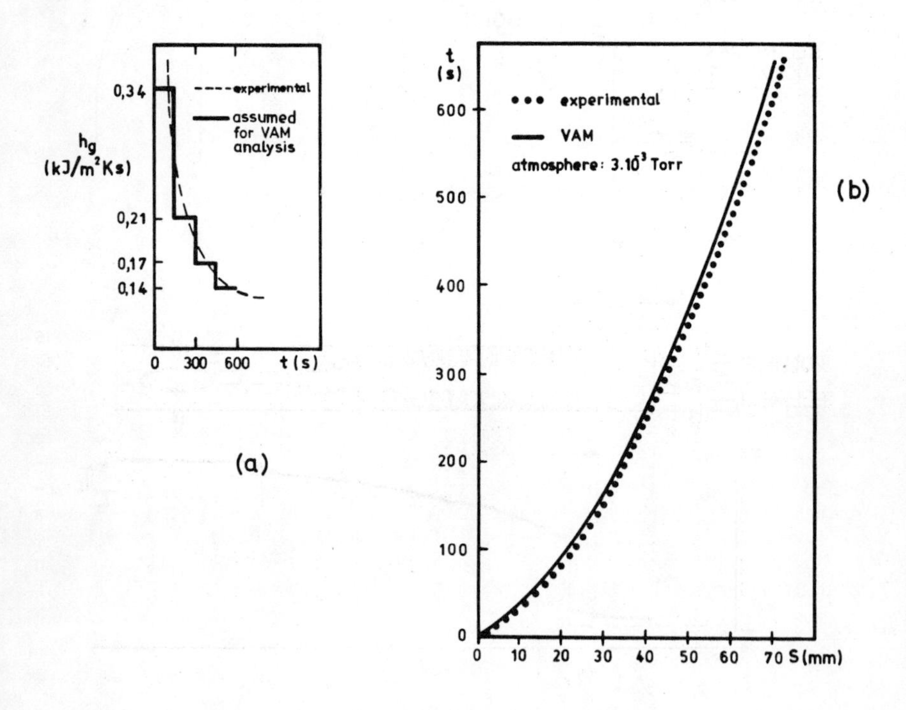

Fig. 2 Experimental data [9] for freezing of steel (in vacuo) against a chill mould: (a) Measured and modelled changes of h with time, (b) Comparison between measured growth behaviour and the predictions of the VAM model.

Moving on to (industrial) CC set-ups, measurements have been made on the variation of heat flux with depth in the mould, an example of which is shown in Fig. 3 (from Volk and Wunnenberg [10]). As the change with depth of the billet surface temperature within the mould is not large (compared with the temperature drop across the interface) the value of h should vary in a broadly similar manner.

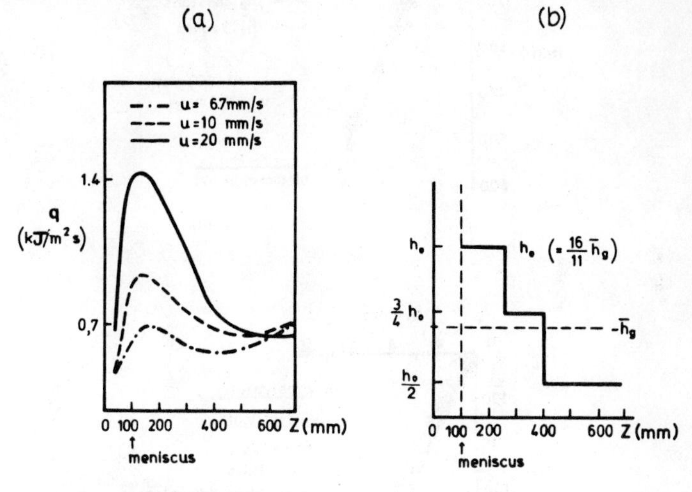

Fig. 3 (a) Experimental measurements [10] of the dependence of the heat flux on depth within the mould (b) Model of expected h_g/depth variation for a high casting speed (see Fig. 4(a)).

A higher casting speed clearly tends to result in greater variations in h and the dependence shown is an attempt to model the behaviour expected with the (fairly high) casting speed for the pool profiles shown in Fig. 4. This shows experimental data obtained by Weinberg et al [11], [12] using radiotracers on a 0.18% C cast. The upper graph illustrates the effect (which is clearly small) of employing the 3-zone variation in h_g when compared with the simple use of a constant value of h_g. In many cases, uncertainties in h_g/depth data are such that it is barely worthwile modelling a variation for purposes of pool profile calculation. The bottom graph compares the predictions of the VAM and Hills models (using constant h_g) with that of a very simple finite difference method.

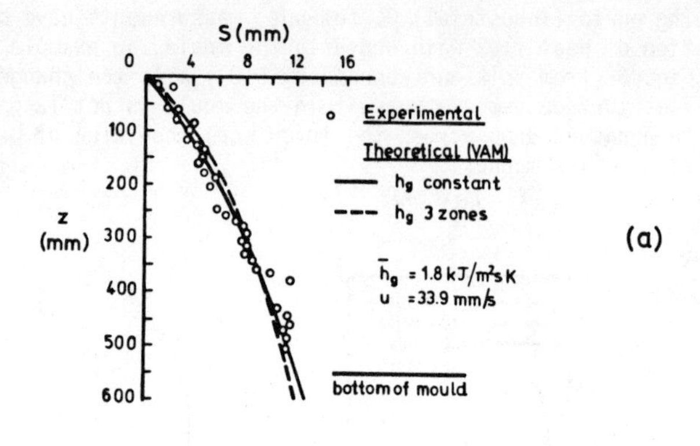

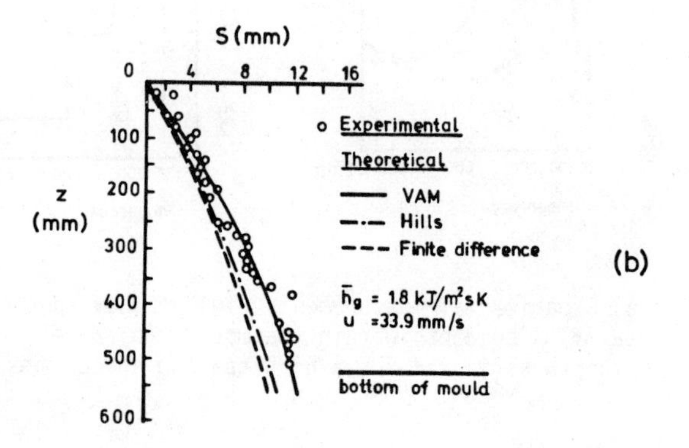

Fig. 4 Experimental [11], [12] and theoretical pool profiles for 0.18% C steel (a) VAM profiles for h_g constant and varying as shown in Fig. 3(a) (b) Curves for 3 models using h_g constant.

Numerical Techniques

Analytical models are clearly subject to limitations, including problems in handling latent heat evolution over an appreciable freezing range, liquid superheat and convection and the effect of finite system dimensions. Good reviews have been published [13], [14] outlining the principles involved in numerical simulation of the solidification process. In the rest of this paper, some results obtained using an explicit finite difference model [15] are presented to illustrate some of the features of interest for CC of steel.

The evolution of latent heat may be modelled either by modifying the heat balance within the mushy zone or by postiterative corrections to the incremental temperature change. (Effectively among the former class are approaches in which the enthalpy is treated as a dependent variable [16]). Postiterative methods are useful in treating discontinuities in the f_s/T curve, while continuous changes are best handled within the algorithm concerned. By combining the two approaches, the present model permits treatment of a generalized f_s/T variation, although discontinuities are in practice not expected for plain carbon steels. The heat balance leading to the finite difference temperature algorithm within the range of continuous change in f_s is detailed in Appendix B, with mention of the bearing that the nature of the solute redistribution has on the f_s/T relationship.

A comparison between experimental (radiotracer) data and model predictions for the pool profile with a 0.1% C steel is presented in Fig. 5. The curves shown represent solute (carbon) redistribution behaviour dictated by the lever rule, which corresponds to the minimum mushy zone width. Although experimental scatter is appreciable (probably caused by local variations in h_g along the length of the billet - associated with the surface undulations observed with this composition), it appears that on average the radiotracers penetrated into the mushy zone to some extent.

Freezing of higher-carbon steels was examined in order to further explore the mushy zone characteristics. The limiting cases of Scheil and lever rule behaviour are illustrated in the f_s/T plots of Fig. 6, which refers to a 0.62% C steel. Also shown are curves corresponding to the Brody and Flemings analysis [17] characterizing incomplete back-diffusion of solute. A value of 0.3 was employed for the back-diffusion parameter α, which is probably of the order expected as a lower limit for steels [18]. Pool profiles corresponding to this case and to the lever rule and Scheil limits are compared in Fig. 7 with measured ("breakout") data from a machine in Switzerland [19].

The experimental points clearly lie near the dendrite tips, indicating that liquid removal techniques do not give rise to significant dendrite remelting (except possibly a little low down in the mould).

Returning to comparisons with radiotracer data from Canada [11], [12] Fig. 8 shows how the fraction solid is predicted to change within the mushy zone for a 0.35% C steel using the Brody and Flemings model. The experimental data are in general consistent with these curves, apart from a slight deviation which may be due to variations in h_g along the mould. In general, it appears that the tracer points probably represent mushy zone penetration down to a depth corresponding to a fraction solid of about 20 - 40%.

Another aspect of some interest is the effect of pouring superheat and liquid convection. In Fig. 9, the experimental (radiotracer) data of Fig. 5 are again presented, this time with pool profile predictions corresponding to an initial superheat of 50 K. It is clear that solidification is then predicted to be significantly retarded, even if the melt remains quiescent.

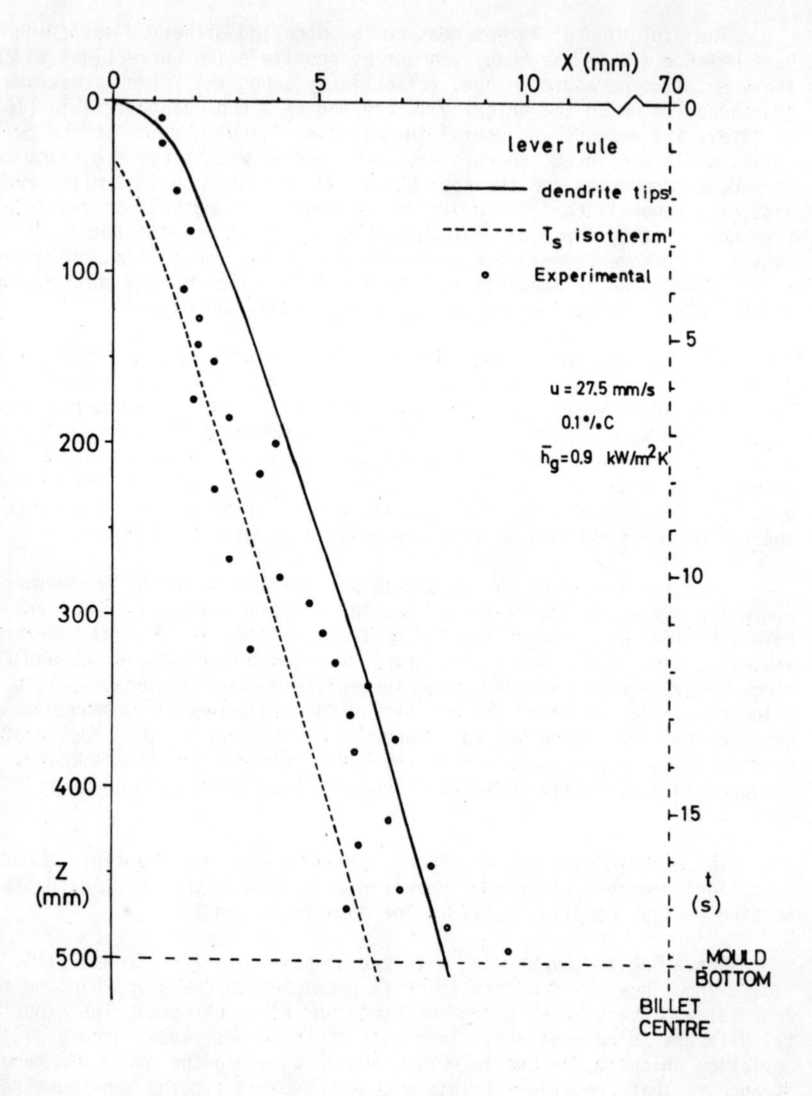

Fig. 5 Experimental [11], [12] and computed pool profiles for 0.1% C steel.

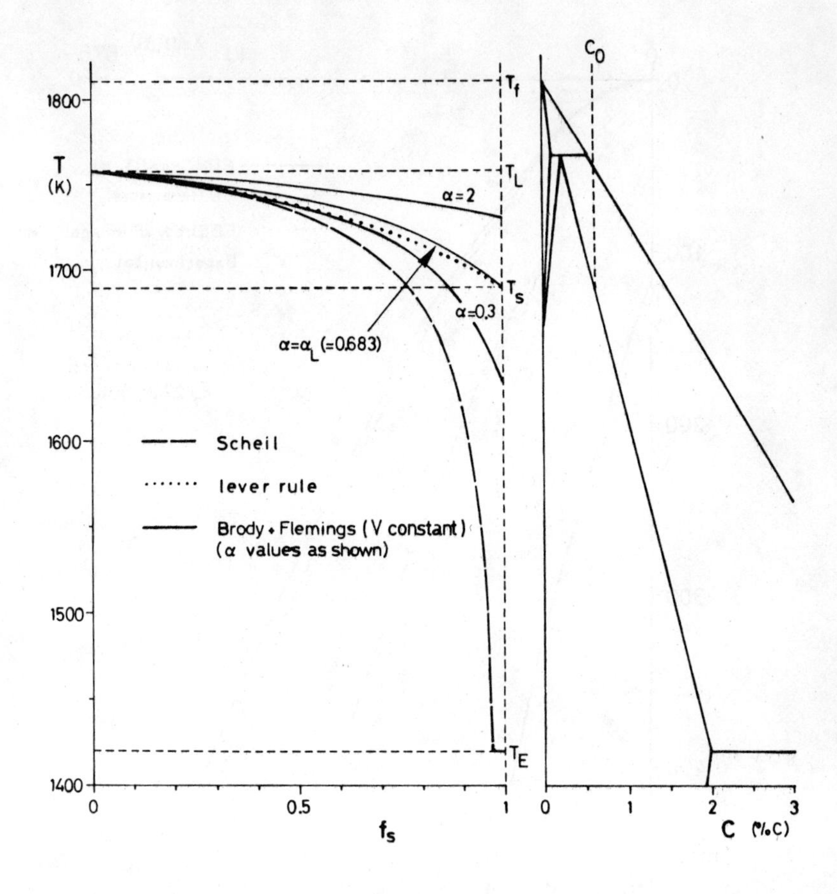

Fig. 6 Theoretical f_S/T curves for a 0.62% C steel freezing under different solute redistribution conditions, together with the phase diagram.

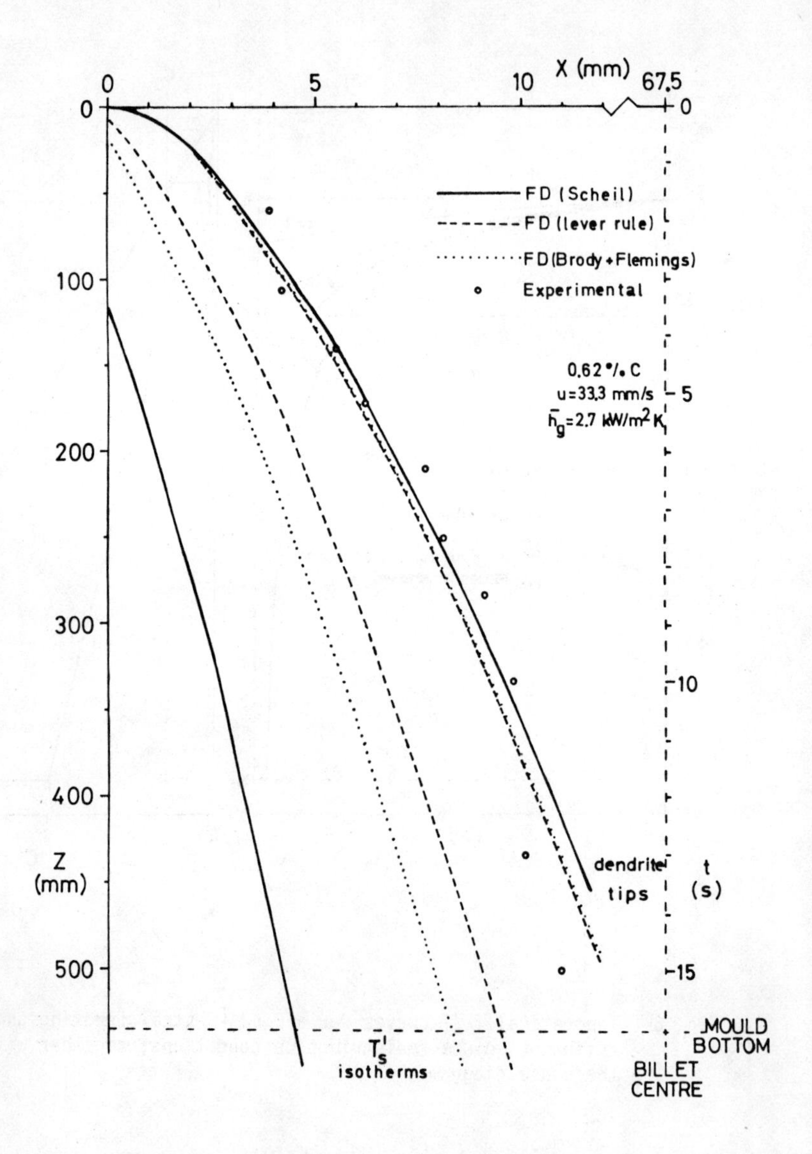

Fig. 7 Comparison between the pool profile measured via a breakout technique [19] for a 0.62% C steel and computed mushy zone limits for different solute redistribution models. (The Brody and Flemings curves correspond to α = 0.3).

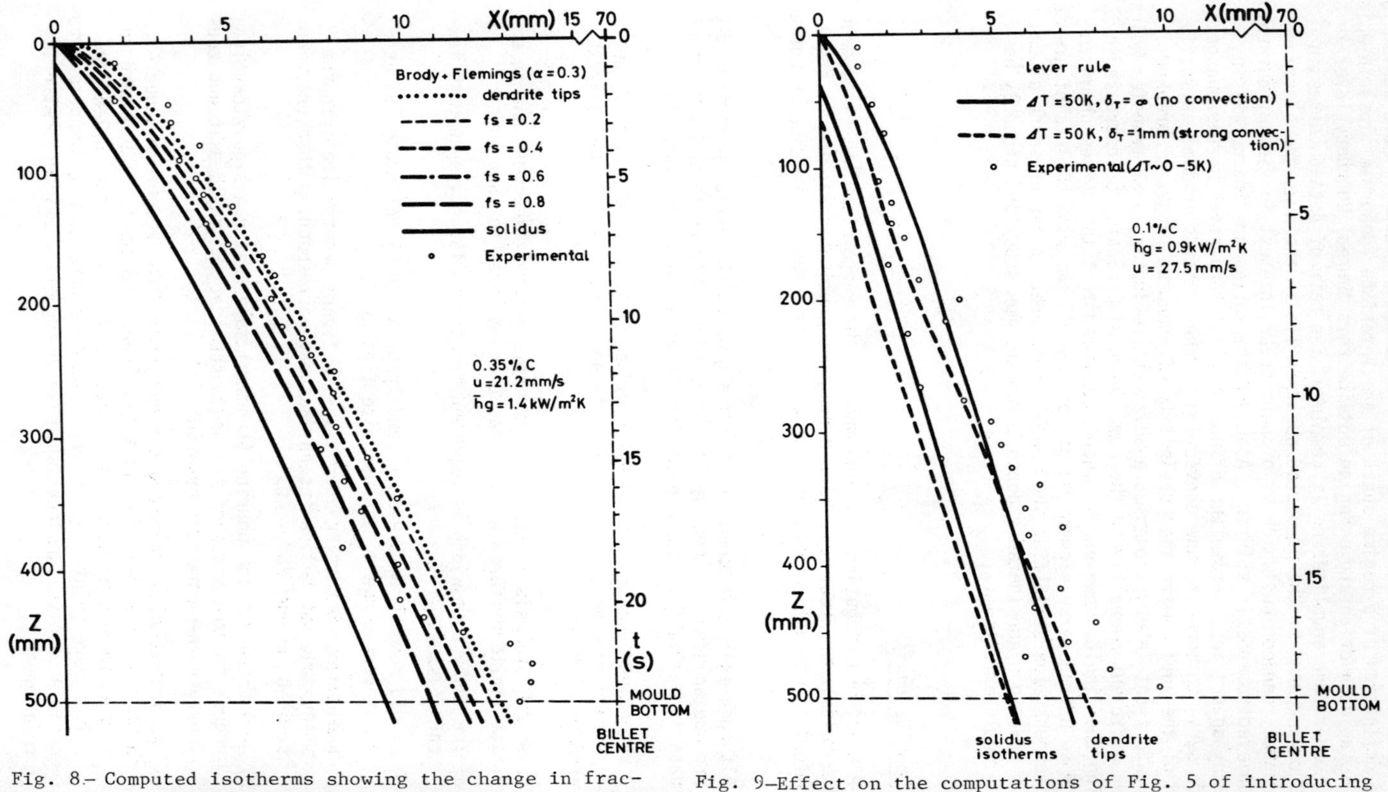

Fig. 8.– Computed isotherms showing the change in fraction solid within the mushy zone using the Brody and Flemings model (α = 0.3), together with radiotracer data [11], [12] for a 0.35% C steel.

Fig. 9–Effect on the computations of Fig. 5 of introducing a pouring superheat of 50 K, with and without (transverse) pool convection. (The curves shown correspond to a billet semithickness of 15 mm).

Among techniques for simulating the thermal effect of liquid convection is the definition of a regime ahead of the interface (dendrite tips) in which there is no convection and a region outside this thermal boundary layer in which temperature equalization is complete. The pair of dotted curves in Fig. 9 corresponds to the predicted pool profile in the presence of strong (horizontal) convection, which is of particular interest in view of current work on electromagnetic stirring. Although the curves are inaccurate in that they do not incorporate the effect of axial convective heat exchange, they do serve to indicate how convection further retards initial solidification of the shell, with the possibility of enhanced danger of breakout. The growth front later becomes more advanced (reflecting heat transfer which is more efficient overall) and the sump depth will clearly be decreased by stirring the liquid. However, it should be noted that the curves shown correspond to a transverse element length of only 15 mm: with billets of greater semithickness, the solidified shell will be thinner in the presence of convection down to a considerable depth below the meniscus (and probably below the mould exit level).

Conclusions

A brief investigation has been made of the range of mathematical techniques available to treat heat flow during the continuous casting of steel. The following observations may be made on the different approaches in current use:

(i) Although empirical equations for the growth and thermal behaviour may provide an adequate description for a given case where shell thickness data is available, they are limited in application and accuracy.

(ii) A recently-developed exact analytical model should prove useful and is rather more versatile and easy to apply than certain approximate analytical models.

(iii) Numerical analysis is essential for more general treatment and an explicit finite difference model has been employed to show that;

a) The mushy zone width is expected to be significant, even with low-carbon steels

b) Liquid removal by breakout (and "pouring out") delineates an isotherm very close to the dendrite tips

c) Radiotracer detection involves significant penetration into the mushy zone and delineates an isotherm corresponding to a fraction solid of around 20% - 40%

d) An increase in the pouring (tundish) superheat can significantly reduce the thickness of the solid shell at the mould exit and thus increase the danger of breakout

e) The introduction of strong convection in the presence of superheat, although resulting in a shallower sump depth, would tend to further reduce the shell thickness at the mould exit (unless the billet were relatively thin). It is thus particularly important to avoid excessive superheat when employing electromagnetic stirring techniques.

Acknowledgements

The authors would like to acknowlege financial support from ALUSUISSE, FAPESP and CONCAST.

NOMENCLATURE

Dimensional Parameters

- a thermal diffusivity (m^2/s)
- B pre-exponential term in empirical expressions (m/s^b)
- c specific heat (J/kg K)
- C composition (wt %)
- d density (kg/m^3)
- D solute diffusivity (m^2/s)
- h heat transfer coefficient (W/m^2 K)
- H latent heat of fusion (J/kg)
- K thermal conductivity (W/m K)
- q heat flux (W/m^2)
- S thickness solidified (m)
- t time (s)
- T temperature (K)
- X distance from metal/mould interface (m)
- Z distance below meniscus level (m)
- β constant (eq. A.3) (s/m)
- γ constant (eq. A.2) (s/m^2)
- λ dendrite arm spacing (m)

Dimensionless Parameters

- b exponent for empirical expressions
- f fraction
- k partition coefficient
- m metal constant = $(K_L\ c_L\ d_L\ /\ K_S\ c_S\ d_S)^{\frac{1}{2}}$
- M metal/mould constant = $(K_S\ c_S\ d_S\ /\ K_m\ c_m\ d_m)^{\frac{1}{2}}$
- n metal constant = $(a_S\ /\ a_L)^{\frac{1}{2}}$
- N metal/mould constant = $(a_S\ /\ a_m)^{\frac{1}{2}}$

α Brody and Flemings [17] back-diffusion parameter (eq. B.5)

δ small increment

∂ partial derivative

ϕ solidification constant in VAM model (eq. A.4)

Ω Clyne and Kurz [18] modified back-diffusion parameter (eq. B.4)

Subscripts

a ambient

f fusion

g global

i interfacial

j of the jth element

L liquid(us)

m mould

o original, zero point

p pouring

s solid(us)

w weighted mean

REFERENCES

[1] H. Jones, "A Comparison of Approximate Analytical Solutions of Freezing from a Plane Chill", J. Inst. Metals, 97 (1969) pp. 38-43

[2] A. Garcia, T.W. Clyne and M. Prates, "Mathematical Model for the Unidirectional Solidification of Metals. II Massive Moulds", Metall. Trans., 10B (1979) pp. 85-92

[3] T.W. Clyne and A. Garcia, "Assessment of a New Model for Heat Flow during Unidirectional Solidification of Metals", Int. J. Heat Mass Transf., 23 (1980) pp. 773-782

[4] A.W.D. Hills, "Simplified Theoretical Treatment for the Transfer of Heat in Continuous-Casting Machine Moulds", J. Iron Steel Inst., 203 (1965) pp. 18-26

[5] R.H. Tien, "Freezing of Semi-infinite Slab with Time-dependent Surface Temperature - An Extension of Neumann's Solution", Trans. Met. Soc. AIME, 233 (1965) pp. 1887-1891

[6] A.J. Pedraza, S. Harriague and D. Fainstein-Pedraza, "On the Use of Analytic Approximations for Describing the Macroscopic Heat Flow During Solidification", Metall. Trans., 11B (1980) pp. 321-330

[7] A.W.D. Hills, "A Generalized Integral-Profile Method for the Analysis of Unidirectional Heat Flow during Solidification", Trans. Met. Soc. AIME, 245 (1969) pp. 1471-1479

[8] T.W. Clyne and A. Garcia, "The Application of a New Heat Flow model to Planar Freezing in Static and Continuous Casting of Steel", submitted to Ironmaking and Steelmaking (Quarterly), (1980)

[9] H. Jacobi, "Heat transfer between ingot and mould during casting of iron under vacuum and various gas atmospheres", Proc. 5th Int. Conf. Vac. Met. & ESR, Munich (1976)

[10] C.R. Taylor, "Continuous Casting Update", Metall. Trans., 6B (1975) pp. 359-375, (citing Volk and Wunnenberg).

[11] S.K. Morton and F. Weinberg, "Continuous Casting of Steel. Part I. Pool profile, liquid mixing and cas structure in the continuous casting of mild steel", J. Iron Steel Inst. 211 (1973) pp. 13-23

[12] J.K. Brimacombe and F. Weinberg, "Continuous Casting of Steel. Part II. Theoretical and Measured Liquid Pool profiles in the mould region during the continuous casting of steel", J. Iron Steel Inst., 211 (1973) pp. 24-33

[13] P.N. Hansen, "Numerical Simulations of the Solidification Process", in "Solidification and Casting of Metals", pp. 350-356, The Metals Soc., London (1979)

[14] W.C. Erickson, "Computer Simulation of Solidification", AFS Int. Cast Metals Res. J., 16 (1980) pp. 30-41

[15] T.W. Clyne, "Numerical Modelling of the Directional Solidification of Metallic Alloys", submitted to Metal Sci., (1980)

[16] N. Shamsundar and E.M. Sparrow, "Analysis of Multidimensional Conduction Phase change via the Enthalpy Model", J. Heat Transfer, Trans. ASME, 97 (1975) pp. 333-340

[17] H.D. Brody and M.C. Flemings, "Solute Redistribution in Dendrite Solidification", Trans. Met. Soc. AIME, 236 (1966) pp. 615-624

[18] T.W. Clyne and W. Kurz, "Solute Redistribution during Solidification with Rapid Solid State Diffusion", submitted to Metall. Trans., (1980)

[19] M.M. Wolf, "Ueber Legierungseinflüsse auf Wärmeabfuhr und Schalenwachstum in der Kokille bei der Erstarrung von Stahlstrangguss", Ph. D. thesis, Swiss Federal Institute of Technology, Lausanne (1978).

APPENDIX A

Principal Equations of the VAM Model

For the generalized case in which h_i is invariant, interface movement is desribed by the equation

$$t = \gamma S^2 + \beta S \tag{A.1}$$

where

$$\gamma = \frac{1}{4 a_s \phi^2} \tag{A.2}$$

$$\beta = \frac{c_s d_s}{\sqrt{\pi}\ \phi \exp(\phi^2)\{M+\text{erf}(\phi)\}\ h_i} \tag{A.3}$$

and the solidification constant ϕ is obtained by iteration from the following condition

$$\frac{\exp(-\phi^2)}{M+\text{erf}(\phi)} = \frac{m\ (T_p-T_f)\ \exp(-n^2\phi^2)}{(T_f-T_o)\{1-\text{erf}(n\phi)\}} + \frac{\sqrt{\pi}\ H\ \phi}{c_s(T_f-T_o)} \tag{A.4}$$

The thermal characteristics of the process are represented by the following equations

$$T_m = T_o + \frac{(T_f-T_o)M}{M+\text{erf}(\phi)}\ \{1 + \text{erf}\ (\phi\ \frac{2\gamma NX\ -\ \beta}{2\gamma S\ +\ \beta})\} \tag{A.5}$$

$$T_s = T_o + \frac{(T_f-T_o)}{M+\text{erf}(\phi)}\ \{M + \text{erf}\ (\phi\ \frac{2\gamma X + \beta}{2\gamma S + \beta})\} \tag{A.6}$$

$$T_L = T_p - \frac{(T_p-T_f)}{1-\text{erf}(n\phi)}\ \{1 - \text{erf}\ (n\phi\ \frac{2\gamma X + \beta}{2\gamma S + \beta})\} \tag{A.7}$$

The case of a perfect heat sink (at T_0) corresponds to $M \rightarrow 0$ and h_i is replaced by h_g, while $T_p \rightarrow T_f$ in the limit of zero superheat. Discontinuous changes in h_i (or h_g) are handled by superposing the appropriate sections of the family of S/t curves corresponding to the series of h_i values in equation [3].

APPENDIX B

Mushy Zone Heat Balance in Finite Difference Model

For the case where the jth element lies within the mushy zone, we have

Heat Extracted from jth element	=	Latent Heat Removed	+	Specific Heat Removed

$$\frac{K_w \, \delta t}{\delta X} (T_{j+1} + T_{j-1} - 2T_j) = H \, d_w \, \delta X \, \delta f_s + c_w \, d_w \, \delta T \, \delta X \quad (B.1)$$

where the subscript w refers to a weighted mean value between those of the solid and liquid phases. It may be noted that the value of H will be negative for solidification. In order to derive an expression for the temperature change, the f_S/T relationship must be such that the differential $(\partial f_S/\partial T)$ can be written down explicitly. Manipulation of eq. (B.1) then leads to the algorithm giving the incremental change in temperature of the jth element during the period δt

$$\delta T = \left(\frac{K_w}{d_w \, c_w}\right) \left(\frac{\delta t}{\delta X^2}\right) \left(\frac{T_{j+1} + T_{j-1} - 2T_j}{1 + \left(\frac{H}{c_w}\right) \left(\frac{\partial f_s}{\partial T}\right)}\right) \quad (B.2)$$

Now, the f_S/T relationship is governed by the assumptions made with regard to the solute redistribution in the mushy zone. A convenient way to represent the general case of this redistribution is to use the recent model of Clyne and Kurz [18], which is effectively a modification of the Brody and Flemings [17] analysis of the case of incomplete back diffusion in the solid. (The new model represents an advance in that it behaves correctly in the lever rule and Scheil equation limits.) The solute redistribution is given by the equation

$$C_s^i = k \, C_o \, [1 - (1-2 \, \Omega \, k) \, f_S]^{(k-1)/(1-2 \, \Omega \, k)} \quad (B.3)$$

where Ω is a constant (characterizing the back diffusion) defined by the expression

$$\Omega = \alpha(1 - \exp(-1/\alpha)) - \tfrac{1}{2} \exp(-1/2\alpha) \quad (B.4)$$

and α is the Brody and Flemings constant, which is related to the solute diffusivity D_S, diffusion path length (dendrite arm spacing) λ and local freezing time t_f by the equation

$$\alpha = \frac{4 \, D_s \, t_f}{\lambda^2} \quad (B.5)$$

For a phase diagram with linear liquidus and solidus lines, this leads to the following f_S/T relationship

$$f_s = \left(\frac{1}{1-2\Omega k}\right) \left[1 - \left(\frac{T_f - T}{T_f - T_L}\right)^{(1-2\Omega k)/(k-1)}\right] \tag{B.6}$$

so that

$$\left(\frac{\partial f_s}{\partial T}\right) = \left(\frac{1}{k-1}\right)(T_f-T_L)^{(2\Omega k-1)/(k-1)}(T_f-T)^{(2-2\Omega k-k)/(k-1)} \tag{B.7}$$

Combination of eqs. (B.2) and (B.7) therefore allows computation of the changing temperature of any given volume element within the mushy zone. Eqs. (B.3), (B.6) and (B.7) reduce to simpler forms in a number of special cases. For example, as $\alpha \to \infty$, $\Omega \to 0.5$ and the lever rule form is then applicable, so that eq. (B.6) reduces to

$$f_s = \left(\frac{1}{1-k}\right)\left(\frac{T_L - T}{T_f - T}\right) \tag{B.8}$$

Similarly, in the limit $\alpha \to 0$, we have $\Omega \to 0$, giving the Scheil equation

$$f_s = 1 - \left(\frac{T_f - T}{T_f - T_L}\right)^{1/(k-1)} \tag{B.9}$$

6th November 1980 / TWC/je

Heat Flow in Welding Processes

MODELING OF HEAT TRANSFER

IN WELDING PROCESSES

P. CACCIATORE

EXXON RESEARCH AND ENGINEERING

FLORHAM PARK, NEW JERSEY

This paper reviews the state-of-the-art in modeling heat transfer mechanisms involved in the welding process. It begins with a brief discussion of the early analytical treatment of temperature transients associated with the welding process, and examines the various assumptions employed to obtain these solutions. The paper then describes development of state-of-the-art numerical approaches to solving nonlinear heat transfer problems. This discussion is combined with an examination of the necessary assumptions and problem areas.

Introduction

An accurate description of the temperature distribution that occurs in a structure subjected to the welding process is essential to predicting distortion and residual stresses. In addition, a knowledge of the heating and cooling rates is useful in predicting metallurgical reactions and microstructure transformation in the weld puddle and heat-affected zone. The generation of detrimental gases and their ability to escape from the weldment, diffuse into the heat-affected zone or remain in some form in the fusion zone, is affected by temperature distributions and cooling rates. Continuous-cooling transformation data from the materials involved can be combined with weld thermal cycles to predict microstructures in the heat-affected zone and weld puddle. This information is essential to evaluating the strength characteristics and life of the structure under its prescribed design loads.

Early Analytical Treatment

A review of the early analytical treatment of temperature transients associated with the welding process, reveals that these efforts were not highly successful in duplicating experimental results. This was due mainly to the number of simplifying assumptions that were necessary to obtain a solution. These assumptions included the following:

a) line or point source to represent the weld arc,
b) two dimensional heat flow (plate thickness ignored),
c) temperature independent material properties,
d) quasi-stationary heat flow,
e) neglect of convection and radiation effects,
f) neglect of heat of transformation.

The heat transfer mechanisms associated with welding are extremely complex and involve all three modes of heat transfer-conduction, convection and radiation. The material undergoes several phase changes in the vicinity of the weld arc and the properties vary rapidly as the temperature increases and decreases with the approach and passage of the weld arc. Not only are properties dependent upon the temperature, but upon the temperature gradients. The heat flow is always three dimensional, but may be approximated in some cases by two dimensional flow (cross sectional analysis).

One by one, as work in this area progressed, each of these assumptions have been removed or relaxed. Researchers have moved from closed form solutions to fairly sophisticated numerical approaches. In many situations, the temperature solutions obtained are adequate to predict distortion, this is particularly true in regions removed from the weld zone. Despite the many advances in the technology, this area still represents a formidable challenge to the engineer searching for a solution for his particular problem. The myriad of welding processes and conditions, and the large number of variables involved make each problem unique. For the engineer interested in the residual stresses and microstructure in the weld zone, the solutions may have to wait until a better understanding of the physical phenomena is obtained and a strong theoretical foundation is laid to model that phenomena.

Basic Assumptions

Perhaps the most fundamental assumption that enters into the numerical analysis of heat transfer during welding is the uncoupling of the heat transfer and thermal stress problems (1). This assumption allows the heat transfer and resulting temperature distribution to be calculated independently of the simultaneously occuring thermal stress response. This assumption simplifies the analysis considerably without introducing any significant error.

A complete mathematical argument for making this assumption is given in Reference 1 . The basic concepts of this argument are:

1) dimensional changes resulting from the mechanical response may be ignored in implementing the thermal boundary conditions,

2) contributions to entropy production via the strain rate effect is far outweighed by the temperate rate term.

Numerical Approaches to Welding Heat Transfer

Two major numerical approaches exist for solving the heat transfer problem-finite differences and finite elements. While both have distinct advantages, the emergence of the finite element method, as the most powerful numerical technique for the solution of heat transfer problems, has overshadowed finite difference developments. The major advantages of the finite element method are:

a) ability to model the most complex and arbitrary geometries,
b) ease of transition from a heat transfer to a thermal stress solution,
c) ability to model complex boundary conditions such as intimate contact.

Governing Equations

The finite element formulation of the heat transfer problem without mechanical coupling has been given by Wilson and Nickell (2) and Becker and Parr (3). Also, finite element solutions of a class of thermoelastic problems were considered by Visser (4), who developed equations based on Gurtin's variational principles for linear initial value problems. A finite element formulation for coupled thermoelastic problems based on Gurtins work is given by Nickell and Sackman in Reference 5 . All of these formulations were based on developing the governing variational principle based on the approach of Gurtin (6).

Oden and Kross (7) presented an alternate formulation for nonlinear coupled thermoelasticity based on energy balances, thereby freeing the development of a finite element model from a dependency on the availability of a suitable variational principle. This fairly comprehensive development approaches the problem from thermodynamic point of view and employs the first law of thermodynamics, entropy concepts and free energy to obtain the governing equations. Several types of boundary conditions are considered including specified temperatures and fluxes, and convective heat transfer.

A third, and much simpler approach for the uncoupled problems of heat transfer is presented in References 8 and 9. The Galerkin weighted residual method is applied to governing differential equation of heat transfer to obtain the finite element equations. This formulation is both elegant and straightforward and will be presented here. All of these various developments yield basically the same equations.

The general class of problems under consideration here are governed by the following differential equation which has to be satisfied in some region R of a continum (10):

$$\frac{\partial}{\partial x}\left[k_x(T)\ \frac{\partial T}{\partial x}\right] + \left[\frac{\partial}{\partial y}\ ky(T)\frac{\partial T}{\partial y}\right] + \frac{\partial}{\partial z}\left[kz(T)\ \frac{\partial T}{\partial z}\right] \tag{1}$$

$$+\ Q - \rho c\ (T)\ \frac{\partial T}{\partial t} = 0$$

where kx, ky, kz = thermal conductivities in three mutually perpendicular directions

ρ = material density

c = material specific heat at constant strain

T = temperature

Q = external heat input per unit volume (sinks and sources).

This equation holds for an orthorhombic molecular structured homogenous anisotropic material. The homogenity is present since the material properties (k, ρ, c) are not considered to be a function of position. Most metals exhibit a cubic crystalline system where the cyclic interchange of the axes of the orthorhombic system is possible and therfore kx(T) = ky(T) = kz(T) = k. The nonlinearity of Equation 1 results from the thermal dependence of the material properties and this can be seen more clearly by rewriting Equation 1 for the cubic system molecule as:

$$k(T)\left[\frac{\partial^2 T}{\partial x^2} + \frac{\partial^2 T}{\partial y^2} + \frac{\partial^2 T}{\partial z^2}\right] + \frac{\partial k}{\partial T}\left[\left(\frac{\partial T}{\partial x}\right)^2 + \left(\frac{\partial T}{\partial y}\right)^2 + \left(\frac{\partial T}{\partial z}\right)^2\right] \tag{2}$$

$$+\ Q\ - \rho c(T)\frac{\partial T}{\partial t} = 0.$$

The governing equation is subject to the following boundary conditions:

$$T = T_B \tag{3}$$

on a portion of the boundary S_1, and

$$k(T)\ \frac{\partial T}{\partial n} + q + h\ (T)\ (T - T_S) = 0 \tag{4}$$

on a portion of the boundary S_2.

In Equation 4

$h(T)$ = boundary surface film coefficient for forced or natural convection; in general this quantity, in addition to being a function of temperature, is a function of the boundary layer which can be expressed in terms of the Grashoff and Prandl flow numbers,

q = boundary surface heat flux vector, which may be specified as a known quantity or may be a function of the temperature (e.g. surface radiation),

T_S = temperature of surrounding medium into which convection is occurring,

$\frac{\partial T}{\partial n}$ = temperature gradient in a direction normal to the boundary S_2; $\frac{\partial T}{\partial t} = 0$ is equivalent to an insulated boundary.

The development of the governing finite element equations is relegated to Appendix A.

Radiation

With the extremely high temperatures encountered in the heat transfer analysis of the welding process, the dominant heat flux loss mechanism in the earlier times is radiation. Radiation in the welding process is assumed to be governed by black body radiation laws, i. e. a body at absolute surface temperature T_S surrounded by a black body at temperature T_o will lose heat at the rate (11):

$$q(T) = \sigma E(T) \, (T_S^4 - T_o^4), \tag{5}$$

where σ is the Stefan-Bolzmann constant and E is the total hemispherical emissivity. The quantity E, which is also a function of temperature, is the ratio of the heat emitted by a body to that emitted by a black body at the same temperature. The emissivity is also a function of surface roughness and level of oxidation. Radiation is in reality a volumetric quantity, i.e. originates within the volume of the body, however, since metals are so opaque to radiation, all but a negligible portion of the radiation leaving the surface originates within a few ten thousandths of an inch of the surface. This is particularly true when a strong temperature gradient exists at and normal to the surface. The radiation geometric shape factor sometimes included in Equation 5 is taken as equal to unity (plane wall, infinite in extent).

The radiation mechanism can be included in our numerical model in one of two distinct ways. The surface temperature of the previous time step is solved with a numerical time stepping procedure and may be used in Equation 5 to furnish an estimate of the heat flux lost at the current time step, the resulting solution is termed a predictor. Normally a corrective (corrector) solution would also be required at the current increment in order to provide sufficient accuracy. To avoid the expense associated with a corrector (two solutions per time step), an extrapolation of the surface temperature can be used. Denoting the previous step by t_n, then

$$T_s(t_{n+1/2}) = \frac{3}{2} T_S(t_n) - \frac{1}{2} T_S(t_{n-1}) \tag{6}$$

This extrapolated mid-interval temperature is a good estimate for the radiation equations (boundary condition) except in cases of extreme temperature changes within an increment.

An alternate approach is contained in Reference 12 where the radiation flux is writtin as;

$$q_r = E\sigma (T_S^4 - T_o^4) = \alpha_r(T)(T_S - T_o) \tag{7}$$

where $\alpha_r(T) = E\sigma (T_S^2 + T_o^2)(T + T_o)$

An algorithm for estimating α_r is obtained by integrating α_r over two time steps and averaging,

$$\alpha_r = \frac{1}{2\Delta t} \int_{t-\Delta t}^{t+\Delta T} E\sigma (T_S^2 + T_o^2)(T_S + T_o)\, dt \tag{8}$$

neglecting the dependence of E on T, Equation 8 yields

$$\begin{aligned}\alpha_r \simeq E\sigma \{ & \left(T_S(t)^2 + T_o(t)^2\right)\left(T_s(t) + T_o(t)\right) \\ & + \frac{1}{4}\left(T_o(t+\Delta t) - T_o(t-\Delta t)\right)^2 \left(T_o(t) - T_S(t)\right. \\ & \left. - \frac{2}{3} T(t-\Delta t)\right) + \frac{1}{3}\left(T_o(t+\Delta T) - T_o(t-\Delta t)\right)\left(T_S(t)\right. \\ & \left. - T_s(t-\Delta t)\right)\left(T_S(t-\Delta t) + T_o(t-\Delta t)\right) + \\ & \left(T_S(t) - T_S(t-\Delta t)\right)^2 \left(T_S(t) + \frac{1}{2} T_o(t+\Delta t) - \frac{1}{6} T_o(t-\Delta t)\right) \} .\end{aligned} \tag{9}$$

Equation 9 involves only known temperatures in the interval from t-Δt to t+Δt and represents a better approximation than the direct calculation of α_r at time t. This method is easily incorporated into the governing equations by replacing h(T) coefficient with $h(T) + \alpha_r$.

Thermal Boundary Conditions

In addition to the boundary conditions already discussed (specified temperature, flux radiation, convection and insulation), two other boundary conditions are useful in the heat transfer analysis of the welding process. The first of these conditions, "intimate contact" (1) was mentioned earlier. This boundary condition is necessary to simulate numerically the addition of molten filler material to the base metal. In the approximate numerical model this condition is modelled as a step-wise addition of finite volumes of filler (several finite elements per time step). After finite element discretization of the filler, the elements of the surfaces being brought instantaneously into intimate contact are taken to be the surfaces of elements of the mesh. The weld deposit process consists of bringing a portion of filler into contact with the base metal in a time interval which is small compared to the overall time scale. The interface surfaces on both the filler and base metal take on the same temperature in that same short time interval. Because of the reduced time scale the heat flux across the surface dominates this portion of the problem.

Neglecting conduction (in the immediate time interval), Equation 6 for a surface becomes (1):

$$\left[C \right] \{\dot{T}\} \Delta t = - \{q\} \Delta t \tag{10}$$

where q = heat flux
Δt = time step.

For two surfaces S^A and S^B brought into contact, the heat flux balance is,

$$\{q\}^A = - \{q\}^B \tag{11}$$

and since the temperatures of corresponding points on the two surfaces must be the same immediately after contact, a linear relationship between nodal temperatures exists;

$$(T^- + \Delta T)^A = S^{AB} (T^- + \Delta T)^B \tag{12}$$

These conditions are introduced into the time stepping scheme to impose the intimate contact condition. The detailed modifications to the equations are given in Reference 18.

A second type of boundary condition that finds some use in welding analyses is the ability to introduce analytic solutions in areas of known behaviour (18). The welding process as a heat transfer problem involves and extremely localized response. At points far removed from the weld itself the heat flow usually adopts a much more regular behavior and the imposition of known analytical solutions as opposed to an extended finite element model should be adequate. A good example of this would be heat flow into a heat sink some distance away from the weld. If at some point the temperature remains constant in time, the relationship between the heat flux and surface remperature (10) is given by

$$q\,(t) = \frac{k}{\sqrt{\pi \kappa t}} \quad (T_S - T_o) \tag{13}$$

where k = thermal conductivity (13)

κ = diffusivity

T_S = surface temperature

t = time

T_o = initial temperature

In this case a large heat sink can be treated as a boundary condition on an adjacent finite body by using a time dependent convective heat transfer coefficient, e. g.

$$h(t) = \frac{k}{\sqrt{\pi \kappa t}} \tag{14}$$

If Equation 13 is interpreted as a boundary layer with a "diffusion thickness" equal to $\sqrt{\pi\kappa t}$, then as $t \to \infty$ the boundary layer becomes so thick as to insulate the surface. Thus Equation 13 is only adequate in the situation where the surface reaches a peak temperature in a short period of time before the diffusion thickness becomes too great. In essence the sink begins to fill and the surface of constant temperature moves into the sink In the case where the surface temperature varies and does not reach a peak rapidly a more rigorous solution is needed. For this situation the boundary flux is specified as

$$q(t) = \frac{-k}{2\sqrt{\pi\kappa}} \int_o^t \frac{T(\tau) - T_o}{(t-\tau)^{3/2}} \, d\tau \tag{15}$$

where the current unknown surface temperature is $T(\tau)$. If extrapolated surface temperatures (based on temperatures at previous times) are used, then Equation 15 becomes an explicit boundary condition for the flux. An implicit form could also be deduced.

With boundary conditions of this type we can utilize many analytical solutions that exist and in the process reduce the cost for the analysis. However if identical numerical models are to be used for the finite element stress analysis, some care must be used in choosing the point of applying the anlytical solution.

Latent Heat (Heat of Transformation)

Heat of transformation is thermal energy released/absorbed upon solid to solid and on solid to liquid transformation. It can occur at discrete temperatures in pure substances or more commonly over a distributed temperature range as in alloys. In many alloys this thermal energy has a significant effect on the temperatures and temperature gradients in the weld zone region.

The earliest treatment of the latent heat (1) used an apparent increase in the specific heat over the specified temperature range to approximate the release and absorption of the addition thermal energy. This typically took the form of a step function change in the ρ vs T curves and difficulties were encountered in obtaining a smooth representation across this peak.

In Reference 13 a more physical approach, based on the observation that the integral of heat capacity with respect to temperature (enthalpy) is a smooth function of temperature even in the range of temperatures where phase change occure.

$$H \text{ (enthalpy)} = \int_{To}^{T} \rho c(T)dT \tag{16}$$

Once a temperature solution is obtained, element nodal enthalpies could be determined from a H vs T curve. An interpolation function, identical to that used for the temperature solution, could be used to express the enthalpy variation over the entire volume of the element:

$$H = \sum_{i=1}^{n} Ni(x,y,z)\ Hi\ (\tau) = N\ H \tag{17}$$

$H_i\ (\tau)$ = nodal enthalpies at time τ

$N_i\ (x,y,z)$ = spatial interpolation functions.

The ρc variable at any point in the element could then be calculated using,

$$\rho c = {}^{dH}/_{dT}. \tag{18}$$

This ρc value is then used as an approximation for the next time step in a transient solution.

A more refined approach is offered by Friedman (14) which utilized the enthalpy concept along with phase change iteration. The enthaply is approximated by:

$$H = \int_{\overline{T}}^{T} \left\{ \rho c(T) + \left[\frac{\rho \ell}{T_L - T_S} - \rho c(T)\right] \left[Z\ (T-Ts) - Z(T-T_L) \right]\right\} dT \tag{19}$$

$$\text{where } Z(z) \pm \begin{matrix} 1 \text{ if } z > 0 \\ 0 \text{ if } z < 0 \end{matrix}$$

ℓ = latent heat of the material,

T_S = temperature at which phase change begins,

T_L = temperature at which phase change is complete,

$\overline{T}$ = initial temperature at start of time step,

T = temperature at the end of the time step.

The expression for H is incorporated into the Fourier heat conduction equation by approximating the $\rho c \frac{\partial T}{\partial t}$ term by $\frac{1}{\Delta t} \rho c\ (T - \overline{T})$ and recognizing,

$$H = \int_{\overline{T}}^{T} \rho c\ (T)\ dT = \rho c\ (T - \overline{T}). \tag{20}$$

An iteration on the temperature solution is performed whenever $\overline{T}$, T or both are within the temperature range bounded by T_s and T_L. The end result is a more accurate description of the temperature distribution as the material passes through the phase transformation.

The method employed for a particular problem is dependent upon the accuracy desired, and type of solution required. For distortion and stresses away from the weld zone, use of the apparent increase in specific heat approach appears to be adequate.

Summary

To summarize our heat transfer model at this point the following features are available;

a) anisotropic homogeneous material with orthorhombic crystalline structure,

b) temperature dependence of the thermal conductivities (kx, ky, kz), specific heat (c), film coefficient (h) and surface emissivity (E),

c) incorporation of material latent heat release during the melting phase change,

d) specified flux, temperature and insulating conditions on element and continuum boundaries,

e) approximation of the "intimate contact" problem and use of analytical solutions in regions of known behavior.

Having defined the governing equations and associated boundary conditions, we can now turn our attention to the technique that will be used to solve these equations.

Time Integration of Governing Equations

Among the numerous choices available for numerically integrating the heat transfer equations, the method used in this study is Crack-Nicolson. In a recent comparison study (15), this method was found to be superior to other difference methods.

The Crank-Nicholson integration operator for the transient heat conduction finite element formulation takes the form (2),

$$\begin{aligned} \left\{ \left[C\right] + \frac{\Delta t}{2} \left[H\right]\right\} \{T(t_{n+1})\} &= \frac{\Delta t}{2} \{Q(t_{n+1})\} + \frac{\Delta t}{2} \{Q(t_n)\} \\ &+ \left\{ \left[C\right] - \frac{\Delta t}{2} \left[H\right]\right\} \{T(t_n)\} \end{aligned} \tag{21}$$

where $\Delta t = t_{n+1} - t_n$ and the matrices $\left[C\right]$ and $\left[H\right]$ are evaluated at the temperature T^a at time t^a ($t^a = \frac{1}{2}(t_n + t_{n+1})$). Equation 2 is obtained by writing Equation A.6 at the middle of the time interval and using a central difference approximation for the time derivative,

$$\dot{T}(t^a) = (T_{n+1} - T_n)/\Delta t. \tag{22}$$

Since neither T^a nor T_{n+1} are known, Equation 21 is a non-linear algebraic system of equations. Quasi-linearization could be achieved by evaluating the matrices $[C]$, $[K]$ at T_n. A modified Crank-Nicolson scheme can be obtained by assuming:

$$\dot{T}_{t+\frac{\Delta t}{2}} = (T_{t+\Delta t} - T_t)/\Delta t$$

$$T_{t+\frac{\Delta t}{2}} = \frac{1}{2} (T_{t+\Delta t} - T_t) \tag{23}$$

and writing Equation A.6 at the middle of the time interval, thus obtaining

$$\left[\frac{2}{\Delta t} [C] + [K] \right] \{T_{t+\frac{\Delta t}{2}}\} = \{Q_{t+\frac{\Delta t}{2}}\} + \frac{2}{\Delta t} \left[C \right] \{T_t\} \tag{24}$$

The matrices $[C]$, $[K]$ and $\{Q\}$ are obtained for a temperature corresponding to a linear extrapolation using the second of Equation 23 written a time step earlier, e. g.

$$\left[C(T_{t+\frac{\Delta t}{2}}) \right] = \left[C(2T_t - T_{t-\frac{\Delta t}{2}}) \right] \tag{25}$$

Characterization Of The Weld Arc

The mechanism involved in the heat transfer from the weld arc to the weld workpiece is complex and not completely understood. The usual assumption made in the work performed to date is that the thermal energy given off by the arc is related to the electrical input.

$$Q = E \ V \ I \tag{26}$$

where
Q = net heat flux
E = efficiency factor
V = electrical input voltage
I = electrical input current.

The efficiency factor is used to account for the thermal losses from the arc and typically will vary from 60% to 95% depending on the welding method.

The flux distribution from the arc is usually assumed to be normal and radially symmetric. For example (15)

$$q = qoe^{-cr2} \tag{27}$$

where qo and c are parameters which are a function of the radius of the welding electrode.

A necessary assumption in most analysis is that the desposition speed (arc velocity) is large compared to the heat transfer rates ahead of the weld arc. This effectively reduces the problem to an analysis of any two dimensional cross section normal to the direction of the weld arc path. This assumption of course embodies the neglect of the end affects of the weld process. In this case the above equation becomes

$$q\ (s,t) = q_o\ e^{-cs^2}\ e^{-cv^2(t-to)^2} \tag{28}$$

v = electrode velocity
to = time of peak heating for the cross section under analysis
s = a measure of the distance on the surface from the centerline of the weld electrode.

The parameters qo and c are obtained from a comparison of the assumed normal flux distribution with the electrode size and establishing an equality between Equation 26 and an integrated value of Equation 28.

Future Work

A great deal of progress has been made in the area of modeling heat transfer in welding. However, a number of serious problems still need to be resolved. The most pressing of these problems is the determination of material properties at the extreme temperature excountered in welding. Usually an extrapolation of relatively low temperature information must be used in lieu of reliable experimental data. Very little data is available to substantiate the extrapolation process. In fact some researchers feel that it will be necessary to resort to treating the weld puddle as a fluid in motion before an accurate assessment of temperatures in this area can be made.

A second problem exists in the characterization of the weld arc. this determines just how much thermal energy is released into the workpiece and how it is distributed over the surface. Obviously this has a significant influence on the width and depth of weld penetration. Additional experimental work is needed in this area to improve the weld arc model.

REFERENCES

1) Hibbit, H. D., "A Numerical Thermo-mechanical Model for the Welding and Subsequent Loading of a Fabricated Structure",Ph.D Thesis, Brown University, June, 1972.

2) Wilson, E. L. and R. E. Nickell, "Application of the Finite Element Method to Heat Conduction Analysis", Nuclear Engineering and Design, Vol. 4, No. 3, Oct. 1966.

3) Becker, E. B. and C. H. Parr, "Application of the Finite Element Method to Heat Conduction in Solids". Technical Report S-117, Rohm and Haas Redstone Research Laboratories, Huntsville, 1967.

4) Visser, W., "A Finite-Element Method for the Determination of Non-Stationary Temperature Distributions and Thermal Deformations", Proceedings Conference on Matrix Methods in Structural Mechanics, AFFLD-TR-66-80, 1966.

5) Nickell, R. E., J. J. Sackman, "Approximate Solutions in Linear Coupled Thermoelasticity", Journal of Applied Mechanics, Vol. 35, No. 2, June, 1968.

6) Gurtin, M. E., "Variational Principles for Linear Initial-Value Problems", Quarterly of Applied Mathematics, Vol. 22, No. 3, 1964.

7) Oden, J. T., and D. A. Kross, "Analysis of General Coupled Thermo-elasticity Problems by the Finite Element Method", Proc. 2nd Conference on Matrix Methods in Structural Mechanics, AFFDL-TR-68-150, October, 1968.

8) Zienkiewicz, O. C., "The Finite Element Method in Engineering Science", McGraw-Hill.

9) Zienkiewicz, O. C. and C. J. Parekh, "Transient Field Problems: Two-Dimensional and Three-Dimensional Analysis by Isoparametric Finite Elements", International Journal for Numerical Methods in Engineer-ing", Vol. 2, 1970.

10) Carslaw, H. S. and J. C. Jaeger, "Conduction of Heat in Solids", Oxford University Press, 1959.

11) McAdams, W. H., "Heat Transmission", McGraw-Hill Book Company, 1954.

12) Comini, G., S. DelGuidice, R. W. Lewis and O. C. Zienkiewicz, "Finite Element Solution of Non-linear Heat Conduction Problems with Special Reference to Phase Change", International Journal for Numerical Methods in Engineering, Vol. 8, 1974.

13) Comini, G., DelGuidice, S., Lewis, R. W., and Zienkiewicz, O.C., "Finite Element Solution of New-linear Heat Conduction Problems with Special Reference to Phase Change", International Journal Numerical Methods in Engineering, Vol. 8, 1974.

14) Friedman, E., "A Direct Interation Method For the Incorporation of Phase Change in Finite Element Heat Conduction Programs", Westing-house Corporation, WAPD-TM-1133, 1974.

15) Wood, W. L., and R. W. Lewis, "A Comparison of Time Marching Schemes for the Transient Heat Conduction Equation", International Journal for Numerical Methods in Engineering, Vol. 9, 1975.

16) Hibbitt, H. D., "Thermomechanical Analysis of Welded Joints and Structures", Fourth Army Materials Technology Conference, Boston, Mass., Sept. 1975.

17) Kazimi, M. S., and C. A. Erdman, "On the Interface Temperature of Two Suddenly Contacting Materials", Journal of Heat Transfer, Nov. 1975.

18) Cacciatore, P. J., "Thermomechanical Finite Element Analysis of the Welding Process", General Dynamics Corporation, Electric Boat Division, P440-76-056, March, 1976.

19) Nickell, R. E., and P. V. Marcal, "Finite Element Analysis of the Welding Process", Division of Engineering, Brown University Report.

Appendix A

FINITE ELEMENT EQUATIONS FOR HEAT TRANSFER

Let the unknown temperature T throughout the solution domain be represented as;

$$T(x_i,t) = \sum_{\ell=1}^{n} N_\ell (x_i) T_\ell(t) \qquad \text{A.1}$$

where $x_i (i=1,2,3)$ = denotes the three dimensional spatial dependence of T

$N_\ell(x_i)$ = shape functions defined piecewise, element by element for each element in the domain,

The simultaneous equations, allowing for the solution of n values of T_ℓ, are obtained typically for point ℓ by equating to zero the weighted and integrated residual resulting from the substitution of Equation A.1 into Equations 1 or 2. Thus for the ℓth equation

$$\int_R N_\ell \left\{ \frac{\partial}{\partial x} \left[k_x (T) \frac{\partial}{\partial x} + k_y (T) \frac{\partial}{\partial y} + k_z (T) \frac{\partial}{\partial z} \right] \sum_1^n N_m T_m(t) + Q - \rho c(T) \frac{\partial}{\partial t} \left(\sum_1^n N_m T_m(t) \right) \right\} dxdydz = 0 \qquad \text{A.2}$$

Integrating the first term by parts we obtain the following;

$$-\int_R \frac{\partial N_\ell}{\partial x} k_x(T) \frac{\partial N_m}{\partial x} + \frac{\partial N_\ell}{\partial y} k_y (T) \frac{\partial N_m}{\partial y} + \frac{\partial N_\ell}{\partial z} k_z (T) \frac{\partial N_m}{\partial z} T_m(t) dV$$

$$+ \int_S N_\ell \left[\sum_1^n k_x(T) \frac{\partial N_m}{\partial x} \ell_x + \sum_1^n ky(T) \frac{\partial N_m}{\partial y} \ell y + \sum k_z (T) \frac{\partial N_m}{\partial z} \ell_z \right] T_m(t)\, dS$$

$$+ \int_R N_\ell\, QdV - \int_R N_\ell \rho c (T) \sum_1^n N_m \dot{T}_m (t) d^V = 0 \qquad \text{A.3}$$

The surface integral only arises on boundaries of type S_2. Substituting Equation 4 into Equation A.3 we obtain;

$$\int_R \left[\frac{\partial N_i}{\partial x} kx(T) \frac{\partial N_j}{\partial x} + \frac{\partial N_i}{\partial y} ky(T) \frac{\partial N_j}{\partial y} + \frac{\partial N_i}{\partial z} kz(T) \frac{\partial N_j}{\partial z} \right] T_j (t) dV$$

$$+ \int_S N_i \left[q + h(T)(T - T_S) \right] ds - \int_R N_i Q dV$$

$$+ \int_R N_i \rho c(T) N_j dV \; \dot{T}_j(t) = 0 \qquad \text{A.4}$$

Since first order derivatives are involved in this integration, only continuity of the interpolation functions N_i (or temperature) need be imposed to insure integrability.

The following matrices are defined for writing Equation A.4 in a more compact form.

$$H_{ij} = \int_V \left[\frac{\partial N_i}{\partial x} kx(T) \frac{\partial N_j}{\partial x} + \frac{\partial N_i}{\partial y} ky(T) \frac{\partial N_j}{\partial y} + \frac{\partial N_i}{\partial z} kz(T) \frac{\partial N_j}{\partial z} \right] dV$$

$$+ \int_S N_i h(T) N_j dS = \text{thermal conductivity}$$

$$C_{ij} = \int_V N_i \; \rho c(T) N_j dV \qquad \text{A.5}$$

$$= \text{thermal heat capacitance}$$

$$\{Q_i\} = - \int_V N_i Q dV + \int_S N_i \left[q - h(T) T_S \right] dS$$

$$= \text{thermal heat vector}$$

$$H \; \{T\} + C \; \{T\} + \{Q\} = o \qquad \text{A.6}$$

Equation A.6 represents n equations in n unknown temperatures. The integrations indicated in Equation A.5 are performed element by element and then assembled in the usual matrix fashion. These equations are nonlinear since the material properties (k,c) are allowed to vary with temperature. The specific heat c can be modified to reflect the latent heat release during phase change. The thermal heat vector is also nonlinear as a result of the temperature dependence of h and the possible nonlinear character of q if radiation occurs.

3-DIMENSIONAL HEAT FLOW DURING FUSION WELDING

Sindo Kou

Department of Metallurgy and Materials Science
Carnegie-Mellon University
Pittsburgh, PA 15213

A computer model was developed to simulate the steady state, three dimensional heat flow during the fusion welding of thick plates. The oblate spheroidal coordinate system was used in conjunction with the enthalpy model in order to facilitate the heat flow calculation. The Gaussian type distribution of the heat flux was assumed. The heat of fusion, the temperature dependence of thermal properties and the surface heat loss were considered. And the effect of the weld pool convection on the heat flow during welding was approximated with an effective thermal conductivity of the liquid phase. Welding heat flows in both 2219 aluminum and 4340 steel were simulated using the computer model developed.

Introduction

The analytical solution to the steady state, 3-dimensional heat flow problem of the bead-on-plate welding of thick sections was first derived by Rosenthal (1) in 1941. The major assumptions made were: 1. a point heat source, 2. no heat of fusion, 3. no surface heat loss due to convection or radiation, and 4. constant thermal properties of the workpiece material. Despite some of these rather unrealistic assumptions, Rosenthal's work in welding heat flow has drawn much attention. Many subsequent investigations (2-14) have been concerned with the refinement or the verification of Rosenthal's original theory. However, no major improvements have been made until the recent application of numerical methods in the simulation of welding heat flow. For example, Hibbitt (15) and Friedman (16) have studied the thermal response of stationary arc welds by using the finite element method. Rosenthal's assumptions were released in these studies since numerical methods are much more powerful in solving moving boundary problems. Friedman (17) and Krutz (18) have also studied the 3-dimensional welding heat flow problem associated with a moving heat source by applying the results of the 2-dimensional welding heat flow problem pertaining to a stationary heat source. These studies have proved to be quite successful. However, 3-dimensional information, such as the weld pool configuration and the temperature distribution, can not be obtained directly from such 2-dimensional simulations.

In the present study, a 3-dimensional welding heat flow model was developed based on the finite difference method and the enthalpy model of Shamsundar (19). The size and the distribution of the heat source, the heat of fusion, the surface heat loss and the temperature dependence of thermal properties were taken into account in such a model. 3-dimensional welding heat flow in 2219 aluminum and 4340 steel were simulated using this model.

Mathematical Model

Shown in Figure 1 is a sketch of the autogenous bead-on-plate welding of a thick plate. The heat source, which can be a tungsten electrode or a plasma torch, is moving at a constant velocity U. As a result of the welding heat input, a weld pool is created under the heat source. Behind the weld pool is the solidified structure of the fusion zone, i.e., the weld bead.

The convection of the weld pool was not simulated in this study due to the complex effects of the electromagnetic force, the imping force of the plasma jet and the surface tension of the liquid metal. Rather, the effective liquid conductivity (20,21) was used to count for the effect of the weld pool convection on the heat flow.

1. Finite Difference Equation

The energy equation, expressed in terms of a coordinate system moving with the heat source at a constant velocity U, can be written as follows:

$$\frac{\partial (\rho H)}{\partial t} = \frac{\partial}{\partial x}\left(k \frac{\partial T}{\partial x}\right) + \frac{\partial}{\partial y}\left(k \frac{\partial T}{\partial y}\right) + \frac{\partial}{\partial z}\left(k \frac{\partial T}{\partial z}\right) + P + U \frac{\partial (\rho H)}{\partial y} \tag{1}$$

where t is time, T temperature, P the source strength per unit volume, H the specific enthalpy and k the thermal conductivity of the workpiece. Assuming the workpiece is of sufficient length so that the steady state heat flow is achieved, except during the initial and the final transients. Therefore, the time-dependent term in Equation (1) can be neglected.

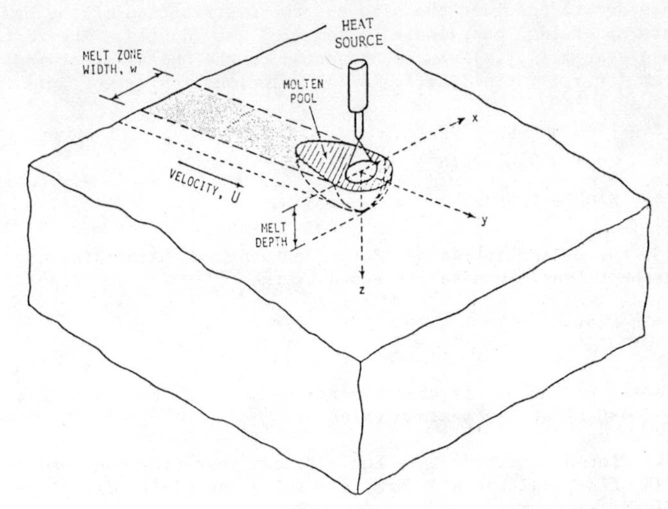

Figure 1. Schematic illustration of a workpiece subjected to a moving heat source of constant speed U. The (x,y,z) coordinate is moving with the heat source at the same speed.

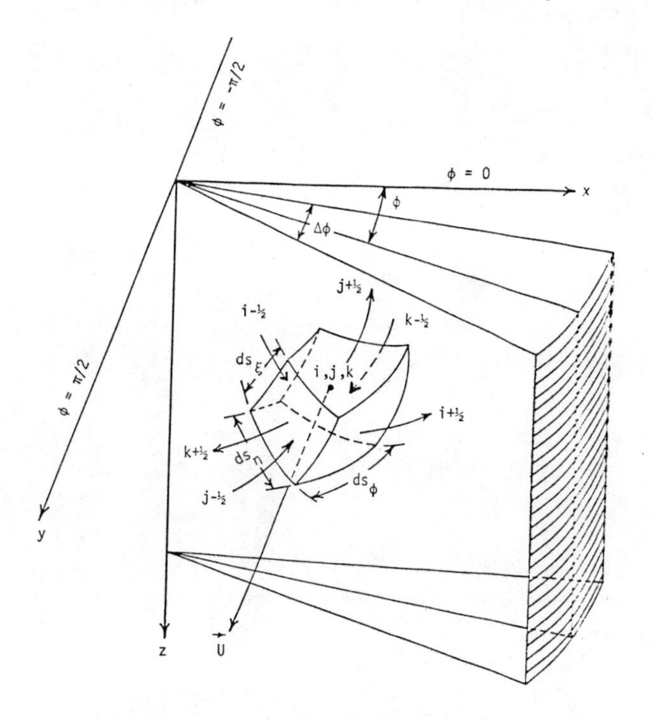

Figure 2. Three-dimensional moving oblate spheroidal coordinate system and problem geometry showing a discritized space domain.

In order to consider the size and the distribution of the heat source, the oblate spheroidal coordinate system (22) was adopted. The rectangular coordinate system (x,y,z) can be converted to the oblate spheroidal coordinate system (η, ξ, ϕ) using the following relationships (see Figure 2):

$$
\begin{aligned}
x &= a \cosh \eta \sin \xi \cos \phi \\
y &= a \cosh \eta \sin \xi \sin \phi \\
z &= a \sinh \eta \cos \xi
\end{aligned}
\qquad (2)
$$

where a is the effective radius of the heat source. According to Pavelic (23), the heat flux q can be expressed by the following equation:

$$q = q_o \exp\left(\frac{-3r^2}{a^2} \right)$$

the effective radius a is thus the radius at which the input heat flux q drops to about 5% of its maximum value q_o.

Substituting Eqn. (2) into Eqn. (1) and rewriting the resultant equation in the finite difference form, the following finite difference equation was obtained:

$$
\begin{aligned}
C_s T_{i,j,k} = {} & C_1 T_{i-1,j,k} + C_2 T_{i+1,j,k} + C_3 T_{i,j-1,k} \\
& + C_4 T_{i,j+1,k} + C_5 T_{i,j,k-1} + C_6 T_{i,j,k+1} \\
& - C_7 H_{i-1,j,k} + C_8 H_{i+1,j,k} - C_9 H_{i,j-1,k} \\
& + C_{10} H_{i,j+1,k} - C_{11} H_{i,j,k-1} + C_{12} H_{i,j,k+1} \\
& + Pa^2(\cosh^2\eta_i - \sin^2\xi_j)/k
\end{aligned}
\qquad (3)
$$

The coefficients in Eqn (3) are:

$$C_{1,2} = \frac{1}{\Delta\eta^2} \mp \frac{1}{2\Delta\eta} \tanh\eta_i$$

$$C_{3,4} = \frac{1}{\Delta\xi^2} \mp \frac{1}{2\Delta\xi} \cot\xi_j$$

$$C_{5,6} = \frac{1}{\Delta\phi^2} \frac{(\cosh^2\eta_i - \sin^2\xi_j)}{\cosh^2\eta_i \sin^2\xi_j}$$

$$C_7 = C_8 = \frac{aU\rho}{2k\,\Delta\eta} \sinh\eta_i \sin\xi_j \sin\phi_k$$

$$C_9 = C_{10} = \frac{aU\rho}{2k\,\Delta\xi} \cosh\eta_i \cos\xi_j \sin\phi_k$$

$$C_{11} = C_{12} = \frac{aU\rho}{2k\,\Delta\phi} \frac{(\cosh^2\eta_i - \sin^2\xi_j)}{\cosh\eta_i \sin\xi_j} \cos\phi_k$$

$$C_s = \sum_{n=1}^{6} C_n = +2 \left[\frac{1}{\Delta\eta^2} + \frac{1}{\Delta\xi^2} + \frac{1}{\Delta\phi^2} \frac{(\cosh^2\eta_i - \sin^2\xi_j)}{\cosh^2\eta_i \sin^2\xi_j}\right]$$

Once temperature T is computed from Eqn. (3), it can be used to calculate the corresponding enthalpy H using standard thermodynamic relationships. This procedure is rather straightforward for either the solid or the liquid phase. For the solid+liquid phase (the mushy zone), however, the following approximation can be used:

$$H = H_s^* (1-f_L) + H_L^* f_L \qquad \text{where } f_L = \frac{T-T_s}{T_L-T_s} \tag{4}$$

In the above equation, H_L^* is the specific enthalpy of the liquid phase at the liquidus temperature T_L, H_s^* is the specific enthalpy of the solid phase at the solidus temperature T_s and f_L is the fraction liquid.

2. Boundary Conditions

The boundary conditions have already been described in detail elsewhere (24). Given below is a summary of the boundary conditions:

(a). $C_1=0$, $C_7=0$, $P=\dfrac{qa \cos \xi_j}{\cosh\eta_i \Delta\eta a^2 (\cosh^2\eta_i - \sin^2\xi_j)}$

at $\eta=0$, $0 \le \xi \le \frac{\pi}{2}, -\frac{\pi}{2} \le \phi \le \frac{\pi}{2}$

(b). $C_5=0$, $C_{11}=0$, $P=0$

at $\eta>0$, $0 \le \xi \le \frac{\pi}{2}$, $\phi = \frac{-\pi}{2}$

(c). $C_6=0$, $C_{12}=0$, $P=0$

at $\eta>0$, $0 \le \xi \le \frac{\pi}{2}$, $\phi = \frac{\pi}{2}$

(d). $T = T_o$

at $\eta \to \infty$, $0 < \xi < \frac{\pi}{2}, \frac{-\pi}{2} < \phi < \frac{\pi}{2}$

(e). $C_3=0$, $C_9=0$, $P=0$

at $\eta > 0$, $\xi = 0$, $\frac{-\pi}{2} < \phi < \frac{\pi}{2}$

(f). $C_4=0$, $C_{10}=0$, $P=0$

at $\eta>0$, $\xi = \frac{\pi}{2}, \frac{-\pi}{2} < \phi < \frac{\pi}{2}$

Results and Discussion

The finite difference equation, Eqn. (3), together with the boundary conditions was solved using the Gauss-Seidel method. A digital Equipment DEC-20 digital computer was used to carry out the computation of the temperature distribution in the workpiece. The thermal properties were updated in each iteration of temperature calculation. The convergence criteria is tested by comparing the new and the old values of $T_{i,j,k}$:

$$\left| T_{i,j,k}(\text{new}) - T_{i,j,k}(\text{old}) \right| \leq 1^{\circ}\text{C} \qquad (5)$$

Both 2219 aluminum and 4340 steel were simulated using the published data of thermal properties (25,26,27). The validity of the heat flow model was verified with Rosenthal's analytical solution under conditions where no liquid phase is present.

Since 3-dimensional heat flow is involved, the results of the calculations were presented in terms of the top view, the side view and the front view. Figure 3 shows the calculated results of 2219 aluminum with the 700°K, the solidus and the liquidus isotherms indicated. The shaded areas indicate the freezing range of the alloy, i.e., the mushy zone. Similar results are given in Figure 4 for 4340 steel.

In order to further verify the validity of the heat flow model welding experiments, including the measurement of thermal cycles and the size of the fusion zone, are being carried out at the Center for the Joining of Materials, Carnegie-Mellon University.

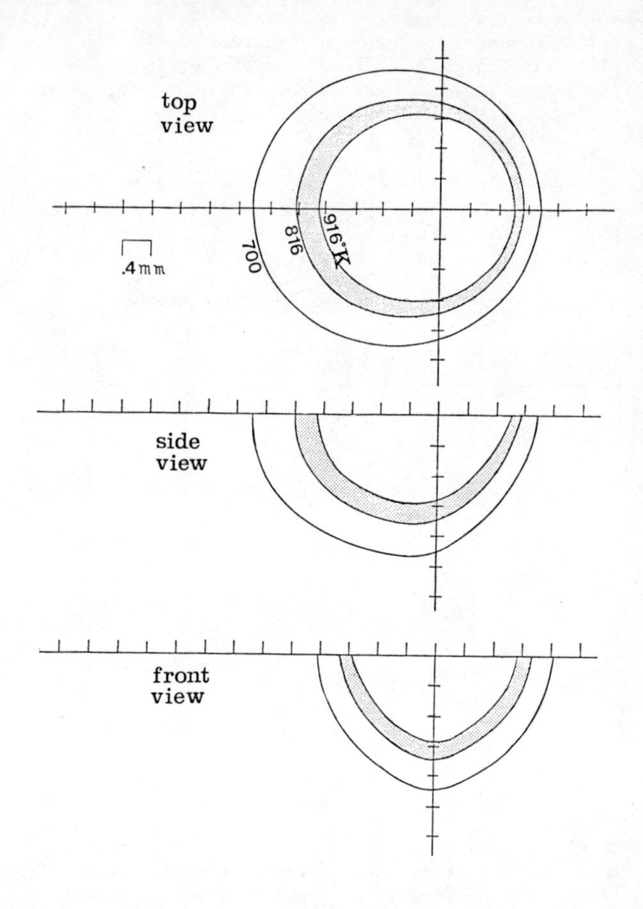

Figure 3. The temperature distribution in a thick plate of 2219 aluminum. The welding velocity is 20 mm/sec, the heat input is 430 Watts.

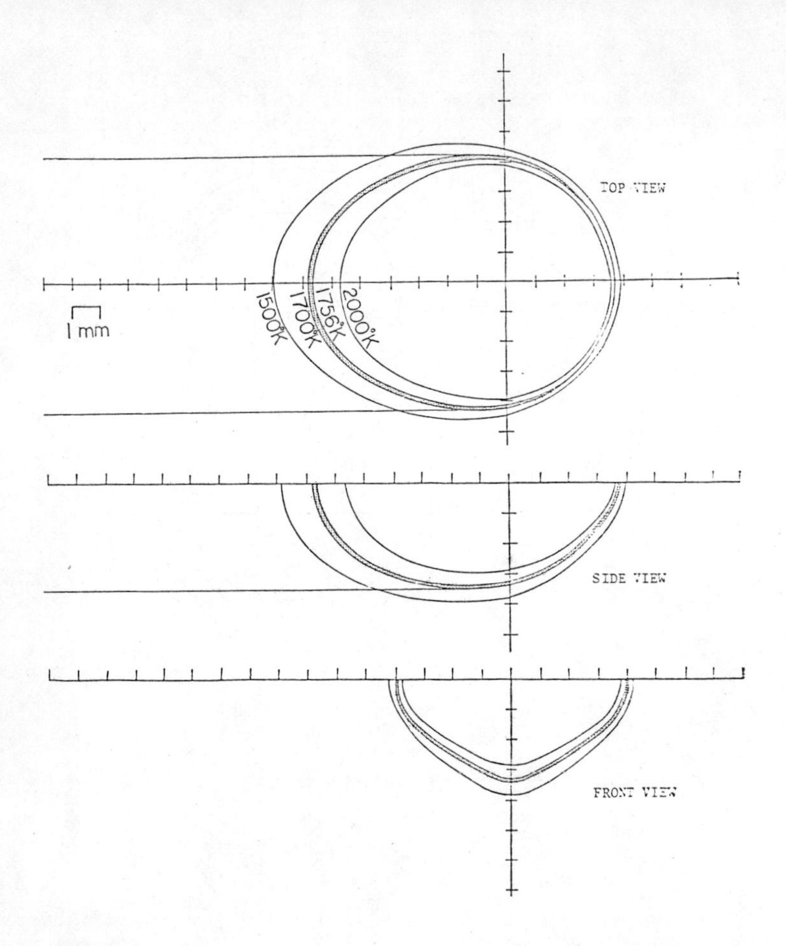

Figure 4. The temperature distribution in a thick plate of 4340 steel. The welding speed is 2 mm/sec, the heat input 1500 Watts.

Acknowledgements

The author gratefully acknowledges support for this study from the Center for the Joining of Materials at Carnegie-Mellon University, under NSF Grant DMR-7824699. The initial support from the Department of Metallurgy and Materials Science is also greatly appreciated.

References

1. D. Rosenthal, Welding Journal,(1941) vol. 20, pp. 220S.

2. W. F. Hess, L. L. Merrill, E. F. Nippes, Jr. and A. P. Bunk, Welding Journal,(1943) vol. 22, pp. 377S.

3. N. N. Rykalin, Calculation of Heat Flow in Welding, Moscow 1951, translated by Z. Paley and C. M. Adams, Jr. for the Army under contract UC-19-066-001-C-3817.

4. R. J. Grosh and Z. A. Trabant, Welding Journal, (1956) vol. 35, pp.396S.

5. C. M. Adams, Jr., Welding Journal, (1958) vol. 37, pp. 210S.

6. P. Jhaveri, W. G. Moffatt and C. M. Adams, Jr., Welding Journal, (1962) vol. 41, pp. 12S.

7. J. M. Barry, Z. Paley and C. M. Adams, Jr., Welding Journal, (1963) vol. 42, pp. 97S.

8. Z. Paley, J. N. Lynch and C. M. Adams, Jr., Welding Journal, (1964) vol. 43, pp. 71S.

9. N. Christensen, V. Davies and K. Gjermundsen, British Welding Journal, February 1965, pp. 54.

10. P. S. Myers, O. A. Uyehara and G. L. Borman, Welding Research Council Bulletin #123, July 1967.

11. D. E. Schillinger, I. G. Betz and H. Markus, Welding Journal, (1970) vol. 49, pp. 410S.

12. E. G. Signes, Welding Journal, (1972), vol. 51, pp. 473S.

13. N. D. Malmuth, W. F. Hall, B. Z. Davis and C. D. Rosen, Welding Journal, (1974) vol. 53, pp. 388S.

14. P. J. Alberry and W. K. C. Jones, paper presented at the 1979 TMS Fall Meeting, Milwaukee, Wis., Sept. 1979.

15. H. D. Hibbitt and P. V. Marcal, Computers and Structures, (1973) vol. 3, pp. 1145.

16. E. Friedman and S. S. Glickstein, Welding Journal, (1976) vol. 55, pp. 408S.

17. E. Friedman, Trans. ASME, J. Pressure Vessel Tech., (1975) vol. 97, Series J, no. 3, pp. 206.

18. G. W. Krutz and L. J. Segerlind, Welding Journal, (1978) vol. 57, pp. 211S.

19. N. Shamsundar and E. M. Sparrow, Journal of Heat Transfer, August, 1975 pp. 333.

20. E. A. Mizikar, Trans. of AIME., (1967) vol. 239, pp. 1747-58.

21. Y. Sharir, A. Grill and J. Pelleg, Metallurgical Transactions, (1980) vol. 11B, pp. 257-65.

22. G. E. Schneider, A. B. Strong and M. M. Yuvanovich, AIAA paper 75-707, AIAA 10th Thermophysics Conference, Denver, CO, May 1975.

23. V. Pavelic, R. Tanbakuchi, O. A. Uyehara and D. S. Myers, Welding Journal, (1969) vol. 48, pp. 295S-305S.

24. S. Kou, S. C. Hsu and R. Mehrabian, "Rapid Melting and Solidification of a Surface Due to a Moving Heat Flux", Met. Trans., vol. 12B (1981), pp. 33.

25. Aluminum, vol. I, Properties, Physical Metallurgy and Phase Diagram, ASM, 1967.

26. Aluminum Standards and Data, the Aluminum Association, 1976.

27. J. Woolman, R. A. Mottram, "The Mechanical and Physical Properties of the British Standard EN Steels", vol. 2,(1966) pp. 73.

HEAT-FLOW SIMULATION OF PULSED CURRENT

GAS TUNGSTEN ARC WELDING

G. M. Ecer and M. Downs
Westinghouse Research and Development Center
Pittsburgh, Pennsylvania 15235

H. D. Brody and A. Gokhale
Department of Metallurgy and Materials Engineering
University of Pittsburgh
Pittsburgh, Pennsylvania 15260

Pittsburgh, PA 15261

ABSTRACT

As one aspect of a program to determine the relations among welding parameters, heat flow, weld zone structure and weld soundness in pulsed arc welding, computer simulations of the heat flow in pulsed welded sheets and plates are being developed. The results of the computer simulations are being compared to thermocouple measurements of the temperature cycles and distributions within the weldment and high speed cinematographic measurements of the variation of the size of the weld pool. The explicit finite difference technique used for the two-dimensional heat flow computations has been found adequate in predicting temperature fields, thermal gradients, the solidus and liquidus movements in and around the weld pool, as well as its size and shape.

INTRODUCTION

The technique of pulsed arc welding combines the desirable features of a high energy input rate process with the advantages of low total heat input. Practical benefits that may be achieved by pulsing include reduction of heat build-up along the weld seam[1], improved weld bead size and shape control[2], improved arc stability[3], reduced sensitivity to disparity in heat sink[2], and overcoming of the gravity effects on the weld pool. Furthermore, in controlling weld metal solidification, pulsed-current arc welding provides more parametric freedom than conventional steady-current methods. The added flexibility results from the new process variables emerging from the cycling of the arc current between high and low currents with the current waveform conforming to a variety of designs. In its simplest form, the arc current follows a square waveshape and introduces six process variables in place of two. The variables of current and voltage in steady-current welding are replaced by high pulse and low pulse currents, their durations (Figure 1) and the corresponding arc voltages. The increased complexity of the weld metal structural control resulting from the introduction of the new variables may be overcome by the utilization of computer-based heat flow simulation techniques.

As one aspect of a program to determine the relations among welding parameters, heat flow, weld zone structure and weld soundness in pulsed arc welding, computer simulations of the heat flow in pulsed welded sheets and plates are being developed. The present paper reports the results of the computer simulations of the pulsed-current GTA welding of thin sheets under two-dimensional heat flow. The predicted temperature fields around a weld pool are compared with those experimentally determined and the simulated movement of the weld pool solidification isotherms is verified through high speed cinematographic measurements.

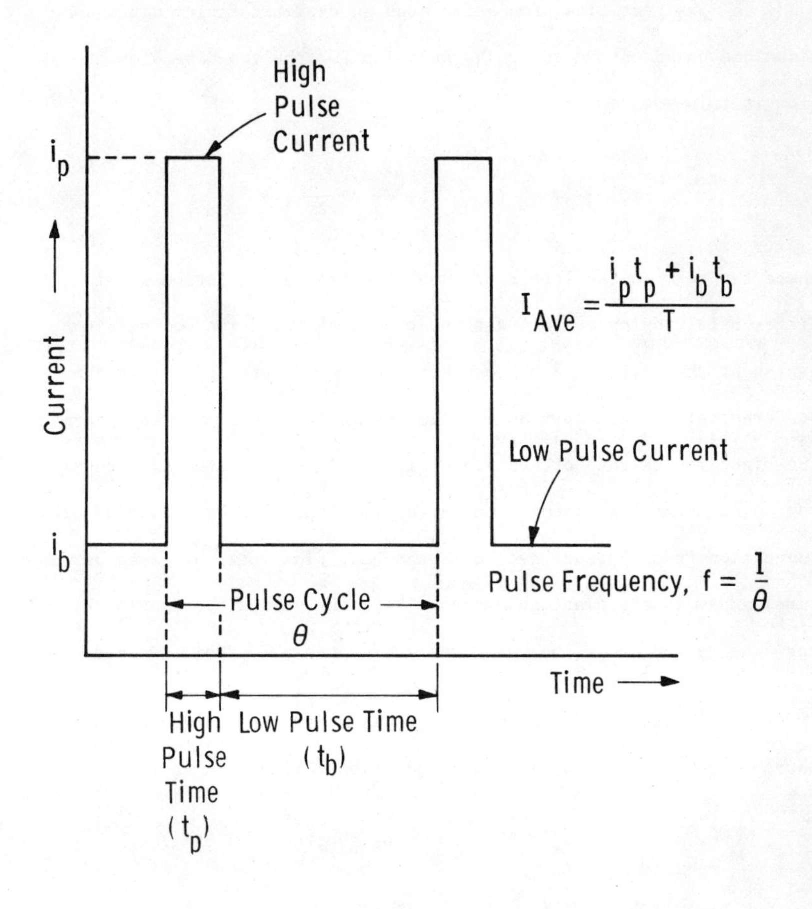

Fig. 1 - An idealized pulsed current waveform and associated definitions.

METHODOLOGY

Computations

The heat flow simulation uses an explicit finite difference technique based on the governing relation for the two-dimensional heat flow in thin sheets

$$\rho C_p \frac{\partial T}{\partial t} = \frac{\partial}{\partial x} (k \frac{\partial T}{\partial x}) + \frac{\partial}{\partial y} (k \frac{\partial T}{\partial y}) + \frac{q}{\omega} \quad (1)$$

where the term on the left represents the heat accumulation within a differential region of interest and the first two terms on the right represent the difference between heat conducted into and out of the differential element (per unit volume per unit time). The last term on the right is the rate of heat input into or loss from the differential region, such as heat input from the arc or heat loss by radiation or convection from the surface. q is the heat flux (per unit area per unit time) and ω is the plate thickness. k is the thermal conductivity, ρ is the density and C_p is the heat capacity of the alloy being welded.

The computational method accounts for the heat loss from top and bottom surfaces by convection and radiation through the expressions

$$q = [2h_r + h_c \text{ (top)} + h_c \text{ (bottom)}] (T - T_o) \quad (2)$$

$$h_r = \sigma\varepsilon \, [T_a^2 + T_{ao}^2][T_{ao}] \quad (3)$$

$$h_c = h' \, (T - T_o)^{0.25} \quad (4)$$

where σ is the Stefan-Boltzman constant, ε is the emissivity, T_a and T_{ao} are the absolute temperature of the surface and ambient, respectively, and h' is a constant in the expression for the heat transfer coefficient for convection.

In the nodal network used, shown in Figure 2, a region of interest near the weld pool is given increased resolution by making the node spacings finer than the surrounding regions. The network itself represents one half of the sheet metal being welded, the other half is assumed the mirror image of the half shown. At the centerline ($x = 0$), no heat flow to the other side, or $dT/dx = 0$, is assumed. The rows and columns of the network are numbered as:

Small grid rows	$1 < J < n$
Small grid columns	$1 < I < m$
Large grid rows	$1 < K < N$
Large grid columns	$1 < L < M$

where n, m, N and M are 72, 22, 85 and 22, respectively, in the version of the computer program whose results are reported here.

The nodal difference equation written for equal increments in x and y within an internal element in the large grid becomes

$$T'_{K,L} = T_{K,L} + \phi\,[T_{K+1,L} + T_{K-1,L} + T_{K,L+1} + T_{K,L-1} - 4T_{K,L} + 8/k\omega] \tag{5}$$

T' and T are the nodal temperatures at $t + \Delta t$ and t, respectively,

$$\phi = \frac{\alpha \Delta t}{\Delta X^2} \tag{6}$$

$$\alpha = \frac{k}{\rho C p} \tag{7}$$

and ΔX is the node spacing. The node spacing is uniform within the two regions and the spacings differ by any, preselected, integral factor. The time step in the explicit finite difference analysis is limited by

$$\Delta t < \frac{\Delta X^2}{2\alpha s} \tag{8}$$

where s = the number of dimensions in the analysis.

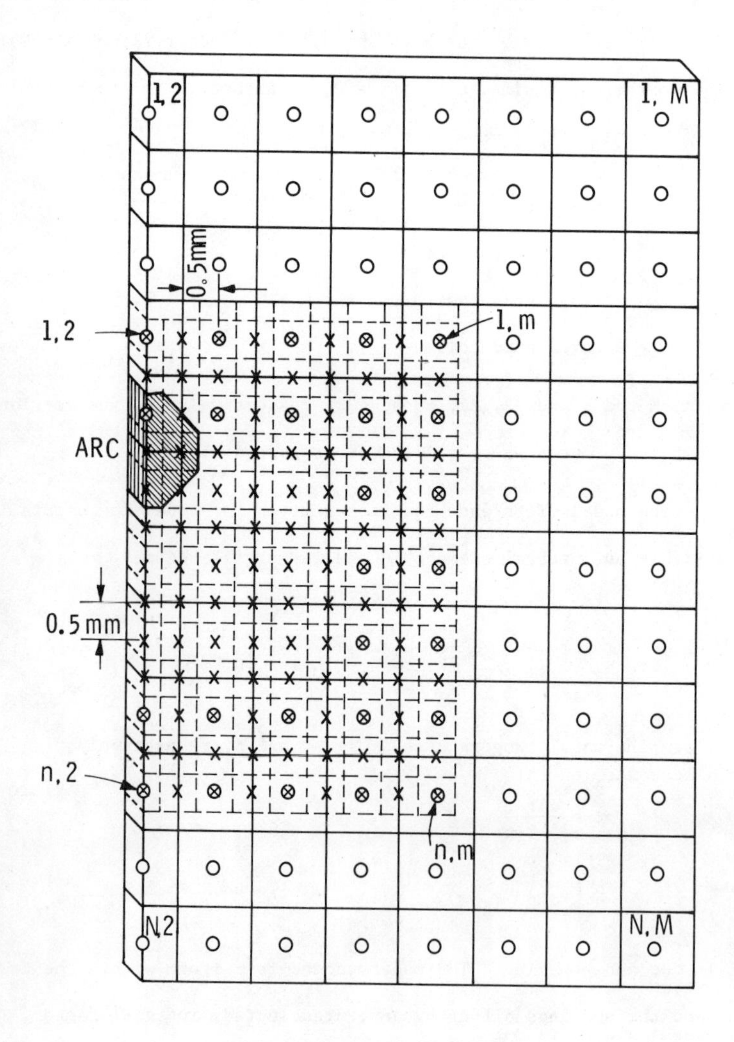

Fig. 2 - A sketch of the element mesh used for simulation of pulsed current tungsten arc welding of sheets.

The contribution of weld pool convection is incorporated into the computations by multiplying the conductivity of the liquid by a factor ranging from 3 to 10. In order to conserve computer time, three zones are set up and a different time step is used in each. In the region with the large node spacing (region 1) a large time step is used. The grid with small node spacing is divided into two zones. The region containing the liquid zone (region 3) is given the largest number of iterations and the smallest time step (due to its small spacing and high assumed thermal conductivity). The rest of the small grid region (region 2) receives an intermediate number of iterations. In the limit the time steps are related by

$$\Delta t_1 = a^2 \Delta t_2 = a^2 b \Delta t_3 \tag{9}$$

where a = ratio of the node spacings, and b = the factor used to increase the thermal conductivity.

The heat input from the arc is spread over several nodes, usually eight. The amount of heat input to each node is weighted and the weighting is changed after each iteration of the time step for region 2 to simulate the movement of the arc. Whenever the arc would have moved the distance of one large node spacing, all nodes are renumbered. The heat input to a node from the arc is calculated by

$$q_p = \frac{1}{4.18} \quad w \mu_p i_p e_p \quad \text{(cal/sec)}$$

$$q_b = \frac{1}{4.18} \quad w \mu_b i_b e_b \tag{10}$$

where w = weighting factor for the node,

μ_b, μ_p = welding arc heat transfer efficiency,

i_p, i_b = arc current for peak and background pulses, respectively,

e_p, e_b = arc voltage for peak and background pulses, respectively.

At the far edges of the simulated sheet heat transfer is assumed to continue in a way to keep a constant curvature for the T vs. x curve, or for $T_{K,M}$ at $x = (M-2)\Delta x$

$$\frac{T_{K,M} - T_o}{T_{K,M-1} - T_o} = \frac{T_{K,M-1} - T_o}{T_{K,M-2} - T_o} \tag{11}$$

A simplified flow diagram for the computer program simulating the pulsed arc welding is shown in Figure 3. A more detailed description of the finite difference method used to approximate heat flow in pulsed current welding was given elsewhere[4]. The input data includes such welding parameters as heat input during t_p and t_b, arc travel speed and values of t_p and t_b and material related properties including solidus and liquidus temperatures, k, ρ, and C_p values for solid, liquid and mushy alloys. Single values for α, k and C_p, selected near the melting point of the alloy, were used in the computations, although they could be introduced as functions of temperature. The latter, however, would have increased the computer time while adding little to the accuracy of the results for the weld pool vicinity which is the zone of interest from the solidification point of view.

The literature sources and the manner of estimating these material properties for the Fe-26Ni alloy used in the present investigation were provided in a previous paper[5]. The material properties used in the computations reported herein are given in Table 1.

Initial temperatures $T_{K,L}$ and $T_{J,I}$ are established by setting them equal to T_o (25°C in most runs). The computer is asked to repeatedly calculate the new times and temperatures until at least 14 seconds elapse. By this time, as experiments indicate, weld size reaches a steady state. Thermal gradients weld pool boundaries and interface velocities may be computed, and time, temperature and position prints for the entire nodal network may be obtained. As a quick check of the size and shape of the weld pool and the temperature field around it printer plots of the melt zones (regions 2 and 3) may be produced. Figure 4 is one such printer plot

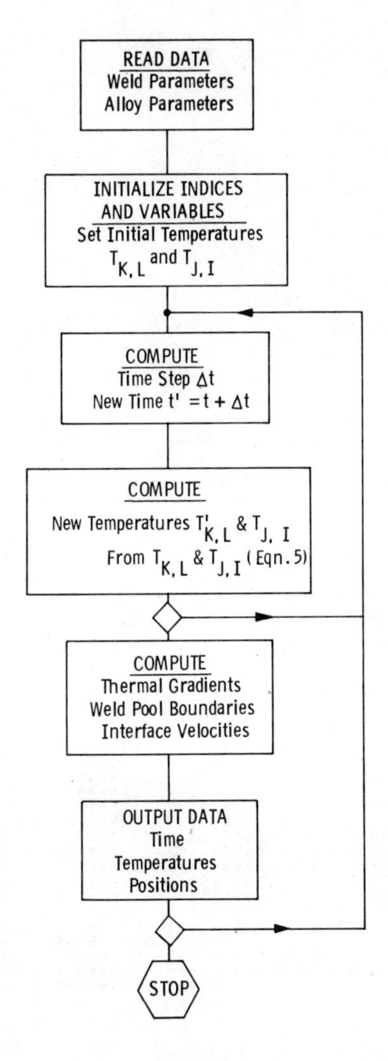

Fig. 3 - Simplified flow diagram for finite difference analysis of pulsed current arc welding.

TABLE 1

PROPERTIES OF THE Fe-26Ni ALLOY USED IN COMPUTATIONS

	Symbol	Units	Value
Solidus Temperature	--	°C	1443
Liquidus Temperature	--	°C	1468
Density	ρ	$g.cm^{-3}$	7.88
Thermal Conductivity	k	$cal.s^{-1}.°C^{-1}.cm^{-1}$	0.1250
Heat Capacity (Solid)	C_p	$cal.g^{-1}.°C^{-1}$	0.1427
Heat Capacity (Liquid)	C_p	$cal.g^{-1}.°C^{-1}$	0.1427
Thermal Diffusivity	$\alpha\ (=\frac{k}{C_p \cdot \rho})$	$cm^2.s^{-1}$	0.1111
Heat Capacity (Mushy Zone)	$C_p(mz)$	$cal.g^{-1}.°C^{-1}$	3.6880

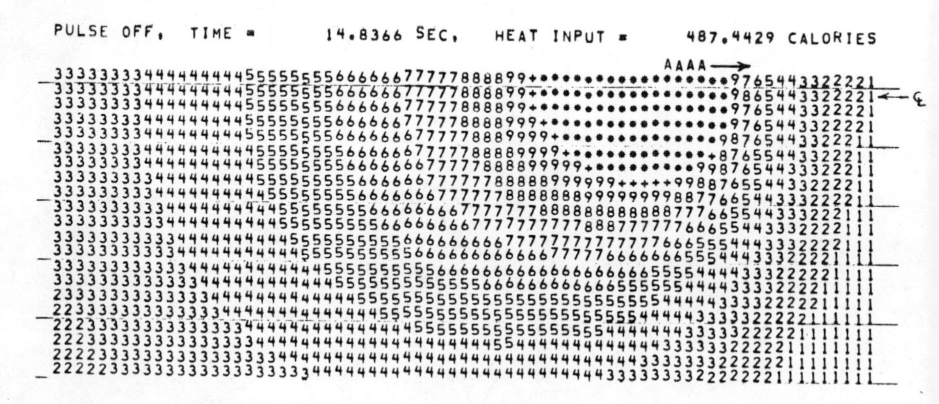

Figure 4 - A typical printer plot of the temperature zones and isothermal lines near the weld pool 14.8366 sec. after the start of the weld No. M1. Each number represents temperatures more than the product of the number and 1/10th of the melting temperature (1468°C) of the alloy. The star and plus signs indicate liquid and mushy zones respectively. Distance between points is 0.5 mm.

where isothermal lines are indicated as boundaries between two adjacent number regions, the mushy zone is shown by (+) signs and the liquid pool is indicated by the (*) signs. The arc position is shown by the four A's.

Experimental

Figure 5 is a schematic of the clamping fixture used to make full penetration welds in 0.08 cm thick sheets of the Fe-26Ni alloy. The welding parameters used for the welds discussed in this paper are listed in Table 2. The tungsten arc moved normal to the plane of the sketch in the direction of the ground connection. The arc length was kept constant at 0.1 in. (0.25 cm). The use of Teflon, with a thermal conductivity of about 250 times less than that of the Fe-26Ni alloy, was to assure two dimensional heat removal from the weld zone.

In addition to the above, the following parameters were kept constant: electrode composition, size and shape were W + 2% ThO_2, 1/8 in. (0.32 cm) diameter, 30° included vertex angle with 0.04 in. (0.1 cm) diameter flat tip; 100% argon as shield gas flowing at a rate of 40 cfh (18.9 ℓ/min) from around the electrode and 4 cfh (1.9 ℓ/min) to the back of the weldment; electrode stick-out from the cup 0.5 in. (1.25 cm) and from the collet 1.16 in. (2.95 cm). All welding was done with direct current electrode negative. Pulsed current waveform remained nearly square shaped within the pulse frequency range of 1 to 100 Hz; an example is shown in Figure 6.

Temperature measurements were taken using 0.02 cm diameter Pt-Pt + 10% Rh thermocouples placed close to and within the fusion zone on the underside of the weldment. The thermocouples were tack welded to the sheet. The thermocouple output was recorded on light sensitive paper using a high speed multi-channel recorder. Arc current transitions from high to low (or vice versa) pulse currents were automatically and instantly marked by the recorder.

TABLE 2

PULSED CURRENT GAS TUNGSTEN ARC WELDING PARAMETERS

USED TO MAKE FULL PENETRATION WELDS IN 0.08 cm THICK SHEETS

OF Fe-26Ni ALLOY

Weld No.	i_p (Amp.)	i_b (Amp.)	t_p (sec.)	t_b (sec.)	Arc Travel (cm/sec.)
G5	140	10	0.167	0.667	0.212
H8	I = 40 Amps, Steady Current				0.212
J4	120	10	0.033	0.133	0.212
J7	120	10	0.208	0.832	0.212
L6	120	10	0.167	0.333	0.605
L8	150	10	0.167	0.333	0.605
M1	150	10	0.038	0.250	0.212

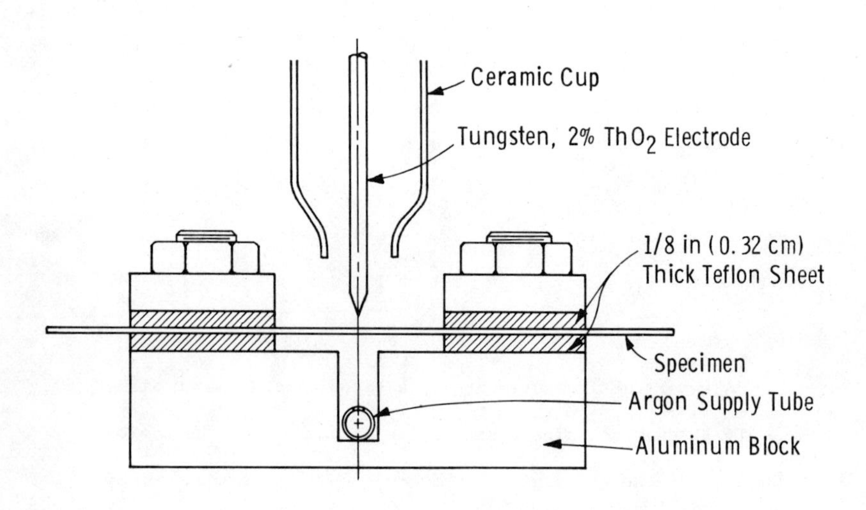

Fig. 5 - Cross-sectional view of the clamping fixture. Torch travel direction was normal to the paper.

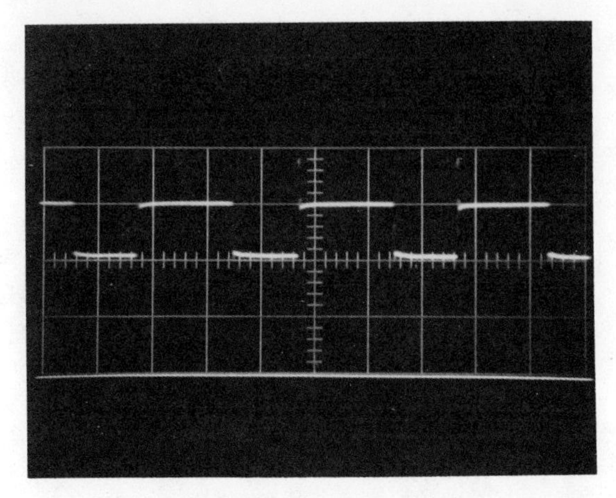

Fig. 6 - A pulsed current waveform typical of those utilized in this study. Here the pulsing parameters were:

i_p = 150A
i_b = 25A
t_p = 0.009s
t_b = 0.006s

High speed motion picture observations of weld pools from the trailing edge direction and from the side were made when the weld puddles reached a steady size. The camera speed increased to about 1000 frames/sec. after the first 0.3 seconds of operation. A pulsating red light within the camera provided markers on the film at a rate of 100 markers per second for more precise measurements of the camera speed. The films produced were analyzed using a professional editing machine with the capacity to record lapsing time and frame sequence electronically. Films were viewed at various speeds as well as frame by frame.

RESULTS AND DISCUSSION

In pulsed welding cycling of the arc current between i_p and i_b and the resulting cyclic variation in heat input can have pronounced effects on the weld geometry. This is illustrated in Figure 7 where the surface appearances of a steady current weld and a pulsed current weld are compared.

Weld No. H8

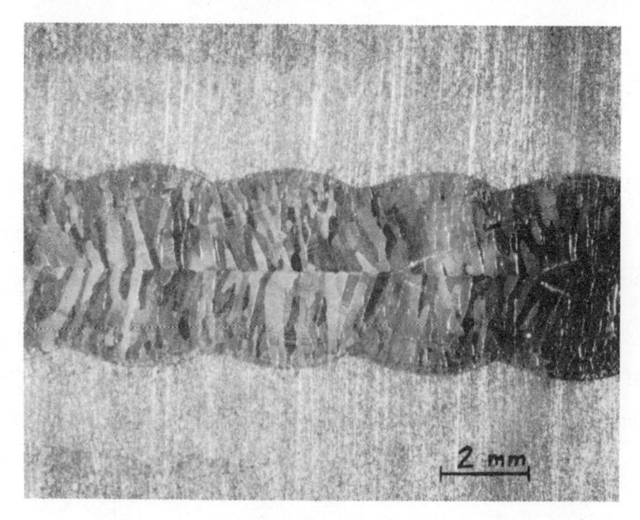

Weld No. L6

Figure 7 - Macrographs comparing the top surfaces of a steady current weld (top) and a pulsed current weld (bottom).

The bottom macrograph is that of the pulsed current weld No. M1 (Table 2) where the pitch, equal to the product of the travel speed and pulse period, was long (P = 3.0 mm) so that the weld appeared as a series of superimposed spot welds. With extremely large pitch, the pool could solidify completely between peak pulses. Smaller pitched pulsed welds, too, could solidify between peak pulses under conditions of low arc speed and large t_b/t_p ratio, thus allowing time to dissipate heat. On the other hand, short pitch and small t_b/t_p ratio pulsed welds would appear and behave as continuous welds.

Temperature cycles experienced by a point within or near the weld zone depends on the sum effect of a number of welding process and material variables that include i_p, i_b, t_p, t_b, arc travel speed, k, α, as well as the distance from the weld centerline. When full penetration welds are desired, under two dimensional heat flow conditions, weld bead width on top and bottom surfaces of the workpiece should be approximately equal. Except for the weld No. L6, all of the welds listed in Table 2 could be considered as 2-D welds, since all showed top to bottom bead width ratios higher than 0.93. In making 2-D welds in thin sheets, one is restricted to a critical maximum heat input per unit length, for a given sheet thickness, above which weld pool surface tension is not sufficient to hold the weld metal together and a weld metal "drop-through" occurs. Thus, all of the process variables have upper and lower limits. Under even such limitations, temperature cycling at points near the weld fusion line can be quite varied and substantial as exemplified by the experimental and computational results presented below.

The temperature vs. time plots of Figures 8 and 11 present the results of computer simulations of the experience of selected locations in the workpiece in comparison with those determined by thermocouple measurements for similar locations. Figure 8 is for a steady current weld; Figures 9-11 are for pulsed current weld. In Figures 8 and 11 the position of the thermocouple locations is displaced slightly from the location of the computer simulation. The locations in Figure 11 were within fusion zone and the thermocouple failed on approach of the arc. The tendency for the experimental temperature plots, for the pulsed current welds, to show somewhat depressed temperature cycling may be a result of the slow response of the thermocouples to sudden changes in temperature. In all four cases computed temperatures are remarkably close to the measured temperatures.

The magnitude of the temperature fluctuations in the liquid and at the interface (mushy zone) may have a bearing on any possible grain multiplication and structure modification that may result from dendrite remelting. In the present study, all of the welds under discussion showed predominantly cellular-columnar solidification structures with small tendency to form dendritic-columnar growth in long pitched, low heat input welds. As a consequence, none of the welds exhibited any degree of equiaxed grain formation. Under three dimensional heat flow conditions for pulsed

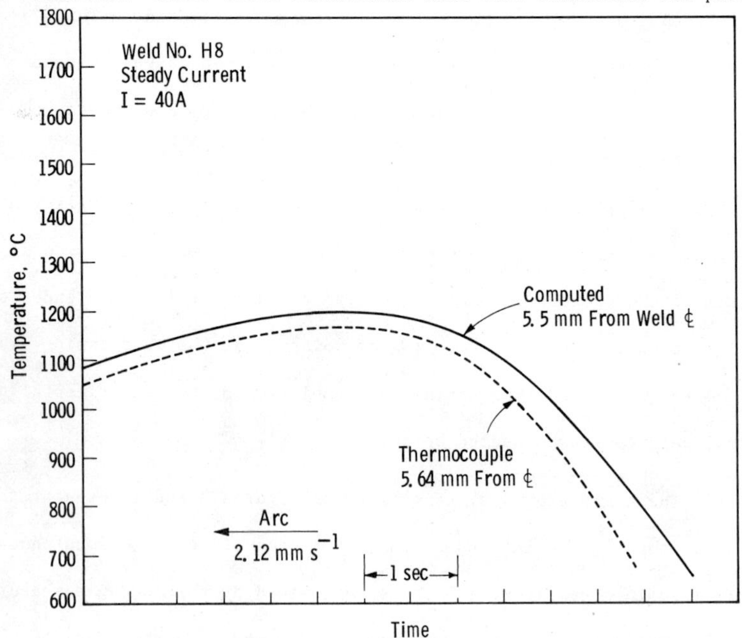

Fig. 8 - Temperature history of a point in the parent metal near the fusion line of a steady current weld. Computed and measured curves are compared. Note the slight difference in distance of the compared locations from the weld centerline.

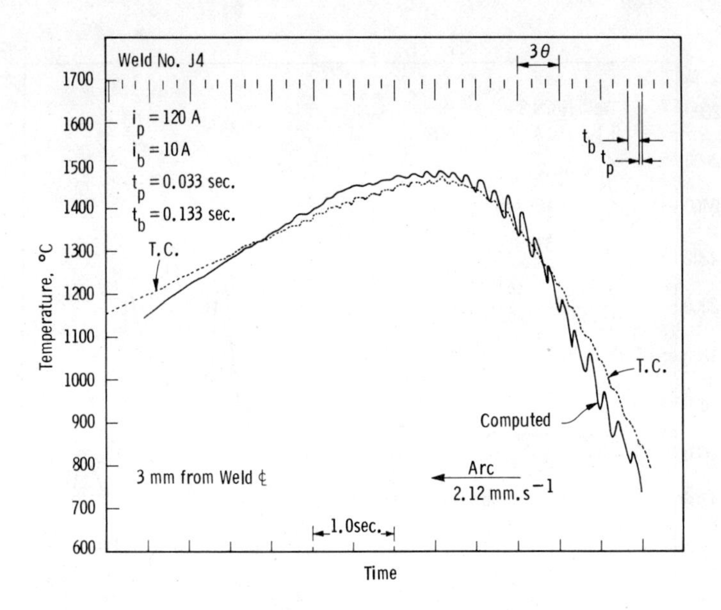

Fig. 9 - Time-temperature profile for a point just within the fusion zone of a pulsed current weld of relatively short pitch (P=0.35 mm).

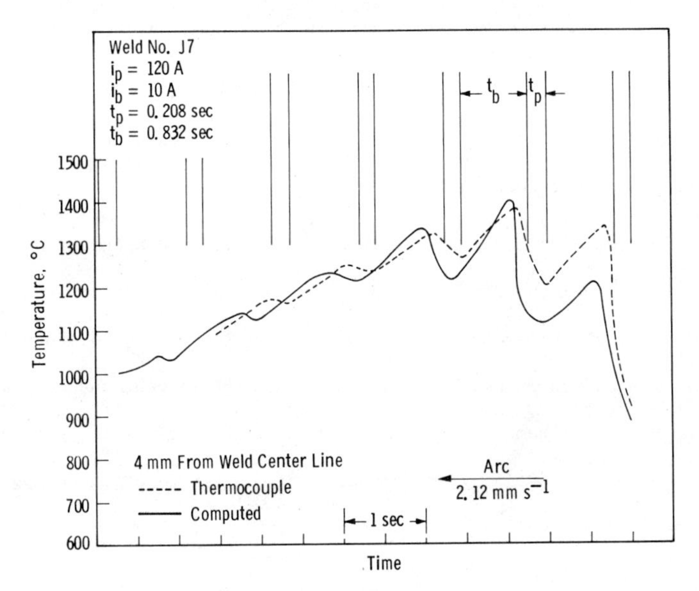

Fig. 10 - Time-temperature profile for a point in the parent metal near the fusion line of a pulsed current weld of moderate pitch (P=2.2 mm).

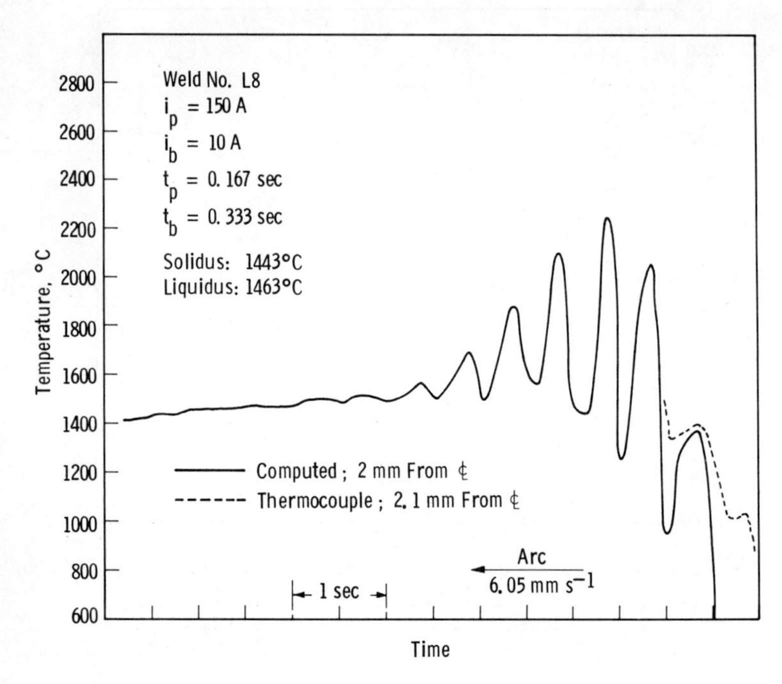

Fig. 11 - Time-temperature profile for a point in the fusion zone of a pulsed current weld of moderately long pitch (P=3 mm). Note the thermocouple malfunction in the melt zone.

welding of thicker plates of the Fe-26Ni alloy and a stainless steel, however, some grain refining effect due to arc current pulsing could be obtained as has been reported elsewhere[6,7].

The computer model predictions of the variations in the weld pool area (A_M), the mushy zone area (A_{MZ}) and length and width of the weld pool within a single pulse period are for a typical pulsed weld shown in Figure 12. The maxima in weld pool width, weld pool length, and weld pool area may be measured from the ripple formation on the weld surface. These three measurements for weld J7 are compared to the maxima on the computed curves. Again, agreement is remarkably good.

As the weld pool expands and contracts within a pulse period, cyclic increases and decreases occur in the thermal gradients along the two major axes as shown for the same weld (No. J7) in Figure 13. The relative constancy of the thermal gradient along the trailing edge is in contrast to the more greatly varying gradient in the transverse direction. It has been noted, elsewhere, freezing rate in the trailing and transverse directions show similarly contrasting behavior[6,7].

The movement of the solidus and the liquidus isotherms can be simulated by the computer anytime during the course of the welding operation. An example of this is shown in Figure 14 where the solidus and the liquidus movements within a single low pulse period, after about 15 seconds of welding, have been plotted. The computed curves are compared to the high speed motion picture observations where the single curve is that perceived by the observer to be a "solid-liquid interface" since the mushy zone boundaries were too diffuse for identification. The graphs, therefore, illustrate the ability of the computational approach in providing detailed information, in this case, more detailed than that obtained through high speed motion picture observations.

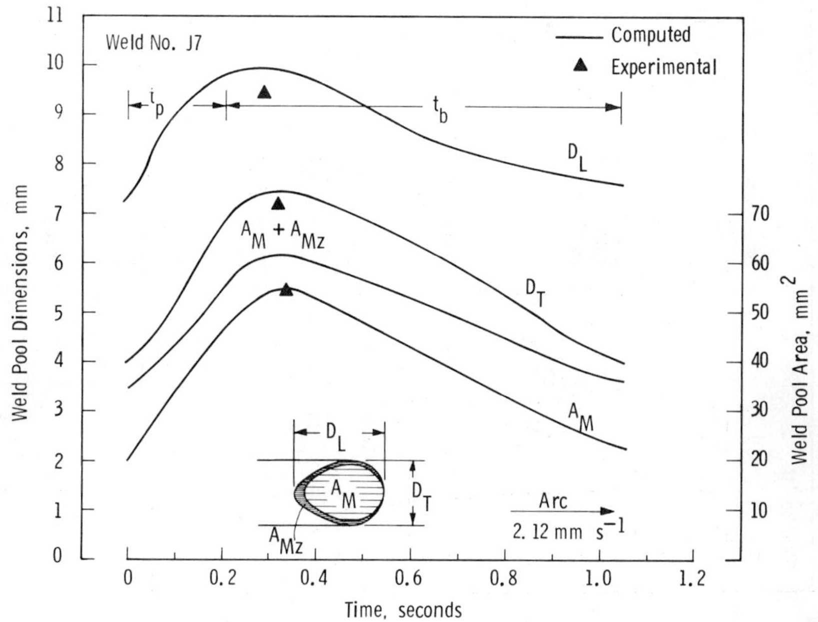

Fig. 12 - Variations of the weld pool length (D_L), width (D_T), mushy zone (A_{MZ}) and the size of the liquid pool within a pulse period.

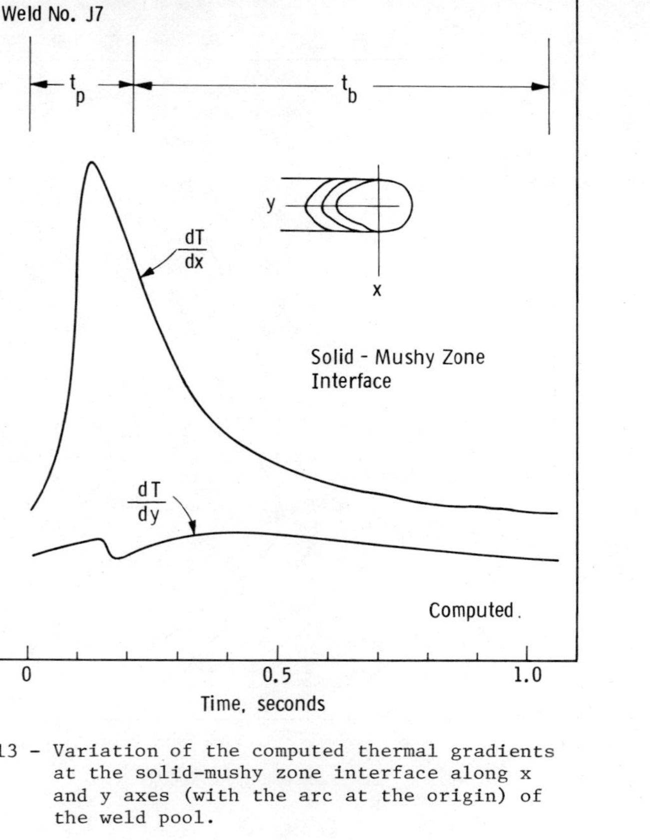

Fig. 13 - Variation of the computed thermal gradients at the solid-mushy zone interface along x and y axes (with the arc at the origin) of the weld pool.

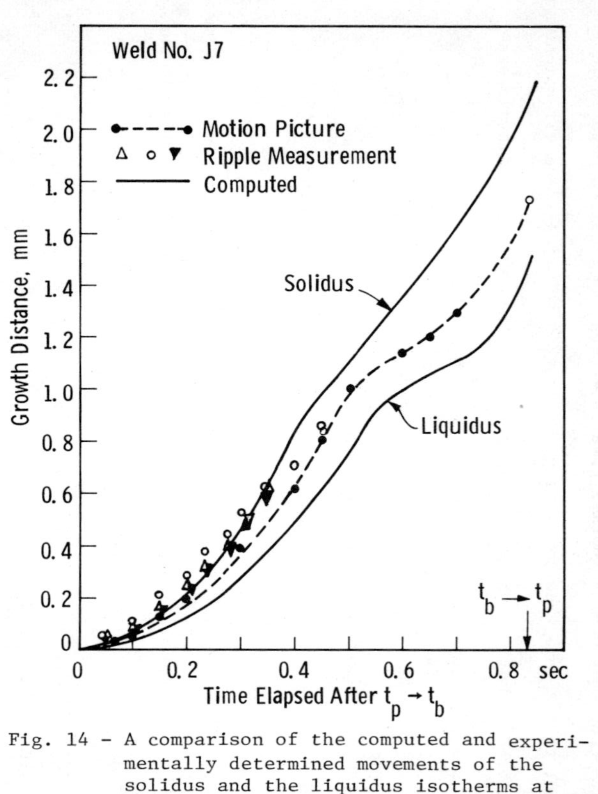

Fig. 14 - A comparison of the computed and experimentally determined movements of the solidus and the liquidus isotherms at the trailing edge of the weld pool. The data are for a single low pulse period about 14 seconds after the start of welding.

SUMMARY AND CONCLUSIONS

A simulation of the pulsed current gas tungsten arc welding of thin sheets of the Fe-26Ni alloy was made. The explicit finite difference technique used for the two-dimensional heat flow computations has been found adequate in predicting temperature fields, thermal gradients, the solidus and the liquidus growth rate in and around the weld pool, as well as its size and shape at anytime during the course of the welding operation. The simplifying assumptions made by selecting constant values for the relevant material thermal properties did not seem to have had a major impact on the accuracy of the computed results as determined by comparisons with experimental findings.

ACKNOWLEDGMENT

This work is being sponsored by the Office of Naval Research, Metallurgy Program, Contract N00014-77-C-0596, Dr. Bruce MacDonald, Scientific Officer.

REFERENCES

1. Boughton, P., "The Pulsed TIG-Welding Process - Part 2 - Potential Applications", in Pulsed TIG Welding Seminar Handbook, Brit. Welding Institute (1973).

2. Needham, J. C., "Pulsed Current Tungsten Arc Welding", in Pulsed TIG Welding Seminar Handbook, Brit. Welding Institute (1973).

3. Ecer, G. M., "Magnetic Deflection of the Pulsed Current Welding Arc", Welding Journal, 59 (6) June 1980, Research Suppl., pp. 183-s to 191-s.

4. Brody, H. D. and Stoehr, R. A., "Computer Simulation of Heat Flow in Casting", Journal of Metals, 32 (9) September 1980, pp. 20-27.

5. Tzavaras, A., Vassilaros, A., Brody, H., and Ecer, G., "Effect of Welding Parameters on Solidification Structures in Pulsed GTA Welded Fe-26%Ni", in Physical Metallurgy of Metal Joining (Eds. R. Kossowsky and M. E. Glicksman), TMS-AIME Conf. Proceedings, 1980, pp. 82-116.

6. Gokhale, A., Brody, H., and Ecer, G., "Relation of Weld Parameters, Heat Flow and Microstructure in Pulsed GTA Welds of 321 Stainless Steel", presented at the ASM Materials and Processes Congress, Oct. 30, 1980, Cleveland, OH.

7. Gokhale, A., "Weld Pool Solidification Structure and Property Control in Pulsed Current Gas Tungsten Arc Welding", M.S. Thesis, Metallurgical Engineering, University of Pittsburgh, 1980.

SIMULATION OF UNSTEADY HEAT-FLOW IN VERTICAL ELECTROSLAG WELDING

A. L. Liby
Rockwell International
Rocky Flats Plant
Golden, Colorado 80401

G. P. Martins and D. L. Olson
Department of Metallurgical Engineering
Colorado School of Mines
Golden, Colorado 80401

SUMMARY

The possibility of controlling the electroslag welding process was examined through experimentation combined with heat flow simulation. The experimentation served to establish the process operating limits and provided input data to the process model. The mathematical model was used to explore a wide range of process conditions.

1. INTRODUCTION

The temperature distribution in a weld as it relates to the final properties of weldments is of long standing interest. The thermal history of electroslag welds differs markedly from that of conventional arc welds. The mass of molten metal and the heat necessary to produce it at any instant are greater for the electroslag process. The heat source moves more slowly during electroslag welding. The predominant effect of this large, and slow moving heat source, is a slower cooling rate of the weld and adjacent base metal for electroslag welding than for the arc welding.

From the metallurgical viewpoint, the electroslag weld has a fusion zone of large grain structure. The orientation of the grains changes from a radial to axial direction with respect to the weld centerline. The volume of parent metal surrounding the electroslag weld that undergoes microstructural change is greater than that experienced by arc welding processes. The gross nature of this heat affected zone surrounding the electroslag weld is of primary concern with regard to the mechanical integrity of these welds. Notable reductions in impact strength of the parent metal adjacent to electroslag welds are attributed to austenitic grain growth and formation of brittle grain boundary fracture during the thermal cycle.

The heat flow analysis associated with electroslag welding may be separated into two parts: generation of heat in the molten pool and conduction through the surrounding parent metal. This investigation is directed primarily towards the second of these aspects.

1.1 Background

1.1.1 Heat Flow in the Weld Pool

The mathematics of heat conduction during solidification processes is inherently complex. Such factors as temperature-dependent physical properties, transient conditions, phase change over a temperature range, and movement of phase boundaries contribute to this complexity. The techniques of analysis of heat conduction with freezing or melting were reviewed by Muehlbauer and Sunderland(1). More recent analysis of systems which melt or freeze over a range in temperature has been made by Tien and Geiger(2) and Cho and Sunderland(3).

Several recent studies have addressed the thermal aspects of the electroslag process. Carvajal and Geiger(4) made an analysis of the temperature distribution and the location of the solidus, mushy and liquidus zones for the electroslag remelting of an Al-4.5% Cu binary alloy. In this analysis a parabolic temperature distribution was applied to the top of the ingot to represent the heat generation in the slag. No direct comparison with experimental data was made in this study, but the results did agree qualitatively with experimental evidence of the shape of the molten pool.

Mitchell and Joshi(5) made a heat balance of a laboratory electroslag remelting furnace. Electrode immersion, slag volume and depth of the molten melt head on the ingot were found to be important factors in determining the heat balance.

A transparent model was developed by Campbell(6) to directly observe phenomena occurring in the slag and metal layers during electroslag remelting. Qualitative observations of the temperature distribution in the molten slag and metal were made. The major means of heat transfer to the molten metal pool was found to be via the droplets which fall from the consumable electrode. The qualitative observations of Campbell were collaborated quantitatively in an experimental investigation by Edwards and Spittle(7). These investigators found that the metal pool depth was proportional to the slag temperature and is determined, to a major extent, by heat transfer to the pool, by molten metal droplets.

Because heat is generated in the electroslag process by resistance heating of the slag, current and voltage profiles in the slag should be important in determining the shape of the molten metal pool. Heat transfer in terms of resistive heating in an axially symmetrical system has been modeled using a resistance network analogue(8). The region of steepest voltage gradient was found to lie below the electrode tip where most of the heat is generated.

A description of the shape of molten metal and slag pool and the amount of penetration into the parent metal during electroslag welding is central to heat flow considerations in both the molten pool and surrounding parent metal. Recently Dilwari, Szekely, and Eagar(9,10) examined the flow behavior in idealized slag and metal pools of rectangular geometry incorporating electromagnetic and thermal buoyancy effects. In addition to a wire electrode, a plate electrode was studied which was claimed to improve the thermal efficiency.

1.1.2 Heat Flow in the Parent Metal

The analysis of heat flow in the parent metal during arc welding has received occasional attention over the last three decades. Most of this work has been based upon the theory of moving sources of heat developed by Rosenthal(11). Linear, two- and three-dimensional flow of heat from moving point, line, and planar sources were examined in this early work. The derivation of a single equation capable of predicting time and rate of cooling for a wide range of material thicknesses, temperature, and welding conditions resulted from this theory. Subsequent investigations(12-14) of heat flow during arc welding have principally involved extension and simplification of Rosenthal's equations, as well as experimental verification.

Previous analytical studies of temperature distribution in the base metal surrounding electroslag welds have also employed exact solutions. Sharapov(15) applied the solution of a line heat-source moving within an infinite plate, to the joining of two parts of cylindrical symmetry, one hollow and one solid, with axes perpendicular. Satisfactory results were reported. A similar theoretical prediction and experimental verification of heat affected zone (HAZ) thermal cycle in heavy plate, butt-welded by the electroslag process, was made by Pugin and Pertsovskii(16). In this case calculation of heat flow in the HAZ was based on replacing the molten weld zone by three moving line heat-sources of different intensities and at different levels.

1.1.3 Scope of Mathematical Models

Exact solutions of heat transfer equations based on moving point or line heat-sources necessitate the assumption of a "quasi-steady" state. This implies that the temperature distribution with respect to the moving source as origin is independent of time. The assumption of a "quasi-steady" state is physically realistic and has been verified experimentally for arc welding processes on most materials. The size of the fused zone of metal caused by an arc traveling along the surface of a plate does not change with time. Prediction of thermal history in arc welding has been refined by considering the experimentally determined shape of the molten pool(17,18).

For the electroslag process the "quasi-steady" assumption does not seem as physically consistent as in the case for arc welding. In electroslag welding the heat source is large and slow moving. The parent metal ahead of the weld is preheated to a high degree. The temperature distribution ahead of the moving source would be expected to change with time for most practical weld geometries. Previous electroslag models have matched the behavior described above by using techniques involving multiple sources and reflection planes.

One of the serious shortcomings of previous heat flow models for electroslag welding is the lack of consideration of the size and shape of the molten slag-metal interface. The heat flow in close proximity to the interface would be expected to depend strongly on its shape and size. Heat flow in this region is considered important in determining the thermal history experienced by the HAZ.

1.2 Statement of the Problem

The following problem is addressed by the work described here: Can the degradation of mechanical properties in the electroslag weld heat-affected zone be controlled within the limits of reasonable process operation? The extent of microstructural change at any point near the weld is a result of the thermal cycle at that point resulting from the flow of heat from the molten weld zone. Control of microstructural change can result only from control of weld heat input and extraction.

The approach to answering the above-stated question was one of experimentation combined with heat flow simulation. The experimentation served a four-fold purpose:

i) to determine limits of process operation

ii) to establish the possibility of metallurgical control by characterizing the microstructures that result at process operation limits

iii) to provide input data to the process model

iv) to provide temperature data, against which the model could be checked

The model was then used to explore a wide-range of process conditions using weld pool shapes, penetrations, and travel speeds established as reasonable by the experiments.

2. EXPERIMENTAL

2.1 Experimental Details

The consumable guide electroslag welding process was used to fill a centerline hole in twelve cylindrical welding specimens. Temperature was monitored continuously at various points in the specimens during each experiment. Specimens were sectioned and examined metallographically following welding. Details are provided in the following section.

2.1.1 Welding Apparatus

Electroslag welds were made using equipment of the consumable guide type. The electroslag welding setup is shown in Figure 1.

Fig. 1 - Experimental Arrangement for Electroslag Welding.

The welding power supply used for experimental welding in this investigation is a Hobart Model RC750. This is a DC constant voltage supply rated at 750 amps at 50 volts with 100% duty cycle. This power supply is designed for remote control from an external panel. The control panel allows adjustment of welding voltage and electrode feed rate. Meters for monitoring welding voltage and current are provided on the control panel.

The guide tube torch and clamp assembly is mounted to a fixed stand and does not move relative to the work during welding. The torch holds the consumable guide tube through which the electrode wire passes and serves as a contact for powering the guide tube and electrode. The electrode wire feed mechanism is powered by a DC electric motor.

The capability for water cooling of specimens was provided by a water jacket made from a 12-in. I.D. sch 40 pipe spool piece as shown at the center in Figure 1. Uniform axial flow of water past the specimen was assured by including a perforated sleeve between the specimen and the inner wall of the water jacket. The specimen was secured to the lid by six bolts on a 3-in. circle. A gasket of 1/8-in. pressed asbestos with SBR binder was used between the specimen and cooling jacket lid.

2.1.2 Welding Specimens

An electroslag welding specimen is shown, attached to the lid of the cooling vessel, with thermocouples in place, in Figure 2 . Electroslag welding is usually applied to full thickness joints between parts of rectangular geometry, such as joining thick plate. However, the cylindrical geometry was chosen for the specimen used in this investigation for two reasons:

Fig. 2 - Cylindrical Electroslag Welding Specimen with Thermocouples in Place.

i) symmetry allows heat flow modeling in two dimensions and

ii) the need for molding shoes is eliminated.

The size of the specimen was chosen in a trade-off between considerations of adequate heat sink and material economy.

2.1.3 Materials

A total of twelve specimens were prepared and used in welding experiments. Cold-rolled 4-in. round AISI 1018 steel was used for three specimens and the remaining nine specimens were made from 4-in. round AISI 8620 steel. The AISI 8620 steel was supplied in the hot rolled and normalized condition.

Flux, wire and guide tube materials were supplied by Hobart Bros. Co.. Two types of electroslag welding flux were supplied. The Hobart PF-203 flux is normally used as a starting flux to bring the process to stable operation quickly. The PF-201 flux is usually added after the process is stable and subsequently throughout the weld to maintain the desired slag bath depth. Major constituents of the PF-201 flux are SiO_2, Al_2O_3, MnO, CaO, MgO, and CaF_2. The PF-203 flux is made up principally of TiO_2 and CaF_2. Smaller amounts of SiO_2, CaO, and MgO are present in the PF-203 flux. The electrical conductivity is raised and the viscosity is lowered as the amounts of CaF_2 and TiO_2 are increased. The PF-203 flux is electrically conducting in the solid state.

The electrode wire used in this investigation was Hobart HB-25P. This wire conforms to E70S-3 classification per AWS specification A5.18-69. The electrode wire was 3/32-in. diameter. A special preservative and lubricant was used on this wire to minimize wire feed problems.

The electrode guide tube was heavy wall tubing of AISI 1018 steel. The outside diameter was 1/2 in. and the inside diameter 1/8..

2.1.4 Process Operation

All electroslag welds were started at approximately 45 volts and 500 amps. The process was started by pressing the start button on the control panel to energize the power supply and begin feeding the electrode wire. An arc is initiated when the electrode wire comes in contact with a wad of steel wool, provided as a starting aid at the bottom of the hole. Once the arc was started, the granulated flux was poured slowly down the hole beside the electrode guide tube. The flux becomes molten from the heat of the arc. When there is a sufficient pool of molten flux, the arc is extinguished and the process enters the electroslag mode. Approximately 75 g. of the flux was poured down the hole during a period of about 30 sec. at the beginning of each weld made in the laboratory. This provided a molten flux pool depth of 1-1/2 to 2 in..

After the arc had extinguished and the process was in the electroslag mode, the wire feed and welding voltage were adjusted to the desired values for each experimental run. Welding voltage, welding current, and electrode feed rate were monitored continuously during each welding experiment. Voltage and current were read from the panel meters on the Porta-Slag control panel. Electrode feed rate was measured by using a stop watch to time the revolutions of the electrode wire spool. The circumference of the wire spool was measured before and after each weld and an average value was used to calculate electrode feed rate.

It was occasionally necessary to add a small amount of flux during the process. This addition made up for flux lost by expulsion from the molten pool and sticking to the side wall of the specimen in areas where the filler metal did not fuse to the parent metal. When the molten slag pool reached the top of the hole in the welding specimen, the process was terminated abruptly by shutting off the power.

2.1.5 Temperature Measurement

Specimen temperature was monitored at twelve different locations during each of the last nine experiments. Temperature measurements were made using Inconel sheathed type K (chromel-alumel) thermocouples. The thermocouple wires were 0.006-in. diameter and the sheath was 0.040-in. outside diameter. The thermocouples were placed in 0.40-in.diameter holes drilled in the specimen as shown in Figure 2 .

The thermocouples were located at four different radial depths ranging from the surface to 1-in., on each of the specimens. The maximum depth of 1-in. for thermocouple placement was dictated by the maximum depth to which the hole could be drilled.

2.1.6 Examination of Electroslag Welds

The electroslag welding specimens were sectioned after welding to determine the character of the weld and the extent of metallurgical change in the unfused parent material. Each specimen was cut into slices transverse to the axis of the cylinder at intervals of 1 to 1-1/2 - in. from top to bottom. The exposed surfaces were etched in the as-cut condition in a solution of 25% HNO_3 in water. These etched surfaces allowed measurement of the amount of penetration of the fusion zone into the parent metal and the width of the heat affected zone surrounding the fusion zone. Selected slices were then sectioned in the direction of the cylinder axis to observe the fusion zone structure, direction of dendritic growth, shape of the metal pool while molten, and length of the fused column of metal.

Etching of all specimens was done in 2% nital for 15-30 sec.. Nital was chosen so as to give good definition of ferrite grain boundaries and good contrast.

2.2 Experimental Results and Discussion

The results of the welding experiments and discussion of the results obtained are now presented in the following sections.

2.2.1 Welding Experiments

The electroslag welding process involves the interaction of a large number of variables. Some of these variables such as weld geometry and choice of flux and other materials are established prior to process operation. Other variables such as welding voltage and electrode feed rate may be controlled while the process is operating.

Welding voltage, current, and electrode feed rate were recorded as a function of elapsed time during each of the experiments. For a given weld, the electrode feed rate solely determines the rate of weld formation or the speed at which the molten pool moves. The ratio of the electrode feed rate (EFR) to the welding speed, V_W, is equal to the ratio of the weld cavity cross section, A_W, to the electrode cross section, A_e:

$$\frac{EFR}{V_w} = \frac{A_w}{A_e} \tag{2.1}$$

The electrode guide tube, area A_{gt}, is also consumed during the welding process. The length of guide tube added to the molten pool is approximately equal to the length of the weld. Therefore:

$$V_w = EFR \times \frac{A_e}{A_w - A_{gt}} \tag{2.2}$$

For the 1-in. diameter weld cavity, 1/2-in. O.D. by 1/8-in. I.D. guide tube, and 3/32-in. diameter electrode wire used in the experiments:

$$V_w = \frac{EFR}{87.1} \tag{2.3}$$

The electrode feed rate is related to welding current through the Ohmic nature of the process and the electrical characteristics of the power supply. Heat for the electroslag welding process is generated by electrical current flowing through the molten slag. The resistance provided by the slag pool is determined by the resistivity of the slag and the length of the paths between the electrode and the molten metal pool. An increase in electrode feed rate reduces the resistance in the welding circuit by reducing the effective path length through the slag pool. The electrical control characteristics of the power supply are such that the welding voltage remains essentially constant as the electrode feed rate is increased. Thus, welding current increases with an increase in electrode feed rate as would be predicted by Ohm's Law.

The experimentally determined relationship between welding current and electrode feed rate for the two different flux compositions used is shown in Figure 3. The PF-201 flux was used in four experiments, but reliable electrode feed rate measurements were made in only two of these experiments. The data plotted in Figure 3 does not represent a constant welding voltage. Welding voltage varied from 36 to 46 volts in the various experiments. Justification for ignoring the effect of variation in welding voltage on the relationship between welding current and electrode feed rate is based on the fact that the change in current with a change in voltage at a constant electrode feed rate is slight.

Figure 3 further illustrates the marked influence that flux chemistry has on the heat produced during an electroslag weld. At a given electrode feed rate, the current used for process operation is roughly 50% higher with the PF-203 flux than with the PF-201 flux. This effect is attributable to the different electrical resistivities of the two fluxes.

The amount of electrical energy consumed for each inch of weld is shown for the two flux compositions in Figure 4. These curves are derived from Figure 3 assuming a constant welding voltage of 43 volts. A decrease in unit energy input with an increase in electrode feed rate is seen for both flux compositions. Stated differently, the proportional rise in welding current is slower than the rise in electrode feed rate. This effect is most likely related to the nature of heat generation at the electrode tip. At low electrode feed rates the current path length through the molten slag is long and the zone in which heat is generated is large and diffuse. As the electrode feed rate increases, the gap between electrode and metal pool becomes

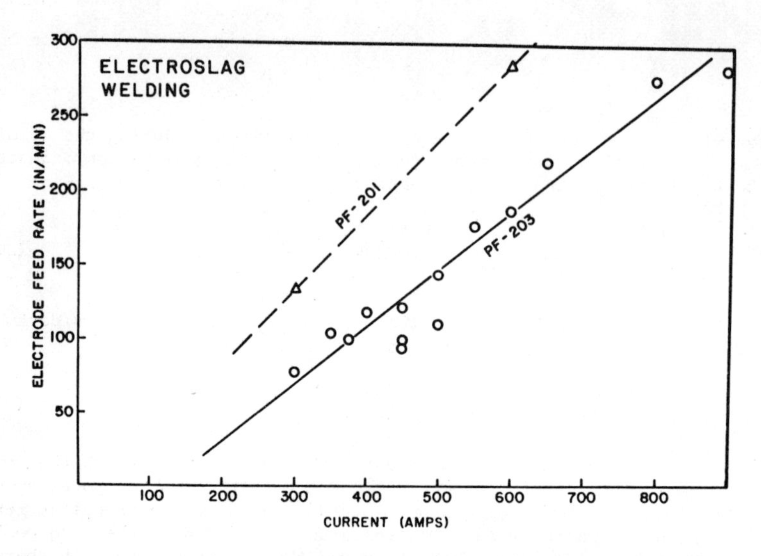

Fig. 3 - Relationship between Welding Current and Electrode Feed Rate for Flux Compositions: PF-201 and PF-203.

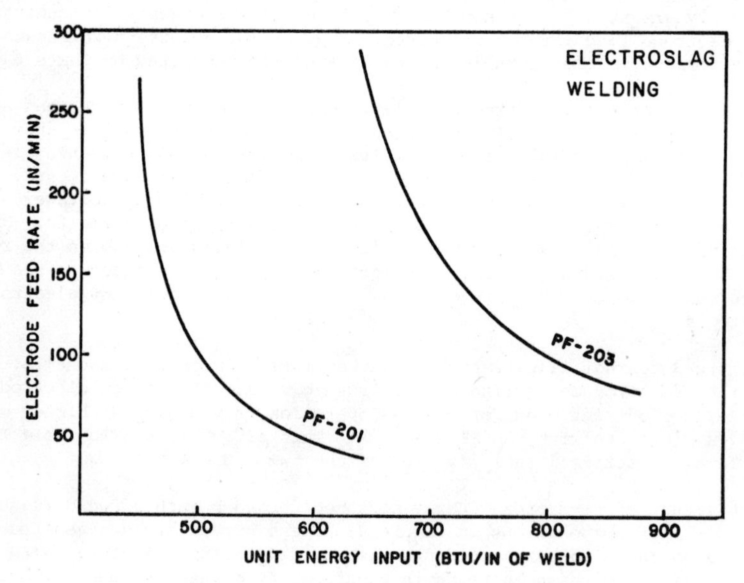

Fig. 4 - Effect of Flux Composition on Energy Input as a Function of Electrode Feed Rate during Electroslag Welding.

smaller and more intense near the electrode tip. Thus, proportionally faster melting of the electrode wire would be expected at higher electrode feed rates.

2.2.2 Temperature Measurement

Time-temperature histories at various points in the electroslag welding specimen were recorded primarily to aid in the formulation and evaluation of the mathematical model for heat flow associated with the moving molten pool. The temperature measurements are also useful in assessing the microstructural changes observed in the parent metal surrounding the weld.

2.2.3 Examination of Welded Specimens

Typical etched macrosections taken perpendicular to the axis of the welding specimen are shown in Figures 5(a) and 5(b). Both of these sections are taken from the same specimen and show an increase in fusion zone size as the weld progresses. This behavior results from the inability to remove heat from the specimen as rapidly as it is input. Material ahead of the moving molten pool is preheated to ever increasing temperatures and thus penetration of the fusion zone into the parent metal can be expected to increase as the weld progresses. The small specimen size used in this investigation enhances the preheating effect. In weldments providing a larger heat sink, the increase in penetration as welding progresses should be less noticeable.

The asymmetry of the weld and heat affected zone shown in Figure 5(b) is caused by the cast of the electrode wire as it comes off the spool. This asymmetry is more apparent during the later stages of welding. Adjustment of a wire straightening device on the welding head helped minimize this problem in subsequent experiments.

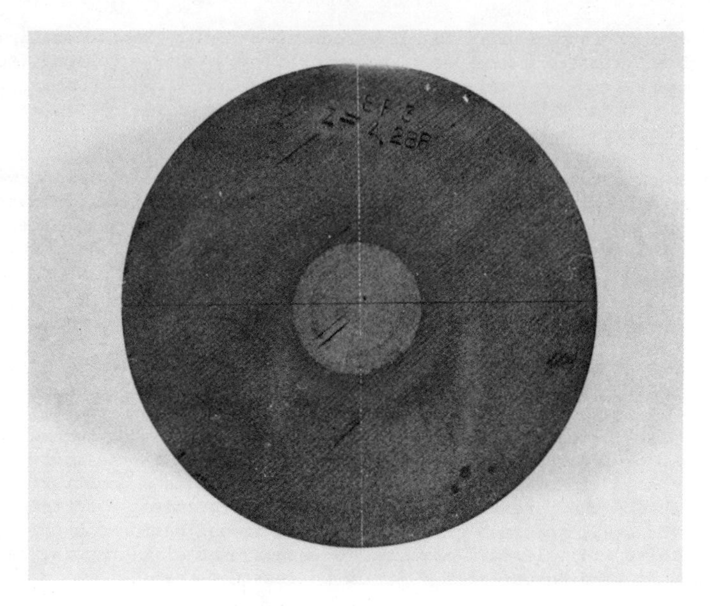

(a)

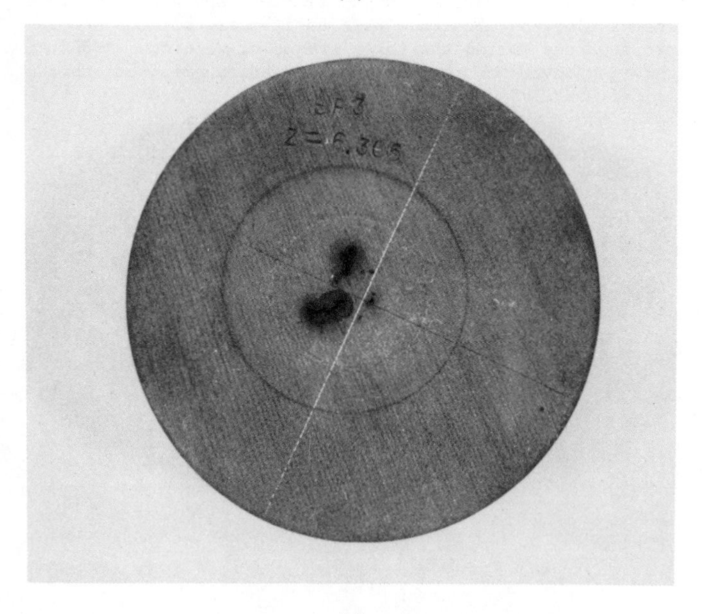

(b)

Fig. 5 - Typical Etched Microstructures Taken at:
(a) 4.3 in. Above Base of Specimen
(b) 6.3 in. Above Base of Specimen.
As welding progresses an increase in the fusion zone is observed.

One of the more important relationships between welding parameters and the nature of weld formation is shown in Figure 6. Here the amount of parent metal penetration at a level 8-in. above the bottom of the specimen is plotted as a function of electrode feed rate. The maximum in this curve is caused by the balance between heat generation, heat removal, and the total amount of energy available to melt the parent metal. Penetration increases as the rate of heat generation outpaces the rate at which heat is removed through the surrounding metal. At higher electrode feed rates penetration decreases with increasing electrode feed rate because of the decrease in unit energy input shown previously in Figure 4.

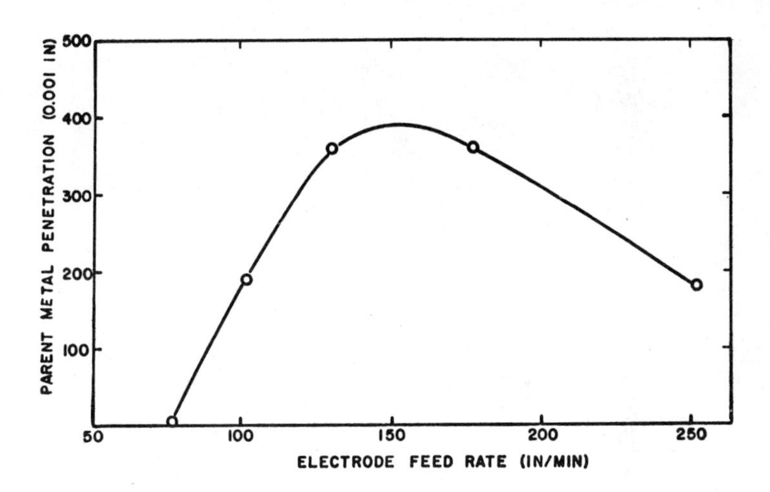

Fig. 6 - The Effect of Electrode Feed Rate on Parent Metal Penetration Using PF-203 Flux, At a Level 8-In. Above Base of Specimen.

Selected welding specimens were examined metallographically to determine the range and extent of microstructural changes in the parent metal surrounding the fusion zone. Figure 7 shows the as-received microstructure of the AISI 8620 steel, (a) near the outside surface and (b) near the center of the 4-in. round. The grain size variation from surface to interior is typical of heavy sections in the normalized condition and is caused by the hot breakdown history and a slower cooling rate at the center of the section during the normalization treatment. In both cases the microstructure is a mixture of ferrite (white areas) and pearlite. In the coarser microstructure found at the center of the section, the prior austenite grains are clearly outlined by the ferrite.

The range of possible changes in parent metal microstructure is illustrated in Figures 8(a) and 8(b). Figure 8(a) is a composite of photomicrographs showing microstructural changes from the weld fusion zone to the unaffected parent metal. This section was taken near the top end of a specimen

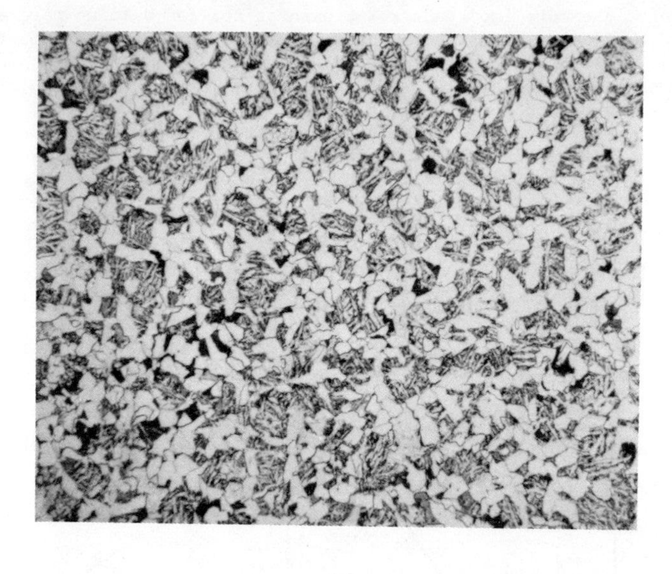

(a)

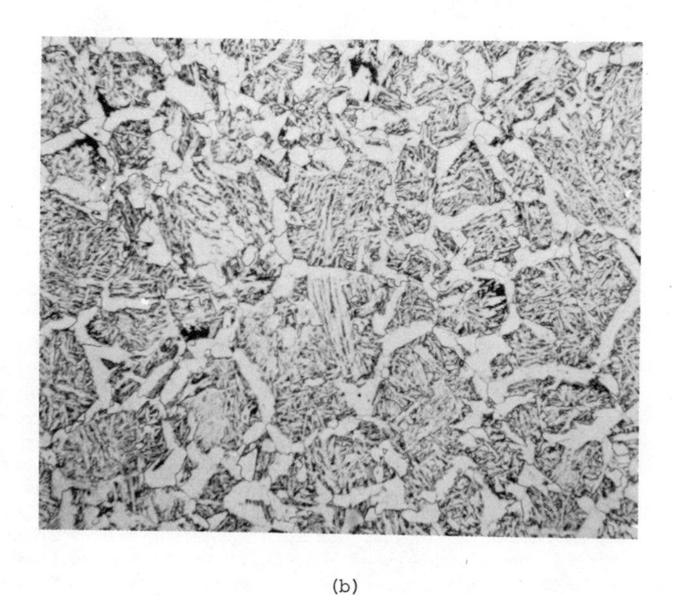

(b)

Fig. 7 - As Received Microstructure of AISI 8620 Steel (100X)
(a) Near Outside Surface
(b) Near Center.

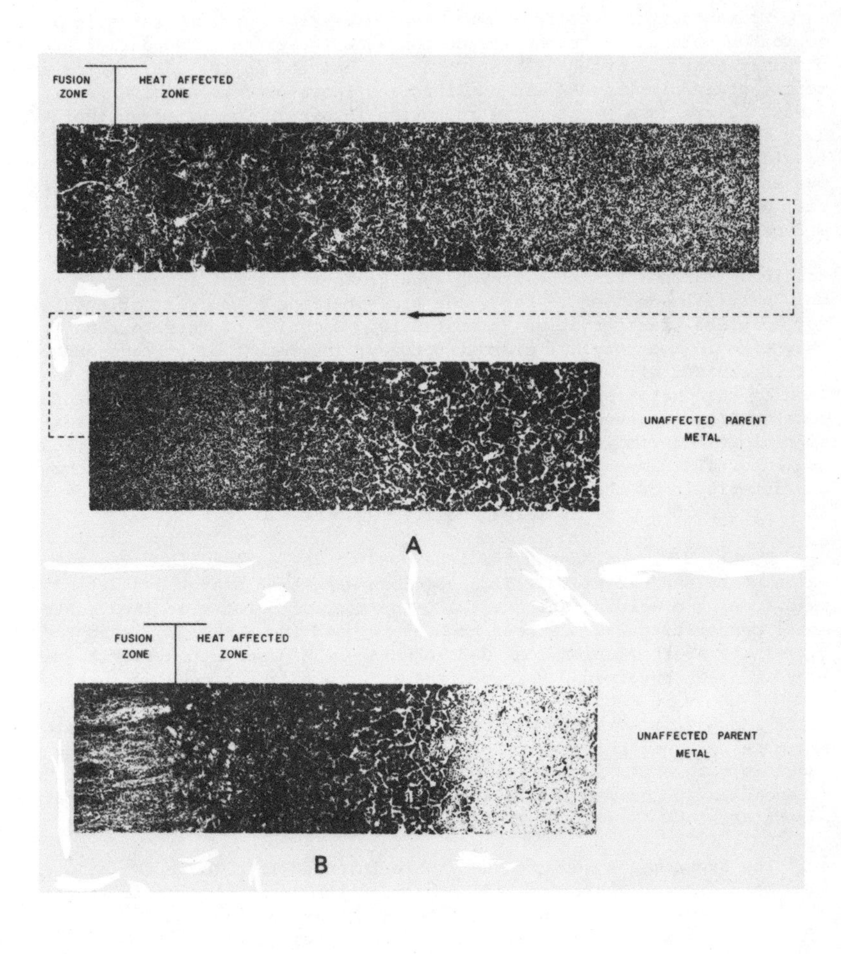

Fig. 8 - Heat-Affected Zone Microstructure for
(a) High Heat Input
(b) Low Heat Input, Electroslag Welds. (30X)

that was welded in air. Cooling of the specimen was only by free convection. Penetration of the weld into the parent metal was about 0.19-in. at this level and the heat-affected zone extended about 0.04-in. beyond the fusion zone. Welding speed was 1.39-in./min.

Widmanstatten ferrite growth, caused by rapid cooling rates, may be observed on both sides of the fusion zone-HAZ interface. The darker areas are probably pearlite in the fusion zone, and bainite in the HAZ. Prior austenite grains in the HAZ are outlined in places by more massive ferrite areas. There is a progressive reduction in size of the prior austenite grains away from the fusion zone. The ferrite morphology changes from Widmanstatten to globular at a distance of about 0.10-in. from the fusion zone. Disappearance of the Widmanstatten ferrite is coincident with successive grain refinement with distance from the weld. A wide zone of very fine ferrite and pearlite may be observed. This zone resulted from complete austenitization during heating followed by decomposition to ferrite and pearlite on cooling. Farther from the fusion zone grain refinement is observed to be only partially complete. Here, the time-at-temperature was not sufficient for complete transformation to austenite. Globules of ferrite remain as they were in the original microstructure. The amount of untransformed ferrite increases with increasing distrance from the fusion zone. In that portion of the photomicrograph showing the structure farthest from the fusion zone, grain refinement is not in evidence. The pearlite in this area shows spheroidization, indicating prolonged heating at a temperature just less than the eutectoid temperature. The microstructural changes described above are characteristic of those which may be expected near electroslag welds in which the molten pool is large and its movement slow.

Figure 8(b) represents the other end of the scale with respect to thermal history and HAZ width. This specimen was taken near the start of the weld made at a welding speed of 2.89-in./min. with water cooling. Parent metal penetration was measured at 0.05-in. and the HAZ width is approximately 0.01-in.. At the fusion zone-HAZ interface a Widmanstatten ferrite morphology is again observed. However, the structure is noticeably finer than in the HAZ shown in Figure 8(a). The zones away from the fusion line in which total and partial transformation to austenite has occurred during welding are much narrower than in the previous case. The dark unresolvable constituent in both zones is bainite or extremely fine pearlite. Only slight spheroidization of pearlite may be observed in a zone next to the unaffected parent metal.

The experiments reported here were intended to allow first hand experience with process operation and to provide the basis for a model that would contribute to the understanding of the electroslag welding process. The relationship between mathematical and experimental results will be discussed next.

3. MATHEMATICAL MODEL

In its simplest form the heat flow analysis associated with electroslag welding may be separated into two parts:

-- generation of heat in the molten pool

-- heat conduction through the surrounding parent metal.

As mentioned previously the present simulation addresses only the second topic. Nevertheless, it is recognized that the heat generation within the slag and molten metal pool and subsequent transport away from this region involve complex phenomena which still have to be tackled. These include:

i) current distribution in the slag

ii) temperature dependent, electrical resistivity of the slag

iii) fluid flow induced by thermally produced buoyancy forces and electromagnetic forces as well as drag forces created by the movement of the electrode and subsequently molten metal drops

iv) radiation heat transfer in the case of transparent molten slags.

3.1 Formulation

The simulation reported in this paper represents a *first attempt* at describing the temperature distribution in a weldment during electroslag welding. The approach examines only the heat conduction in the solid regions surrounding the slag and molten metal pool. The geometry of this pool is known *a priori* and its movement is determined by the rate of weld metal addition according to the size and speed of the wire and given by Equation 2.2 or 2.3. In addition, the interface between the pool and base metal is assumed isothermal. Property values are also assumed to be independent of temperatures. This model was then applied to a cylindrical weld specimen in which the coaxial electroslag weld was performed. The temperature history during welding may thus be simulated.

The physical situation at some instant during the weld is represented in Figure 9(A). The molten pool is approximated by a parabola in section. This paraboloid, comprising the slag and molten metal pool was confirmed from preliminary experimental observations. Its characteristic dimensions are:

R_1 -- the radius of the co-axial hole in which the electroslag weld was made

R_2 -- the radius of the maximum penetration into the parent metal

R_3 -- the outer radius of the specimen

L -- the length of the specimen

$Z_1(r,t)$ -- the height of the pool within the radial limits:

$$R_1 \geqslant r \geqslant o$$

$Z_2(r,t)$ -- the height of the pool within the radial limits:

$$R_2 \geqslant r \geqslant R_1$$

r,z,t -- the radial, axial, and time coordinate respectively:

$$L \geqslant z \geqslant o$$

These dimensions define five regions within the solid where the heat conduction equation may be applied. This is illustrated in Figure 9(B).

The unsteady heat conduction equation for each region reduces to two dimensions because of axial symmetry and is:

$$\frac{\partial T_i}{\partial t} = \alpha\left(\frac{\partial^2 T_i}{\partial r^2} + \frac{1}{r}\frac{\partial T_i}{\partial r} + \frac{\partial^2 T_i}{\partial z^2}\right) \; ; \quad 5 \geqslant i \geqslant 1 \qquad (3.1)$$

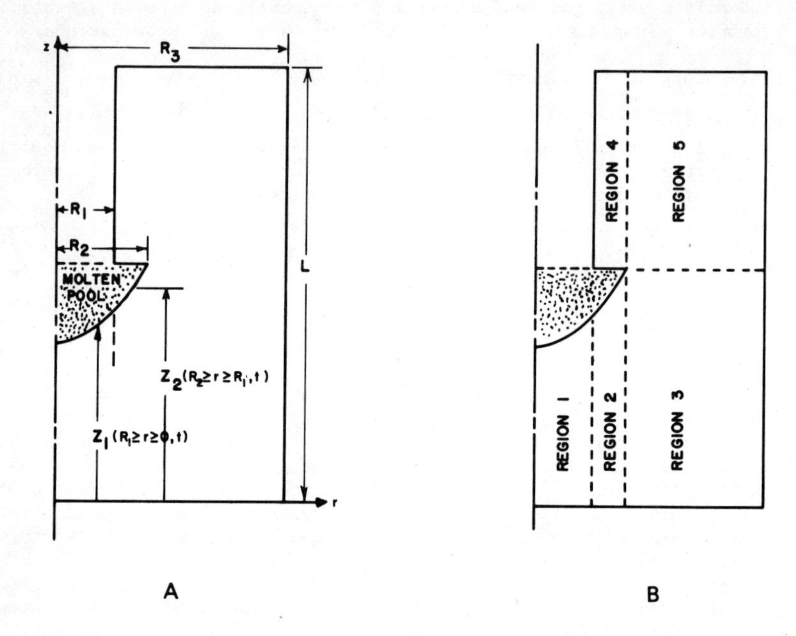

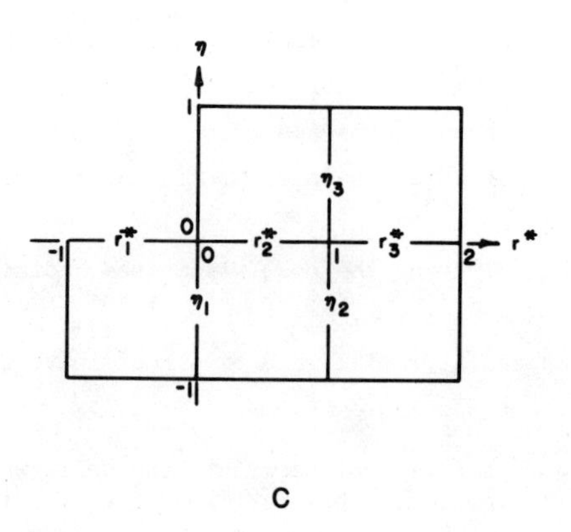

Fig. 9 - (A) Geometry of Weld Specimen
(B) Regions Delineated in Solid
(C) Transformed Coordinates

where $T_i = T_i(r,z,t)$ and initially the sample is at uniform temperature, T_o, which is the same as that of the bulk cooling medium. In symbols:

$$T(r,z,t = o) = T_o \tag{3.2}$$

According to the value of *i*, the following limits on the spatial coordinates and accompanying boundary conditions apply.

Region 1: $Z_1(r,t) \geqslant z \geqslant o$

$$R_1 \geqslant r \geqslant o$$

$$\left.\frac{\partial T_1}{\partial r}\right|_{o,z,t} = o \tag{3.3}$$

$$\left.\frac{\partial T_1}{\partial r}\right|_{R_1,z,t} = \left.\frac{\partial T_2}{\partial r}\right|_{R_1,z,t} \tag{3.4}$$

$$-k \left.\frac{\partial T_1}{\partial z}\right|_{r,o,t} = h_1 \left(T_1 \Big|_{r,o,t} - T_o \right) \tag{3.5}$$

$$T_1(r,Z_1,t) = T_m \tag{3.6}$$

Region 2: $Z_2(r,t) \geqslant z \geqslant o$

$$R_2 \geqslant r \geqslant R_1$$

$$\left.\frac{\partial T_2}{\partial r}\right|_{R_2,z,t} = \left.\frac{\partial T_3}{\partial r}\right|_{R_2,z,t} \tag{3.7}$$

$$T_2(R_1,z,t) = T_1(R_1,z,t) \tag{3.8}$$

$$-k \left.\frac{\partial T_2}{\partial z}\right|_{r,o,t} = h_1 \left(T_2 \Big|_{r,o,t} - T_o \right) \tag{3.9}$$

$$T_2(r,Z_2,t) = T_m \tag{3.10}$$

Region 3: $Z_2(R_2,t) \geqslant z \geqslant o$

$R_3 \geqslant r \geqslant R_2$

$$T_3(R_2,z,t) = T_2(R_2,z,t) \tag{3.11}$$

$$-k \left.\frac{\partial T_3}{\partial r}\right|_{R_3,z,t} = h_2 \left(T_3\Big|_{R_3,z,t} - T_o\right) \tag{3.12}$$

$$-k \left.\frac{\partial T_3}{\partial z}\right|_{r,o,t} = h_1 \left(T_3\Big|_{r,o,t} - T_o\right) \tag{3.13}$$

$$\left.\frac{\partial T_3}{\partial z}\right|_{r,Z_2(R_2,t),t} = \left.\frac{\partial T_5}{\partial z}\right|_{r,Z_2(R_2,t),t} \tag{3.14}$$

Region 4: $L \geqslant z \geqslant Z_2(R_1,t)$

$R_2 \geqslant r \geqslant R_1$

$$T_4(R_2,z,t) = T_5(R_2,z,t) \tag{3.15}$$

$$\left.\frac{\partial T_4}{\partial r}\right|_{R_1,z,t} = o \tag{3.16}$$

$$T_4(r,Z_2(R_2,t),t) = T_m \tag{3.17}$$

$$\left.\frac{\partial T_4}{\partial z}\right|_{r,L,t} = o \tag{3.18}$$

Region 5: $L \geqslant z \geqslant Z_2(t)$

$R_3 \geqslant r \geqslant R_2$

$$\left.\frac{\partial T_5}{\partial r}\right|_{R_2,z,t} = \left.\frac{\partial T_4}{\partial r}\right|_{R_2,z,t} \tag{3.19}$$

Region 5:
(Cont.)

$$-k \left. \frac{\partial T_5}{\partial r} \right|_{R_3,z,t} = h_2 \left(T_5 \Big|_{R_3,z,t} - T_o \right) \tag{3.20}$$

$$T_5 \Big|_{r,Z_2(R_2,t),t} = T_3 \Big|_{r,Z_2(R_2,t)t} \tag{3.21}$$

$$\left. \frac{\partial T_5}{\partial z} \right|_{r,L,t} = o \tag{3.22}$$

Equation 3.3 expresses the condition of symmetry of the temperature distribution at the center line of the specimen in the solidified region. Equations 3.4, 3.7, 3.14, and 3.19 provide for continuity of the heat fluxes at the interfaces between the solid regions being considered. Similarly, Equations 3.8, 3.11, 3.15, and 3.21 assure continuity of temperatures across the interfaces in the regions where these equations are applied. The interphase heat transport (between specimen and cooling medium) is expressed by Equations 3.5, 3.9, and 3.13 for the plane base of the specimen and by Equations 3.12 and 3.20 for the curved surfaces of the cylinder. Equations 3.6, 3.10, and 3.17 state that the pool/solid interface is at a unique melting point, T_m, of the solid--the inclusion of a melting range was not felt to be justified at this stage. Finally, it is assumed that the heat flux across the interface of the unfilled part of the hole in the specimen and also at the top surface of the specimen is negligibly small. These conditions are expressed by Equations 3.16, 3.18, and 3.22, respectively. The moving pool/solid metal interface was described by the following equations which represent a constant axial velocity of the parabola.

$$Z_1(r,t) = bt + ar^2 \; ; \; R_1 \geqslant r \geqslant o \tag{3.23}$$

$$Z_2(r,t) = bt + ar^2 \; ; \; R_2 \geqslant r \geqslant R_1 \tag{3.24}$$

The quantity b in Equations 3.23 and 3.24 is the velocity of welding given by Equation 2.3. The parameter a describes the shape of the parabola with origin about the center line of the weld specimen and is related to the form factor, FF, defined as the ratio of the pool diameter to the depth. Thus:

$$a = 2(\mathrm{FF} \times R_2)^{-1} \tag{3.25}$$

In order to solve the equations numerically it was found convenient to transform each region to a non-dimensional coordinate system and at the same time bring the moving parabolic boundary to rest. The transformed coordinates shown in Figure 9(C) are defined as follows and delineated by the regions indicated.

The transformed equations are provided in the Appendix.

$$r_1^* = \frac{r - R_1}{R_1} \quad ; \quad o \geqslant r_1^* \geqslant -1 \tag{3.26}$$

$$r_2^* = \frac{r - R_1}{R_2 - R_1} \quad ; \quad 1 \geqslant r_2^* \geqslant 0 \tag{3.27}$$

$$r_3^* = 1 + \frac{r - R_2}{R_3 - R_2} \quad ; \quad 2 \geqslant r_3^* \geqslant 1 \tag{3.28}$$

$$\eta_1 = \frac{z - Z_1(r,t)}{Z_1(r,t)} \quad ; \quad o \geqslant \eta_1 \geqslant -1 \tag{3.29}$$

$$\eta_2 = \frac{z - Z_2(r,t)}{Z_2(r,t)} \quad ; \quad o \geqslant \eta_2 \geqslant -1 \tag{3.30}$$

$$\eta_3 = \frac{z - Z_2(R_2,t)}{L - Z_2(R_2,t)} \quad ; \quad 1 \geqslant \eta_2 \geqslant o \tag{3.31}$$

TABLE 3.1

Coordinates Appropriate to Transformed Regions

Region	Coordinates
1	r_1^* , η_1
2	r_2^* , η_2
3	r_3^* , η_2 $(r = R_2)$
4	r_2^* , η_3
5	r_3^* , η_3

In addition a non-dimensional temperature and time were introduced thus:

$$U_i = \frac{T_i - T_o}{T_m - T_o} \tag{3.32}$$

$$\tau = \frac{\alpha t}{R_1^2} \tag{3.33}$$

Although the transformed equations are complex in form, it is felt that the additional work (differentiation and algebra) is worthwhile in terms of the utility it brings to effecting the numerical solution.

The property values used in the simulations are shown in Table 3.2. A single heat transfer coefficient, h, was assigned to the base and curved sides of the weld specimen. Preliminary estimates indicated that a value of 5 Btu hr^{-1} ft^{-2} $^{o}F^{-1}$ was reasonable and an order of magnitude increase in this value was investigated to test the sensitivity of this parameter.

Further details may be obtained from Reference 19.

TABLE 3.2

Property Values Used in Heat-Flow Calculations

Property	Symbol	Value
Thermal conductivity	k	25 Btu hr^{-1} ft^{-1} $^{o}F^{-1}$
Density	ρ	490 lb ft^{-3}
Specific heat capacity	C_p	0.11 Btu lb^{-1} $^{o}F^{-1}$
Thermal diffusivity	α	0.48 ft^2 hr^{-1}
Surface heat transfer coefficient	h	5.0 Btu hr^{-1} ft^{-2} $^{o}F^{-1}$
Melting temperature of specimen	T_m	2760°F
Initial temperature and bulk temperature of cooling system	T_o	70°F

3.2 Results and Discussion

Isotherms generated by the simulation are shown in Figures 10 and 11. Each figure shows three different conditions for a fixed welding speed -- 3.69 in./min. and 1.39 in./min., respectively.

In Figure 10 the effect of the form factor of the weld pool on the isotherm distribution is shown. The penetration into the parent metal is fixed at 0.150 in.. As the form factor is increased (a shallower pool), the isotherms in the range 2000 to 1400°F are confined to a narrower region, both axially and radially -- Figures 10A and 10C.

In Figure 11 isotherm distributions are shown for different parent metal penetrations with a fixed form factor of 2. The penetration of the isotherms increases rapidly as the penetration is first increased by a factor of 10 -- Figure 11A versus Figure 11B. The effect is particularly dramatic when the penetration is increased to 0.15 in. in Figure 11C.

Four simulated transient temperature responses, at a point in the specimen are shown in Figures 12A, B, C, and D. In Figures 12A and 12C, the simulated responses are compared with those recorded by the thermocouple in the specimen at the location indicated. The predicted peak temperature in Figure 12A is about 75 deg. F higher than that measured.

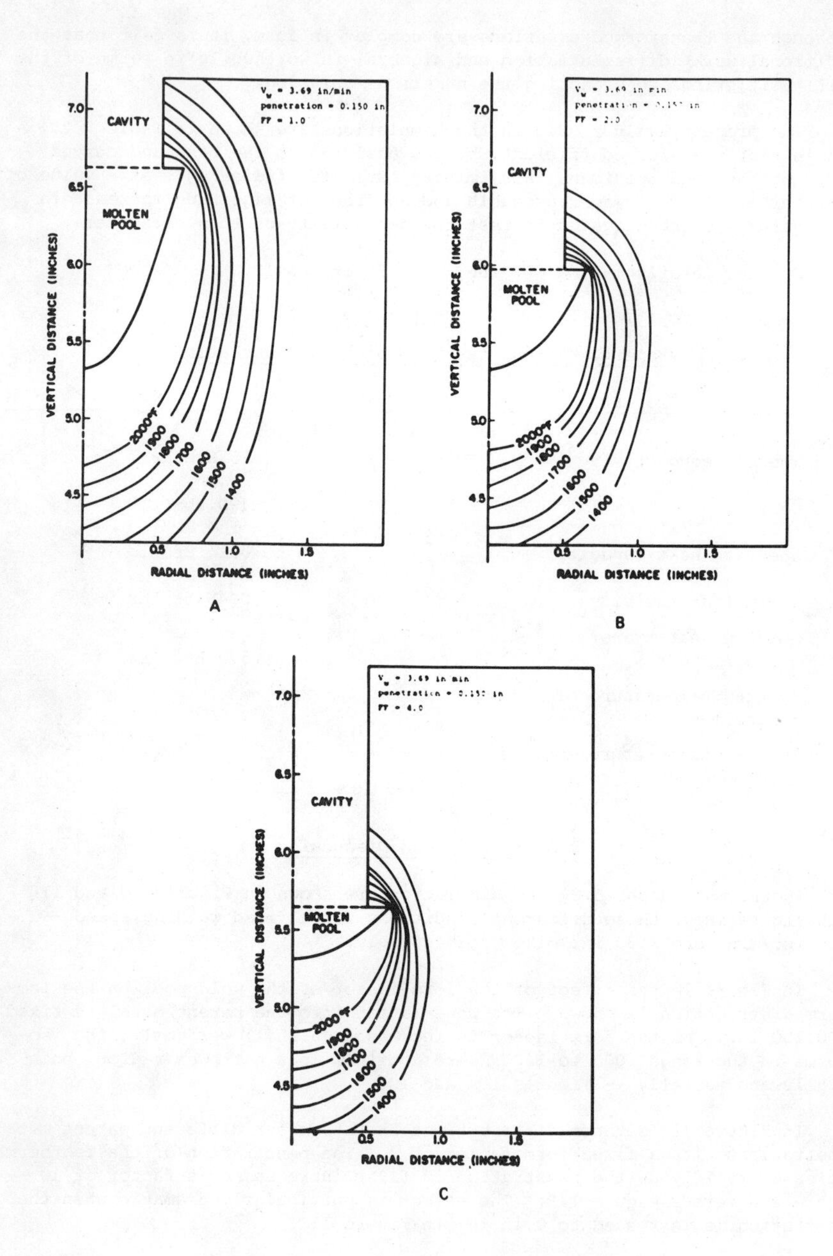

Fig. 10 - Predicted Isotherms. Effect of Form Factor.
Welding Speed: 3.69 in./min.. Penetration: 0.150 in..
Form Factors: A - 1.0; B - 2.0; C - 4.0.

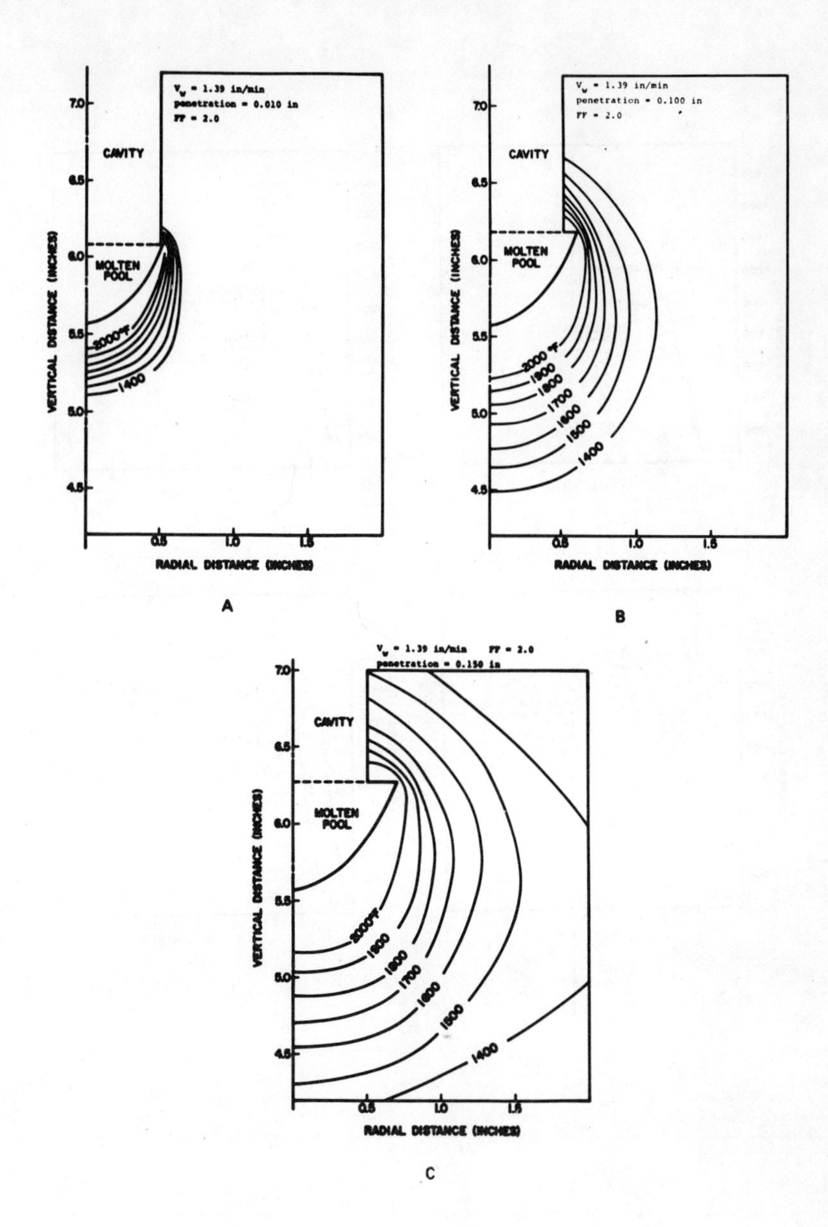

Fig. 11 - Predicted Isotherms. Effect of Penetration.
Welding Speed: 1.39 in./min.. Form Factor: 1.0.
Penetrations: A - 0.010 in.; B - 0.100 in.; C - 1.150 in..

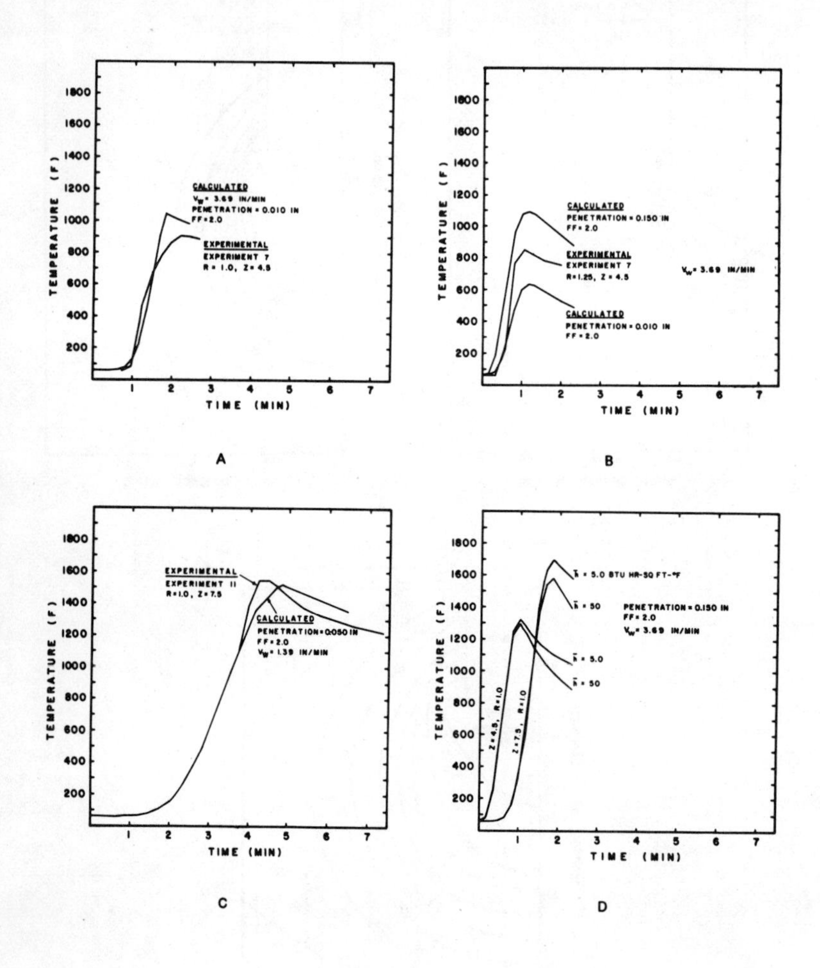

Fig. 12 - Transient Temperatures at a Point in the Weld Specimen.
A, B, and C: Experimental versus Predicted.
D: Sensitivity of Heat Transfer Coefficient.

The simulation shown in Figure 12C, at the lower welding speed, produced closer agreement with the measured temperature history.

In the other two plots (12B and 12D), the temperature-time response for the same weld at two different axial positions and slightly different radial positions -- 1.25 in. versus 1 in. -- are shown. The predicted results are seen to be quite sensitive to the penetration into the parent metal, according to Figure 12B. The value of this quantity was bracketed because it was found that experimental welds were, in fact, asymmetric.

The simulated temperature response higher up the specimen (7.5 in.) lags that at the lower value of 4.5 in., as may be seen in Figure 12D. Also included in this figure is the effect of the heat transfer coefficient. In both cases it is evident that conduction is the limiting heat transfer mechanism during the time investigated. This is manifested by the relative insensitivity of the temperature response, when the magnitude of h is increased by a factor of 10.

As a final point in this discussion it is worthwhile mentioning that a heat balance was carried out at the molten pool-solid interface in order to assess how this compared with actual heat input rates. This was obtained by integrating the heat flux over this boundary. In two cases studies, for welding speeds of 1.39 in./min. and 3.69 in./min., the predicted heat transfer rate varied from 500 to 200 Btu/min. over a period of 5 minutes and 2 minutes, respectively. The average heat input rates in the experiments corresponding to these conditions were approximately 900 to 1800 Btu/min., respectively. Thus the fraction of heat reporting into the specimen appears to be reasonable in light of the present model formulation.

4. CONCLUSION

The experimental phase of this investigation has shown that slag chemistry is important in determining the operating characteristics of the electroslag welding process. This arises because the slag constituents have a direct bearing on its electrical as well as its interfacial characteristics. The resistivity of the molten slag determines the power consumed in relation to the speed of welding for the process.

The observations and analyses of the metallurgical changes of the parent metal adjacent to the experimental welds have confirmed that the nature and degree of metallurgical change in the heat affected zone are dependent on the welding parameters. This reinforces the contention that metallurgical control of electroslag welds is, indeed, possible.

The mathematical model for transient heat flow from the moving slag-metal pool, based on a symmetrical parabolic isothermal molten-solid interface that moves upward at the speed of welding, was successful within the limits of its intended use. The model successfully simulated experiments performed in the laboratory and provided results to a wide range of other possible conditions within the operability of the process. Analysis of results of the computer simulations allowed useful conclusions with regard to process operation and insight into controlling the metallurgy of the electroslag weld.

The more important conclusions that may be drawn from the experimental work and the computer simulations are:

i) The chemical composition of the flux used in the process plays an important part in the formation of the electroslag weld. The resistivity of the molten slag determines the amount of heat generated at a given welding speed. The amount of heat generated will determine the overall size of the molten metal pool and the potential for metallurgical changes in the unfused metal adjacent to the weld.

ii) Metallurgical changes in the heat affected zone vary according to the chosen welding parameters. Metallurgical control of electroslag welds by controlling welding parameters is possible.

iii) The welding speed is important with regard to control of the metallurgy of electroslag welds. A high welding speed minimizes the effects of other welding variables.

iv) The degree of parent metal penetration is another important factor in determining the thermal experience of the metal near the fusion zone. For small values of parent metal penetration the effects of other variables are minimized.

v) The computer simulation can predict the thermal history of the weld, provided *a priori* knowledge of the weld pool shape and parent metal penetration are known.

vi) Not withstanding the limitation indicated in v), above, the model can be used to investigate the sensitivity of the process to the welding parameters.

vii) The technique of transforming the geometry of the physical system should prove to be useful for further model development where the equation of motion of the solid-molten pool interface is incorporated.

5. ACKNOWLEDGEMENTS

This work was supported in part by the Department of Energy. The welding equipment used was provided by Hobart Brothers Company. Appreciation is expressed for these contributions.

6. REFERENCES

1. Muehlbauer, J. C., and Sunderland, J. E., 1965, "Heat Conduction with Freezing or Melting," Appl. Mech. Rev., v. 18, no. 12, pp. 951-959.

2. Tien, R. H., and Geiger, G. E., 1967, "A Heat-Transfer Analysis of the Solidification of a Binary Eutectic System," Am. Soc. Mech. Engineers Trans., Heat Transfer Jour., v. 67, Aug., pp. 230-234.

3. Cho, S. H., and Sunderland, J. E., 1974, "Phase Change Problems with Phase Change Thermal Conductivity," Trans. ASME, v. 96, May., pp. 214-217.

4. Carvajal, L. F., and Geiger, G. E., 1971, "An Analysis of the Temperature Distribution and the Location of the Solidus, Mushy, and Liquidus Zones for Binary Alloys in Remelting Processes," Metallurgy Trans., v. 2. no. 8, pp. 2087-2092.

5. Mitchell, A., and Joshi, S., 1973, "The Thermal Characteristics of the Electroslag Process," Met. Trans., v. 4, no. 3, pp. 631-642.

6. Campbell, J., 1970, "Fluid Flow and Droplet Formation in the Electroslag Remelting Process," Jour. Metals, v. 22, no. 7, pp. 23-35.

7. Edwards, K. P., and Spittle, J. A., 1972, "Temperature Distributions in the Slag and Metal Remelting of Aluminum Alloys," Inst. Metals Jour., v. 100, pp. 244-248.

8. Peover, M. E., 1972, "Electroslag Remelting: A Review of Electrical and Electrochemical Aspects," Inst. Metals Jour., 1. 100, pp. 97-106.

9. Dilawari, A. H., Szekely, J., and Eagar, T. W., "Electromatnetically and Thermally Driven Flow Phenomena in Electroslag Welding," Met. Trans. B, 1978, v. 9B, pp. 371-381.

10. Dilawari, A. H., Eager, T. W., and Szekely, J., "An Analysis of Heat and Fluid Flow Phenomena in Electroslag Welding," Welding Jour., 57 (1) 1978, Research Suppl., pp. 24-s to 30-s.

11. Rosenthal, D., 1949, "The Theory of Moving Sources of Heat and Its Application to Metal Treatments," Am. Soc. Mech. Engineers Trans., v. 68, no. 11, pp. 849-866.

12. Adams, C. M., Jr., 1958, "Cooling Rates and Peak Temperatures in Fusion Welding," Welding Jour., v. 37, no. 5, pp. 210-s - 215-s.

13. Schillinger, D. E., Betz, I. G., and Markus, H., 1970, "Simplified Determination of Thermal Experience in Fusion Welding," Welding Jour., v. 49, no. 9, pp. 410-s - 418-s.

14. Signes, E. C., 1972, "A Simplified Method for Calculating Cooling Rates in Mild and Low Alloy Steel Weld Metals," Welding Jour., v. 51, no. 10, pp. 473-s - 484-s.

15. Sharapov, Y. V., 1965, "The Heat Distribution During the Electroslag Welding of Thick-Walled Structures," Automatic Welding, v. 18, no. 6, pp. 38-45.

16. Pugin, A. I., and Pertsovskii, G. A., 1963, "Calculation of the Thermal Cycle in the HAZ When Welding Very Thick Steel by Electroslag Process," Automatic Welding, v. 16, no. 6, pp. 12-21.

17. Rykalin, N. N., and Beketov, A. A., 1967, "Calculating the Thermal Cycle in the Heat Affected Zone From the Two Dimensional Outline of the Molten Pool," Welding Production, no. 1, pp. 42-47.

18. Pavelic, V., and others, 1969, "Experimental and Computed Temperature Histories in Gas Tungsten-Arc Welding of Thin Plates," Welding Jour., v. 48, no. 7, pp. 295-s - 305-s.

19. Liby, A. L., 1975, "Metallurgical Control of Electroslag Welds," Ph.D. Thesis, Colorado School of Mines.

7. APPENDIX

The transformed equations which were used to obtain the finite difference equations are presented here. The Crank—Nicolson technique, together with the Gauss-Seidel method for solving the resultant system of algebraic equations, was employed in the numerical computations. The computer program was written in Fortan IV.

Heat-Flow Equations

Region 1:

$$\frac{\partial U_1}{\partial \tau} - (1+\eta_1)\frac{\partial \ln Z_1}{\partial \tau}\frac{\partial U_1}{\partial \eta_1} = \frac{\partial^2 U_1}{\partial r_1^{*2}} - (1+\eta_1)\frac{\partial \ln Z_1}{\partial r_1^*}\frac{\partial^2 U_1}{\partial \eta_1 \partial r_1^*}$$

$$+ (1+\eta_1)\left[2\left(\frac{\partial \ln Z_1}{\partial r_1^*}\right)^2 - \frac{\partial^2 \ln Z_1}{\partial r_1^{*2}}\right]\frac{\partial U_1}{\partial \eta_1}$$

$$+ \left[(1+\eta_1)\frac{\partial \ln Z_1}{\partial r_1^*}\right]^2 \frac{\partial^2 U_1}{\partial \eta_1^2}$$

$$+ \frac{1}{1+r_1^*}\left[\frac{\partial U_1}{\partial r_1^*} - (1+\eta_1)\frac{\partial \ln Z_1}{\partial r_1^*}\frac{\partial U_1}{\partial \eta_1}\right] + \frac{R_1^2}{Z_1^2}\frac{\partial^2 U_1}{\partial \eta_1^2} \qquad (A.1)$$

Region 2:

$$\frac{(R_2-R_1)^2}{R_1^2}\left[\frac{\partial U_2}{\partial \tau} - (1+\eta_1)\frac{\partial \ln Z_2}{\partial \tau}\frac{\partial U_2}{\partial \eta_2}\right] = \frac{\partial^2 U_2}{\partial r_2^{*2}}$$

$$- 2(1+\eta_1)\frac{\partial \ln Z_2}{\partial r_2^*}\frac{\partial^2 U_2}{\partial \eta_2 \partial r_2^*}$$

$$+ (1+\eta_1)\left[2\left(\frac{\partial \ln Z_2}{\partial r_2^*}\right)^2 - \frac{\partial^2 \ln Z_2}{\partial r_2^{*2}}\right]\frac{\partial U_2}{\partial \eta_2}$$

$$+ \left[(1+\eta_1)\frac{\partial \ln Z_2}{\partial r_2^*}\right]^2 \frac{\partial^2 U_2}{\partial \eta_1^2}$$

$$+ \frac{1}{r_2^* - \frac{R_1}{R_2-R_1}}\left[\frac{\partial U_2}{\partial r_2^*} - (1+\eta_1)\frac{\partial \ln Z_2}{\partial r_2^*}\frac{\partial U_2}{\partial \eta_1}\right]$$

$$+ \frac{(R_2-R_1)^2}{Z_2^{\;2}} \frac{\partial^2 U_2}{\partial \eta_2^{\;2}} \tag{A.2}$$

Region 3:

$$\frac{(R_3-R_2)^2}{R_1^{\;2}} \left[\frac{\partial U_3}{\partial \tau} - (1+\eta) \frac{\partial \ln Z_2}{\partial \tau} \frac{\partial U_3}{\partial \eta_2} \right] = \frac{\partial^2 U_3}{\partial r_3^{*2}}$$

$$+ \left[\frac{1}{1+r_3^*} + \frac{R_2}{R_3-R_2} \right] \frac{\partial U_3}{\partial r_3^*} + \frac{(R_3-R_2)^2}{Z_2^{\;2}} \frac{\partial^2 U_3}{\partial \eta_2^{\;2}} \tag{A.3}$$

Region 4:

$$\frac{(R_2-R_1)^2}{R_1^{\;2}} \left[\frac{\partial U_4}{\partial \tau} + (1+\eta_3) \frac{\partial \ln(L-Z_2)}{\partial \tau} \frac{\partial U_4}{\partial \eta_3} \right] =$$

$$\frac{\partial^2 U_4}{\partial r_2^{*2}} + \left[\frac{1}{r_2^* + \frac{R_1}{R_2-R_1}} \right] \frac{\partial U_4}{\partial r_2^*} + \frac{(R_2-R_1)^2}{(L-Z_2)^2} \frac{\partial^2 U_4}{\partial \eta_3^{\;2}} \tag{A.4}$$

Region 5:

$$\frac{(R_3-R_2)^2}{R_1^{\;2}} \left[\frac{\partial U_5}{\partial \tau} + (1-\eta_3) \frac{\partial \ln(L-Z_2)}{\partial \tau} \cdot \frac{\partial U_5}{\partial \eta_3} \right] =$$

$$\frac{\partial^2 U_5}{\partial r_3^{*2}} + \left[\frac{1}{(r_3^*-1) + \frac{R_2}{R_3-R_1}} \right] \frac{\partial U_5}{\partial r_3^*}$$

$$+ \frac{(R_3-R_2)^2}{(L-Z_2)^2} \frac{\partial^2 U_5}{\partial \eta_3^{\;2}} \tag{A.5}$$

Equations 3.23 and 3.24, for the movement of the pool, become:

$$Z_1(r_1^*,\tau) = \frac{bR_1^{\;2}\tau}{\alpha} + aR_1^{\;2}(r_1^*+1)^2 \tag{A.6}$$

$$Z_2(r_2^*,\tau) = \frac{bR_1^{\;2}\tau}{\alpha} + a(r_2^*(R_2-R_1) + R_1)^2 \tag{A.7}$$

In addition, the derivatives involving Z_1 and Z_2 for the various regions are as follows:

Region 1:

$$\frac{\partial \ln Z_1}{\partial \tau} = \frac{bR_1^2}{\alpha Z_1} \tag{A.8}$$

$$\frac{\partial \ln Z_1}{\partial r_1^*} = \frac{2aR_1^2(r_1^*+1)}{Z_1} \tag{A.9}$$

$$\frac{\partial^2 \ln Z_1}{\partial r_1^{*2}} = \frac{2aR_1^2}{Z_1} - \left(\frac{\partial \ln Z_1}{\partial r_1^*}\right)^2 \tag{A.10}$$

Region 2:

$$\frac{\partial \ln Z_2}{\partial \tau} = \frac{bR_1^2}{\alpha Z_2} \tag{A.11}$$

$$\frac{\partial \ln Z_2}{\partial r_2^*} = \frac{2a(r_2^*(R_2-R_1) + R_1)\ (R_2-R_1)}{Z_2} \tag{A.12}$$

$$\frac{\partial^2 \ln Z_2}{\partial r_2^{*2}} = \frac{2a(R_2-R_1)^2}{Z_2} - \left(\frac{\partial \ln Z_2}{\partial r_2^*}\right)^2 \tag{A.13}$$

Regions 3, 4 and 5:

$$\frac{\partial Z_2}{\partial \tau} = \frac{bR_1^2}{\alpha Z_2(R_2,\tau)} \tag{A.14}$$

It should be noted that in these regions Z_2 is independent of either r_2^* or r_3^*.

Initial and Boundary Conditions

The initial condition for all regions is:

$$U_i(r_j^*, \eta_k, \tau) = 0 \tag{A.15}$$

where subscript i refers to the region of interest, and j and k are the appropriate coordinates as given in Table 3.1.

The boundary conditions may be classified into five types:

Continuity of Temperature

These boundary conditions take the general form:

$$U_i = U_{\substack{i-1 \\ (i-2)}} \tag{A.16}$$

and correspond to Equations 3.8, 3.11, 3.15 and 3.21, for i = 2, 3, 4 and 5 with the bracketed subscript used when i = 5.

Symmetry or No-flux Conditions

The relevant equations to be transformed are 3.3, 3.16, 3.18, and 3.22. These become, respectively:

$$\left[\frac{\partial U_1}{\partial r_1^*} - (1+\eta_1)\frac{\partial \ln Z_1}{\partial r_1^*}\frac{\partial U_1}{\partial \eta_1}\right]\Bigg|_{-1,\eta_1,\tau} = 0 \tag{A.17}$$

$$\left.\frac{\partial U_4}{\partial r_2^*}\right|_{o,\eta_3,\tau} = 0 \tag{A.18}$$

and for the last two,

$$\left.\frac{\partial U_i}{\partial \eta_3}\right|_{r_{i-2},1,\tau} = 0 \tag{A.19}$$

where i = 4 and 5.

Continuity of Heat Fluxes

Equations 3.4, 3.7, 3.14, and 3.19 become:

$$\left.\frac{\partial U_2}{\partial r_2^*}\right|_{o,\eta_2,\tau} = \frac{R_2-R_1}{R_1}\left[\frac{\partial U_1}{\partial r_1^*} - (1+\eta_1)\frac{\partial \ln Z_1}{\partial r_1^*}\frac{\partial U_1}{\partial \eta_1}\right]\Bigg|_{o,\eta_1,\tau} + (1+\eta_1)\left.\frac{\partial \ln Z_2}{\partial r_2^*}\frac{\partial U_2}{\partial \eta_2}\right|_{o,\eta_2,\tau} \tag{A.20}$$

$$\left.\frac{\partial U_3}{\partial r_3^*}\right|_{1,\eta_2,\tau} = \frac{R_3-R_2}{R_2-R_1}\left[\frac{\partial U_2}{\partial r_2^*} - (1+\eta_1)\frac{\partial \ln Z_2}{\partial r_2^*}\frac{\partial U_2}{\partial \eta_2}\right]\Bigg|_{1,\eta_2,\tau}$$

..... (A.21)

$$\left.\frac{\partial U_3}{\partial \eta_2}\right|_{r_3^*,o,\tau} = \frac{Z_2(o,\tau)}{L-Z_2(o,\tau)} \left.\frac{\partial U_5}{\partial \eta_3}\right|_{r_3^*,o,t} \tag{A.22}$$

$$\left.\frac{\partial U_5}{\partial r_3^*}\right|_{1,\eta_3,\tau} = \frac{R_3-R_2}{R_2-R_1} \left.\frac{\partial U_4}{\partial r_2^*}\right|_{1,\eta_3,\tau} \tag{A.23}$$

respectively.

Interphase Heat-Flow

The convective heat transport in the system is described by Equations 3.5, 3.9, 3.12, 3.13, and 3.20 which, when transformed, become:

$$\left.\frac{\partial U_1}{\partial \eta_1}\right|_{r_1^*,-1,\tau} = \frac{Z_1(r_1^*,\tau)h_1}{k} \left.U_2\right|_{r_1^*,-1,\tau} \tag{A.24}$$

$$\left.\frac{\partial U_2}{\partial \eta_2}\right|_{r_2^*,-1,\tau} = \frac{Z_2(r_2^*,\tau)h_1}{k} \left.U_2\right|_{r_2^*,-1,\tau} \tag{A.25}$$

$$\left.\frac{\partial U_3}{\partial r_3^*}\right|_{2,\eta_2,\tau} = -\frac{(R_3-R_2)h_2}{k} \left.U_3\right|_{2,\eta_2,\tau} \tag{A.26}$$

$$\left.\frac{\partial U_2}{\partial \eta_2}\right|_{r_3^*,-1,\tau} = \frac{Z_2(o,\tau)h_1}{k} \left.U_3\right|_{r_3^*,-1,\tau} \tag{A.27}$$

$$\left.\frac{\partial U_5}{\partial r_3^*}\right|_{2,\eta_3,\tau} = -\frac{(R_3-R_2)h_2}{k} \left.U_3\right|_{2,\eta_3,\tau} \tag{A.28}$$

It should be noted that the quantities on the right hand side multiplying the U_i's is of the form of a Biot number. Its value, when referenced to a characteristic dimension in the system (this would be needed in Equations A.24, A.25, and A.27, since Z_1 and Z_2 are time dependent), may be used to assess the relative importance of conduction and convection in the system.

Isothermal Surfaces

The pool-solid interface temperature is at the melting point as indicated by Equations 3.6, 3.10, and 3.17. These become:

$$U_i \Big|_{r_j^*, \eta_k, \tau} = 1 \tag{A.29}$$

where i = 1, 2 and 4, and j and k depend on the coordinate system associated with Region i, according to Table 3.1.

MATHEMATICAL MODELING OF THE TEMPERATURE PROFILES

AND WELD DILUTION IN ELECTROSLAG WELDING OF STEEL PLATES

T. Deb Roy, J. Szekely*, and T. W. Eagar*

Metallurgy Section
Department of Materials Science and Engineering
The Pennsylvania State University
University Park, PA 16802

*Department of Materials Science and Engineering
Massachusetts Institute of Technology
Cambridge, MA 02139

This paper describes a calculation procedure for the detailed prediction of temperature profiles and weld dilution in the electroslag welding of mild steel plates. The temperature profiles in the liquid slag and the liquid metal regions are calculated in three dimensions under steady state conditions. The values of the heat generation patterns needed in the temperature profile calculation are computed by solving a three dimensional electric field equation in the liquid slag region. The weld dilution is computed as a function of time by solving a transient three dimensional heat flow equation for the base plate.

Good agreement achieved between the predicted temperature profiles in the base plate and the available measurements illustrate the capabilities of the model. Reasonable agreement has also been obtained between the measurements and the predicted values of dilution as affected by changes in the plate gap and the weld speed.

Introduction

In recent years, the electroslag welding process has been a subject of growing interest to engineers and research scientists because of its potential attractiveness for the welding of thick plates. At present there is some reservation in the application of this fast and economical welding process for critical applications. In the current electroslag welding practice the size of the heat affected zone is relatively large and the fracture toughness of the welds is often poor.

The work described in this paper is a part of a research program aimed at developing a fundamental understanding of the heat and fluid flow phenomena in the electroslag welding process with the ultimate objective of providing guidelines for operation so as to achieve improved weld properties.

Most of the earlier work on the electroslag welding of steels was concerned with the physical characterization of the weld by examining its microstructure and by conducting other suitable tests to determine fracture toughness and other mechanical properties of the weld for a variety of welding conditions. Some examples of these studies are the papers by Paton,[1] Eagar and Ricci,[2] des Ramos, Pense and Stout,[3] Liby and Olson,[4] and by Jackson.[5] So far relatively less work has been done on the prediction of important weld characteristics such as the size of the heat affected zone, grain growth, and the weld dilution.

The first modeling work on the ESW process, presented by Dilawari, Szekely, and Eagar,[6] deals with the calculation of heat and fluid flow in the liquid slag and liquid metal regions in idealized two dimensional systems. It is found that the computed results are significantly influenced by the system geometry. Subsequently, a computation scheme has been presented[7] to calculate the temperature fields and the heat generation patterns in truly three dimensional systems. The work establishes that a reduction in the heat input, a crucial objective in the electroslag welding process, can be achieved by using narrow plate gaps or by using closely spaced multiple electrodes.

This previous paper,[7] while providing useful information (such as the effect of electrode asymmetry on the heat generation pattern) about the process, did not deal with some of the important aspects of the welding process, such as the weld dilution. Also, the temperature profiles were calculated for the liquid metal and the liquid slag regions only and the calculation of transient heat flow in the base plate was not carried out.

These problems are addressed in the present paper. Some of the attractive features of the earlier computation scheme, such as the computation of heat generation patterns and temperature profiles in the liquid region in a truly three dimensional situation are retained in the present model. However, the computation has been extended to account for the three dimensional transient heat flow in the base plate and the weld dilution.

In the following sections, we shall outline the mathematical model and the computational procedure. This is followed by a section on the comparison between the computed results and the available experimental data and the conclusions from this study.

Mathematical Modeling

In the electroslag welding system, a diagram of which is shown in Figure 1, the current passes through the consumable electrode, through the molten slag, to the base plate, either directly or indirectly through the molten metal pool. This results in the establishment of voltage profiles and spatially distributed heat generation patterns in the liquid slag. As a result of the heat generation, the consumable electrode melts, forming a pool of liquid metal underneath the liquid slag. The solidification of the liquid metal pool forms the weld. The nonuniform current density in the slag results in the establishment of an electromagnetic force field which causes circulation in the liquid region. Heat generated in the liquid slag is transported to the base plate by conduction and by convection. Part of this heat is utilized for the melting of the base plate and is responsible for the weld dilution while the remainder is transferred into the base plate and is responsible for the thermal cycle in the base plate, the establishment of a heat affected zone, and the resultant grain growth.

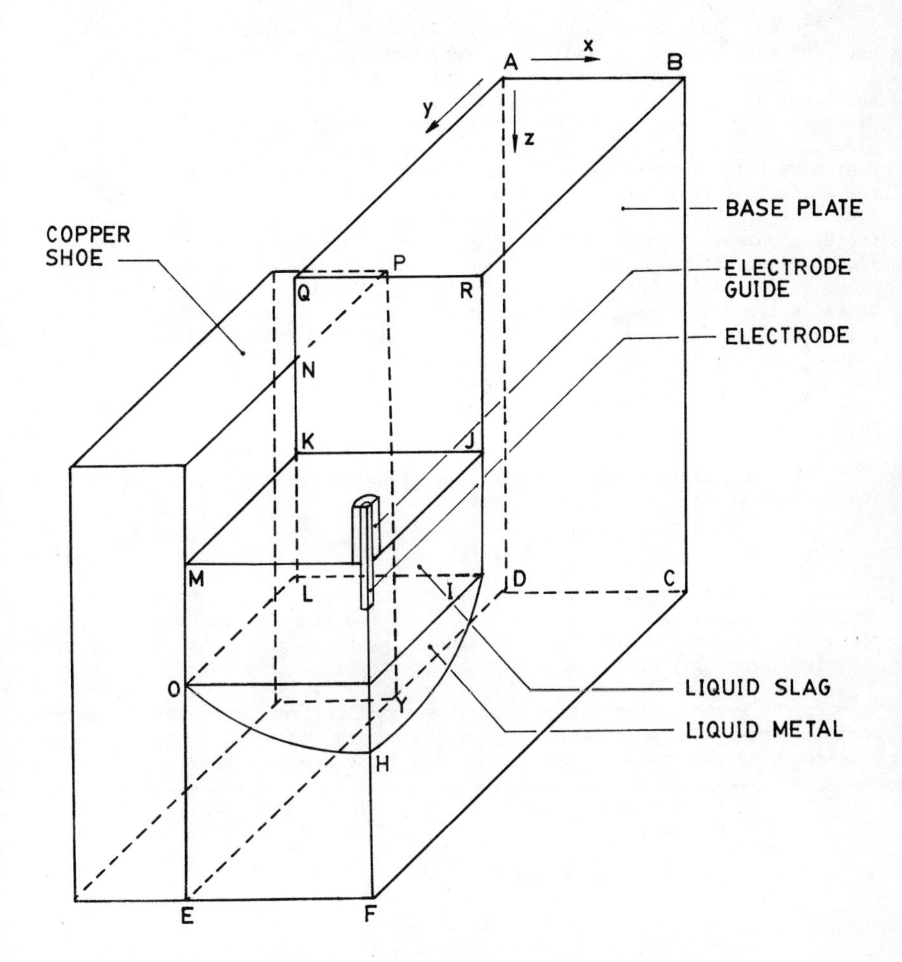

Figure 1. Sketch of the ESW Process

The following physical processes are to be considered in order to simulate the welding process:

(a) The passage of current from the electrode through the molten slag and metal pools to the base plate results in the establishment of a spatially distributed heat generation pattern.

(b) Heat transfer in the liquid metal and liquid slag takes place by conduction, convection, and by the transport of the liquid metal drops from the slag to the metal phase.

(c) Thermal energy is transferred in the base plate by transient conduction and the time dependent position of the boundary separating the molten and the solid metal regions is determined by a dynamic heat balance.

The calculation procedure adopted here for the detailed prediction of heat generation patterns and temperature profiles in the liquid region of the electroslag welding process has been presented recently.[7] In short, the modeling equations are represented by statements of the appropriate electric field equations together with Ohm's law in three dimensions in order to calculate the spatially distributed heat generation pattern. The heat generation pattern is calculated only for the liquid slag phase since the electrical resistance of the slag phase is several thousand times higher than that of the metal phase and practically all of the heat is generated in the slag phase. Having calculated the heat generation patterns for the liquid slag, attention is focused on the calculation of the temperature profiles in the liquid slag and liquid metal. Estimates of the temperature profiles are made by solving the heat flow equations using effective thermal conductivities to account for the convection in the liquid metal and liquid slag phases. The calculation scheme[7] accounts for the transport of heat from the slag to metal phase by the liquid metal drops, the energy loss due to electrolysis, and the energy required for the heating of the cold slag charge.

The temperature profile in the base plate is represented by the following relationship:

$$\nabla(k\nabla T) = \rho C_p \frac{\partial T}{\partial t} \tag{1}$$

The boundary conditions used for the solution of the above equation are presented in Appendix A. It is noted that both the thermal conductivity and the specific heat are allowed to be temperature dependent, but this temperature dependence is determined solely by the nature of the material.

In the present computation the property values for low carbon steel are used and the following relationships are employed to describe the temperature dependence of the thermal conductivity,[8] and the specific heat.[9]

$$\begin{aligned} k &= 9.2 \times 10^{-2} - 5.86 \times 10^{-5}T \quad \text{for } T < 1050\ \text{K} \\ &= 3.05 \times 10^{-2} \quad \text{for } T > 1050\ \text{K} \end{aligned} \tag{2}$$

k is expressed in kW/mK

$$\begin{aligned} C_p &= 0.46 + 4.18 \times 10^{-4}T \quad \text{for } T < 1000\ \text{K} \\ &= 0.88 \quad \text{for } T > 1000\ \text{K} \end{aligned} \tag{3}$$

C_p in the above equation is expressed in kW/kgK

The melting rate of the base plate can be calculated from the following equation:

$$m_b = (Q_{s\ell p\ell} - H_{s\ell p\ell})/\Delta H \qquad (4)$$

where $Q_{s\ell p\ell}$ and $H_{s\ell p\ell}$ (in kW) are the rate of heat transfer from the liquid slag to the base plate and the rate of heat transfer from the melting base plate to the interior of the base plate. These quantities are calculated from the temperature field by integrating the local values of heat fluxes, and ΔH is the latent heat of fusion in kJ/kg. The amount of base metal dilution is readily obtained from the following expression:

$$\%\ \text{Dilution} = m_b/(m_b + m) \qquad (5)$$

where m is the electrode melting rate calculated during the heat flow calculation in the liquid slag region.[7]

Computational Method

Tasks

The computational scheme described in the present section is designed to perform the following tasks:

(1) Solution of the electric field equation in the slag phase to calculate the voltage profiles and heat generation patterns in three dimensions.

(2) Solution of the steady state, three dimensional heat flow equations in the liquid regions with due consideration for the transport of heat from the slag to the metal phase by the liquid metal drops, the energy loss due to electrolysis, and the energy required for the heating of the cold slag charge.

(3) Calculation of the three dimensional temperature profiles in the base plate as a function of time by solving a transient heat conduction equation.

Sequence of Operations

As has been discussed earlier, the electric field equations and the heat flow equations are not explicitly coupled. The calculation of the temperature profiles in the liquid slag and the liquid metal regions requires knowledge of the heat generation pattern which in turn requires solution of the electric field equation in the slag region. However, the physical dimension (width) of the region is not known a priori since the width of this region is related to the weld dilution--a quantity which can be calculated only after the heat flow equations are solved for the liquid regions and for the base plate. Therefore, an iterative computational scheme, outlined in Figure 2, has been adopted.

In short, an iterative scheme is used in which the computation is started with an assumed value of dilution. The electric field equation and the modified heat conduction equations in the liquid regions are solved

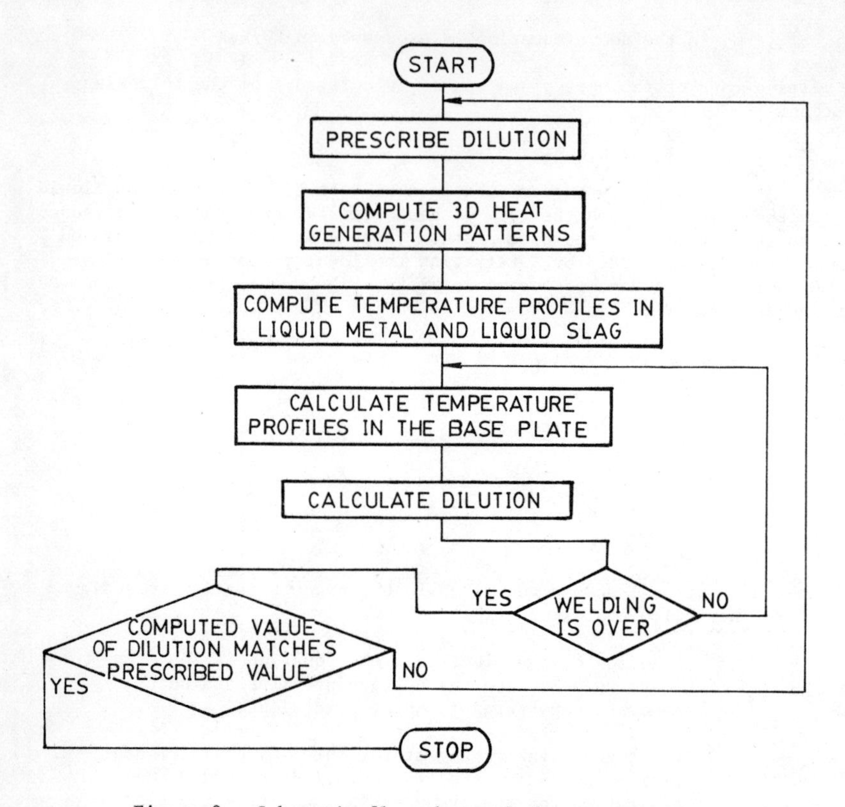

Figure 2. Schematic flow sheet of the computer program

sequentially. This is followed by the solution of the transient heat conduction equation for the base plate to compute the temperature profiles and the weld dilution as a function of time. The time average value of the weld dilution is then compared with the initially assumed value of dilution. The procedure is repeated until the assumed and the computed values of weld dilution agree within 2%.

Details of the Computational Work

Use of Effective Conductivity. In the present computations effective or enhanced values of thermal conductivity are used for the computations relating to the bulk of the liquid slag and the liquid metal phases. Near the solid walls (grid nodes closest to the solid walls) relatively lower values of effective thermal conductivity presented in Table I are used. The effects of using different values of effective conductivity have been examined in our earlier paper.[7] In the present computations, the near wall values are used only for the grid nodes next to the solid walls.

Grids. The electric field equation and the heat flow equations are solved in finite difference forms using nonuniformly spaced rectangular grids. Irregular boundaries are approximated by judicious choice of grids.

Table I. Values of Effective Thermal Conductivity (kcal/mKs)

	bulk	near wall
slag	0.6	0.075
metal	0.3	0.15

For example, the circular cross section of the electrode is represented by a regular octagon, The locations of the grid nodes are presented in Table II.

Table II. Details of the Finite Difference Grid Distribution

Direction	Location (cm)
Grids in the Liquid Region	
x	0, 0.05, 0.18, 0.34, 0.50, 0.66, 0.83, 1.0, 1.08, 1.15, 1.23, and 1.27 (12 nodes)
y	60.3, 60.38, 60.54, 60.82, 61.1, 61.38, 61.66, 61.94, 62.22, 62.38, 62.46, and 62.5 (12 nodes)
z	0, 0.2, 0.63, 1.07, 1.51, 1.94, 2.37, 2.80, 3.0, 3.2, 3.5, 3.8, 4.1, 4.4, and 4.5 (15 nodes)
Fixed Grids in the Base Plate (cm)	
x	0, 0.16, 0.48, 0.80, 1.12, and 1.27 (6 nodes)
y	0, 15.0, 35.0, 44.0, 50.0, 53.95, 55.22, 56.49, 57.12, 57.76, 58.4, 59.03, 59.48, 59.85, 60.10, 60.30, 60.80, 61.40, 62.00, and 62.50 (20 nodes)
Examples of Time Dependent Flexible Grid Locations in the z-Direction (cm)	
t = 203 s	0, 5.6, 11.4, 17.1, 22.8, 28.5, 34.2, 36.9, 39.3, 39.9, 41.1, 42.3, 42.9, 43.8, 44.4, 44.9, and 45.8 (17 nodes)
t = 390 s	0, 5.6, 11.4, 17.1, 22.8, 28.5, 31.2, 33.6, 34.2, 35.4, 36.6, 37.2, 38.1, 38.7, 39.2, 39.9, and 45.8 (17 nodes)

In the base plate, temperature variations are maximum in the vicinity of the base plate/liquid slag interface. To achieve a reasonably high accuracy in the computations, closer grid spacings are needed in the vicinity of the melt/solid boundary. This is achieved by choosing sufficiently small grid sizes (nonuniformly spaced, fixed) in the x and y directions (in the horizontal plane) and by using time dependent flexible grid locations in the vertical z-direction (welding direction). Since the welding speed is related to the electrode melting rate and is calculated prior to the computations in the base plate, the z-location of the top of the slag surface can be determined with reference to a fixed origin at any given time. During the computations, the time steps are calculated in such a manner than the top of the slag surface coincides with one of nine, fixed predetermined locations in the z-direction. In addition to these nine grids, eight flexible grids are stationed on both sides of the "top slag surface"--two of these being above this surface and six below it. The total number of working grids at any given time in the z-direction was, therefore, 17 stationed flexibly with due consideration to the location of the plane representing the top slag surface, The locations of the working z-grid nodes in the base plate at two time instances are presented in Table II, along with the locations of the x and y nodes in the base plate.

The total number of grids at any given time is, therefore, 6 x 20 x 17 = 2040 in the base plate and 12 x 12 x 15 = 2160 in the liquid region. A typical calculation takes about 500 sec of central processing time on MIT's IBM 370 digital computer.

Solution of the Finite Difference Equations. The finite difference equations are solved by an iterative scheme involving a line-by-line solution procedure. The scheme known as tridiagonal matrix algorithm (TDMA) is a particular version of Gaussian elimination technique in which the values of a variable, such as voltage or temperature at all the grid points on a line, can be updated at a time. This method is appreciably faster than the point-by-point iteration scheme in which the value of a variable is updated only at a single grid node. To achieve fast convergence, the algorithm TDMA is used in the vertical direction where the variation in the value of a variable such as the voltage is significant. The computation starts with some assumed values of a variable at all the grid nodes. The assumed values are then updated along a chosen line and the boundary conditions are applied at both ends of the line. Once TDMA is applied for a line adjacent to a boundary surface, variable values on the nearest vertical line on this surface are updated by applying the appropriate boundary condition. After one cycle of iteration, the absolute values of the difference between the "old" and the updated values of the variable at all grid nodes are compared with a certain prescribed small quantity to check for the convergence. For the voltage calculations needed to compute heat generation patterns, the maximum allowable difference $|\phi - \phi_{old}|$, is prescribed to be less than 0.00001 volt. This specified difference in voltage amounts to only 0.00025% of the voltage difference between the electrode and the base plate. The accuracy of the computations is also ensured by means of independent checks. For example, for the voltage calculations, the numerically integrated value of heat generation pattern over the 3D region is compared with the product of the overall current and the voltage difference between the electrode and the base plate. For the calculation of temperature profiles, a maximum difference of 0.01 K was used as a convergence criteria to ensure sufficient accuracy in computations.

Computed Results and Discussion

The computed results provide information on the spatially distributed heat generation patterns, on the temperature fields in the molten regions, on the temperature fields within the base plate, and on the weld dilution.

In the model, the spatially distributed heat generation pattern is calculated by solving the electric field equation and is not taken as an adjustable parameter. A detailed study of the heat generation patterns and the temperature profiles in the liquid metal has been presented in an earlier paper.[7] The modeling work presented in this paper is an extension of the earlier work. Here the temperature distribution in the base plate has been treated as a transient three dimensional problem. The weld dilution is accounted for in the computations. The computed values of the temperature profiles and the weld dilution are compared with the available experimental data.

The data used for the computations are the typical values of different parameters encountered for the ESW of mild steel plates. These are presented in Table III.

Table III. Values of Various Parameters Used for the Calculation of Temperature Fields

Plate thickness	2.54×10^{-2} and 5.08×10^{-2} m
Initial plate gap	1.9×10^{-2} m & 3.0×10^{-2} m
Width of copper shoe	1.02×10^{-1} m
Electrode radius	1.2×10^{-3} m
Metal drop radius	1.2×10^{-3} m
Depth of the slag pool	3.0×10^{-2} m
Electrode immersion in the slag	1.5×10^{-2} m
Length of the base plate	0.30 and .61m
Electrode voltage	37V
Current	calculated in the program, roughly 500A
Molten slag electrical conductivity	375 $(\text{ohm.m})^{-1}$
Molten slag thermal conductivity	1.05×10^{-3} kJ/(ms ^{o}K)
Molten metal thermal conductivity	2.1×10^{-2} kJ/ms ^{o}K)
Heat transfer coefficient at the slag/cooling shoe interface	1.67 kJ/m^{2}s ^{o}K)
Heat transfer coefficient at the base plate/cooling shoe interface	0.13 kJ/(ms^{o}K)
Convective heat transfer coefficient at the base plate surfaces	0.084 kJ(ms^{o}K)
Specific heat of the electrode	0.84 kJ/kg
Specific heat of the liquid metal drops	0.84 kJ/kg
Specific heat of the slag	1.05 kJ/kg
Latent heat of fusion of the electrode	272 kJ/kg
Emissivity of the free slag surface	0.6
Viscosity of the slag	1.0×10^{-2} kg/(ms)
Density of the liquid metal	7×10^{3} kg/m^3
Density of liquid slag	2.7×10^{3} kg/m^3

In the base plate transient three dimensional temperature profiles are calculated. Since the quantity of the data generated is voluminous, only a representative selection is presented here. However, the computer isotherms exhibit the typical behavior expected from a moving distributed heat source, the strength of which is calculated by solving electric field equations.

In Figure 3, the computer values of temperature at two different monitoring locations is shown as a function of time for the experimental conditions reported by Benter, et al.[10] Their experimental data are also presented in this diagram to enable a direct comparison between the predictions

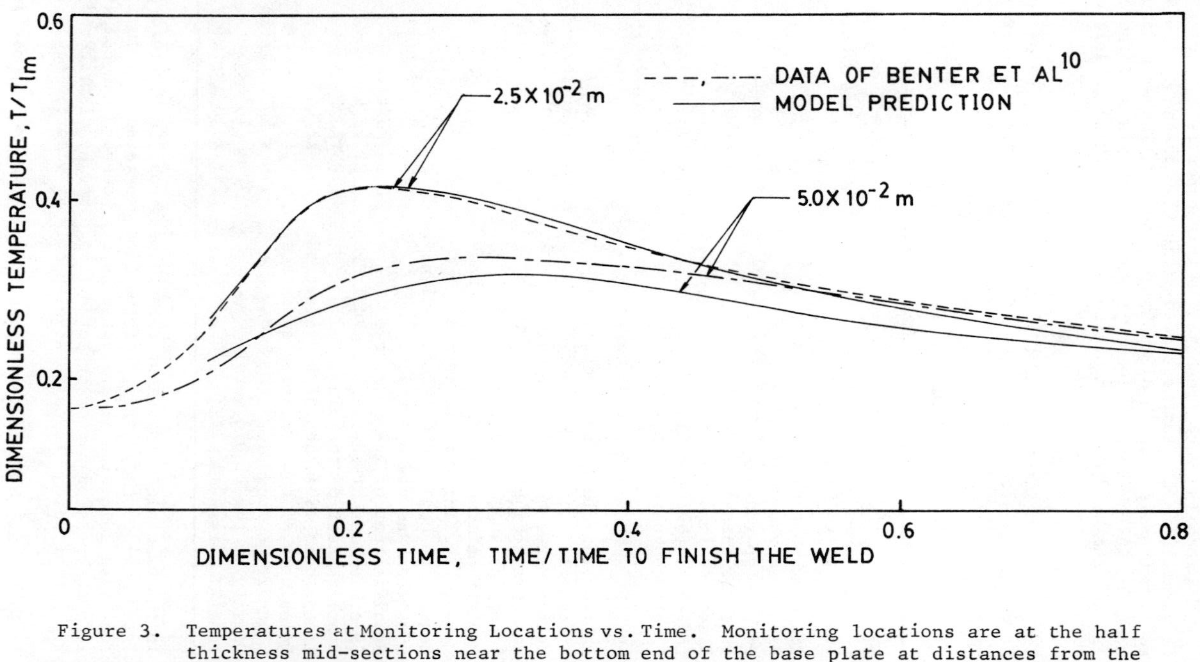

Figure 3. Temperatures at Monitoring Locations vs. Time. Monitoring locations are at the half thickness mid-sections near the bottom end of the base plate at distances from the weld face indicated on the graphs. Plate dimensions: 0.47m long, 0.61m wide, and 0.02m thick. Welding voltage = 37V, Plate gap = 0.03m, Electrode immersion = 0.015m. Welding speed = 3.1×10^{-4}m/s.

based on the model and the measurements. It is noted that the model did not incorporate the effect of the starting tab and, for this reason, one minute "starting delay" was made to represent this phenomenon.

Inspection of Figure 3 shows very good agreement between the predictions and the measurements.

The transient temperature profiles in the base plate establish the size of the heat affected zone and influence the grain growth in this region. Both these parameters are important in determining the mechanical properties of the weld. The roles played by different process parameters on the size of this zone and the grain growth phenomena are being studied at this time and will be reported in due course. However, it is seen from Figure 3 that relatively high temperatures are achieved at one inch distance from the welding face of the plate at the botton end of the vertical mid-section. Since the size of the heat affected zone is thought to be related to the peak temperature, the relatively high temperature value at the said monitoring location is consistent with a relatively large size of the heat affected zone commonly encountered in the ESW process. Furthermore, the data shown in Figure 3 underline the fact that electroslag welding is a transient process which has two important implications. One of these is that the properties of the weld may exhibit significant spatial variations; the other is that great care has to be taken in the selection of the test piece geometry, in order to reproduce faithfully the conditions in production scale operations.

As noted in the earlier section, the weld dilution is computed iteratively. The weld dilution and the solidification profile define the weld form factor, an important parameter in determining the weld quality. For this reason, it is of interest to examine the implications of the computer results regarding the weld dilution.

Figure 4 shows a plot of the percentage dilution, defined earlier in Eq. (5), against a dimensionless time. It is seen that initially the dilution is low, it attains a more-or-less steady value over a major length of the weld, and finally the dilution increases as the top of the plate is being approached. This behavior seems reasonable because during the initial stages most of the thermal energy is being used to preheat the plate, while a quasi-steady type heat balance is being maintained subsequently in the process, corresponding to the horizontal part of the curve. It is noted that the initially low level of dilution is usually compensated for, by using a starting tab.

Figure 5 shows a selection of experimental data reported by Benter et al,[10] relating the percentage dilution to the weld speed. It is seen that the data exhibit appreciable scatter; this is in part attributable to the difficulties in assessing dilution accurately; moreover the data represent measurements taken from several sections. The data indicate that dilution tends to decrease slightly, with increasing welding speed. The full line represents predictions based on the model, which seems to be in reasonable agreement with the measurements.

Figure 6 shows experimental measurements of Ricci[2] concerned with the effect of the plate gap on the percentage dilution. The theoretical predictions based on the model are also shown for the sake of comparison.

Here again the experimental measurements do exhibit appreciable scatter; nonetheless the predictions based on the model seem to be in reasonable agreement with the measurements.

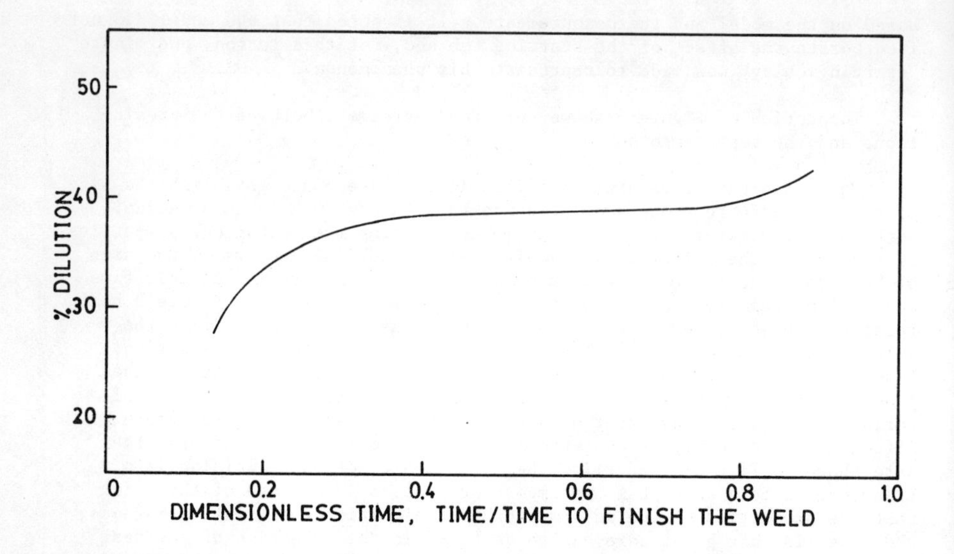

Figure 4. Dilution vs. Time. Plate dimensions = 0.47m long, 0.61m wide, and 0.025m thick. Welding voltage = 37V, Plate gap = 0.03m, Electrode immersion = 0.015m. Weld speed = 3.1×10^{-4}m/s.

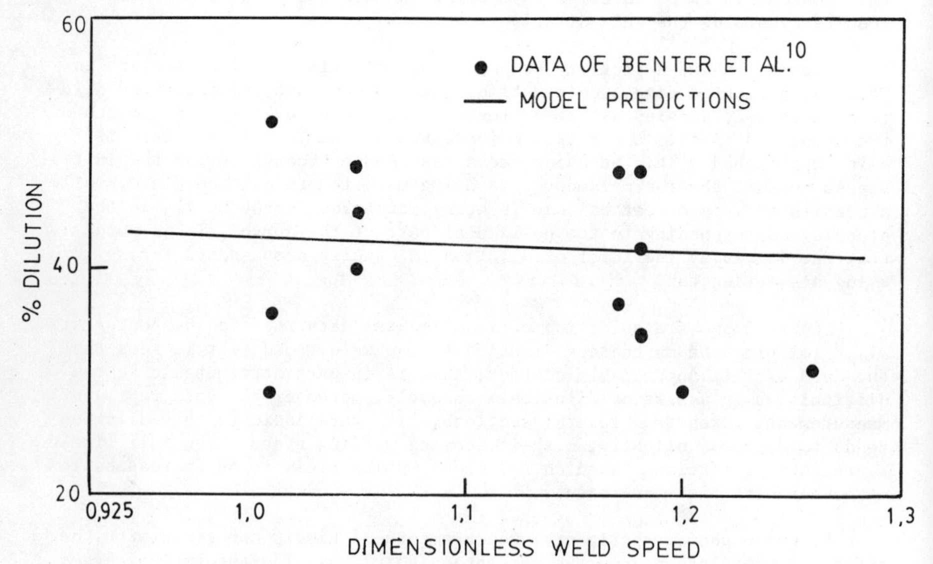

Figure 5. Dilution vs. Dimensionless Weld Speed. Plate dimensions: 0.47m long, 0.61m wide, and 0.025m thick. Welding voltage = 37V, Plate gap = 0.03m. Electrode immersion = 0.015m. Reference weld speed = 3.1×10^{-4}m/s.

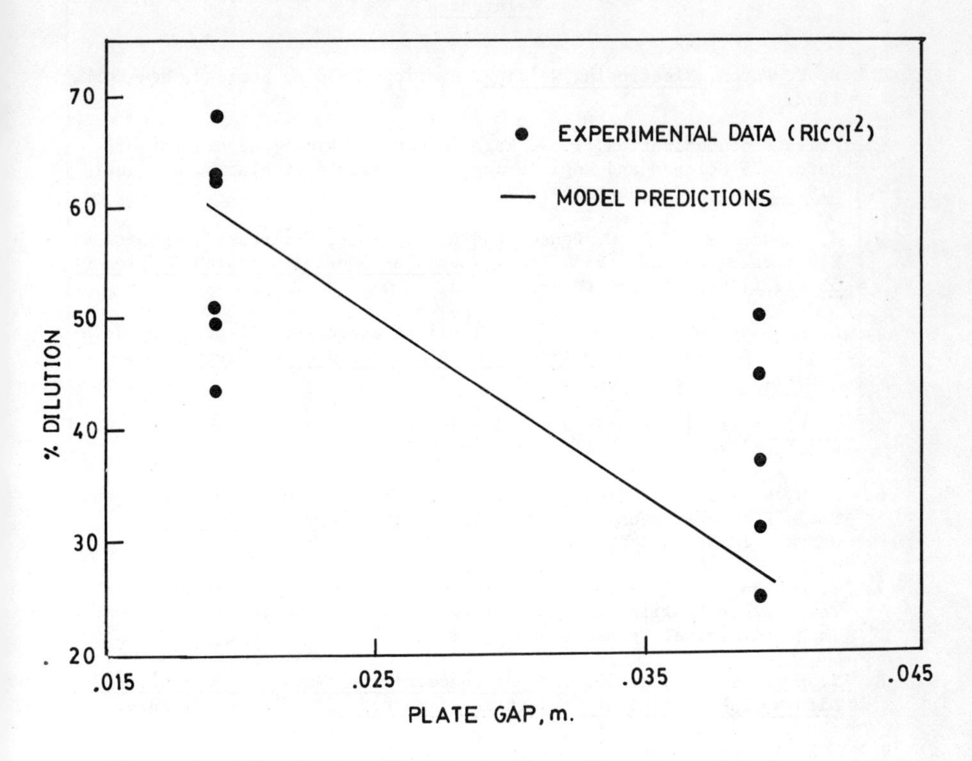

Figure 6. Dilution vs. Plate Gap. Plate dimensions: 0.3m long, 0.3m wide, and 0.05m thick. Welding voltage = 45V, Electrode immersion = 0.095m.

Conclusions

A mathematical representation of the ESW process has been used to compute the transient three dimensional temperature profile in the base plate and the weld dilution. The computed values of these quantities are found to be in good agreement with the available experimental data. The computed results indicate that electroslag welding is a transient process and that the properties of the weld may exhibit significant spatial variation. The weld dilution is found to be influenced by both the plate gap and by the weld speed. The dilution increases strongly with a decrease in the plate gap and increases mildly with a decrease in the weld speed.

References

1. B. E. Paton, Electroslag Welding, American Welding Society, New York, 1962.

2. Private communication, T. W. Eagar and W. S. Ricci, Department of Materials Science and Engineering, MIT, Cambridge; also W. S. Ricci, M.S. Thesis, MIS, September 1979.

3. J. B. des Ramos, A. W. Pense, and R. D. Stout, "Fracture Toughness of Electroslag Welded A 5376 Steel," Welding Journal Research Supplement, 55 (1) (1976) pp. 1-s to 4-s.

4. A. L. Liby, D. L. Olson: "Metallurgical Aspects of Electroslag Welding: A Review," Quarterly of the Colorado School of Mines, V 69 (1974).

5. C. E. Jackson: "Fluxes and Slags in Welding," Bulletin No. 190, Welding Research Council, December 1973.

6. A. H. Dilawari, J. Szekely, and T. W. Eagar, "An Analysis of Heat and Fluid Flow Phenomena in Electroslag Welding," Welding Journal, January 1978, p. 245.

7. T. Deb Roy, J. Szekely, and T. W. Eagar, "Heat Generation Patterns and Temperature Profiles in Electroslag Welding," accepted for publication in Metallurgical Transactions, B.

8. Y. S. Touloukin, et al, Thermal Conductivity, Metallic Elements and Alloys, Thermophysical Properties of Matter, IFI/Plenum, NY 1970.

9. Y. S. Touloukin, et al, Specific Heat, Metallic Elements, and Alloys, Thermophysical Properties of Matter, IFI/Plenum, Ny 1970.

10. W. P. Benter, Jr., P. J. Konkol, B. M. Kapadia, A. K. Shoemaker, and J. F. Sovak, "Acceptance Criteria for Electroslag Weldments in Bridges," Phase I. Final Report, U.S. Steel Corporations, April 1977.

Acknowledgement

The authors gratefully acknowledge the support from the Department of Energy under contract No. ER-78-S-02-4799.A001.

APPENDIX A

Boundary Conditions for the Calculation of Temperature Profiles in the Base Plate

The following boundary conditions are used for the computation of temperature profile in the base plate. Figure 1 shows the different places mentioned in this section.

(i) Vertical surface of the plate, ABCD plane:

$$k \frac{\partial T}{\partial y} = h_c(T - Ta) \qquad (A1)$$

(ii) Bottom end of the plate, CDEF plane:

$$-k \frac{\partial T}{\partial z} = h_c(T - Ta) \qquad (A2)$$

(iii) Mid-section symmetry plane, RJI HFCB plane:

$$\frac{\partial T}{\partial x} = 0 \qquad (A3)$$

(iv) Plate-cooling shoe interface, PYEOLN plane:

$$k \frac{\partial T}{\partial x} = h_c(T - Ta) \qquad (A4)$$

(v) Vertical surface of the plate, ADYPNQ plane:

$$k \frac{\partial T}{\partial x} = h_c(T - Ta) \qquad (A5)$$

(vi) Top surface of the plate, ABRQ plane:

$$k \frac{\partial T}{\partial z} = h_c(T - Ta) \qquad (A6)$$

(vii) Mid-section symmetry plane, EFHO plane:

$$\frac{\partial T}{\partial y} = 0 \qquad (A7)$$

(viii) The curved surface at the liquid metal/plate surface, OLIH plane:

$$T = T_{\ell m} \qquad (A8)$$

(ix) Liquid slag/plate interface, KJIL plane:

$$T = T_{\ell m} \qquad (A9)$$

(x) Vertical surface of the plate QRJK plane:

$$K \frac{\partial T}{\partial y} = \sigma F_{s\rightarrow \blacksquare}(\varepsilon_{s\ell} T_{s\ell av}^{4} - \varepsilon T^{4}) - h_c (T - Ta) \qquad (A10)$$

Where σ is the Stephen-Boltzman constant, $F_{s\rightarrow \blacksquare}$ is the view factor, and $T_{s\ell av}$ is the average temperature of the top slag surface.

List of Symbols

C_p - Specific heat of the material of the base plate, kJ/kgK

hc - Heat transfer coefficient at the base plate surface, kJ/m^2sK

$H_{s\ell p\ell}$ - Total amount of heat transported from the liquid slag to the base plate, kH/s

k - Thermal conductivity, kJ/msK

m - Electrode melting rate, kg/s

m_b - Melting rate of the base plate, kg/s

t - Time, s

T, Ta, $T_{\ell m}$ - The symbol T refers to temperature in the computation domain; subscripts a and ℓm refer to ambient temperature and the melting point of metal, K

x, y, z - Distance coordinates, m

ρ - Density, kg/m^3

$\varepsilon_{s\ell}$ - Emissivity of the slag surface

ε - Emissivity of the base plate

ϕ - Updated value of voltage

ϕ_{old} - Value of voltage prior to iteration

Stress and Deformation: Castings and Weldments

MODELS OF STRESSES AND DEFORMATIONS IN SOLIDIFYING BODIES

O. Richmond
U.S. Steel Corporation
Monroeville, Pennsylvania

Optimum design of casting processes depends, at least in part, on a knowledge of the stresses and deformation that develops in the solidifying body. Excessive stresses and deformations can lead to air gaps and associated loss of heat transfer, cracks at the solidification front and associated segregation, other surface and internal cracks, and even shell breakouts. Examples are given of past attempts to develop mathematical models of the mechanical behavior of solidifying bodies, especially in the case of unidirectional solidification, and an attempt is made to assess current trends and research opportunities. It is pointed out that a complete research effort involves four parts: (1) accurate measurement of mechanical behavior at appropriate temperatures and strain rates, (2) development of elastoviscoplastic constitutive models that represent this behavior, (3) development of computational methods for treating the appropriate boundary value problems, and (4) experimental evaluation of predictions of the model under controlled laboratory conditions.

Introduction

Modeling of the solid-mechanical aspects of casting (and welding) processes, ie., the stresses and strains in the solidifying body, is by no means an established art. Some rather crude'strength-of-materials' approaches have been used for estimating the conditions needed for causing massive air gaps in molds, or for estimating support spacings and bending moments required for guiding partially-solidified strands in continuous casters. However, the capability to accurately predict the detailed stresses and deformations in the solid behind the solidification front is only now being developed. The more crude models will not be considered here. Rather, the scope will be limited to the more detailed models, because the time seems right for their rapid development, and because it is these which show promise of significantly improving design and analysis methods for casting processes.

Some past developments will be reviewed very briefly, and then an assessment of current trends and opportunities will be attempted. Most prior work has assumed that the thermomechanical problem is uncoupled. That is, the heat generated by inelastic deformation during solidification is assumed to have negligible effect on the solidification rates and temperature fields, and also the mechanical deformation of the solidifying shell is assumed to have negligible effect on the heat transfer. Thus, the geometrical development of the solid and its associated temperature fields, that is the thermal problem, is determined first. Then the mechanical problem is solved, taking account of the temperature-dependence of mechanical properties but utilizing the already known temperature distribution. Most of the results which have been obtained to date consider only single-phased pure metals so that effects such as those caused by solid state transformations and mushy zones are not taken into account. Also most of the results have been obtained for rigid molds with simple geometries, in many cases dealing only with unidirectional solidification. Thus effects due to mold distortion and air gap formation are usually ignored.

Selected Results on Unidirectional Solidification

The detailed modeling of solidification problems may be said to have begun with the work of Weiner and Boley[1] sponsored by the U.S. Steel Research Laboratory, and published in 1963. These investigators considered the case of a casting with a constant surface temperature in a square mold, as illustrated in Figure 1a. The mechanical behavior of the solid was assumed to be rate-independent, non-hardening, with temperature-dependent flow stress, but temperature-independent elastic moduli. Effects of melt fluid pressure and of creep were neglected so that the casting remained square as it shrunk, and an air gap, though this was not discussed, formed from the beginning of solidification. The stresses for the chosen cooling history, constant surface temperature, were compressive at the outer surface and tensile near the solidification front.

More than a decade later, in 1971, Richmond-Tien[2] published new results for a rectangular mold. Viscous (creep) effects were included in their model as well as temperature-dependent elastic moduli. Also the effect of

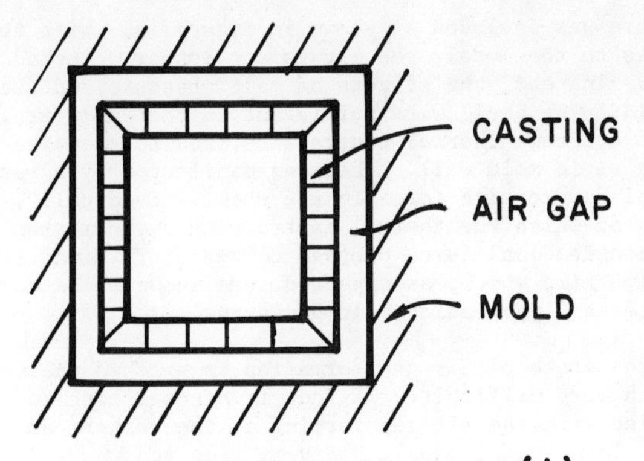

a. WEINER-BOLEY PROBLEM (1)

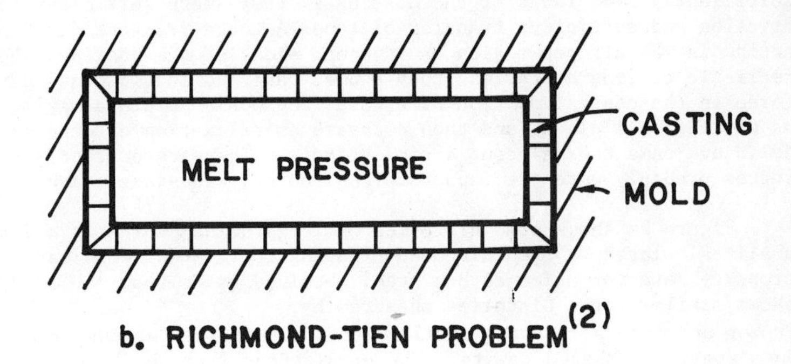

b. RICHMOND-TIEN PROBLEM (2)

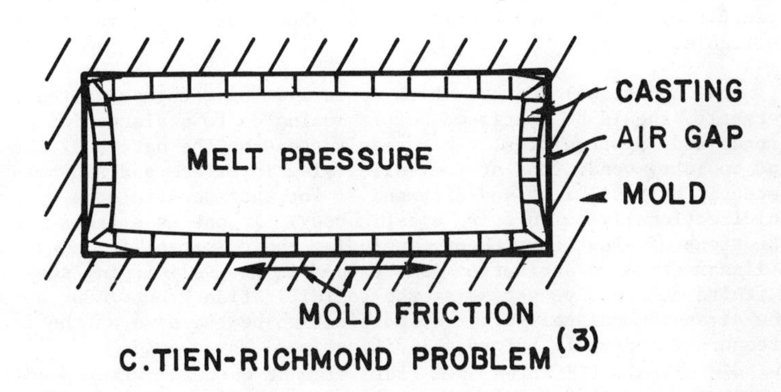

C. TIEN-RICHMOND PROBLEM (3)

FIG. 1 ILLUSTRATION OF SOLIDIFACATION PROBLEMS

melt pressure was included as shown in Figure 1b. With these improvements to the model, the air gap no longer occurred at the beginning of cooling. Instead, the effects of melt pressure and the creep response in the solidifying shell were sufficient in the early stages of solidification to overcome thermal contraction, and to keep the shell pressed against the rigid mold wall. This was manifested by a resultant longitudinal compressive force in the shell. Eventually, however, a time was reached at which the thermal contraction overcame the creep and the resultant longitudinal force dropped to zero. This was taken as the air gap formation time which, as expected, was shown to be delayed by decreasing cooling rates and increasing melt pressures.

Once the stage of air gap formation is reached the mathematical problem becomes much more difficult. In fact it becomes both two-dimensional and uncoupled with the air gap forming at the corners as illustrated in Figure 1c. In a recent paper, however, Tien and Richmond(3) have argued that their solution, which involves no lateral contraction, could continue to apply in the central portion of sufficiently wide faces of the casting if there were sufficient mold friction and sufficient transfer of tensile force from the fluid pressure acting in the air gap region on adjacent sides of the casting. For realistic cooling histories, they showed that the longitudinal tensile force in the central portion of a wide face would attain a maximum value as cooling took place, and then decrease as relaxation effects in the solid overcame thermal contraction effects. They argued that such tensile forces probably were the cause of longitudinal wide-face cracks.

Figure 2a shows some histories of longitudinal force in a constrained shell calculated by Tien-Richmond(3) using their model with material property data for a low carbon steel obtained by Feltham(4). Figure 2b shows similar force histories measured by Frober and Oeters(5) for a similar steel constrained against contraction in a special I-shaped cavity. The predictions from the model and the experimental observations are seen to be quite similar, but more specific evaluations of predictions by experiments of this type are very desirable.

One additional result, which appears to have escaped attention until recently, should be mentioned before moving on to a discussion of research trends and opportunities. This result concerns the nature of the initial and boundary conditions at the solidifying interface and has been treated recently by Vasillicos and Richmond(6) for the specific case of a unidirectionally-solidifying elastic body. If one uses the standard equations of elasticity theory including the compatability equations, a linear stress distribution is obtained in the solidifying shell, with a finite compressive stress at the solidification front. But shouldn't the stress immediately after solidification be the same as the fluid pressure immediately before solidification? Why should the act of solidification of a fluid element onto an existing solidified body cause a sudden large compressive stress? The answer is that it should not. The standard compatability equations apply to a body which is uniformly stress-free all at once. The elements of a solidifying body are stress-free at different times. Thus, the solution of a solidifying-body problem is fundamentally different from a fixed-body problem. This important point seems not to have been appreciated and a number of solutions in the literature are therefore incorrect. The solution given by Richmond and Tien(2) is incorrect but has been corrected in Reference 3. The correct solution for the problem illustrated in Figure 3 is shown along with the incorrect one. The stress for the correct case is seen to be zero at the solidification front.

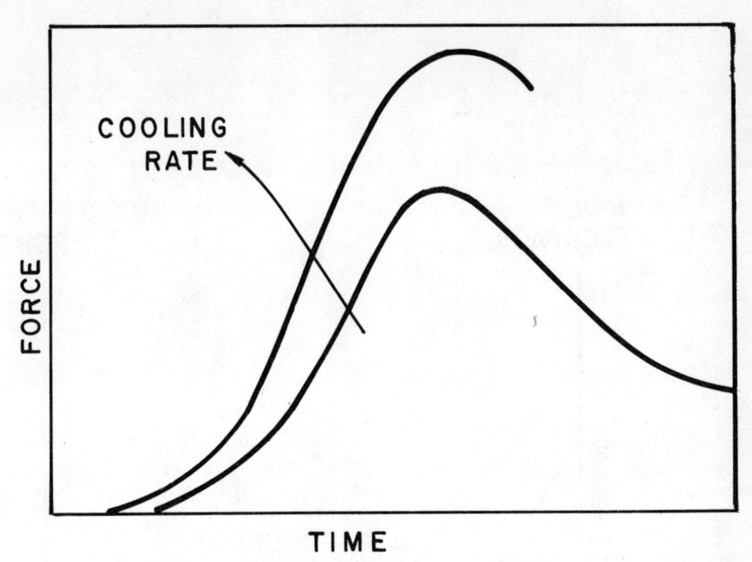

a. CALCULATED(3)

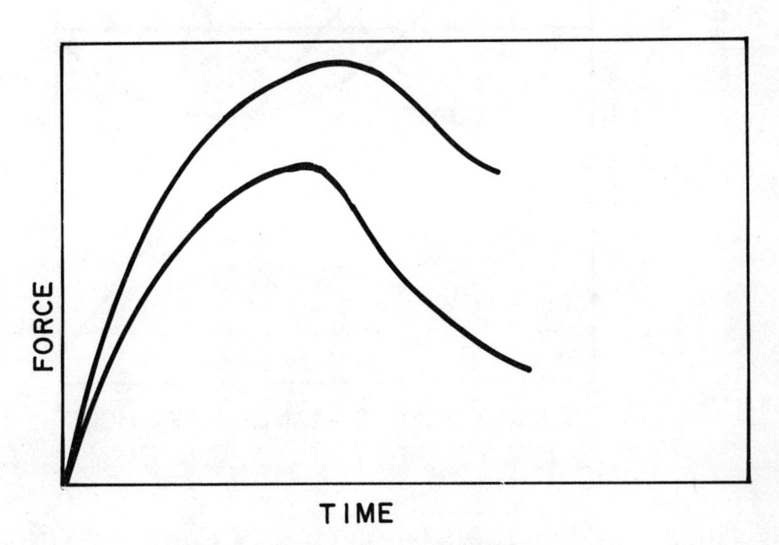

b. MEASURED(5)

FIG. 2 LATERAL FORCE IN A CONSTRAINED SOLIDIFYING SLAB

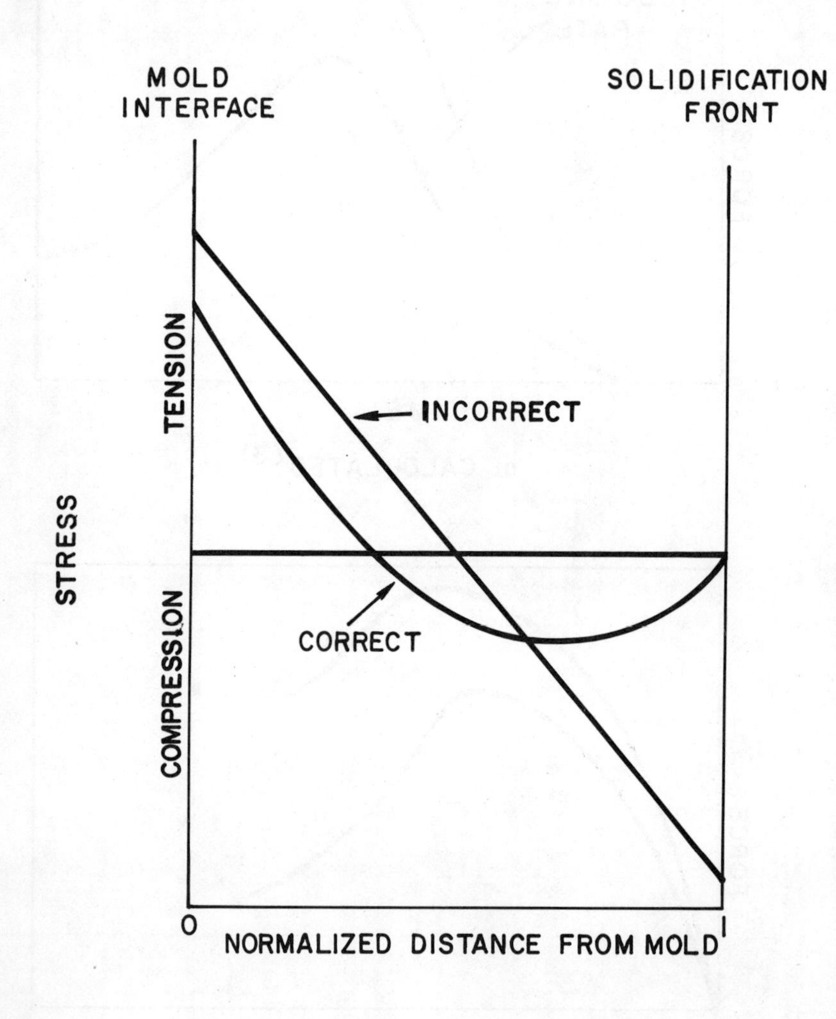

FIG. 3 STRESS DISTRIBUTION IN A SOLIDIFYING ELASTIC SLAB (6)

Research Trends and Opportunities

Before describing some of the current trends and opportunities envisioned for research on the mechanical behavior of solidifying bodies, the four principal elements of model development are listed: (1) accurate measurements of the mechanical behavior of the solid material under carefully controlled conditions of uniform temperatures and strain rates, (2) development of elastoriscoplastic constitutive equations that adequately represent these measurements; (3) development of computational methods for predicting mechanical behavior of materials obeying these constitutive equations and capable of solving initial and boundary conditions appropriate for solidification problems; and (4) evaluation of these models and methods by comparing quantitative predictions with measurements obtained in controlled solidification experiments.

In the area of measurements of mechanical behavior, the work of Wray[7,8] on iron-based alloys is particularly noteworthy. In the area of constitutive model development, the internal-variable models based on a single internal variable[9-11] are particularly attractive. Figure 4a from Anand[12] shows the results of such a model compared with some of Wray's data, and Figure 4b shows its predicted behavior in strain-rate increment tests. The behavior appears to be very realistic. In the area of computational methods, the tremendous progress in both finite-element and boundary-element methods, as well as finite-difference methods, offers great promise for two-and three-dimensional problems. Finally in the area of experimental evaluations, the type of experiment performed by Frober and Oeters,[5] referred to earlier, on constrained unidirectional solidification is representative of the type of work that should be done.

It seems clear that there will be considerable progress in the years immediately ahead in the development of theoretical tools which can predict the stresses and deformations in solidifying bodies. The discussion here has been limited primarily to uncoupled, unidirectional models in pure, single-phase metals, but considerable progress can be expected in two-and three-dimensional solutions, in the treatment of pure metals with solid-state transformations as well as alloys with mushy zones, and even in coupled problems.

Acknowledgment

I have benefited greatly, over a number of years from interactions with my colleagues R.H. Tien and P.J. Wray. I also am grateful to my more recent collegues, L. Anand and A. Vassilicos for the use of some of their unpublished work.

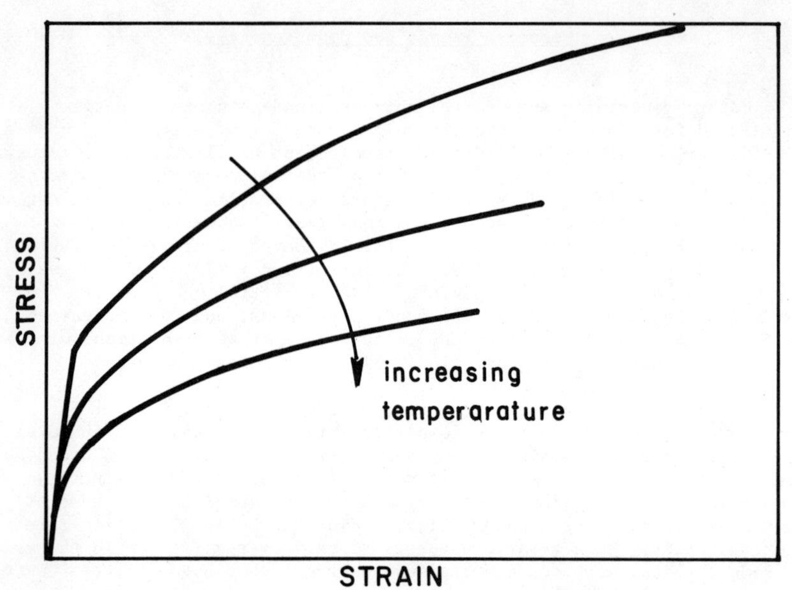

a. CONSTANT STRAIN RATE TESTS

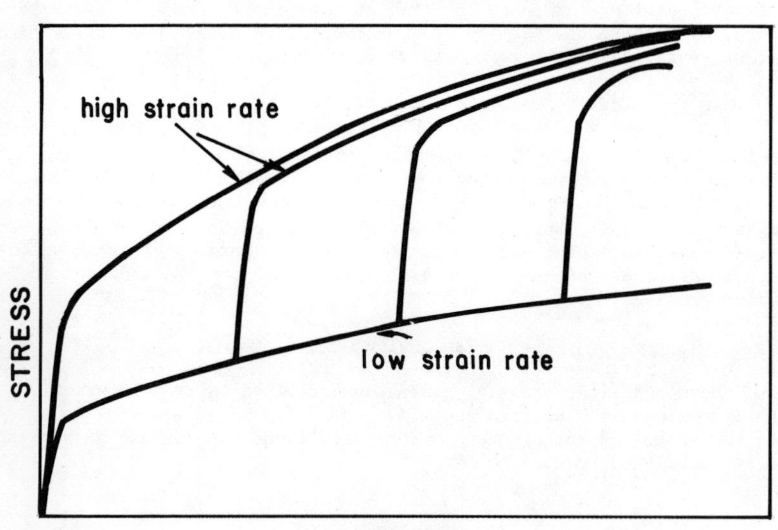

b. STRAIN-RATE INCREMENT TESTS

FIG. 4 MECHANICAL BEHAVIOR OF INTERNAL VARIABLE MODEL

References

1. J. H. Weiner and B. A. Boley, "Elastoplastic Thermal Stresses in a Solidifying Body," J. Mech. Phys. Solids, 11, (1963), pp. 145-154.
2. O. Richmond and R. H. Tien, "Theory of Thermal Stresses and Air-Gap Formation during the Early Stages of Solidifcation in a Rectangular Mold," J. Mech. Phys. Solids, 19, (1971), pp. 273-284.
3. R. H. Tien and O. Richmond, "Theory of Maximum Tensile Stresses in the Solidifying Shell of a Constrained Rectangular Casting," To be published in J. Appl. Mechs.
4. P. Feltham, "The Plastic Flow of Iron and Plain Carbon Steels above the A_3 Point", Proc. Roy Soc., 66, (1953), pp. 865-883.
5. J. Fröber and F. Oeters, "On the Mechanical Behavior of Steel During Solidification," Arch. Eisenhuttenwes. 51, (1980), pp. 43-48.
6. A. Vassilicos and O. Richmond, "Stresses in a Solidifying Elastic Plate," To be published.
7. P. J. Wray and M. F. Holmes, "Plastic Deformation of Austenitic Iron at Intermediate Strain Rates," Metall Trans., 6A, (1975), pp. 1189-1196.
8. P. J. Wray, "Plastic Deformation of Delta-Ferritic Iron at Intermediate Strain Rates," Metall. Trans., 7A, (1976), pp. 1621-1627.
9. E. W. Hart, "Constitutive Relations for the Nonelastic Deformation of Metals," ASME J. Eng. Matts. ad Tech., 98, (1976), pp. 193-202.
10. J. R. Rice, "Continuum Mechanics and Thermodynamics of Plasticity in Relation to Microscale Deformation Mechanisms," Constitutive Equations in Plasticity, A. S. Argon, ed., The MIT Press, Cambridge, (1975), pp. 23-79.
11. S. R. Bodner and Y. Partom, "Constitutive Equations for Elastic-Visco-plastic Strain-Hardening Materials," ASME J. Appl. Mech., 42, (1975), pp. 385-389.
12. L. Anand, "Constitutive Equations for the Rate-Dependent Deformation of Metals at Elevated Temperatures," Submitted for publication.

MODELS OF STRESSES AND DEFORMATION DUE TO WELDING - A REVIEW

Koichi Masubuchi

Department of Ocean Engineering
Massachusetts Institute of Technology
Cambridge, Massachusetts

This paper first discusses the use of computers in welding. Applications of computers to welding started around 1965 and they have grown significantly since then. The paper then discusses development of studies on residual stresses and distortion during welding. Since the late 1960's a number of investigators in the world have developed computer programs for analyzing transient thermal stresses and metal movement during welding.

Systematic research efforts have been carried out at M.I.T. on residual stresses and distortion in weldments. A series of computer programs have been developed to calculate residual stresses and distortion in various types of welds. Experimental data have been generated on weldments in various materials and processes. Some experimental data on steel weldments are presented in this paper.

Introduction

For a long time welding research has been primarily empirical based upon experience and experimental data. One of the major reasons is that it is rather difficult to analytically simulate welding phenomena, especially when computations must be carried out manually. With the development of modern computers it is possible to execute complex calculations with reasonable time and cost. This paper presents a review of the present state-of-the art of analytical modeling of stresses and deformation due to welding.

Use Of Computers In Welding

Before discussing the use of computers for analyzing residual stresses and distortion, I would like to discuss briefly uses of computers in welding fabrication in general. We, researchers at M.I.T., became interested in this subject about ten years ago, and a survey was made to determine how computers were used in studies on welding, the results of which were published in technical journals(1). Review journals including ASM Review of Metal Literature, U. S. NASA Scientific and Technical Aerospace Reports, and Applied Mechanics Reviews were used to identify publications on welding. The survey also covered published articles on other fabrication technologies, including forming and machining, extraction and refining, metal production, and foundry.

Results are summarized in Figure 1, which shows numbers of papers which mention the word "computer" in either titles or abstracts. For all fabrication technologies investigated, the numbers of papers which mention the word "computer" have increased significantly since around 1965. Also shown in parentheses are total numbers of papers published in 1970 on these fabrication technologies. There were about 1200 papers on forming and machining, about 620 papers on extraction and refining, and so on. These numbers indicate that these technologies are similar not only in their natures of work but also in numbers of publications.

As far as the welding technology is concerned, there were about 950 articles published in 1970. During the 10 year-period from 1960 to 1970, about 1000 articles on welding were published each year in technical journals in major industrial nations. Of these papers, only a few mentioned the word "computer" in either titles or abstracts. There was only one in 1965, and the number increased steadily since then. In terms of the total number of papers the welding technology was the second largest, only after forming and machining. As far as the use of computers is concerned, however, the welding technology was behind all of the technologies surveyed, and the foundry technology was slightly ahead of the welding technology. The reason for fewer computer applications to welding and foundry technologies than to the other technologies is perhaps due to the fact that welding and foundry involve complex phenomena which are difficult to express analytically.

From this modest beginning which started around 1965 the use of computers in welding has grown significantly. Recent papers by Masubuchi(2,3) discuss the present state-of-the-art of computer applications to welding with special emphasis to numerical analyses of various phenomena involved in welding. During the last 15 years a number of investigators in the world have developed many computer programs covering various subjects including (a) heat flow, (b) solidification, solid-state transformation and other phenomena in welding metallurgy, (c) residual stresses and distortion, and (d) service behaviors of welded structures. In the remainder of this paper I will discuss recent developments in the analysis of thermal stresses, residual stresses, and distortion in weldments.

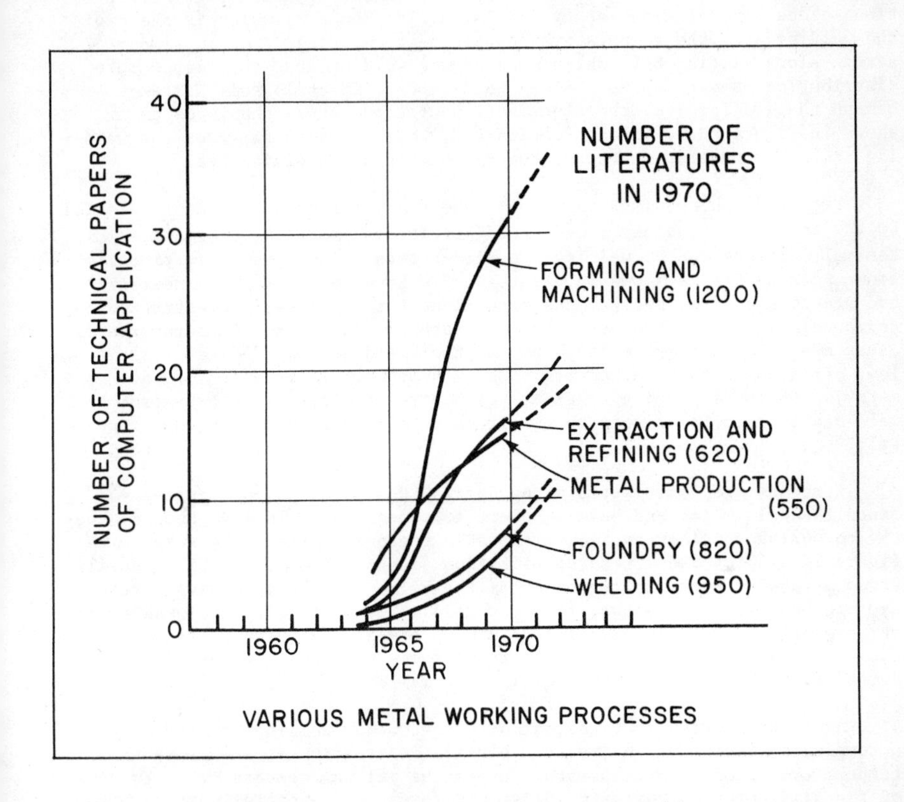

Fig. 1 - Computer applications in various metal working processes.

Development Of Studies On Residual Stresses And Distortion In Weldments

Because a weldment is locally heated by the welding heat source, its temperature distribution is not uniform and changes as welding progresses. During the welding heat cycle, complex strains occur in the weld metal and the base-metal regions near the weld. The strains produced during welding are accompanied by plastic upsetting. As a result, residual stresses remain after welding is completed. Shrinkage and distortion are also produced. These stresses and distortion cause many problems in welding fabrication. Thermal stresses during welding often cause cracking. High tensile residual stresses near the weld may promote fractures of the weldment. Compressive residual stresses in the base plate, often combined with distortion, may reduce the buckling strength. Correcting unacceptable distortion is extremely costly and in some cases impossible.

Figure 2 shows schematically changes of temperature and stresses during welding. A weld bead is being laid along the x-axis of a plate. The welding arc, which is moving at speed v, is presently located at the origin, O, as shown in Figure 2a. Figure 2b shows temperature distributions along several cross sections. Along Section A-A, which is ahead of the welding arc, the temperature change due to welding, ΔT, is almost zero. Along Section B-B, which crosses the welding arc, the temperature distribution is very steep. Along Section C-C, which is some distance behind the welding arc, the distribution of temperature change is as shown in Figure 2B-3. Along Section D-D, which is very far from the welding arc, the temperature change due to welding again diminishes.

Figure 2c shows distributions of the thermal stress component parallel to the x-axis, σ_x, along these cross sections.* Along Section A-A, thermal stresses due to welding are almost zero. The stress distribution along Section B-B is shown in Figure 2c-2. Stresses in regions near the arc are compressive, because the expansion of these areas is restrained by surrounding areas heated at lower temperatures. Since the temperatures of areas near the weld are quite high and the yield strength of material is low, stresses in these areas are as high as the yield strength of the material at corresponding temperatures. Stresses in regions away from the weld are tensile to balance the compressive stresses in regions near the weld.

Stresses are distributed along Section C-C as shown in Figure 2c-3. Since the weld metal and base-metal regions near the weld have cooled, they try to shrink causing tensile stresses in regions close to the weld. Figure 2c-4 shows the stress distribution along Section D-D. High tensile stresses are produced in regions near the weld, while compressive stresses are produced in regions away from the weld. The distribution of residual stresses that remain after welding is completed as is shown in this figure. The cross-hatched area, M-M in Figure 2a shows the region where plastic deformation occurs during the welding thermal cycle.

As shown in Figure 2, thermal stresses during welding are produced by a complex mechanism which involves plastic deformation at a wide range of temperatures from room temperature up to the melting temperature. Because of the difficulty in analyzing plastic deformation, especially at elevated temperatures, analyses using manual computations were limited to very simple cases such as spot welding.

With the development of modern computers it has become possible to analyze with reasonable cost and time transient thermal stresses and metal movement during welding. The first significant attempt to use a computer in the analysis of thermal stresses during welding was done by Tall in 1961(4). He developed a simple program on thermal stresses during bead welding along the center line of a strip. The determination of the temperature distribution was treated as a two-dimensional problem. For the stress analysis, the longitudinal stress, σ_x, was treated as a function of the lateral distance y only and σ_y and τ_{xy} were assumed to be zero. In this paper, such an analysis designated one-dimensional. In 1968, Masubuchi, et al(5) developed a FORTRAN program, based upon Tall's analysis,

---

* In the two-dimensional plane stress case, as shown in Figure 2, three stress components including σ_x , σ_y and τ_{xy} exist.

for the one-dimensional thermal stresses during welding.

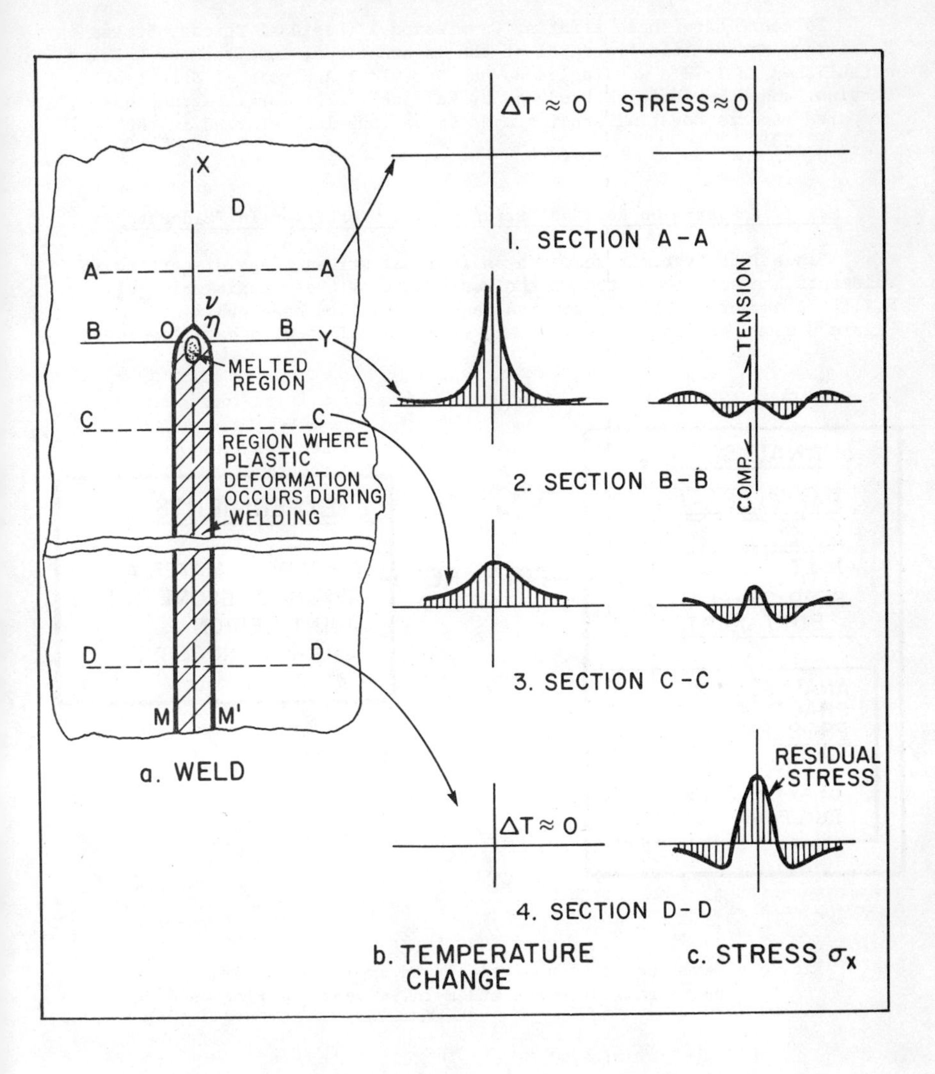

Fig. 2 - Schematic representation of changes of temperature and stresses during welding.

Since 1970, the computer analysis of transient thermal stresses during welding has become more common. At M.I.T. and several other laboratories around the world, investigators are currently developing computer programs related to, or specifically for, this kind of analysis. For example, finite-element programs for calculating welding thermal stresses have been developed at Battelle Memorial Institute (6,7), Brown University (8), Osaka University (9,10), and the University of Tokyo (11,12) as well as M.I.T.*

To coordinate these efforts, Commission X (Residual Stress, Stress Relieving, and Brittle Fracture) of the International Institute of Welding established in 1972 a working group on "Numerical Analysis of Stresses, Strains, and Other Effects Produced by Welding". The working group has prepared reports covering studies made in various laboratories in the world (3,13).

M.I.T. Research On Residual Stresses And Distortion In Weldments

Since 1968 systemic research on residual stresses and distortion in weldments has been conducted at the Department of Ocean Engineering of M.I.T. A number of theses, reports, and papers have been published. Figure 3 shows how the analytical and experimental studies have developed.

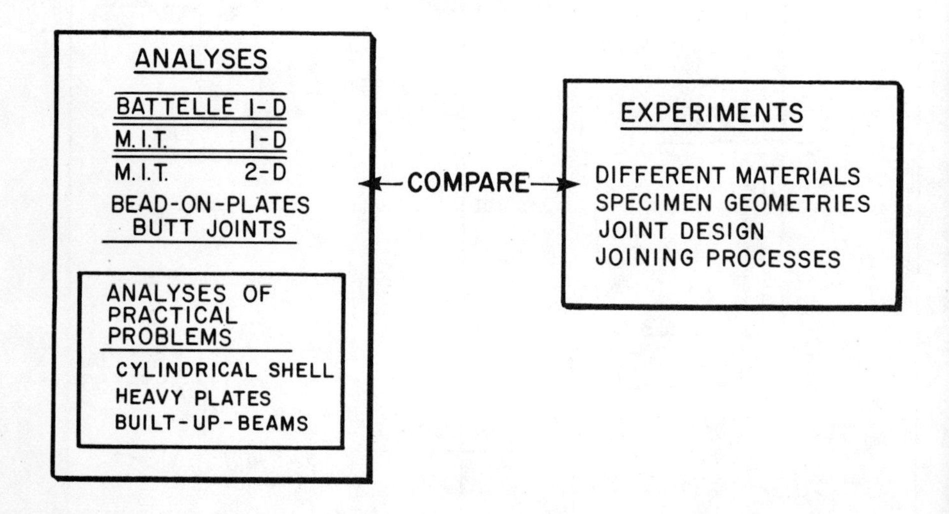

Fig. 3 - Development of analytical and empirical studies at M.I.T. on thermal stresses and metal movement during welding.

Analytical Studies

The first to be developed were the one-dimensional programs. The original Battelle program developed by Masubuchi, et al (5) was improved

*References (6) through (12) represent only some publications, primarily early publications. Many more articles have been written.

during a study completed in 1972 (14). The M.I.T. program, developed by Andrews, et al (14), was an improvement over the Battelle program in the following ways:

1. In the Battelle program the material was assumed to be perfectly plastic; in the M.I.T. program, strain hardening of the material was taken into consideration.

2. Calculation of total mechanical strains, important when comparing the results of the theoretical study with experimental data, was included in the M.I.T. program.

Details of the M.I.T. program are described in the NASA Contract Report CR-61351 (14). More recently the program has been further improved and it can now handle various practical problems, such as stresses during multipass welding (15).

The development of two-dimensional finite element programs was the next step. The initial effort was made by Iwaki (16), who developed a finite element program for analyzing thermal stresses during bead-on-plate welding on a plate. Muraki (17) further improved the programs to cover thermal stresses and metal movement in two-dimensional plane stress and plane strain conditions. Toshioka (18) made an effort to include the effects of metallurgical transformation on transient thermal stresses that form during welding. After the heat flow is analyzed, the metal properties are determined using the continuous cooling transformation (CCT) diagram, and dimensional changes due to mettallurgical transformation are taken into consideration as well as thermal strains. The latest work, which has been sponsored by the Office of Naval Research, involves analysis of thermal stresses and residual stresses in heavy weldments in HY-130 steel. The analysis includes the effects of metallurgical transformation on thermal stresses and residual stresses in weldments.

Analyses also have been expanded to cover some pratical problems including:

1) Stresses produced during girth welding of a thin-walled cylindrical shell

2) Longitudinal stresses in heavy weldments

3) Stresses and distortion during welding fabrication of built-up beams.

Experimental Studies

In studying the thermal stresses and metal movement that occur during welding, it is important to compare analytical predictions with experimental results. During the last 10 years a series of experiments were conducted at M.I.T. for various sponsors. Most of the experiments were carried out by graduate students. The experiments conducted so far cover the following conditions:

1) Materials. Low-carbon steel, high-strength steels, stainless steel, aluminum alloys, titanium alloys, columbium and tantalum, and zircaloy.

2) Plate Thickness. 0.3 to 25mm.

3) Joint Types. Bead-on-plate and butt welds in single pass and multipass (up to 20 passes), fillet welds, and built-up beams.

4) Processes. Shielded metal-arc, gas metal-arc, gas tungsten-arc, electron beam and laser beam processes, and flame heating.

In these experiments thermocouples were used to measure temperatures, while electrical resistance strain gages were used to determine strain changes. Experimental results have been reported in various reports, theses, and papers. Experimental results have been compared with analytical predictions using various computer programs. Some results on high-strength steels are presented in this paper.

Integration Effort

An additional important effort at M.I.T. has been to integrate results of analytical and experimental studies performed at M.I.T. and other laboratories. The integration effort has been carried out primarily by Professor K. Masubuchi. For example, he wrote the Welding Research Council Bulletins No. 149 and No. 174 covering the state-of-the-art of residual stresses, distortion, and their effects on service behaviors of welded structures. A comprehensive monograph entitled "Analysis of Welded Structures - Design and Fabrication Considerations" was developed under a three-year contract for the Office of Naval Research (20). The monograph has been further modified and a book entitled "Analysis of Welded Structures - Residual Stresses and Distortion and Their Consequences" has been published recently (21). This book covers many subjects related to residual stresses, distortion, and their effects on service behaviors of welded structures including brittle fracture, fatigue, hydrogen embrittlement, stress corrosion cracking, and buckling strength of welded structures.

Examples Of Experimental And Analytical Results On Steel Weldments

In order to show examples of experimental data and analytical predictions, results obtained by Hwang are presented in this paper. Hwang(22) studied thermal stresses and residual stresses in weldments in low-carbon steel, high-strength steels, 308 stainless steel, and a titanium alloy (Ti-6Al-2Cb-1Ta-1Mo). Table I shows experimental conditions of steel weldments. The steels used include low-carbon steel and quenched and tempered steels (ASTM A517 and HY-80). Data on stainless steel and titanium weldments are not included here. Figure 4 shows bead-on-edge and butt welds used in the experiments. Details of the results are given in Hwang's thesis (22). This paper shows only a portion of the results.

Figure 5 shows changes of temperature and longitudinal strains at 1 inch from the welding edge on Specimen No. 2, gas tungsten-arc (GTA) welding of low-carbon steel. Experimental data agreed well with calculated values. The one-dimensional program was used for calculating longitudinal strains.

Figure 6 shows similar results on Specimen No. 4, GTA welding of ASTM A517 steel. Although temperature data agreed well with calculated values, strain data were quite different from calculated values. Similar results were obtained in a number of welds made in high-strength steels (19,21,23). In the many experiments conducted so far analytical predictions agreed reasonably well with experimental data obtained on most materials.

But in some welds in high-strength steels analytical predictions had rather poor agreement with experimental data. We suspect that the effect of metallurgical transformation could be a major cause of the discrepancy. An effort is being made to modify the analysis to include the effect of metallurgical transformation on thermal stresses during welding.

Table I. Conditions of Experiments by Hwang

Material	Type of Weld and (Specimen Size)	Process	Filler Metal	Voltage	Ampere	Speed
Low-Carbon Steel	Bead On Edge (48" x 6")	GMA	A 675	25	380	.322
Low-Carbon Steel	Bead On Edge (48" x 6")	GTA	None	12	295	.268
Q & T Steel (A517)	Bead On Edge (48" x 6")	GMA	A 675	25	420	.333
Q & T Steel (A517)	Bead On Edge (48" x 6")	GTA	None	12	295	.255
HY-80 Steel	Butt Weld	GMA	A 675	25	300	.483

Additional experiments were made on Bead-on-edge welds in 308 stainless steel and titanium alloy (Ti-6Al-2Cb-1Ta-1Mo).

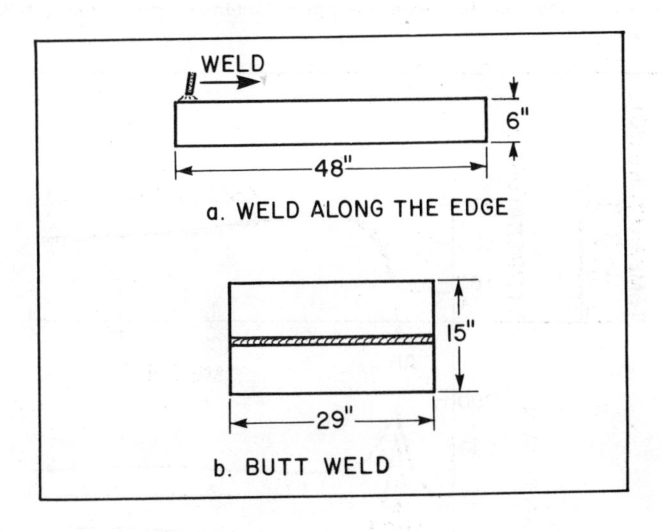

Fig. 4 - Test specimens used by Hwang.

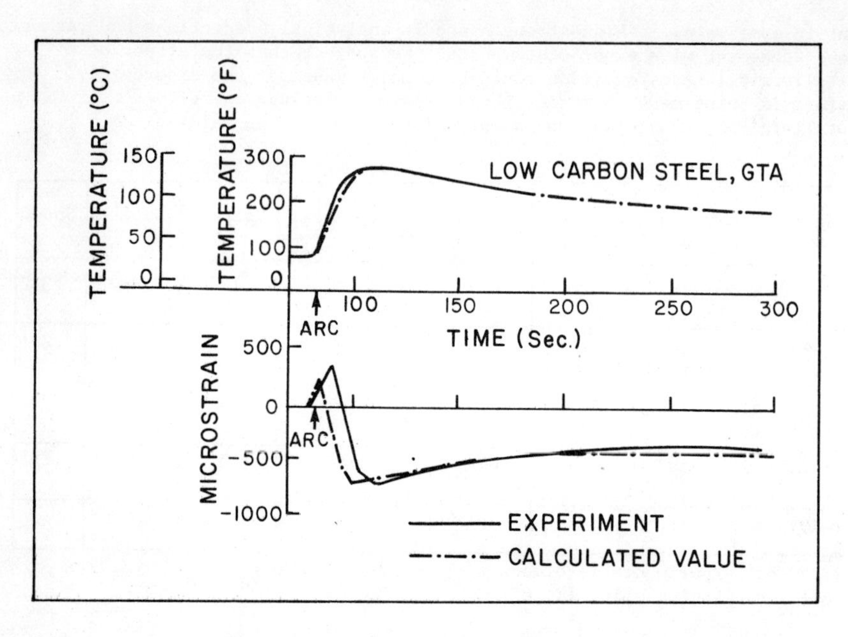

Fig. 5 - Longitudinal strain and temperature at 25.4 mm (1.0") from welding edge, specimen #2 (low-carbon steel, GTA).

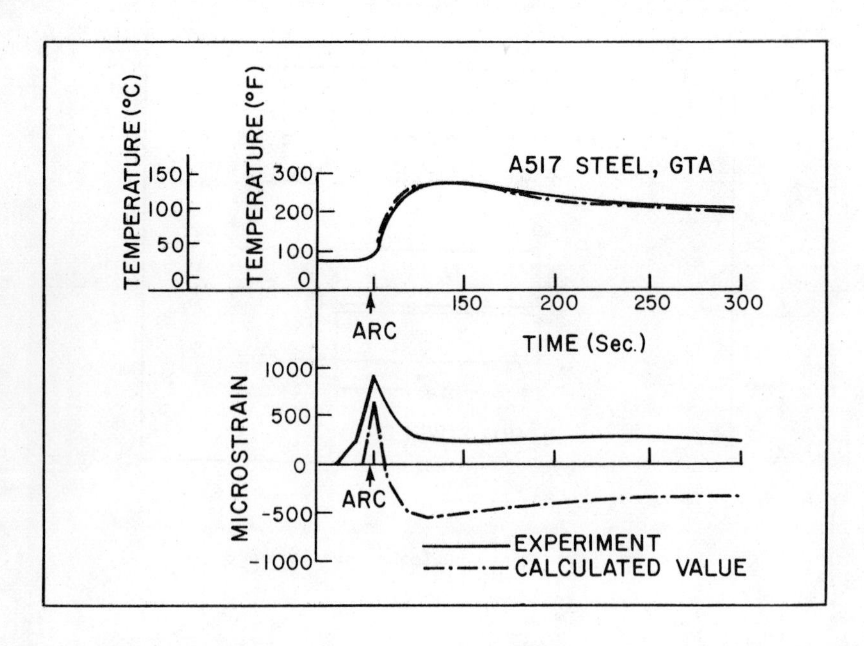

Fig. 6 - Longitudinal strain and temperature at 25.4 mm (1.0") from welding edge, specimen #4 (A517 GTA).

Figure 7 shows the relationship between the lateral distance from weld, y, and longitudinal residual stress, σ_x, on Specimen No. 2, GTA welding of low-carbon steel. Experimental data agreed well with analytical predictions.

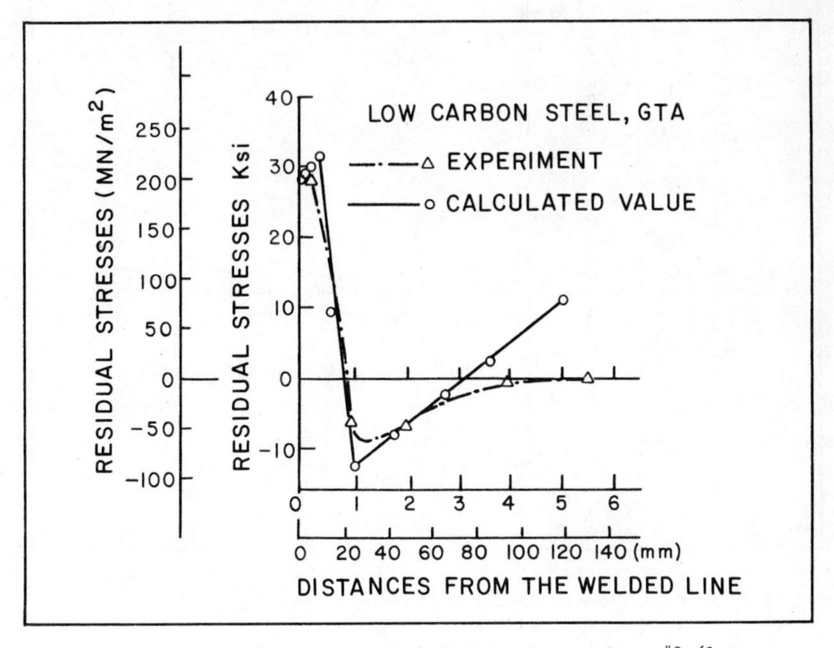

Fig. 7 - Residual stresses distribution in specimen #2 (low-carbon steel, GTA).

Figure 8 shows similar results on Specimen No. 4, GTA welding of ASTM A517 steel. Although the analysis predicted concentrated high tensile stresses in small areas near the weld, experimentally determined residual stresses near the weld were considerably lower than the analytical predictions.

It has been well established that for welding ordinary low-carbon steel with the yield strength of 35 - 40 ksi, the peak residual stress usually approaches the yield stress value. Curve 0 of Figure 9 shows schematically a typical distribution of longitudinal residual stress, or stresses acting in the direction parallel to the weld line, in a butt weld in low-carbon steel. Residual stresses are tensile in regions near the weld, while they are compressive in regions away from the weld. However, based upon experimental data and analytical results obtained by various investigators there are indications that this is not the case for high-strength steels. This subject is discussed in detail in a recent book by Masubuchi (21). Curves 1, 2, and 3 show schematically three possible distributions of longitudinal residual stresses in a butt weld in high-strength steels such as HY-130 and HY-180 steels (22-24).

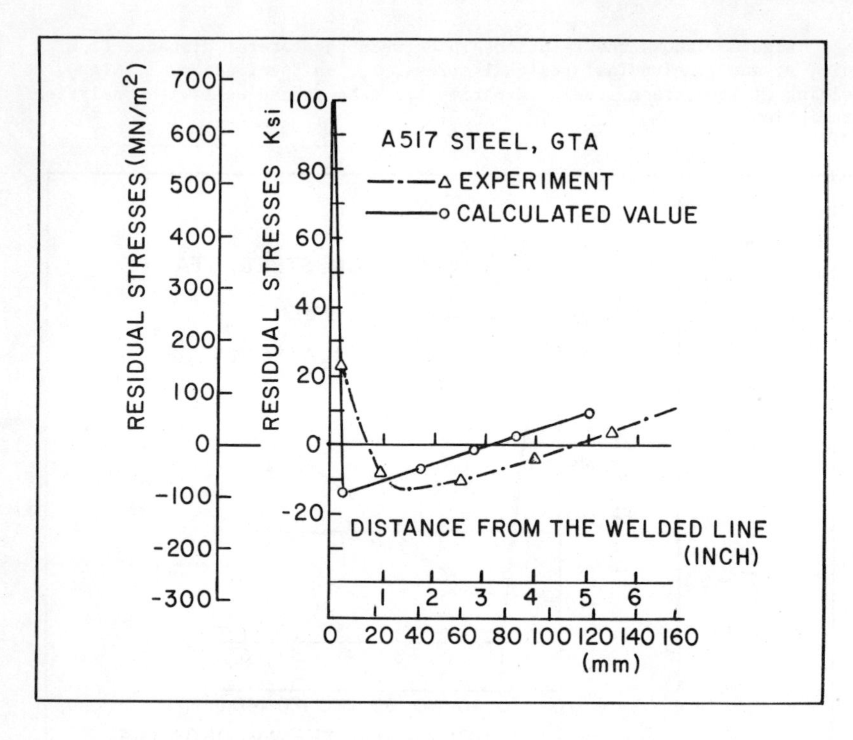

Fig. 8 - Residual stresses distribution in specimen #4 (A517 steel, GTA).

If it is assumed that the maximum residual stress is as high as the yield stress, the distribution would be given by Curve 1. In such a case, the residual stress and distortion would cause severe problems in the fabrication of welded structures using high-strength steels. In Curve 2, the high tensile residual stresses are confined to small areas. In such a case, the distortion would be significantly less, but cracking due to high tensile residual stresses would be a problem. The stress distribution given by Curve 3 is one that would cause few problems related to residual stresses.

Neither experimental nor analytical data support Curve 1. This possibility can be ruled out completely. Experimental data obtained with the use of strain gages tend to support curve 3. On the other hand, analytical data neglecting the effects of phase transformation tend to support Curve 2. Recent experimental data obtained at M.I.T. on heavy section HY-130 steel weldments have shown that the peak residual stress can be as high as the yield stress value.

On the basis of the experimental and analytical information obtained thus far, it is safe to assume that the residual stress disbritubtion in an actual structure is a cross between Cruves 2 and 3. If the x-ray diffraction techniques are used for measuring residual stresses in the weld metal and adjacent base-metal regions, it is likely that widely scattered results ranging from the yeild stress value to the much lower values will be obtained even in compression in some regions. However, the average value of the residual stresses in these regions would be similar to those shown by Curve 2 indicating that stresses are slightly higher than those in low-carbon steel weldments.

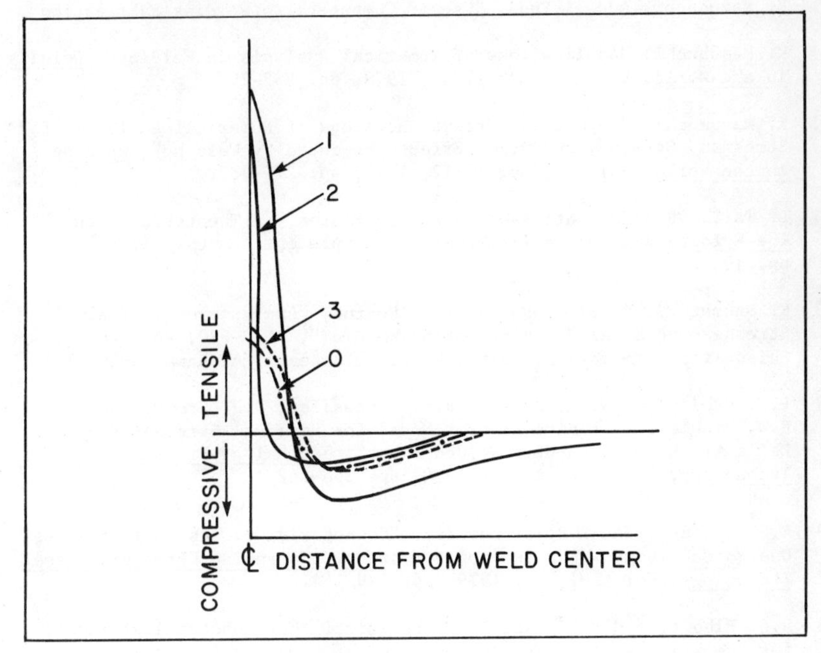

Fig. 9 - Possible distributions of longitudinal residual stresses in a butt weld in high strength steel.

Closing Remark

At the end of this paper I would like to point out the need for obtaining basic data on material properties at elevated temperatures close to the melting temperatures. Very little data are presently available on material properties above creep temperatures. This is a major problem in analyzing transient thermal stresses during welding. I strongly recommend that research be done to generate more data on material properties at elevated temperatures.

References

(1) K. Masubuchi, and T. Yada, "Use of Computers in Welding Fabrication".

(2) K. Masubuchi, "Applications of Numerical Analysis in Welding", Welding in the World, Vol. 17, No. 11/12, 1979, pp. 268-291.

(3) K. Masubuchi, "Report on Current Knowledge of Numerical Analysis of Stresses, Strains, and Other Effects Produced by Welding", Welding in the World, Vol. 13, No. 11/12, 1975, pp. 271-287.

(4) L. Tall, "Residual Stresses in Welded Plates - A Theoretical Study", The Welding Journal, 43(1), Research Supplement, January 1964, pp. 10s-23s.

(5) K. Masubuchi, B. Simmons, and R.E. Monroe, "Analysis of Thermal Stresses and Metal Movement During Welding", RSIC-820, Redstone Scientific Information Center, Redstone Arsenal, Alabama, July, 1968.

(6) E.F. Rybicki, D.W. Schmueser, R.B. Stonesifer, J.J. Groom, and H.W. Mishler, "A Finite Element Model for Residual Stresses and Deflections in Girth-Butt Welded Pipes", Journal of Pressure Vessel Technology, Vol. 100, August 1978, pp. 256-262.

(7) E.F. Rybicki, and R.B. Stonesifer, "Computation of Residual Stresses Due to Multipass Welds in Piping Systems," Journal of Pressure Vessel Technology, Vol. 101, May 1979, pp. 149-154.

(8) H.D. Hibbit, and P.V. Marcel, "A Numerical Thermo-Mechanical Model for the Welding and Subsequent Loading of a Fabricated Structure", Department of the Navy, NSRDC Contract No. N00014-67-A-019-0006, Technical Report No. 2, March 1972.

(9) Y. Ueda, and T. Yamakawa, "Analysis of Thermal Elastic-Plastic Stress and Strain During Welding", Document X-616-71, Commission X of the International Institute of Welding, 1971.

(10) K. Satoh, Y. Ueda, and S. Matsui, "1972-73 Literature Survey on Welding Stresses and Strains in Japan", Document X-699-73, Commission X of the International Institute of Welding, 1973.

(11) T. Nomoto, "Finite Element Analysis of Thermal Stresses during Welding", PH.D. Thesis, University of Tokyo, 1971.

(12) Y. Fujita, K. Terai, S. Matsui, H. Matsumura, T. Nomoto, and M. Otsuka, "Studies on Prevention of End Cracking in One-Side Automatic Welding, Part 3", Journal of the Society of Naval Architects of Japan, Vol. 136, Dec. 1974, pp. 459-465.

(13) K. Masubuchi, "Activities of Working Group on Numerical Analysis of Stresses, Strains, and Other Effects Produced by Welding", Document X-786-75, Commission X of the International Institute of Welding, 1975.

(14) J.B. Andrews, M. Arita, and K. Masubuchi, "Analysis of Thermal Stresses and Metal Movement During Welding", NASA Contractor Report NASA CR-61351, prepared for the G.C. Marshall Space Flight Center, December 1970 (for sale by the National Technical Information Service, Springfield, Virginia 22151).

(15) M. Nishida, "Analytical Prediction of Distortion in Welded Structures", M.S. Thesis, M.I.T., March 1976.

(16) K. Masubuchi, and T. Iwaki, "Thermo-elasto-plastic Analysis of Orthotropic Plates by the Finite Element Method", Journal of the Society of Naval Architects of Japan, Vol. 130, 1971, pp. 195-204.

(17) T. Muraki, J.J. Bryan, and K. Masubuchi, "Analysis of Thermal Stresses and Metal Movement During Welding, Part I: Analytical Study and Part II: Comparison of Experimental Data and Analytical Results", Journal of Engineering Materials and Technology, ASME, January 1975, pp. 81-84 and 85-91.

(18) Y. Toshioka, "Effects of Material Properties on Residual Stresses and Deformation in Weldments", M.I.T., June 1974 (unpublished).

(19) V.J. Papazoglou, and K. Masubuchi, "Study of Residual Stresses and Distortion in Structural Weldments in High-Strength Steels", Technical Progress Report under Contract No. N00014-75-0469 (M.I.T. OSP #82558) to the Office of Naval Research, M.I.T., November 1979.

(20) V.J. Papazoglou, and K. Masubuchi, "Development of Analytical and Empirical Systems for Parametric Studies of Design and Fabrication of Welded Structures", Final Report under Contract No. N00014-75-C-0469 (M.I.T. OSP #82558) to the Office of Naval Research, M.I.T., November 1977.

(21) K. Masubuchi, Analysis of Welded Structures - Residual Stresses and Distortion and their Consequences, Pergamon Press, Oxford and New York, 1980.

(22) J.S. Hwang, "Residual Stresses in Weldments in High-Strength Steels", M.S. Thesis, M.I.T. January 1976.

(23) K.M. Klein, "Investigation of Welding Thermal Strains in Marine Steels", M.S. Thesis, M.I.T. May 1971.

(24) K. Masubuchi, "Thermal Stresses and Metal Movement During Welding Structural Materials, Especially High Strength Steels", International Conference on Residual Stresses in Welded Construction and Their Effects, London, November 15-17, 1977, The Welding Institute, London.

GEOMETRIC FEATURES OF A CHILL-CAST

SURFACE AND THEIR THERMOMECHANICAL ORIGIN

Peter J. Wray

United States Steel Corporation
Research Laboratory
Monroeville, Pennsylvania

(The full text of this paper appears in Metallurgical Transactions, 12B (1981). An extended summary is presented here.)

When a liquid is cast against a smooth mold wall, the normal expectation is the cast surface will merely assume the smoothness of the mold wall. In the case of rapidly cooled castings, however, the cast surface will often exhibit geometric features that are a consequence of the thermomechanical behavior of the solidifying shell and have little to do with the geometry of the mold wall. Such a surface is generally considered to depreciate the quality of a casting, as well as to reduce the heat transfer between the solidifying shell and the mold. The purpose of this study was to document the cast-surface features, establish their dependence on casting conditions, and determine their origin.

In the experimental part of the study, lead was chosen as a representative casting material—observation of the clean cast surface was straightforward, and the relatively low melting point permitted easy control of the casting process. The castings were made by pumping the liquid lead into a mold chamber and decanting the liquid to leave a solidified shell against the vertical mold wall. The casting speed, taken as the steady speed of the ascending meniscus, was varied in the range 0.025 to 17 cm/s. To obtain a variety of chilling conditions, in one type of mold chamber, plane-faced copper-mold plates of different thicknesses were used. The other mold chamber consisted simply of copper or pyrex cylindrical tubes.

Fig. 1 - External surface of lead cast against a copper plate at a casting speed of 0.67 cm/s. The width of the casting is 10 cm.

Three types of cast-surface features were recognized. The first type consisted of fine horizontal corrugations regularly spaced about one mm apart (Fig. 1). The average spacing decreased gradually with increasing casting speed until at the higher speeds some shell surfaces were smooth and highly reflective. There was no apparent influence of the chilling condition on the spacing. Further characteristics of the corrugations are they are present close to the leading edge of the shell, indicating they form at an early stage of solidification, and large dendritic surface grains may extend across many of them.

In an attempt to understand what mechanism may be responsible for the formation of the corrugations, the thermoelastic deformation of a solidifying shell in the vicinity of its leading edge was examined by using the ANSYS finite-element computer program. This is a plane-strain beam problem in which the parabolically shaped beam is initially resting on a foundation AB as shown in Figure 2.

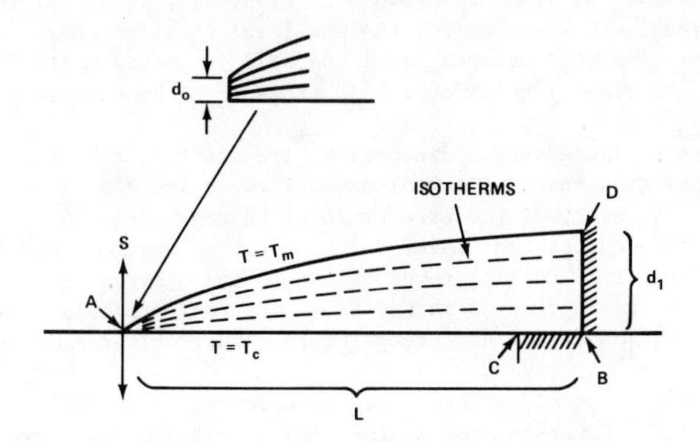

Fig. 2 - Geometry of simulated shell with the solid-liquid interface AD described by a parabolic relation and with a uniform temperature T_c at the shell/mold interface AB.

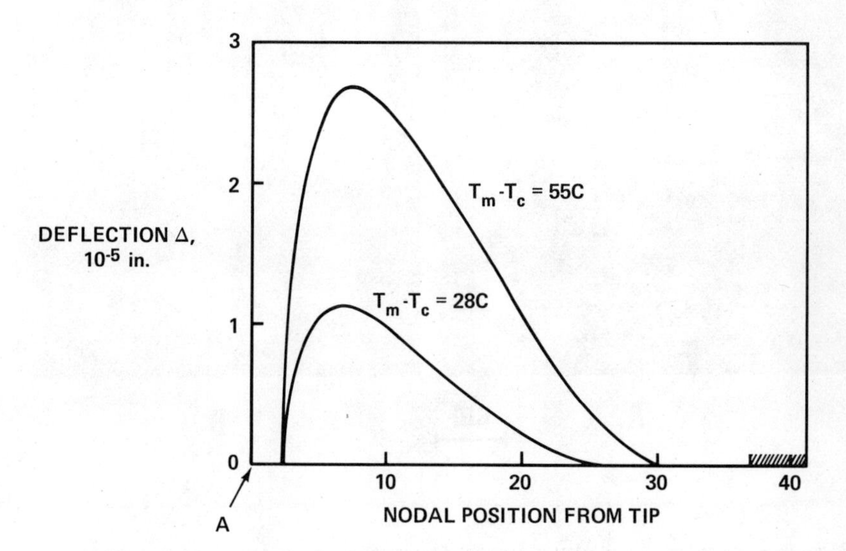

Fig. 3 - Computed separation of the shell from the mold wall, as a function of the distance from leading edge A. The two curves are for different values of temperature T_c.

With the imposition of thermal gradients, the shell distorts and produces a gap or deflection Δ between it and the mold wall. The variation of the gap distance along the shell/mold interface, AB, depends on the choice of the thermal conditions. For the condition of a uniform temperature at AB, the gap is largest in the vicinity of the leading edge A, as shown in Figure 3, although contact is always maintained at the edge itself. With increasing casting speed, the shell thickness decreases and the maximum gap distance between shell and mold increases until the shell becomes sufficiently thin for liquid pressure to be countereffective.

These predicted shell distortions are elastic, and removal of the temperature gradient during cooling will cause the gap at the mold wall to diminish. However, if the strains involved are sufficient for plastic yielding to occur at the elevated temperatures, undulations of the shell could remain after cooling to give rise to the observed corrugations. Theoretical examination of this possibility awaits consideration of the dynamic behavior of the high-temperature elastic-plastic deformations.

A second type of surface feature, observed at the higher speeds and illustrated in Figure 4, was in the form of wrinkles that were spaced at

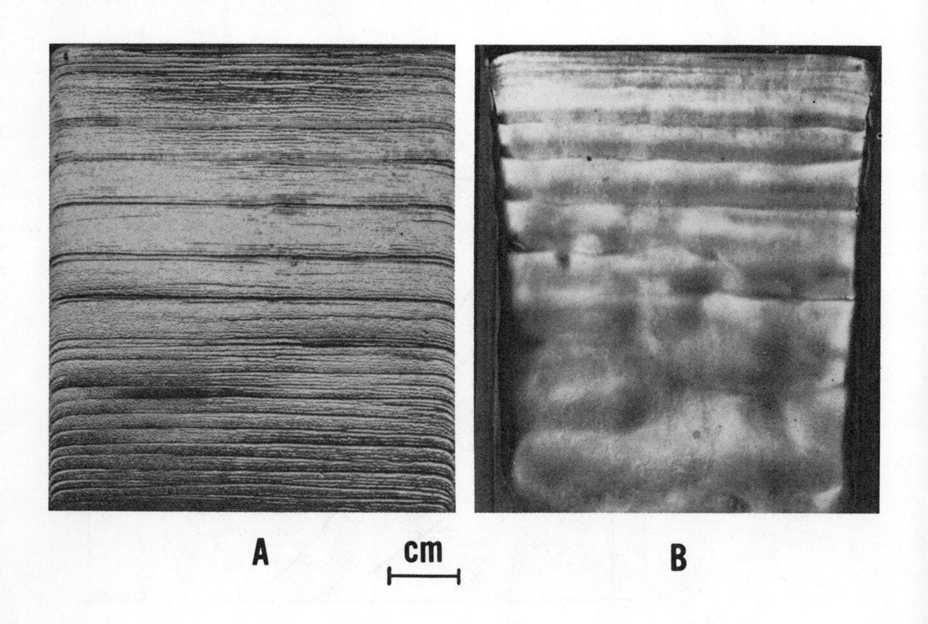

Fig. 4 - Decanted lead plate formed by casting at a speed of 1.6 cm/s. The exterior surface (A) shows the fine corrugations plus the wider-spaced wrinkles. The uneven growth of the interior surface and its correspondence to the wrinkles is illustrated in B.

multiples of the fine corrugations. Furthermore, with increasing casting speed, the spacing of the wrinkles increased. It was observed that these wrinkles are associated with the uneven thickening of the solidifying shell as shown in Figure 4B. Earlier unpublished work by O. Richmond and N. C. Huang provided some theoretical basis for the wrinkle formation when they found that perturbations of the temperature at the mold surface, combined with the temperature gradients associated with heat flow through the solidifying plate, could produce separation of the plate from the mold in critical areas. Prediction of the actual pattern of this lift-off is an unsolved instability problem that must accommodate the interference of the finer spaced corrugations.

The third class of surface features was observed only at slow casting speeds. The spacing of the laps shown in Figure 5 increased with decreasing speed. Examination of the meniscus behavior during the slow-speed casting revealed that periodically the meniscus itself would begin to freeze, causing the slowly rising liquid to form a liquid drop at the center of the freezing "cap." The liquid drop then expanded over the cap, made contact with the mold surface, and repeated the freezing cycle to form a series of laps. This type of cyclic solidification is favored by a low degree of superheat in the liquid and a high rate of heat extraction.

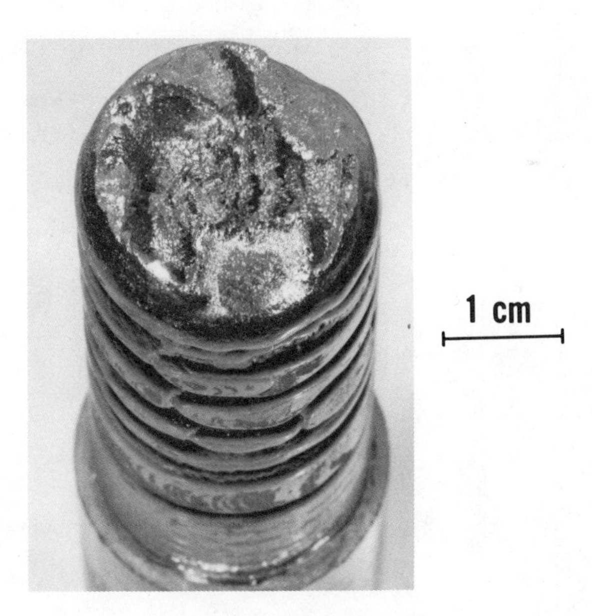

Fig. 5. - Upper view of lead cast in pyrex tube at slow speed of 0.027 cm/s, showing the partially frozen meniscus associated with the formation of the surface laps.

MECHANICAL, PHYSICAL AND THERMAL DATA FOR MODELING

THE SOLIDIFICATION PROCESSING OF STEELS

Peter J. Wray

United States Steel Corporation
Research Laboratory
Monroeville, Pennsylvania

For the purpose of modeling the solidification behavior and calculating the stresses in steel castings, welds, etc., references to the best available measurements of the relevant mechanical, physical, and thermal properties of iron and plain carbon steels have been compiled. The amount of property data available is judged sufficient to enable substantial extension and refinement of the present thermomechanical models.

As is clearly demonstrated in these Proceedings, the physical basis and the mathematical procedures available for the modeling of solidification processes are being steadily improved. However, these improvements generally require an ever more discriminating choice of the material property values to be used in the models. To show that at least in the case of plain carbon steels the modeling is not yet limited by data availability, a compilation has been made of references to the best measurements of each property for iron, iron-carbon alloys, and plain carbon steels.

The properties are dealt with in alphabetical order. In most cases the important values of a property, such as the value at the melting point, have been quoted. For a more complete documentation of the measurements and a discussion of their accuracy, etc., the interested reader should refer to the original papers.

Density

For pure iron in the solid state, Lucas[1] gives the temperature dependence of the specific volume ν. Adopting the form of his equations but slightly modifying the coefficients, we have

$$\nu_{\alpha} = 0.1270 + 5.08 \times 10^{-6} (T-20) \times 9.81 \times 10^{-10} (T-20)^2 \quad (1a)$$

$$\nu_{\gamma} = 0.1225 + 9.45 \times 10^{-6} (T-20) \quad (1b)$$

$$\nu_{\delta} = 0.1234 + 9.38 \times 10^{-6} (T-20) \quad (1c)$$

for the α-ferrite, austenite and δ-ferrite phases, respectively. The units of specific volume are cm^3/g and of temperature are °C. The relative volume change accompanying the δ-ferrite to austenite phase change at 1400°C is a decrease of 0.60 percent and the volume change for the austenite to α-ferrite transformation at 910°C is an increase of 1.03 percent. These values are in close agreement with the values of 0.54 and 1.05 percent obtained by Richter.[2] For comparison, at the melting point of 1536°C the relative contraction for liquid to δ-ferrite is 3.17 percent.

For the volume increase of austenite with increasing carbon content, the results of Ridley and Stuart[3] suggest the relation

$$\nu = \nu_{\gamma} + 7.688 \times 10^{-3} (C) \quad (2a)$$

where ν_{γ} is the specific volume of pure iron given in Eq. 1b and C is the carbon content in weight percent. Similarily, for α-ferrite,

$$\nu = \nu_{\alpha} + 8.154 \times 10^{-3} (C) \quad (2b)$$

Using Eqns. 1 and 2 the density of the solid and the solidus of Fe-C alloys can be determined. The results are given in Table I, together with the solidus temperatures taken from Hansen.[4]

Table I. Density of Iron-Carbon Alloys at the Solidus and Liquidus

Carbon wt pct		Solidus Temp, °C	Solid Density, g/cm^3	Liquidus Temp, °C	Liquid Density g/cm^3	β_M pct
0	δ	1536	7.265	1536	7.035	3.17
0.05	δ	1523	7.268	1533	7.033	3.23
0.10	δ	1490	7.284	1530	7.031	3.47
0.16	δ	1490	7.312	1527	7.028	3.88
0.60	γ	1440	7.306	1490	7.007	4.01
1.00	γ	1355	7.305	1466	6.995	4.24
1.50	γ	1252	7.307	1428	6.993	4.30
2.06	γ	1153	7.314	1385	6.998	4.32
2.50	γ	1153	7.308	1342	7.010	4.08
3.00	γ	1153	7.302	1294	7.022	3.86
3.50	γ	1153	7.296	1242	7.039	3.52
4.00	γ	1153	7.289	1182	7.062	3.12
4.32	γ	1153	7.286	1153	7.069	2.84

The variation of the density of the liquid phase with carbon was measured by Lucas [5]. His evaluation of the liquidus temperatures and the density at those temperatures is included in Table I. The volume changes associated with the solidification of Fe-C alloys have been discussed by Wray [6]. Estimates of the relative contraction β_M accompanying freezing are given in Table I, where

$$\beta_M = \frac{V_L - V_S}{V_L} = 1 - \frac{\rho_L}{\rho_S} \qquad (3)$$

with ρ_L being the density of the liquid at the liquidus temperature and ρ_S the density of the solid at the solidus temperature.

Elastic Moduli

Measurements of the elastic properties of iron at elevated temperatures are presented in Table II.[7-11] A general examination of the literature revealed that at the melting point the elastic moduli are about one quarter of their low-temperature value, as indicated in Table II. Because of relaxation due to structural features such as grain boundaries, particle surfaces, etc., at elevated temperatures, the measured values of the moduli are a function of the rate of straining.[12] The modulus values given in Table II from Ref. 10 are unrelaxed values.

The only known high temperature measurements of Young's modulus made since 1962 are those of Arnoult and McLellan.[11] These authors were primarily interested in the effect of carbon, and found that at 1000°C the modulus decreased about 0.7 percent per atomic percent of carbon (1 w/o C = 4.6 a/o C, 2 w/o C = 8.8 a/o C).

Table II. Elastic Moduli of Iron at Elevated Temperatures

Refs	(7)	(8)		(9)		(10)	(11)
Temp, °C	E	E	E	G	ν**	E	E
RT	218*	212	203	78	0.275	171	-
100	211	206	200	77	0.275	168	-
200	204	200	194	75	0.275	166	-
300	194	192	188	73	0.280	163	-
400	183	184	180	68	0.285	158	-
500	170	176	171	66	0.300	154	-
600	156	167	157	60	0.310	144	-
700	144	152	134	51	0.320	131	-
800	-	125	108	39	0.375	107	-
900	-	106	85	29	0.450	93	-
1000	-	101	96	35	0.370	99	111
1050	-	-	-	-	-	-	101
1100	-	-	-	-	-	93	-
1200	-	-	-	-	-	87	-
1300	-	-	-	-	-	78	-
1400	-	-	-	-	-	49	-
1440	-	-	-	-	-	44	-

* E and G are in units of GPa where 1 GPa = 0.145×10^6 psi.

** Poisson's ratio.

Emissivity

The emissivity of steel was discussed recently by Pehlke and Turner.[13] Their review showed that the emissivity increases with surface roughness from about 0.5 for a polished steel surface to greater than 0.9 for a rough ingot surface. The emissivity also increases gradually with increasing temperature, but this effect is probably negligible given the uncertainty of surface roughness, oxide thickness, etc., encountered in practice. Values varying from 0.75 to 0.88 are generally used for engineering calculations.

Measurements of the emissivity of steels in inert atmospheres are useful in that they may reveal trends. For example, Shvarev, et al.[14] have shown that for Fe-C alloys containing up to 1.8 C a sharp drop in emissivity occurs at the freezing point, but beyond 3.0 C, the solidification is accompanied by a sharp increase in emissivity. With the conditions in this experiment, the emissivity of solid pure iron at the melting point was approximately 0.29.

Latent Heat of Transformation

The thermodynamic functions for solid iron were reviewed by Orr and Chipman.[15] From enthalpy data, they estimated the heats of transformation to be 215, 200, and 3300 cal per g-atom for the $\alpha \to \gamma$, $\gamma \to \delta$, and $\delta \to L$ transformations respectively.

Phase Diagram of the Fe-C System

The phase diagram according to Hansen [4] and modified by Lucas'[5] liquidus data should suffice for initial representation of plain carbon steel behavior. However, with the development of more detailed solidification models and their application to a wider range of steels, the determination of actual solidus and liquidus temperatures will be necessary.

Plastic Flow Behavior

The plastic deformation of the austenite and δ-ferrite phases of iron was measured by Wray and Holmes[16] and Wray,[17] respectively, over the range of strain rates commonly encountered in continuous casting, namely 5×10^{-6} to 2×10^{-2} sec^{-1}.

For the same intermediate range of strain rate, the deformation of plain carbon steels has recently been measured by Palmaers[18] and Wray.[19] In the absence of a valid theory of viscoplasticity, it is not possible at present to describe the measured deformation behavior simply, briefly, and at the same time accurately. However, some general features of the behavior can be illustrated and even described in an approximate fashion. For example, the effects of carbon content and temperature on the flow stress of 0.8 Mn, 0.25 Si plain carbon steels at a strain of 0.10 are shown in Figure 1. The observation that the logarithm of the flow stress σ varies linearly with temperature leads to the equation

$$\sigma = A \exp(-BT) \tag{4}$$

where A depends on carbon content and strain and B depends primarily on strain rate. Using this relationship for extrapolation to higher temperatures, the flow stress at the solidus can be estimated as shown in Figure 2 for strain rates of 2.5×10^{-5} and 2.3×10^{-2} sec^{-1}.

The variation of the flow stress with strain rate at a strain of 0.05 is illustrated in Figure 3. Using a hyperbolic sine relation for the strain-rate dependence and a temperature dependence different from Equation (4), data such as provided in Figure 3 can be described by

$$\dot{\varepsilon} = K_{\varepsilon} [\sinh(\alpha\sigma)]^{n} \exp(-Q_F/RT) \tag{5}$$

where the coefficients K_{ε}, α, n and Q_F vary as shown in Table III and R is the gas constant.

In hypoeutectoid steels containing less than 0.8 percent carbon, a two-phase $\alpha + \gamma$ structure may exist at temperatures above 720°C. The plastic deformation behavior of such mixtures has recently been examined by Wray[20] as a function of carbon content.

Measurements of the plastic deformation of steels at strain rates greater than 2×10^{-2} sec^{-1} and temperatures greater than 1000° C are scarce.

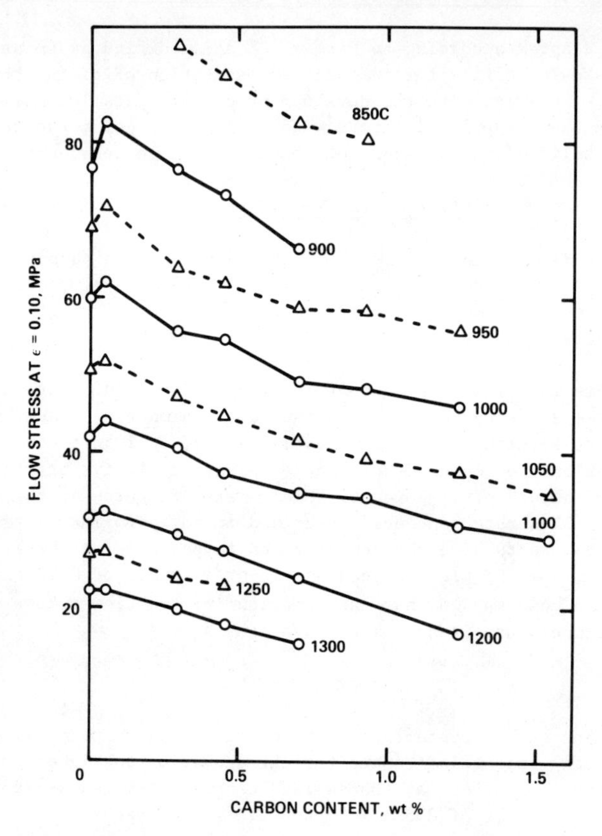

Fig. 1 - Illustration of the decrease in flow stress at $\varepsilon = 0.10$ with increasing carbon content at various temperatures.

The available data due to Nadai and Manjoine,[21] Cook,[22] and Grant and co-workers[23-26] do not extend above 1200°C.

For strain rates below 10^{-5} sec^{-1} the extensive creep data obtained by Feltham[27] for a series of plain carbon steels may be useful. At strain rates below 10^{-8} sec^{-1} and temperatures above 1250°C, it has been established that the deformation is Newtonian viscous.[29] In these experiments, not only is the viscosity of the solid measured, but so is the surface energy, as noted below.

Specific Heat

Elevated temperature determinations of this property were reviewed by Lange.[30] From his collection of data, it is concluded that the specific heat of the austenite phase varies linearly with temperature from about

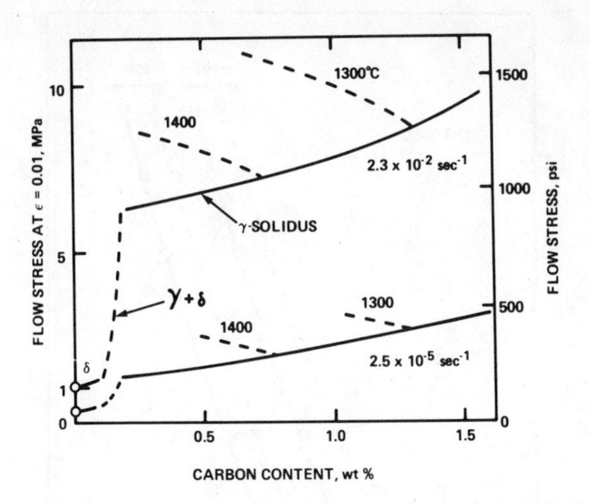

Fig. 2 - Variation of the flow stress at the austenite solidus, for two different strain rates.

0.155 $cal \cdot g^{-1} \cdot °C^{-1}$ at 910°C to 0.17 $cal \cdot g^{-1} \cdot °C^{-1}$ at 1390°C. For the δ-ferrite the specific heat varies from about 0.175 $cal \cdot g^{-1} \cdot °C^{-1}$ at 1390°C to 0.18 $cal \cdot g^{-1} \cdot °C^{-1}$ at the melting point. Estimates of the specific heat of the liquid phase vary, but a temperature independent value of 0.19 $cal \cdot g^{-1} \cdot °C^{-1}$ might be reasonable.

Surface Energy

The solid-vapor interface energy for pure iron was given as 2.5 ± 0.32 J/m^2 for the austenite phase and 1.95 ± 2.0 J/m^2 for the δ-ferrite phase.[28] In the case of a ferritic Fe-3Si alloy, the value was 1.65 ± 0.20 J/m^2.[29] The literature dealing with the measurement of the energy of the liquid-vapor surface of iron is copious. There seems to be reasonable agreement to a value of 1.78 ± 0.01 J/m^2 at the melting point. This value decreases with increasing carbon content,[31] while the temperature dependence is in dispute because of adsorption effects.[32]

Direct measurement of the energy of solid-liquid interfaces is very difficult, and has not been attempted for ferrous materials. From consideration of homogeneous nucleation of solid particles, Turnbull[33] indirectly estimated a value of 0.204 J/m^2 and Chadwick[34] quoted a value of 0.234 J/m^2. However, these values are probably too low. It is reasonable to expect the energy of the solid-liquid interface to be higher than that of the solid-solid interface, yet the grain-boundary energy for δ-ferrite is 0.47 J/m^2,[28] and for austenite it is evaluated at 0.77 and 0.87 J/m^2 according to two different procedures.[28,35]

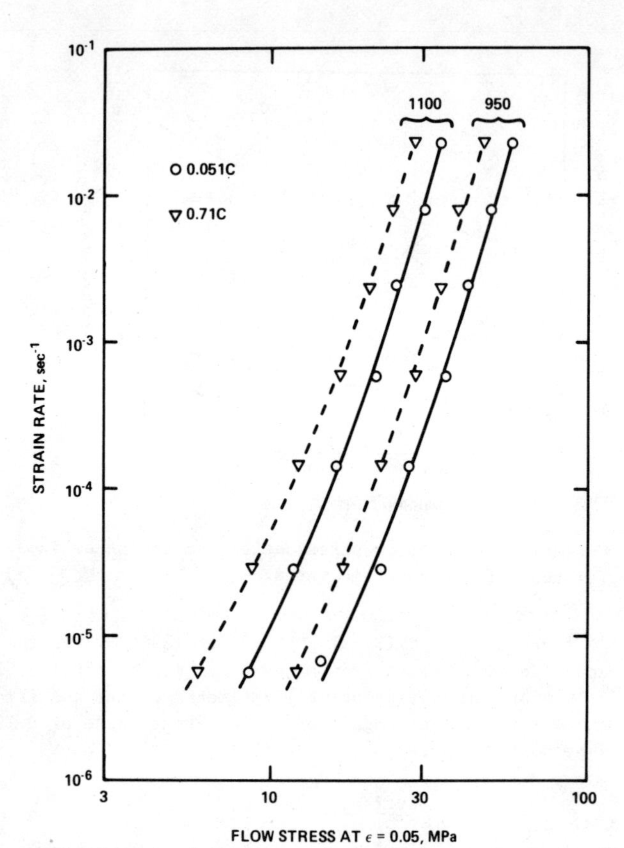

Fig. 3 - Strain-rate dependence of the flow stress at $\varepsilon = 0.05$ for two steels at 950 and 1100°C.

Table III - Evaluation of the Material Parameters of Equation 5, Using the Flow Stress Data of Wray(19)

Carbon Content wt pct	True Strain	K_ε 10^{10} sec^{-1}	α $(MPa)^{-1}$	n	Q_F kJ/mole	Number of Data Points
0.051	0.002	0.420	0.0741	5.98	329	20
0.051	0.05	50.8	0.022	5.64	342	18
0.29	0.002	45.4	0.135	4.11	420	19
0.29	0.05	45.2	0.0431	4.30	380	15
0.46	0.002	907.	0.0356	6.90	341	17
0.46	0.05	62.6	0.0262	5.30	346	17
0.93	0.002	590.	0.0900	5.41	414	22
	0.05	7.98	0.0290	4.72	323	24
1.25	0.002	392.	0.0849	5.64	401	18
	0.05	165.	0.0351	5.02	368	15
1.54	0.05	6.63	0.0639	4.51	371	21

Gas constant R = 8.131×10^{-3} kJ/mole K.

Thermal Conductivity

The thermal conductivity of iron at elevated temperatures was reviewed by Lange,[30] and documented by Touloukian et al.[36] In the same year. Above 700°C, the influence of alloying content is insignificant, and the thermal conductivities of iron and steels merge into a single narrow band of values as indicated in Table IV. At the liquid-to-solid transformation, the thermal conductivity of iron is estimated to decrease by about 15 percent.

Table IV - Recommended Values for the Thermal Conductivity of Iron and AISI 347 Stainless Steel.

Temp, °C	Thermal Conductivity Watt cm^{-1} K^{-1} Iron	AISI 347
27	0.803	0.148
127	0.694	0.165
227	0.613	0.179
327	0.547	0.192
427	0.487	0.205
527	0.433	0.219
627	0.380	0.232
727	0.326	0.246
827	0.297	0.259
927	0.282	0.273
1027	0.299	0.286
1127	0.309	0.300
1227	0.318	0.313
1327	0.327	0.327
1427	0.336	-
1527	0.345	-
1627	0.415	-
1727	0.426	-

1 Watt cm^{-1} K^{-1} = 57.78 Btu hr^{-1} ft^{-1} F^{-1}.

Thermal Expansion

A thorough examination of the thermal expansion of iron at elevated temperatures was reported by Richter.[2] His measurements of the linear coefficient of thermal expansion, presented in Table V, agree well with the results of earlier workers. For austenite the value of 22.0 x 10^{-6}/°C is independent of temperature and is about 40 percent greater than for ferrite.

Table V - Linear Thermal Expansion Coefficient for Iron

Temperature, °C	Mean Coefficient $\frac{1}{L_o}\frac{L(T) - L_o}{T - T_o}$	Differential Coefficient $\frac{1}{L(T)}\frac{L(T + \Delta T) - L(T)}{\Delta T}$
	10^{-6}/°C	
0	11.4	11.5
100	12.3	13.1
200	13.2	14.3
300	13.8	15.5
400	14.3	16.2
500	14.7	16.4
600	15.0	16.0
700	15.1	15.4
800	15.1	14.9
900	15.3	15.4
1000	-	22.0
1100	-	22.0
1200	-	22.0
1300	-	22.0
1400	-	16.5
1500	-	16.6

Viscosity

A concise and useful review of the viscosity of liquid iron and steels is included in the recent paper by Iida and Morita.[37)] At the melting point of iron the viscosity is 5.2 $\pm$ 0.6 mPa·s, decreasing to about 4.0 mPa·s at 1750°C. The reader may note that the viscosity of solid iron at 1515°C is 5.9 x 10^{18} mPa·s.[(28)] (Note that 1 mPa·s = 1 centipoise).

The viscosity of liquid iron is sensitive to impurity content. For example, the addition of 0.05 percent sulfur lowers the viscosity 20 percent.[(38)]

Conclusions

Knowledge of the properties of iron and plain carbon steels at elevated temperatures is sufficiently well advanced that the modeling of solidification processing should not be impeded. Furthermore, in many cases properties such as density, thermal conductivity and expansion etc., depend only slightly on composition. Therefore values obtained for iron and plain carbon steels should be satisfactory for modeling of a wide variety of steels.

The major deficiency is the lack of a general constitutive equation to adequately describe the viscoplastic behavior. For the interim, approximate descriptions can be developed according to the particular application. A deficiency of lesser current import is the unavailability of accurate liquidus and solidus measurements for a wide variety of steels.

References

1. L. D. Lucas, "Density of Metals at High Temperatures (In Solid and Liquid States), "Mem. Sci. Rev. Met., 69, (1972) pp. 479-92.

2. F. Richter, "Dilation of Armco Iron and Pure Iron Between -190 and 1450°C," Arch. Eisenh., 41 (1970) pp. 709-714.

3. N. Ridley and H. Stuart, "Partial Molar Volumes From High-Temperature Lattice Parameters of Fe-C Austenite," Met. Sci. J., 4 (1970) pp. 218-22.

4. M. Hansen, Constitution of Binary Alloys, 2nd edition, McGraw-Hill Book Company, New York, 1958.

5. L. D. Lucas, "Specific Volume of Liquid Metals and Alloys at High Temperatures, Part II," Mem. Sci. Rev. Met., 61 (1964) pp. 97-116.

6. P. J. Wray, "Predicted Volume Change Behavior Accompanying the Solidification of Binary Alloys," Metall, Trans., 7B (1976) pp. 639-646.

7. R. Kimura, "On the Elastic Moduli of Ferromagnetic Materials Part II, The Change in Young's Modulus Due to Magnetization and Temperature," Proc. Phys. Maths. Soc. Japan, 21, (1939) pp. 786.

8. W. Koster, "Elastic Modulus and Damping of Iron and Iron Alloys, Arch. Eisenh., 14 (1940) pp. 271-

9. R. I. Garber and A. I. Kovalev, "Investigation of the Temperature Dependence of the Modulus of Elasticity of Iron," Zavodsk. Lab., 24 (1958) pp. 539.

10. D. R. Hub, "Measurement of Velocity and Attenuation of Sound in Iron Up to the Melting Point." Paper No. 551 in Proc. IVth Intern. Cong. Acoustics, Copenhagen, 1962.

11. W. J. Arnoult and R. B. McLellan, "Variation of the Young's Modulus of Austenite With Carbon Concentration," Acta Met., 23 (1975) pp. 51-56.

12. C. Boulanger, C. Crussard, "Study of Mechanical Properties at Very High Temperatures", Rev. Metall., 53, (1956) pp. 715-728.

13. R. D. Pehlke and R. H. Turner, "Emissivity of Low Alloy Steel Powders During Sintering," Metals Eng. Quart., 15 (1975) pp. 37-42.

14. K. M. Shvarev, V. S. Gushchin, B. A. Baum, and P. V. Gel'd, "Integral Radiation Capacity of Fe-C Alloys at Elevated Temperatures," Russian Metallurgy, 4/1976, pp. 64-68.

15. R. L. Orr and J. Chipman, "Thermodynamic Functions of Iron," Trans. AIME, 239 (1967) pp. 630-33.

16. P. J. Wray and M. F. Holmes, "Plastic Deformation of Austenitic Iron at Intermediate Strain Rates," Metall. Trans., 6A (1975) pp. 1189-1196.

17. P. J. Wray, "Plastic Deformation of Delta-Ferritic Iron at Intermediate Strain Rates," Metall. Trans., 7A (1976) pp. 1621-27.

18. A. Palmaers, "High Temperature Mechanical Properties of Steel as a Means for Controlling Continuous Casting," C. R. M. Metallurgical Reports, No. 53, November 1978, pp. 23-31.

19. P. J. Wray, "Effect of Carbon Content on the Plastic Flow of Plain Carbon Steels at Elevated Temperatures." Submitted for publication.

20. P. J. Wray, "Plastic Flow and Fracture of Plain Carbon Steels in the Ferrite plus Austenite Region." Submitted for publication.

21. A. Nadai and M. J. Manjoine, "High-Speed Tension Tests at Elevated Temperatures, Parts II and III," J. App. Mechs., 63 (1941) pp. A77-91.

22. P. M. Cook, "True Stress-Strain Curves for Steel in Compression at High Temperatures and Strain Rates for Application to the Calculation of Load and Torque in Hot Rolling," pp. 86-97 in Proc. Conf. on Properties of Materials at High Rates of Strain, Inst. Mech. Eng., London, 1957.

23. R. Nordheim, T. B. King, and N. J. Grant, "The Effect of Phosphorus on the Deformation and Fracture Characteristics of Iron From 1600 to 2000°F," Trans. AIME, 218 (1960) pp. 1029-32.

24. S. Y. Ogawa, T. B. King, and N. J. Grant, "Deformation and Fracture Characteristics of Fe-S, Fe-S-O, and Fe-S-Mn Alloys at High Strain Rates and Temperatures," Trans. AIME, 224 (1962) pp. 12-18.

25. W. R. Lawson, R. N. Embriaco, T. B. King, and N. J. Grant, "The Effect of High Strain Rates on Cast Fe-C Alloys at Hot-Working Temperatures," pp. 235-255 in Continuous Casting, Interscience Publishers, New York, 1962.

26. B. E. Lindblom and N. J. Grant, "Hot Plasticity of an Nb-treated Carbon Steel at High Strain Rates," Jernk. Ann., 155 (1971) pp. 595-600.

27. P. Feltham, "The Plastic Flow of Iron and Plain Carbon Steels Above the A_3 Point," Proc. Roy. Soc., 66 (1953) pp. 865-83.

28. A. T. Price, H. A. Holl, and A. P. Greenough, "The Surface Energy and Self-Diffusion Coefficient of Solid Iron Above 1350 C," Acta Met., 12 (1964) pp. 49-58.

29. H. Jones and G. M. Leak, "The Effect of Surface Adsorption on Zero Creep Measurements in Iron-Silicon Alloys," Acta Met., 14 (1966) pp. 21-27.

30. K. W. Lange, "On the Thermal Conductivity of Iron," Arch. Eisenh., 41 (1970) pp. 559-562.

31. U. Mittag and K. W. Lange, "Experimental Results for the Interfacial Tension of Fe-C, Fe-Si, Fe-Cr and Fe-N Alloys," Arch. Eisenh., 47 (1976) pp. 65-69.

32. M. E. Fraser, W-K. Lu, A. E. Hamielec, and R. Murarka, "Surface Tension Measurements on Pure Liquid Iron and Nickel by an Oscillating Drop Technique," Metall. Trans., 2 (1971) pp. 817-823.

33. D. Turnbull, "Formation of Crystal Nuclei in Liquid Metals," J. Appl. Phys., 21 (1950) pp. 1022-28.

34. G. A. Chadwick, "Some Aspects of the Solidification of Ferrous Alloys," pp. 207-214 in Chemical Metallurgy of Iron and Steel, ISI Publication No. 147, 1971.

35. L. H. Van Vlack, "Intergranular Energy of Iron and Some Iron Alloys," Trans. AIME, 191 (1951) pp. 251-259.

36. Y. S. Touloukian, R. W. Powell, C. Y. Ho, P. G. Klemens, Thermophysical Properties of Matter. Volume I. Thermal Conductivity: Metallic Elements and Alloys, Plenum, New York, 1970.

37. T. Iida and Z-I. Morita, "Estimation of Some Physical Properties of Liquid Steel Near the Liquidus Temperature," pp. 104-110 in Proc. Third Intern. Iron Steel Cong., ASM, Metals Park, Ohio, 1979.

38. M. G. Frohberg and T. Cakici, "Measurement of the Viscosity of Liquid Iron With Special Consideration of the Temperature Control," Arch. Eisenh., 48 (1977) pp. 145-49.

Fluid Flow: Castings and Ingots

A REVIEW OF OUR PRESENT UNDERSTANDING OF

MACROSEGREGATION IN AXI-SYMMETRIC INGOTS

S. D. Ridder and R. Mehrabian
Metallurgy Division
National Bureau of Standards
Washington, D.C.

S. Kou
Department of Metallurgy and Materials Science
Carnegie-Mellon University
Pittsburgh, Pa.

Our present understanding of the mechanisms responsible for certain types of macrosegregation occurring in ESR, VAR and continuous cast ingots are reviewed. Experimental observations on both a high temperature alloy Ni-27 wt.% Mo and low temperature Sn-Pb alloys are compared to theoretical predictions. The mathematical models developed extend previous work by coupling the convective heat and fluid flow in the fully liquid metal pool above the liquidus isotherm to the interdendritic fluid flow responsible for macrosegregation.

Introduction

The macrosegregation patterns found in ingots produced by Electroslag Remelting (ESR), Vacuum Arc Remelting (VAR), or continuous casting, although improved over more conventional methods, continue to limit ingot size, composition and casting speed. It is, therefore, important to understand the mechanisms responsible for these segregation patterns and, furthermore, to develop predictive theoretical models that can be used for on-line control of process variables for the economic production of ingots with acceptable homogeneity.

In a number of papers (1-11), macrosegregation in castings has been correlated with the interdendritic fluid flow present during solidification. In these papers, a few of the possible driving forces which can lead to interdendritic fluid flow were considered. Specifically, solidification contraction and density driven convection were noted as the main causes of the interdendritic fluid flow. The research reviewed herein goes on to consider a third effect: convective flow in the fully liquid metal ahead of the liquidus isotherm and, at the same time, incorporates a permeability coefficient that varies with dendrite arm spacing (12).

Theoretical calculations were verified on a small scale (200 mm diameter) ESR furnace and on a low-temperature (Sn-Pb alloys) casting apparatus capable of producing 90 mm diameter cylindrical ingots. Both systems work under axi-symmetric heat flow conditions and all the reported data are from the steady-state region of the ingots.

Theory

The basic "solute redistribution" equations derived in previous papers (1,2,5) have been used to quantitatively predict the extent of macrosegregation in an ingot for certain heat and fluid flow conditions during solidification. For a binary alloy the "solute redistribution" equation describing the effect of interdendritic fluid flow on volume fraction solid is (2,5,8):

$$\frac{1}{g_L}\frac{\partial g_L}{\partial t} = -\frac{\rho_L}{\rho_s(1-k)}\left(1 - \frac{\vec{n}\cdot\vec{v}}{\vec{n}\cdot\vec{u}}\right)\frac{1}{C_L}\frac{\partial C_L}{\partial t} \qquad (1)$$

where

$$-\frac{\vec{n}\cdot\vec{v}}{\vec{n}\cdot\vec{v}} = \frac{\vec{v}\cdot\Delta T}{\frac{\partial T}{\partial t}}$$

g_L = volume fraction liquid,
C_L = liquid composition,
k = equilibrium partition ratio,
$\vec{n}$ = unit vector normal to isotherms,
t = time,
T = temperature,
$\vec{u}$ = isotherm velocity,
$\vec{v}$ = interdendritic fluid flow velocity,
ρ_L = liquid density, and
ρ_s = solid density.

Derivation of Equation (1) was based on the assumptions that a) no significant undercooling occurs before solidification, b) equilibrium at the liquid-solid interface exists throughout solidification, c) diffusion in the liquid between each individual small interdendritic channels is complete, d) diffusion in the solid is negligible, e) solid density is constant, and f) no pores form.

The important parameter in Equation (1) is the component of interdendritic fluid flow velocity that is perpendicular to the isotherms, i.e., $\vec{n} \cdot \vec{v}$. This leads to the following macrosegregation criteria (8):

$$1 - \frac{\vec{n} \cdot \vec{v}}{\vec{n} \cdot \vec{u}} \gtreqless \frac{g_S \rho_S + g_E(\rho_{SE} - \rho_S)}{\rho_L g_L} \quad (2)$$

where g_S is volume fraction solid, g_E is volume fraction eutectic and ρ_{SE} is the density of the eutectic solid.

While casting cylindrical, axi-symmetric ingots, where heat is extracted in both the axial and radial directions, "mushy" zones with roughly parabolic liquidus and solidus surfaces are produced as shown schematically in Figure 1.

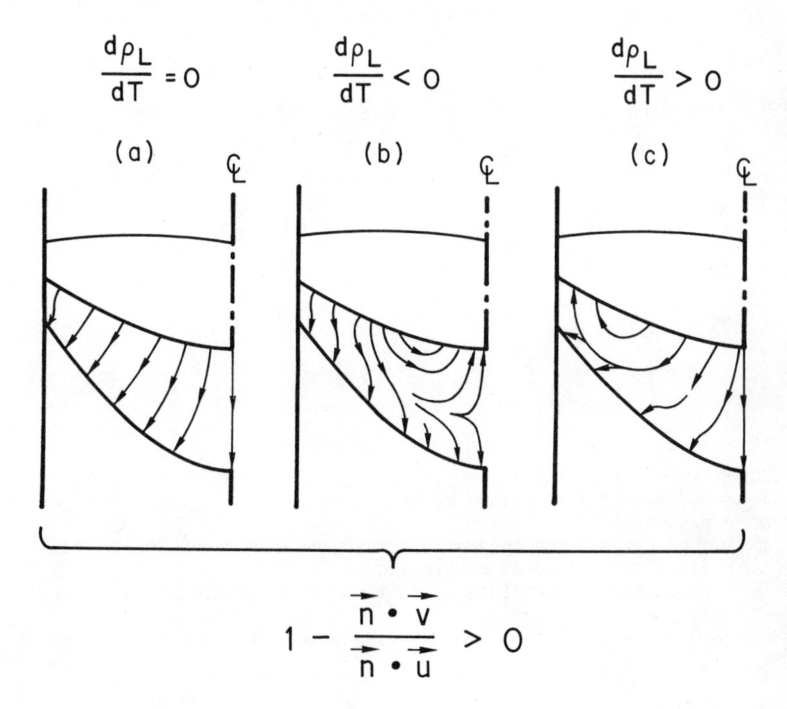

Fig. 1. Schematic illustration of possible interdendritic fluid flow in axi-symmetric ingots. (a) Flow resulting in negative segregation at ingot centerline, (b) flow resulting in positive segregation, (c) flow resulting in enhanced negative segregation. From reference (12).

If liquid density in the "mushy" zone is relatively constant from the solidus to liquidus isotherms (Fig. 1a), or if this density decreases during solidification (Fig. 1c), the flow of liquid metal outward from the center of the "mushy" zone will produce conditions such that the left side of Equation (2) is greater than the right side at the centerline, the opposite being true at the mold wall. In this case, negative segregation is expected at the centerline and positive segregation at the mold wall.

If liquid density in the "mushy" zone increases with decreasing temperature, a driving force exists to reduce the outward flow shown in Figure 1a and may cause it to turn inwards as in Figure 1b. Whether or not the flow turns inwards depends on the time available for flow, the resistance of the porous interdendritic medium, and the liquid density gradient in the "mushy" zone. Figure 1b shows a flow reversal near the top-center of the "mushy" zone. In this case, the left side of Equation (2) becomes smaller than the right side and positive segregation is expected at the ingot centerline.

The theoretical results to be discussed below required the calculation of the interdendritic fluid flow before Equation (1) could be used to compute the macrosegregation. This was done using the following equations derived in a previous paper (8).

$$\vec{v} = -\frac{K}{\mu g_L}(\nabla P - \rho_L \vec{g}) \tag{3}$$

and

$$\nabla^2 P + \left(\frac{2}{g_L}\nabla g_L + \frac{1}{\rho_L}\nabla\rho_L + a_1\frac{\nabla C_L}{C_L} + \frac{1}{\gamma}\nabla\gamma\right)\cdot\nabla P =$$

$$\left(\frac{2}{g_L}\nabla g_L + \frac{2}{\rho_L}\nabla\rho_L + a_1\frac{\nabla C_L}{C_L}\right)\cdot\ \rho_L\vec{g}$$

$$-\frac{\mu}{\gamma g_L}\left(\vec{u}\cdot\nabla C_L\right)\left(\frac{1}{\rho_L}\frac{d\rho_L}{dC_L} + \frac{a_1}{C_L}\right) \tag{4}$$

where

a_1 = $(\rho_s - \rho_L)/\rho_s(1 - k)$,
K = permeability of porous medium,
P = pressure,
$\vec{g}$ = acceleration due to gravity,
μ = visocity of interdendritic liquid, and
γ = permeability coefficient, defined by the following equation:

$$K = \gamma g_L^2 \tag{5}$$

The boundary conditions at the centerline and mold wall in a cylindrical and axi-symmetric ingot are such that v_r (radial component of velocity) is zero ($\partial\rho/\partial r = 0$) (8).

At the solidus isotherm liquid metal flows to feed shrinkage:

$$\vec{v} = -\frac{(\rho_{sE} - \rho_{LE})}{\rho_{LE}} \vec{u}_E \tag{6}$$

where

$\vec{u}_E$ = solidus (eutectic) isotherm velocity perpendicular to the isotherm, and
ρ_{LE} = eutectic liquid density

At the liquidus isotherm two different boundary conditions were tried. First, we assumed that there was no penetration of the convective flow from the fully liquid region above the liquidus into the "mushy" zone:

$$P|_{liquidus} = f(r) = P_a + \rho_{Lo}\, gH \tag{7}$$

where

P_a = pressure at the top of the liquid metal pool
H = height of the liquid metal pool, a function of r,
ρ_{Lo} = density of the liquid at the liquidus isotherm, and
g = $|\vec{g}|$

In the second case, penetration of convective flow from the fully liquid region was considered. The pressure along the liquidus was modified using the following equation:

$$P|_{(r,\, z_L)} = \int_o^r \frac{\partial P}{\partial r}\, dr + \int_{z_L}^o \frac{\partial P}{\partial z}\, dz + C \tag{8}$$

where

r = radial distance
z_L = locus of liquidus isotherm measured from the top of the liquid metal pool, and
C = a constant of integration, evaluated by assuming the pressure is atmospheric at the top of the liquid metal pool.

The pressure gradients in the fully liquid zone were calculated from the velocity distribution in this region.

To compute the convective flow velocities in the fully liquid region a method developed by Gosman et al. (13) was used, resulting in the following differential equations:

$$\frac{\partial}{\partial z}\left(\frac{\xi}{r}\frac{\partial \psi}{\partial r}\right) - \frac{\partial}{\partial r}\left(\frac{\xi}{r}\frac{\partial \psi}{\partial z}\right) + \frac{\partial}{\partial z}\left[\frac{\nu}{r}\frac{\partial}{\partial z}(r\xi)\right] + \frac{\partial}{\partial r}\left[\frac{\nu}{r}\frac{\partial}{\partial r}(r\xi)\right] - \zeta g \frac{\partial T}{\partial r} = 0 \tag{9}$$

$$\frac{\partial}{\partial z}\left(\frac{1}{r}\frac{\partial \psi}{\partial z}\right) + \frac{\partial}{\partial r}\left(\frac{1}{r}\frac{\partial \psi}{\partial r}\right) - \xi = 0 \tag{10}$$

$$\frac{\partial}{\partial z}\left(T\frac{\partial \Psi}{\partial r}\right) - \frac{\partial}{\partial r}\left(T\frac{\partial \psi}{\partial z}\right) + \frac{\partial}{\partial z}\left(\alpha_L\, r\, \frac{\partial T}{\partial z}\right) + \frac{\partial}{\partial r}\left(\alpha_L\, r\, \frac{\partial T}{\partial r}\right) = 0 \tag{11}$$

where

ζ = coefficient of volume expansion = $-\frac{1}{\rho_L}(\frac{\partial\rho_L}{\partial_T})p$, $\simeq 1 \times 10^{-4}\,°K^{-1}$,
ν = kinematic visocity = μ/ρ_L
ξ = vorticity = $\partial v_r'/\partial_z - \partial v_z'/\partial_r'$, and
ψ = stream function described by

$$v_r' = \frac{1}{r}\frac{\partial\psi}{\partial z} \tag{12}$$

$$v_z' = -\frac{1}{r}\frac{\partial\psi}{\partial r} \tag{13}$$

$\alpha_L = \frac{\kappa_L}{\rho_L C_p}$ = thermal diffusivity

κ_L = thermal conductivity of the liquid metal = 30.1 and 19.3 $Wm^{-1}°K^{-1}$ for Sn-rich and Pb-rich alloys, respectively,

C_p = heat capacity per unit mass at constant pressure = 0.24 and 0.19 kJ $kg^{-1}°K^{-1}$ for Sn-rich and Pb-rich alloys, respectively.

The symbol $\vec{v}'$ is used instead of $\vec{v}$ to indicate the difference between the liquid velocity in the "mushy" zone and the liquid velocity in the liquid metal pool. The velocity of the liquid in the "mushy" zone, $\vec{v}$, is referenced to the isotherms while the velocity of the liquid metal pool, $\vec{v}'$, is the total fluid velocity. The relationship between $\vec{v}$ and $\vec{v}'$ is given by:

$$v_r = v_r' \tag{14}$$

$$v_z = v_z' - R \tag{15}$$

where R = the steady state casting speed.

Equations (9-11) were solved simultaneously by the relaxation method of Gosman et al. (13) on a grid of points delineating the "mushy" and liquid regions of the casting. This grid pattern was adjusted to insure a smooth fit at the solidus and liquidus boundaries by allowing the radial and/or axial spacings to vary in dimension in a progressive manner. An example of one such grid pattern is shown in Figure 2. This variable spacing mesh provides a closer fit between the theoretical and the experimental "mushy" zone size and shape, and helps in achieving a more stable solution to the differental Equations (4) and (9-11).

Apparatus and Experimental Procedure

High Temperature Ni-Mo Studies

The alloys used in the research reviewed in this paper were chosen from both low and high temperature binary model systems. For the high temperature studies a 200 mm diameter ESR furnace (schematically illustrated in Figure 3) was used to cast ingots from a Ni-27 wt.% Mo alloy (8).

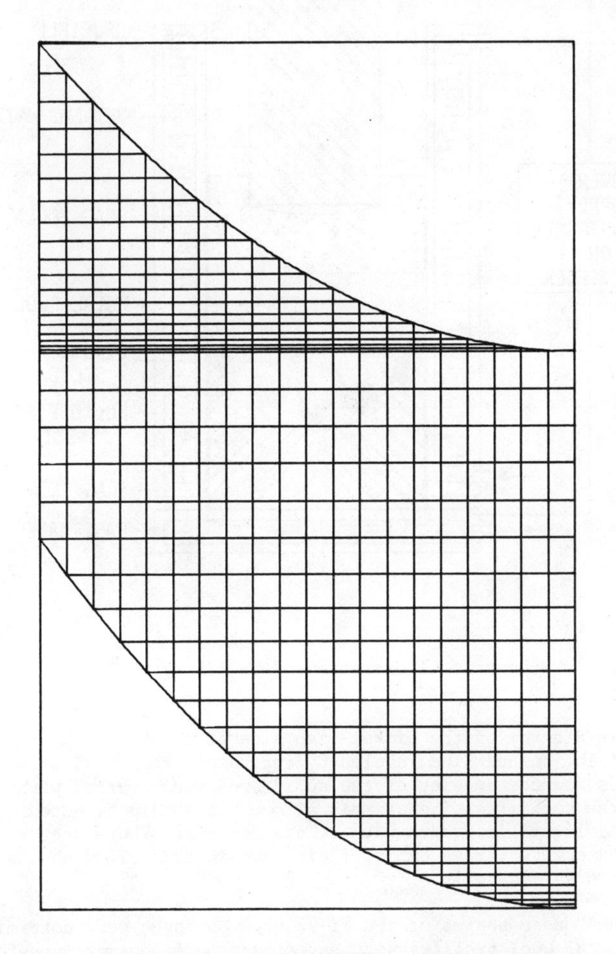

Fig. 2. Typical set of grid lines used to compute fluid velocities in the "mushy" region. From reference (12).

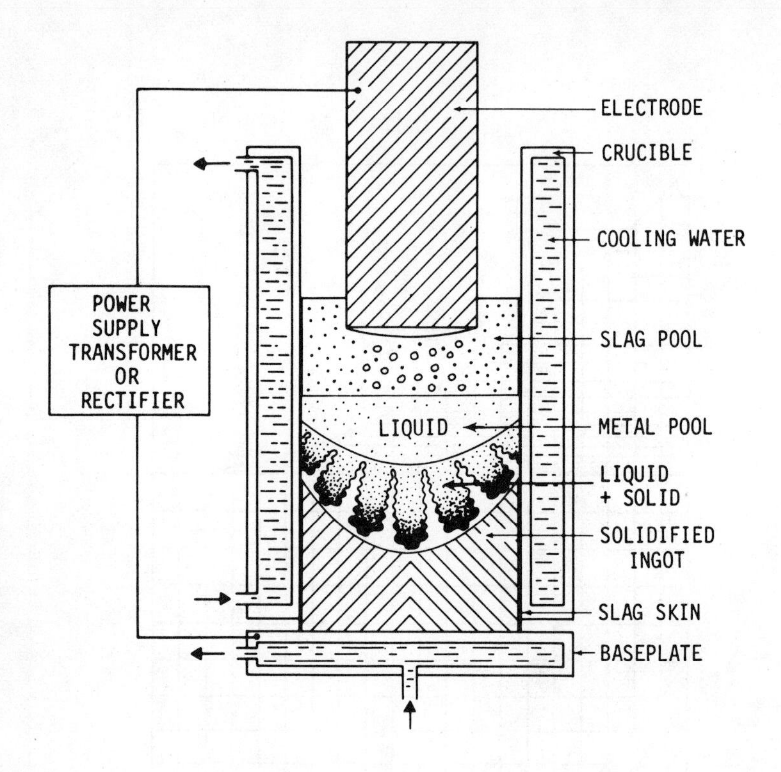

Fig. 3. Schematic illustration of an ESR system.

The phase diagram of the nickel-rich side of the Ni-Mo system is shown in Figure 4. The dashed line labeled C_o represents the Ni-27 wt.% Mo alloy. Figure 5 shows how the density of the solid and liquid varies with composition during the solidification of this alloy. It should be noted that the density of the liquid increases with solute content. Therefore, provided sufficient convective forces exist, fluid flow in this alloy should behave like that shown in Figure 1b.

The "mushy" zone shapes of the Ni-27 wt.% Mo ingot were determined by marking the metal pool profiles with successive additions of tungsten powder and calculation of the cooling rate versus radial position as determined from the measured secondary dendrite arm spacings (DAS) using the following equations:

$$DAS = a \left(\frac{\partial T}{\partial t}_{avg} \right)^{-n} \quad (16)$$

$$z_L - z_E = \Delta T \cdot R \cdot (DAS/a)^{1/n} \quad (17)$$

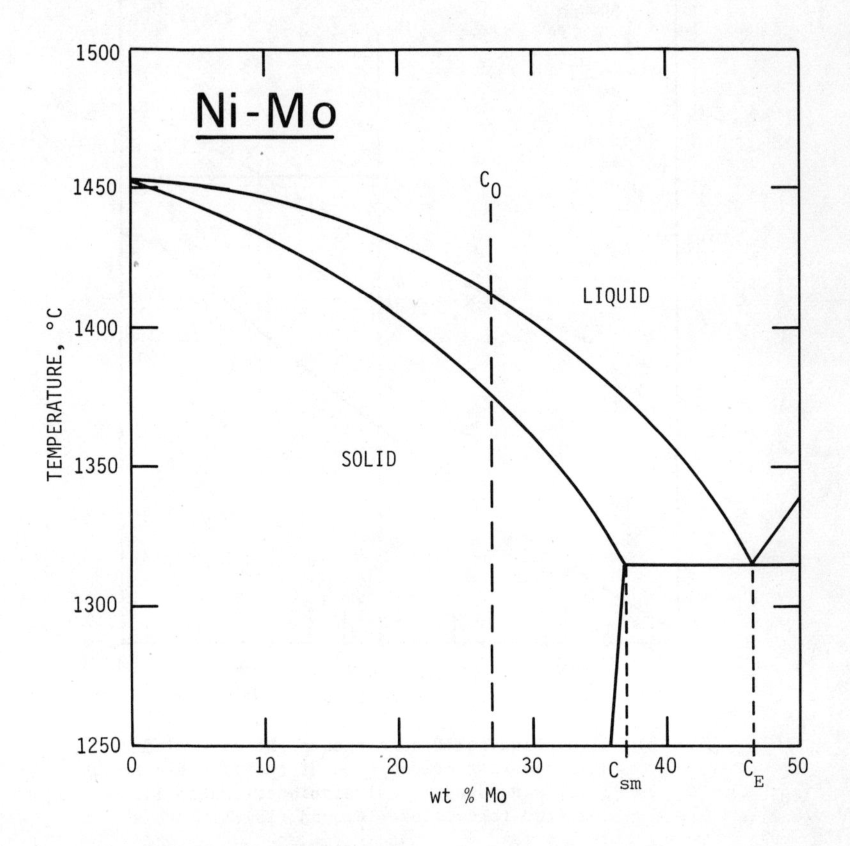

Fig. 4. Nickel-rich corner of the Ni-Mo phase diagram. From reference (13).

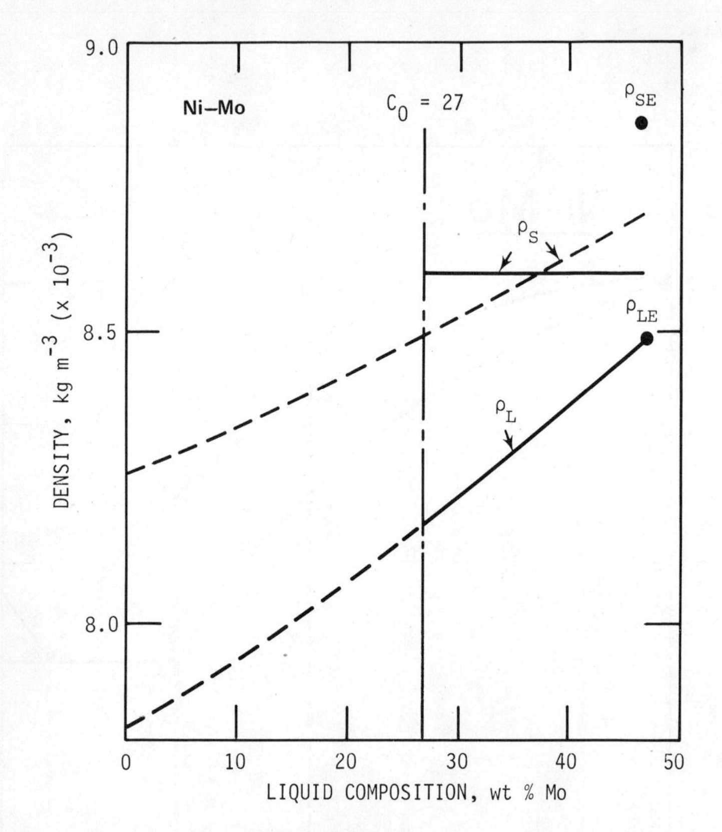

Fig. 5. Liquid and solid densities of Ni-rich Ni-Mo alloys versus composition of the liquid. Solid lines were used in calculations, dashed lines are from references (14) and (15).

where

z_L-z_E = height of the "mushy" zone at a given distance from the mold wall, r;
ΔT = liquidus minus the solidus temperature of the alloy;
a = 43.6
n = 0.35

The constants _a_ and _n_ were determined empirically from experiments on a unidirectional ingot (8).

Figure 6 shows the pool profiles in a cross section of the Ni-27 wt.% Mo ingot produced by adding tungsten powder to the slag pool in 10 minute internals while the ingot was being cast.

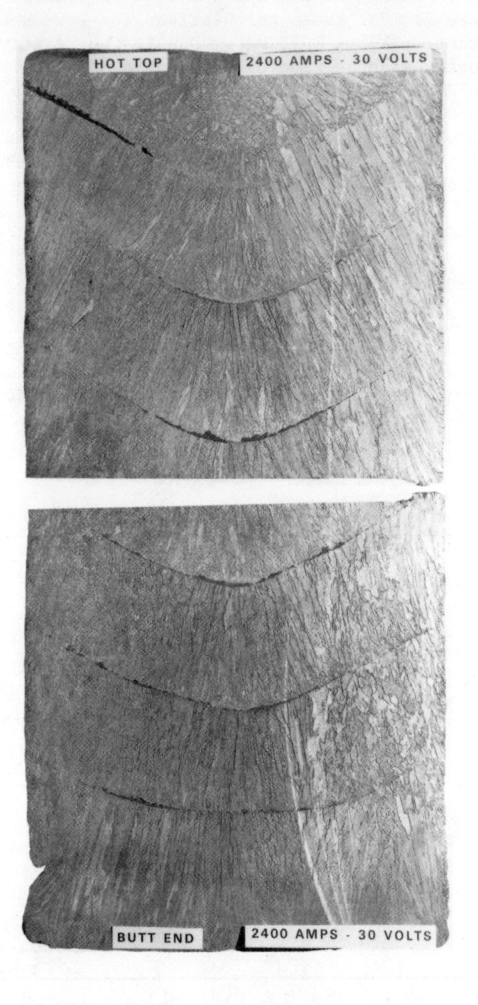

Fig. 6. Metal pool profiles in the Ni-27% Mo alloy ESR ingot marked by tungsten powders. Ingot and electrode diameters, ingot height, slag thickness, melt rate and average vertical pool velocity were 196mm, 152mm, 400mm, 57mm, 26.2g/sec and 0.09mm/sec, respectively. From reference (8).

The calculated "mushy" zone shape and flow velocities for this ingot are shown in Figure 7. For these calculations the permeability coefficient, γ, was held constant at 1.5×10^{-5} mm^2. Due to the relatively small density gradients and short residence time of the liquid metal in the "mushy" zone, convective forces did not have sufficient time or strength to produce a velocity turnaround at the centerline. The resulting outward flow of interdendritic liquid, although weak, is similar to that shown in Figure 1a and produces a slight negative segregation at the ingot centerline. The experimental and theoretical segregation curves supporting this is shown in Figure 8.

Low Temperature Studies With Sn-Pb Alloys

A schematic illustration of the low temperature laboratory apparatus is shown in Figure 9. The apparatus was designed to permit controlled solidification of Sn-Pb alloy ingots with predetermined temperature gradients (e.g., metal pool profiles, "mushy" zone sizes and shapes, and isotherm velocities, etc.).

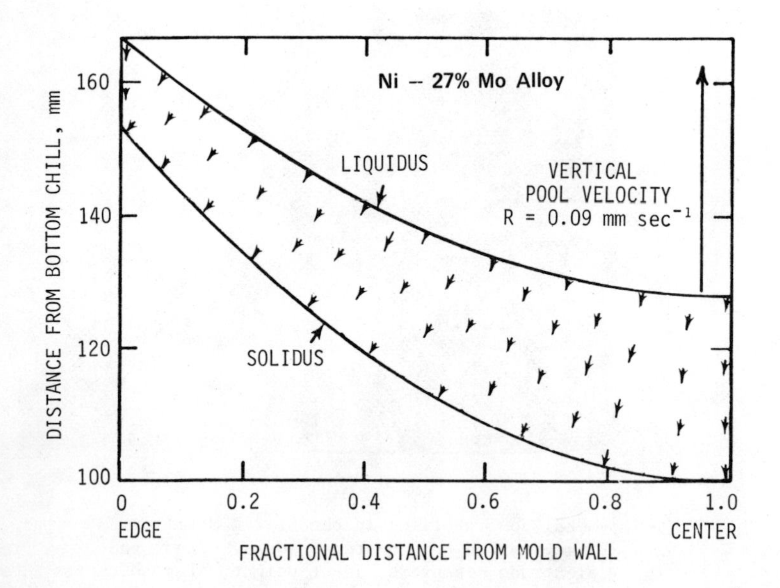

Fig. 7. Calculated flow velocities in the Ni-27% Mo ingot. The arrow's orientation and length designate the direction and magnitude of the interdendritic fluid flow, respectively. Note the direction of flow does not turn around and go from the cooler to the hotter regions. The slight fanning outward of the flow is expected to lead to positive segregation at the mold wall and negative segregation at the centerline. From reference (8).

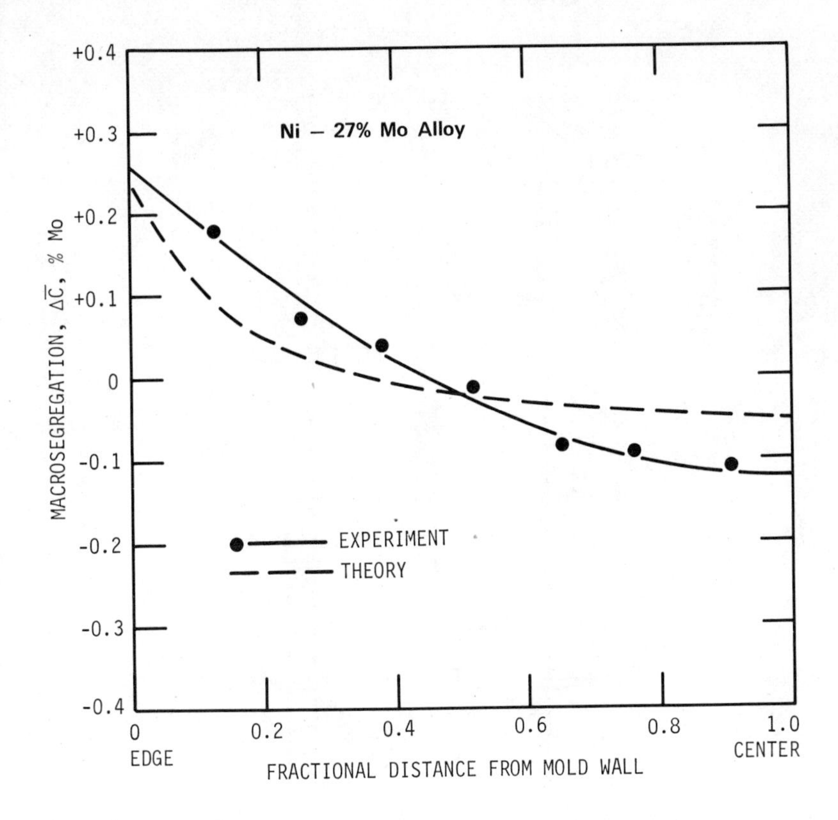

Fig. 8. Comparison of experimental and theoretical segregation profiles in the Ni-27% Mo ingot. Calculated flow velocity distribution for this ingot is shown in Fig. 7. From reference (8).

Variation of metal bath temperature, flow rate of the liquid metal from the coaxial feed system and the oil temperature were combined to control liquid metal pool depths, "mushy" zone size and shape and isotherm velocities during solidification.

The phase diagram of the Sn-Pb system is shown in Figure 10 with the alloys on interest marked by the two shaded areas. The solid and liquid densities of the two sides of the eutectic are plotted versus liquid composition and temperature in Figure 11. In the Sn-rich alloy, the density of the liquid increases with increasing solute content (as in the Ni-27 wt.% Mo alloy discussed earlier), whereas, in the Pb-rich alloy, the opposite is true. It was anticipated that the effect of gravity induced interdendritic fluid flow, hence macrosegregation, from this driving force would be opposite in the two alloys solidified under similar conditions.

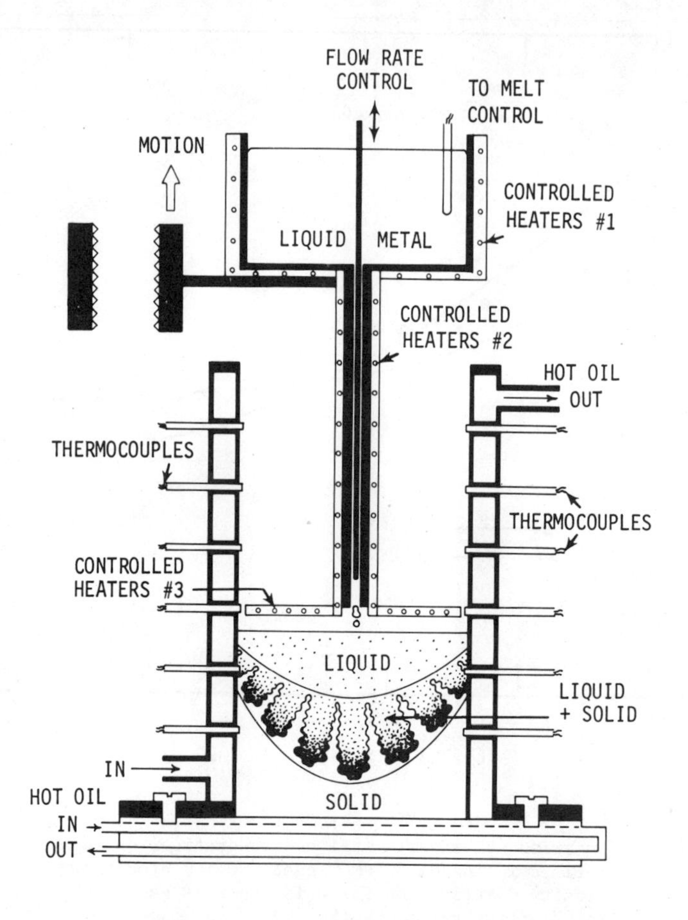

Fig. 9. Schematic illustration of the laboratory apparatus for simulation of macrosegregation.

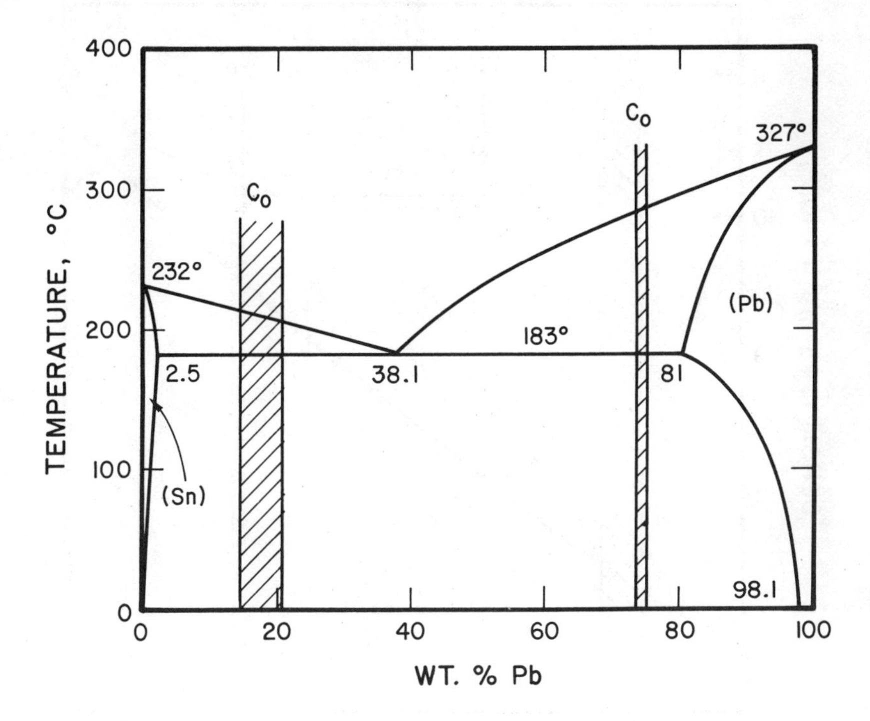

Fig. 10. Phase diagram of the Sn-Pb system. Shaded areas represent alloys studied in this work. From reference (13).

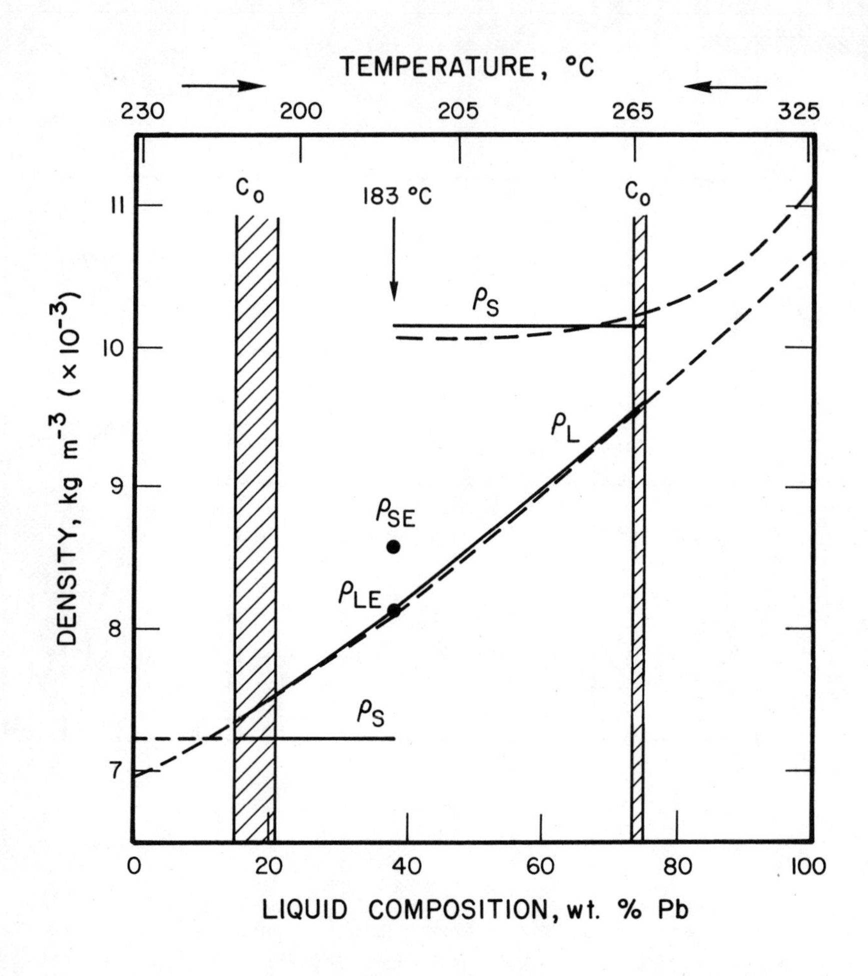

Fig. 11. Liquid and solid densities of Sn-Pb alloys on both sides of eutectic versus composition and temperature of the liquid. ρ_s, ρ_L, ρ_{sE} and ρ_{LE} designate the densities of the solid, liquid, eutectic solid and eutectic liquid, respectively. Solid lines were used in calculations, dashed lines are from references (14) and (15) and shaded areas are alloys studied in this work.

Utilizing the low temperature casting apparatus discussed above, a series of tin and lead-rich ingots were cast. Temperature profiles, derived from the data generated by thermocouples placed in each ingot while it was cast, were used to plot liquidus and solidus isotherm shapes, as well as the depth of the liquid metal pool above the liquidus isotherm. The ingots were subsequently sectioned and examined by x-ray fluorescence to determine the radial composition profile in the steady-state region of each casting.

Figure 12 shows the steady-state isotherms in the Sn-16% Pb ingot. The flow fields computed using Equations (3), (4), and (9-11) are shown in Figures 13 and 14.

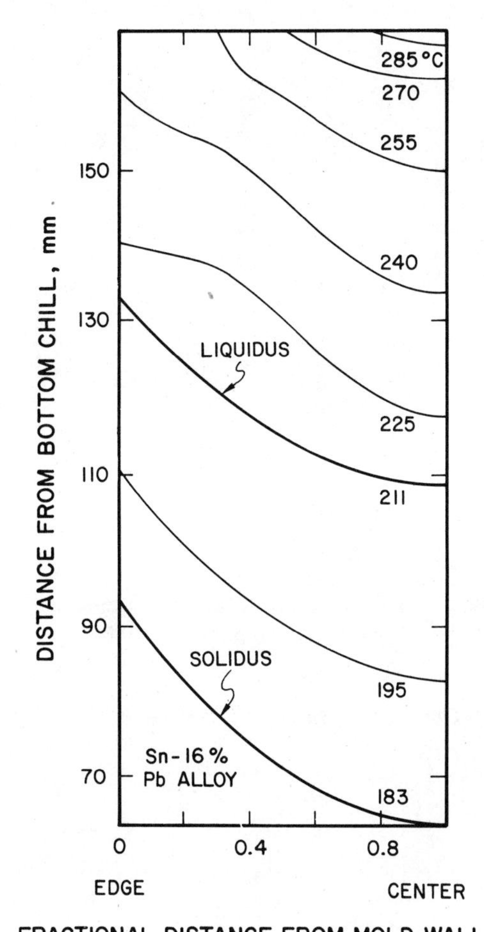

Fig. 12. Isotherms in the Sn-16% Pb ingot. From reference (12).

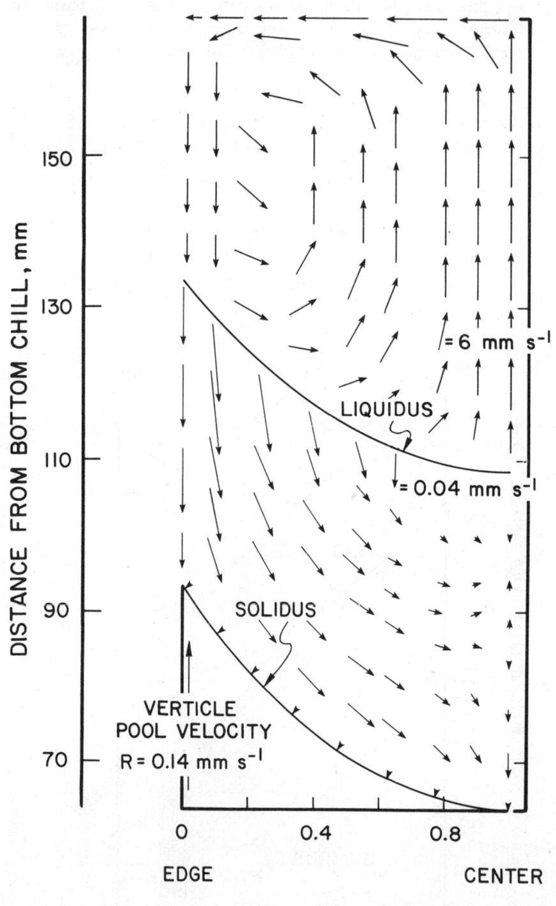

Fig. 13. Calculated flow velocities in the Sn-16% Pb ingot. The resulting segregation profile is shown in Fig. 15. From reference (12).

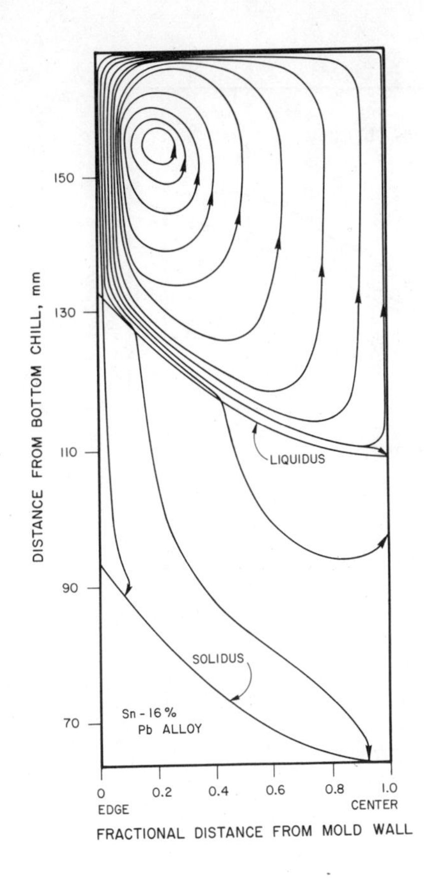

Fig. 14. Streamlines in the liquid metal pool and in the "mushy" zone of the Sn-16% Pb ingot. From reference (12).

In Figure 13 velocities are plotted as vectors in both the "mushy" region and in the liquid metal pool using a different scale for each. In the liquid metal pool velocities are on the order of 6 mm/sec while in the "mushy" zone this drops to less than 0.04 mm/sec. A streamline plot is shown in Figure 14 to help clarify the nature of the flow pattern.

The flow pattern in the interdendritic region of this ingot turns around near the centerline, corresponding to the condition shown in Figure 1b. This should result in positive segregation at the centerline and, consequently, negative segregation at the mold wall. The experimental findings are compared with the theoretical predictions in Figure 15. This figure shows the segregation profiles that are theoretically predicted for both the case of no fluid flow penetration into the "mushy" zone from the liquid metal pool, and the case in which flow due to convection in the bulk liquid above the liquidus isotherm penetrates the "mushy" zone. The permeability coefficient, γ_o, was given the value of 9×10^{-5} mm^2 at the centerline and allowed to vary according to the following equation:

$$\gamma_j = \gamma_o \, (DAS_j/DAS_o)^2 \tag{18}$$

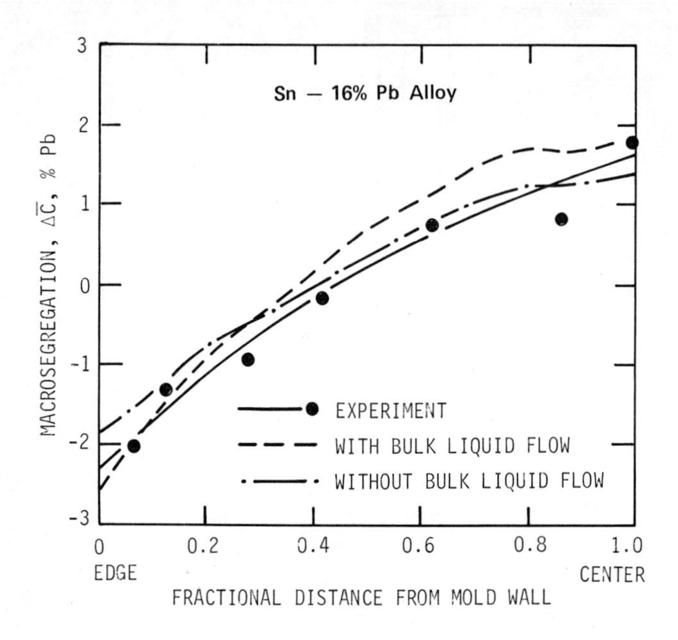

Fig. 15. Comparison of experimental and theoretical segregation profiles in the Sn-16% Pb ingot. Calculated flow velocities distributions for this ingot are shown in Figs. 13 and 14. From reference (12).

where

γ_j = the value of γ at some radial position j,
γ_o = the value of γ at the centerline,
DAS_j = the dendrite arm spacing at some radial position j, and
DAS_o = the dendrite arm spacing at the centerline.

Since the flow in the upper liquid metal pool is in the same direction as the flow in the "mushy" zone, a reinforcement of the interdendritic flow occurs. As noted in Figure 15, this reinforcement is small, resulting in a slight increase in the magnitude of segregation over that which is predicted when convective flow in the liquid metal pool is ignored.

The steady-state isotherms in the Pb-26.5% Sn ingot are shown in Figure 16 and the resulting flow velocities are shown in Figure 17 and 18, respectively. The data shown in Figures 16-18 are based on the solution of Equations (3-4) and (9-11). Experimental data was used for initial values and boundary conditions where applicable, and a value of 2.3×10^{-5} mm^2 was assigned to the permeability coefficient, γ, at the centerline.

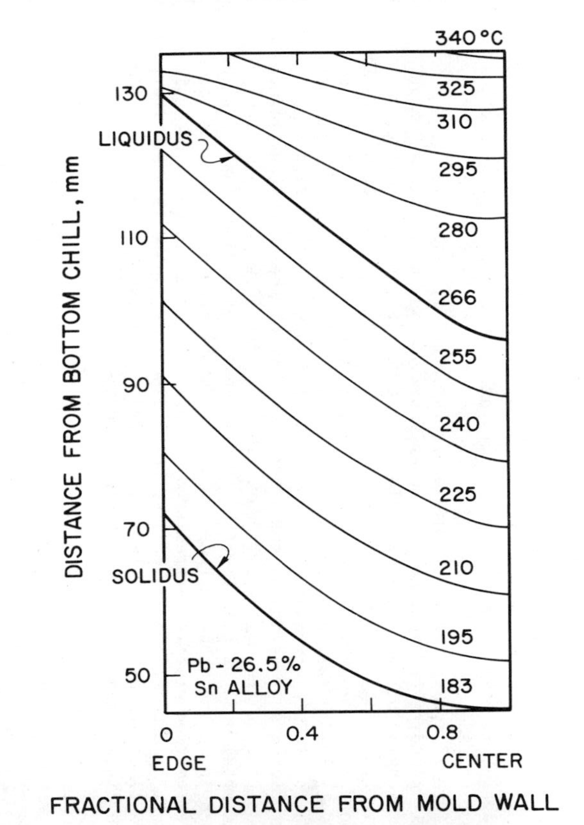

Fig. 16. Calculated isotherms in the Pb-26.5% Sn ingot. From reference (12).

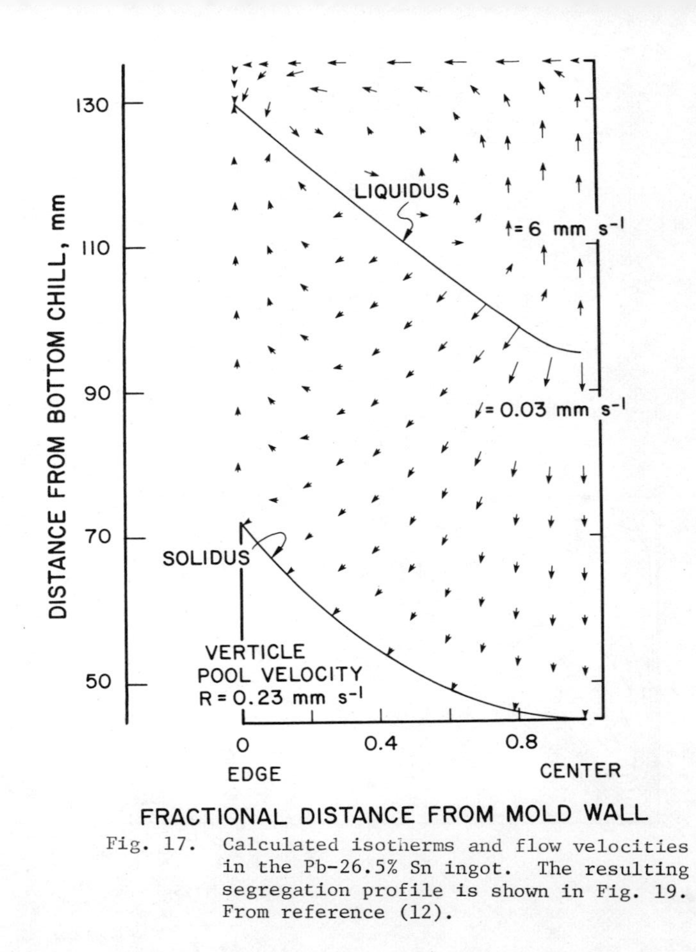

Fig. 17. Calculated isotherms and flow velocities in the Pb-26.5% Sn ingot. The resulting segregation profile is shown in Fig. 19. From reference (12).

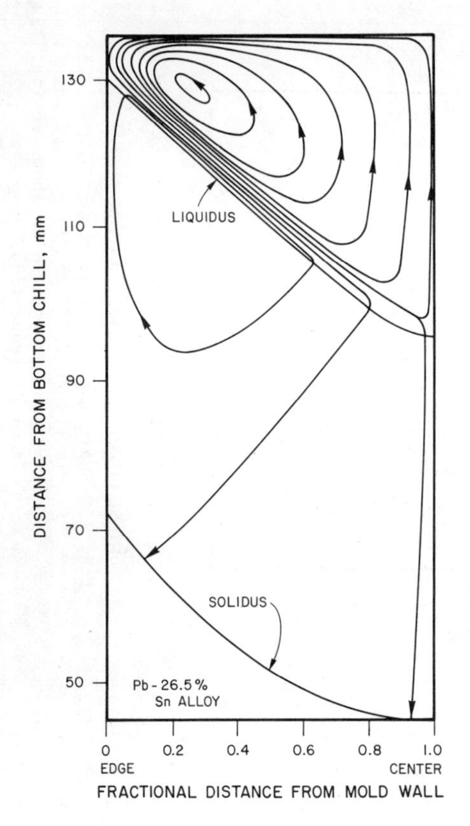

Fig. 18. Streamlines in the liquid metal pool and in the "mushy" zone of the Pb-26.5% Sn ingot. From reference (12).

Again, γ, was allowed to vary according to Equation (18). As in the case of the Sn-16 wt.% Pb ingot discussed above, the flow velocities shown in Figure 17 are scaled to two different reference velocities. As previously noted, the density of the interdendritic liquid decreases during solidification of the lead-rich alloy, whereas, the opposite is true for the tin-rich alloy discussed above. As seen in Figures 17 and 18, the tin enriched liquid in the cooler regions of the "mushy" zone rises along the mold wall due to the buoyancy forces which arise in the solidification of this alloy. This corresponds to the flow condition shown in Figure 1c. The predicted segregation is therefore, negative at the ingot centerline and positive at the mold wall. However, since the flow pattern in the liquid metal pool of this Pb-rich ingot is in the same direction as in the Sn-rich ingot and opposing the flow in the "mushy" region, a decrease in the segregation profile is predicted if this flow is included in the calculations. Again, this is a small effect, as seen in Figure 19 where the two dashed lines represent the theoretical predictions, one including the flow in the liquid metal pool and the other ignoring this flow.

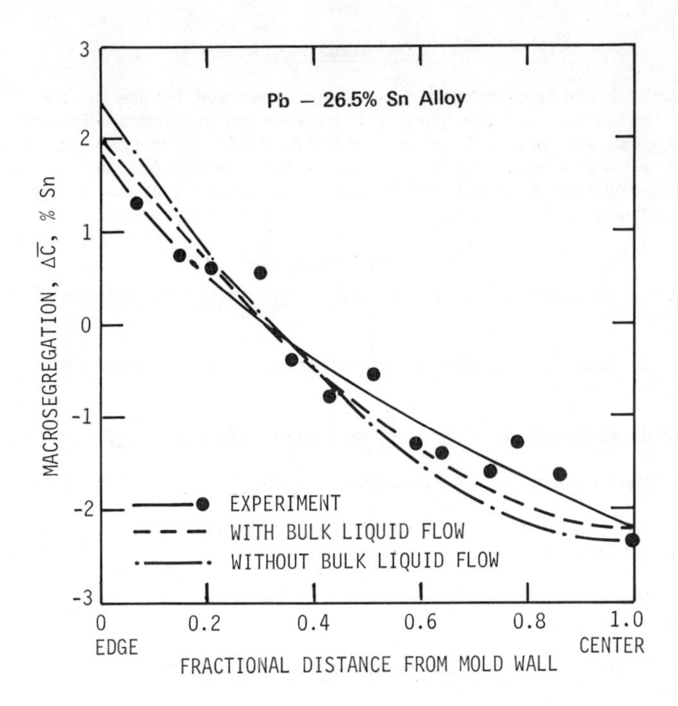

Fig. 19. Comparison of experimental and theoretical segregation profiles in the Pb-26.5% Sn ingot. Calculated flow velocity distributions for this ingot are shown in Figs. 17 and 18. From reference (12).

Conclusions

In this paper we have reviewed the extension of the previous macrosegregation theory for flow in the "mushy" zone from one-dimensional heat flow and two-dimensional fluid flow to two-dimensional heat and fluid flow in axi-symmetric ingots by utilizing variable mesh spacing in the finite difference equations. This incorporation of variable mesh spacing has allowed a smooth transition near the solidus and liquidus boundaries, resulting in more stable solutions of the macrosegreation equations. At the same time, a variable permeability coefficient, defined by Equation (18), results in a closer agreement between the theoretical and experimental segregation profiles.

Furthermore, a mathematical model has also been presented that couples the flow in the liquid metal pool above the liquidus isotherm to the interdendritic flow in the "mushy" zone during solidification. Results of this work to date, on the experimental apparatus described in this paper, indicate that the natural convection in the liquid metal pool has little effect on the interdendritic fluid flow and hence macrosegregation.

Acknowledgements

Most of the work reviewed here was sponsored by the National Science Foundation as part of the USA-USSR Cooperative program in Electrometallurgy. This program was under grant number DMR76-03682. Technical monitor of the project is Dr. Robert Reynik. This program was carried out at the University of Illinois at Urbana-Champaign, Illinois and was part of a joint Cabot Corporation-University of Illinois research program.

References

1. M. C. Flemings and G. E. Nereo, Trans. TMS-AIME, 239 (1967) pp. 1449-1461.

2. M. C. Flemings, R. Mehrabian and G. E. Nereo, Trans. TMS-AIME, 242 (1968) pp. 41-49.

3. M. C. Flemings and G. E. Nereo, Trans. TMS-AIME, 242 (1968) pp. 50-55.

4. R. Mehrabian and M. C. Flemings, Trans. TMS-AIME, 245 (1969) pp. 2347.

5. R. Mehrabian, M. A. Keane and M. C. Flemings, Met. Trans., 1 (1970) pp. 1209-1220.

6. R. Mehrabian, M. A. Keane and M. C. Flemings, Met. Trans., 1 (1970) pp. 3238-3241.

7. M. C. Flemings, Scandiavian J. of Metallurgy, 5 (1976) p. 1.

8. S. D. Ridder, F. C. Reyes, S. Chakravorty, R. Mehrabian, J. D. Nauman, J. H. Chen and H. J. Klein, Met. Trans. B, 9B (1978) pp. 415-425.

9. S. Kou, D. R. Poirier and M. C. Flemings, "Proceedings of the Electric Furnace Conference", Iron and Steel Society of AIME, 35 (1977) pp. 221-228.

10. S. Kou, D. R. Poirier and M. C. Flemings, Met. Trans. B, 9B (1978) pp. 711-719.

11. D. R. Poirier, M. C. Flemings, H. J. Klein, Advances in Metals Processing, Proceedings of Sagamore Army Materials Research Conference, 1978, J. J. Burke, R. Mehrabian, J. Weiss, eds., Plenum Publishing Corp., New York, 1981.

12. S. D. Ridder, S. Kou and R. Mehrabian, accepted for publication in Met. Trans. B.

13. A. D. Gosman, W. M. Pun, A. K. Runchal, D. B. Spalding and M. Wolfshtein, Heat and Mass Transfer in Recirculation Flows, pp. 18-112, Academic Press, London and New York, 1969.

14. D. P. Desai, T. D. Hawkins, M. Gleiser and K. K. Kelly, Selected Values of the Thermodynamic Properties of Binary Alloys, p. 1267, ASM, 1973.

15. H. R. Thresh, A. F. Crawley and D. W. G. White, Trans. TMS-AIME, 242 (1968) p. 819.

16. H. R. Thresh and A. F. Crawley, Met. Trans., 1 (1970) p. 1531-1535.

SOME EFFECTS OF FORCED CONVECTION ON MACROSEGREGATION

D. N. Petrakis, M. C. Flemings and D. R. Poirier

D. N. Petrakis, former graduate student at The University of Arizona, is Manufacturing Engineer at Lockheed Missile and Space Company in Sunnyvale, CA. M. C. Flemings is Ford Professor of Engineering, Department of Materials Science and Engineering, Massachusetts Institute of Technology, Cambridge, MA. D. R. Poirier is Associate Professor of Metallurgical Engineering, The University of Arizona, Tucson, Arizona.

Laboratory-scale apparatus was used to study effects of rotating the liquid pool on the macrosegregation in remelted ingots. The apparatus was used to make a series of Sn-Pb ingots (9∿12% Pb) with solidification rates which varied from 1.7×10^{-3} to 9.2×10^{-3} cm/sec and with rotations which varied from 10 to 60 rpm. Analyses, based upon computer calculations, were also included to determine the effect of the forced convection on the flow of interdendritic liquid in the mushy zone during solidification. Rotation of the liquid pool was produced by rotating a set of immersed heating elements. The rotation had the effect of changing the shape of the mushy zone at the center of the ingot where the liquidus isotherm took a convex shape, as opposed to the more usual concave shape without rotation. Depending on the rate of rotation in the liquid pool and the shape of the liquidus isotherm, rotation can be used to minimize macrosegregation. It was also observed that with rotational speeds greater than 40 rpm, the tangential nature of flow was sufficiently disturbed by thermocouple tubes to cause severe localized macrosegregation. The tubes, which extended from top to bottom of the ingots, were used for the purpose of obtaining temperature measurements, but it was found that they also produced a localized pressure drop located behind the tube positions, which altered the macrosegregation. Macrosegregation is calculated by a computer model based on the equations that describe the flow of interdendritic liquid in the mushy zone. The effect of convection in the liquid pool is simulated by changing the prescribed pressure at the liquidus isotherm which is used as a boundary condition for the flow field. Using the model, relationships among the rotation of the liquid pool, solidification rate, and shape of the mushy zone are demonstrated, as they all influence the formation of macrosegregation.

INTRODUCTION

In previous work by Kou et al.[1], experimental ingots (8 cm dia.) of Sn-15% Pb were solidified at approximately 5 x 10^{-3} cm/s in order to study macrosegregation comparable to that which can be observed in large commercial remelted ingots. In the experimental ingots, severe segregation was produced including a large variation of composition across the ingots, and, in extreme cases, "freckles". Both types of macrosegregation are caused by gravity induced flow of interdendritic liquid within the mushy zone during solidification. This flow was modelled and agreement between calculated segregation, based upon this flow, and the segregation observed in the experimental ingots was excellent.

In a second study[2], the use of centrifugal force as a method of controlling and reducing macrosegregation was presented. For example, Fig. 1 shows the effect of rotating the mold during solidification. In Fig. 1a, with no rotation, there is severe positive segregation, including freckles, centrally located. By employing modest rotation (ω = 8.7 rad/s, Fig. 1b) freckles are eliminated and macrosegregation is reduced. At a greater speed, Fig. 1c, macrosegregation is significantly reduced, but with excessive speed (ω = 15 rad/s, Fig. 1d), the centrifugal force predominates over the gravity force and causes outward flow resulting in freckles at the surface of the ingot.

In this study, we present effects produced by a different method of employing rotation in the system. Rather than rotating the mold during solidification, only the liquid pool above the mushy zone is rotated. One effect of the rotation is to alter the temperature field in the liquid pool such that the shape of the liquidus isotherm changes from concave to convex. This observation was also made by Edwards and Spittle[3] who used electromagnetic stirring to study the effects of externally applied forces to control heat and fluid flow in experimental ESR ingots of Al-3% Cu alloy. In stirred ESR ingots, they observed an increase in pool depth and a decrease in the temperature gradient ahead of the solid-liquid interface. Also, they observed macrostructures which indicated that a convex-shaped solid-liquid interface had been established during stirring.

The effects of rotation of the liquid pool are shown in Fig. 2 in which rotation was 10 rpm during solidification of the lower half of an experimental ingot of Sn-10 pct Pb alloy and 40 rpm for the upper half of the same ingot. Details which pertain to these experiments are given later in this paper. In the lower half of the ingot, the rotational speed has little effect on macrosegregation. The shape of the mushy zone is comparable to that observed for similar ingots in previous works.[1,2] Since solidification rate is relatively slow for the lower half of the ingot (3.7 x 10^{-3} cm/s), and the density of the interdendritic liquid increases with

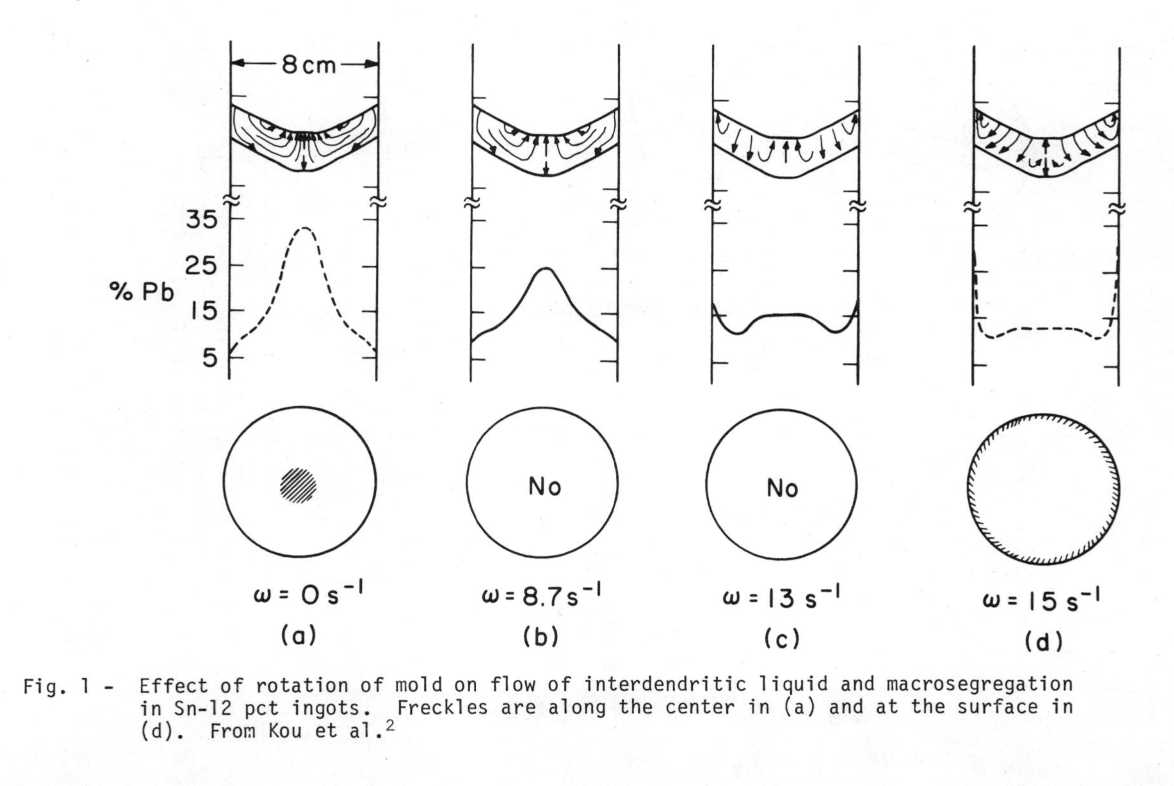

Fig. 1 - Effect of rotation of mold on flow of interdendritic liquid and macrosegregation in Sn-12 pct ingots. Freckles are along the center in (a) and at the surface in (d). From Kou et al.[2]

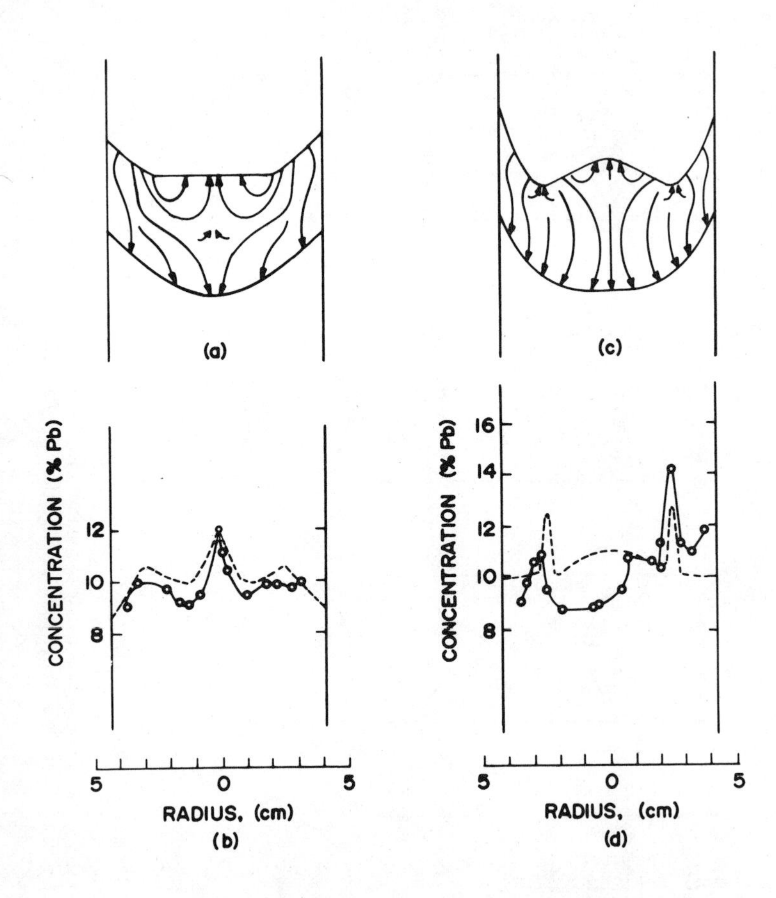

Fig. 2 - Effects of rotation of the liquid pool above the mushy zone: (a) and (c) shape of mushy zone and calculated flow pattern with rotations of 10 rpm and 40 rpm, respectively; (b) and (d) macrosegregation across the ingot with rotations of 10 rpm and 40 rpm, respectively. For Ingot 5 in Table 1.

decreasing temperature during solidification, then there is severe segregation across the ingot. This segregation is predictable based upon the results and analyses for comparable ingots with no rotation of the liquid.[1]

In the upper half of the ingot, however, significant effects caused by the increased rotation of the liquid are apparent. Here the forced convection has altered the shape of the mushy zone and produced the convexity of the liquidus isotherm in the central region. The strong positive segregation along the ingot axis is now replaced by segregation peaks at approximately mid-radius positions, and the segregation is no longer symmetrical. In the presence of a stagnant liquid pool, the convex liquidus should cause negative segregation in the central region, but, in this liquid pool which is rotated at 40 rpm, positive segregation in the central region is expected. The segregation peaks at approximately mid-radius are predictable and are due to a stationary and vertical tube located at that position. As the rotating liquid flows past the tube, there is a localized reduction of pressure which is sufficient to cause flow of solute-rich interdendritic liquid towards the region of the reduced pressure.

In the past fourteen years, it has been shown that flow of interdendritic liquid causes almost all types of segregation in castings and ingots. It has been demonstrated that interdendritic liquid convects as a result of solidification shrinkage and/or the force of gravity; the latter is extremely important when solidification rate is relatively slow (less than 10^{-2} cm/s). In previous work on macrosegregation in remelted ingots, the effect of convection in the liquid metal pool, above the liquidus isotherm, on the flow of interdendritic liquid was not considered, even though flow in the bulk liquid can be appreciable especially in the presence of an electro-magnetic force field as in the ESR process.[1,4] At the time of those works, any effect of convection in the bulk liquid on the flow of interdendritic liquid and macrosegregation was thought to be small since the penetrating bulk liquid encounters a dense dendritic network, which greatly impedes the bulk flow, at a short distance behind the advancing dendrite tips. In the light of another paper in this conference[5], as well as this paper, this assumption can now be more critically examined.

Many investigators in their attempt to determine the nature of columnar to equiaxial transition and crystal multiplication mechanisms, have studied the effects of forced convection on ingot structure and the manipulation of convection by externally applied forces. To suppress convection, homogeneous magnetic fields and rotation have been employed. Much of this work has been reviewed by Cole[6]. On the other hand, rotating magnetic fields and oscillation enhance convection and equiaxial grain growth during solidification is enhanced.[6,7]

Remelted ingots differ from static ingots not only in the chemical changes that take place during melting but also in the heat transfer and liquid pool convection conditions under which solidification takes place. Growth takes place in the presence of a positive temperature gradient ahead of the mushy zone as heat is constantly introduced in the liquid pool. Besides natural convection in the liquid pool there is mixing due to the liquid entering the pool.[8,9], and, in ESR ingots, convection in the liquid pool is induced by the strong convection of the slag.[10,11] Campbell attributed the difference in macrostructures observed between ESR and VAR ingots to the fact that in ESR divergence of electric current occurs primarily in the slag layer which causes rapid stirring in the slag and very little stirring in the metal pool, whereas in VAR vigorous stirring occurs in the metal pool, which aids dendrite fragmentation.

Takahashi et al.[12,13] studied the effects of forced convection on macrosegregation. In their experiments the melt was contained between two concentric cylinders with solidification from the inner cylinder which was cooled and rotated. They showed that there was an increase of the deflection angle of dendrites with an increase in the rate of rotation of the inner cylinder and a decrease in the solidification rate. In their analysis of segregation, the "washing effect" of solute due to the presence of convection was considered; however, the effect of the gravity field and the centrifugal force on the convection of interdendritic liquid within the mushy zone was neglected.

CONVECTION IN THE LIQUID POOL

In order to calculate macrosegregation it is necessary to determine the velocity field of interdendritic liquid in the mushy zone. This flow is assumed to be bidirectional (radial and axial directions in cylindrical coordinates) and is calculated in the same manner as in previous works by Kou et al.[1] and by Ridder et al.[4] with the exception of the boundary condition which is specified along the liquidus isotherm. In the calculations presented here, interdendritic flow velocity is approximated by treating the liquid above the liquidus isotherm (i.e. bulk liquid) and the liquid below (i.e., interdendritic liquid) separately. The effect of convection in the bulk liquid is assessed by computing the pressure along the liquidus isotherm and implementing this pressure as a boundary condition for flow of interdendritic liquid in the mushy zone.

For the system shown in Fig. 3, the major component of flow in the bulk liquid is in the Θ-direction. Since the magnitudes of v_r and v_z are small (when compared to v_Θ) we assume that they are zero. As such, the three components of the Navier-Stokes equation reduce to

$$\frac{d}{dr}\left[\frac{1}{r}\frac{d}{dr}(rv_\Theta)\right] = 0, \qquad (1)$$

$$\frac{\partial P}{\partial r} - \rho \frac{v_\Theta^2}{r} = 0 \quad , \tag{2}$$

and

$$\frac{\partial P}{\partial z} - \rho g_z = 0 \quad , \tag{3}$$

where v_Θ is tangential velocity, r is radial coordinate, P is pressure, ρ is density of the bulk liquid, z is axial coordinate and g_z is the component of gravitational acceleration in the axial direction.

In the annulus between the concentric cylinders, the velocity satisfies these boundary conditions:

$$v_\Theta = KR\omega \text{ at } r = KR \tag{4}$$

and

$$v_\Theta = 0 \quad \text{at } r = R \quad , \tag{5}$$

where KR is the radius of the inner cylinder, R is the radius of the outer cylinder, and ω is the angular velocity of the inner cylinder. Equations (1), (4) and (5) give the velocity in the Θ-direction as

$$v_\Theta = \frac{\omega K^2 R^2}{1-K^2} \left(\frac{1}{r} - \frac{r}{R^2} \right) \quad . \tag{6}$$

The pressure within this region must satisfy Eqs. (2) and (3) and the total differential,

$$dP = \frac{\partial P}{\partial r} dr + \frac{\partial P}{\partial z} dz \quad . \tag{7}$$

After substituting Eqs. (2), (3) and (6) into Eq. (7) and integrating with $P(R,0) = P_o$ as the ambient pressure, the pressure distribution in the region $KR \leq r \leq R$ becomes

$$P = \frac{\rho}{2} \left(\frac{\omega K^2 R}{1-K^2} \right)^2 \left(\xi^2 - \frac{1}{\xi^2} - 4 \ln \xi \right) + \rho g z + P_o \tag{8}$$

where $\xi = r/R$. We assume that this pressure also applies for $z < z_L$.

In the region $0 \leq r \leq KR$ (i.e., directly under the bottom of the rotating cylinder), it is assumed that

$$v_\Theta = \omega r \quad . \tag{9}$$

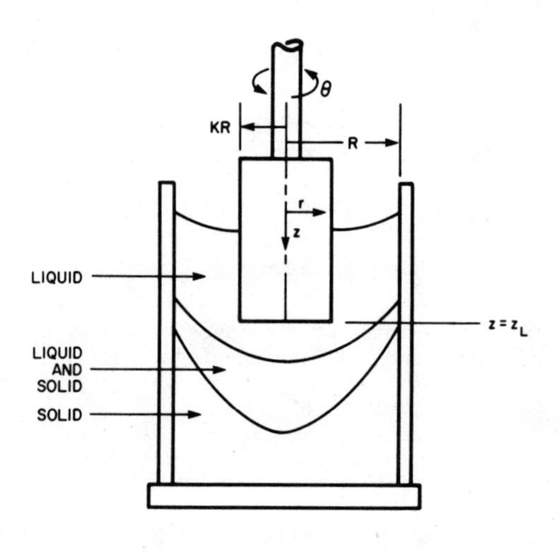

Fig. 3 - Solidification with rotation of liquid pool.

Equations (2), (3) and (9) are substituted into Eq. (7) for integration; the constant of integration is evaluated at $r = KR$ for any z greater than z_L, and P is given by Eq. (8). The result is the pressure in the region $0 \leq r \leq KR$ which is

$$P = \frac{\rho\omega^2 R^2}{2}\left[(\xi^2 - K^2) + {\frac{K^2}{1-K^2}}^{2}\left(K^2 - \frac{1}{K^2} - 4\ln K\right)\right] + \rho g z + P_o . \quad (10)$$

Finally, Fig. 4 shows the dynamic pressure (i.e., Eqs. (8) and (10) without the last two terms) with respect to ξ.

To arrive at Fig. 4 a number of assumptions have been made; these include no turbulence, $v_r = v_z = 0$, and no drag induced by the leading tips of the dendrites along the liquidus isotherm. Therefore, Fig. 4 is highly idealized, but it is representative of the actual dynamic pressure at the mold wall ($\xi = 1$) and at the center ($\xi = 0$). The importance of Fig. 4 is that the effect of rotating the bulk liquid is to produce an increase in pressure (at a given value of z) with an increase in r; i.e., as the mold wall is approached, pressure increases.

Now consider the case in which the rotating bulk liquid encounters a cylinder, with an axis parallel to the axis of rotation and located between $r = KR$ and $r = R$. The pressure reduction, due to the flow past the cylinder, relative to the dynamic pressure of the free stream velocity is given by Fig. 5 in terms of the dimensionless group

$$\frac{P - P_o'}{\frac{1}{2}\rho V_\infty^2} \quad (11)$$

where P is the stagnation pressure calculated by Eq. (8), P_o' is the static pressure (i.e., $P_o + \rho g z$) and V_∞ is the free stream velocity which is approximated as v_Θ in Eq. (6).

EFFECT OF CONVECTION ON MACROSEGREGATION

Figures 6 and 7 summarize calculations which show: (1) the effect of a convex liquidus isotherm on macrosegregation, (2) the effect of rotation of the bulk liquid, and (3) the effect of a localized pressure reduction at the liquidus isotherm (as caused by flow past a cylinder). The geometry of the mushy zone used for these calculations corresponds to an experimental ingot (Ingot 1).

In order to determine the effect of a convex liquidus isotherm, the flow pattern and segregation were calculated by ignoring convection in the liquid pool, Figures 6a and 6b. Only the pressure due to the metallostatic head was used as boundary condition at the liquidus isotherm. Other conditions and parameters used in calcula-

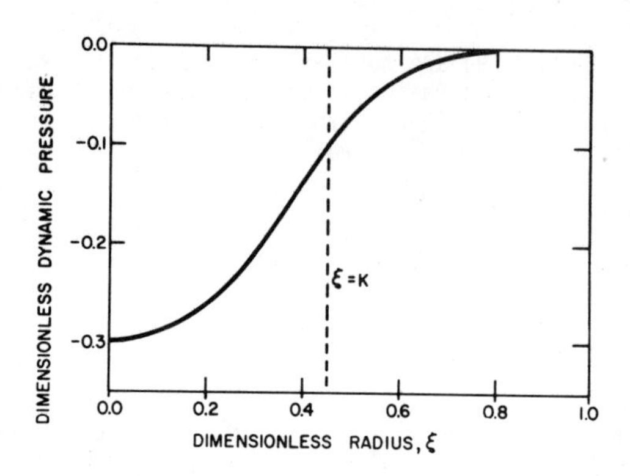

Fig. 4 - Dynamic pressure as a function of dimensionless radius.

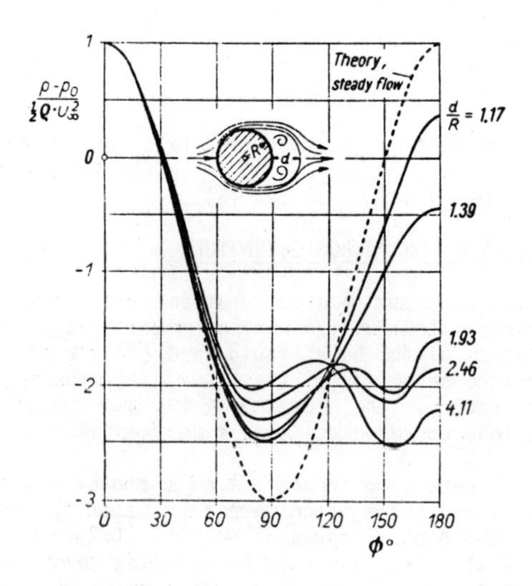

Fig. 5 - Pressure distribution around a circular cylinder for free flow past a cylinder. From Schlichting.[14]

tions for Figs. 6 and 7 are summarized in Table 1. The effect of the convex liquidus isotherm is to cause downward flow from the center which fans outward, i.e. flow is from hotter to cooler regions, except in the upper region near the wall. Interdendritic flow velocity is larger near the center of the ingot, resulting in negative segregation in this region. If the liquidus isotherm had the more usual concave shape, the segregation would be positive at the centerline as shown in previous works.[1,2] Where the flow reverses direction near the abrupt change in slope of the liquidus isotherm, there is slight positive segregation.

In the flow pattern of Fig. 6c the effect of convection in the bulk liquid is taken into account by specifying pressure along the liquidus isotherm according to Eqs. (8) and (10). Due to the rotation (60 rpm), the pressure near the mold wall is now great enough to cause upward flow of interdendritic liquid in the central region in the upper half of the ingot, resulting in positive segregation at the center, Fig. 6d.

When the effects of convection and the localized pressure decrease due to flow past a cylinder are both taken into account, the calculated flow pattern appears almost the same as if the presence of thermocouple tubes is neglected, Fig. 6e. However, there is some localized flow towards the region of the tube which causes an additional peak of positive segregation, Fig. 6f.

From the calculated flow patterns it is seen that the effect of strong convection (60 rpm) in the liquid pool is to alter the flow pattern of interdendritic liquid, causing a complete flow reversal in the upper central region of the mushy zone which results in the centerline segregation. With this strong convection, a localized pressure reduction at the position of cylinder causes the formation of a peak of positive segregation, Fig. 6f.

To show the effect of a reduced rotation, the pressure in the bulk liquid is calculated assuming that the inner cylinder rotates at 43 rpm. The calculated flow pattern of interdendritic liquid and segregation profile are shown in Figs. 7a and 7b, respectively. The flow pattern is identical to Fig. 6a, but the magnitude of velocity of interdendritic liquid which moves towards the liquidus isotherm is less than in Fig. 6a. Consequently the segregation at the centerline is significantly reduced.

Figures 6 and 7 illustrate that rotation of the bulk liquid can be advantageous in reducing macrosegregation in remelted ingots. Figure 6d shows positive segregation at the center when rotation is relatively strong (60 rpm), but when the rotation employed is reduced to (43 rpm) and the convex liquidus isotherm is maintained, then the segregation is decreased significantly (Fig. 7b). With no rotation and a concave liquidus, such an ingot would exhibit substantial positive segregation in the central portion of the ingot.[1]

TABLE 1

Experimental Variables and Solidification Parameters

Ingot	Solidification Rate, cm/s	Rotation rpm	Composition wt. pct. Pb	DAS* Microns	γ_o, cm^2	$\frac{P-P'_o}{(\rho V_\infty^2/2)}$
1	9.2×10^{-3}	0	10.3	35-40	5×10^{-7}	
2-lower	7.3×10^{-3}	0	11.5	36-39	2.5×10^{-7}	
2-upper	9.0×10^{-3}	23	11.4	36-40	1×10^{-7}	
3-lower	9.2×10^{-3}	0	11.0	38-42	1×10^{-7}	
3-upper	10.0×10^{-3}	25	11.8	37-42	1×10^{-7}	
4-lower	5.6×10^{-3}	0	10.2	37-43	1×10^{-7}	
4-upper	5.9×10^{-3}	18	10.1	37-44	2.5×10^{-7}	
5-lower	3.7×10^{-3}	10	9.7	43-53	7×10^{-7}	
5-upper	5.4×10^{-3}	40	10.5	40-50	2.5×10^{-7}	-2.50
6	4.5×10^{-3}	10	9.0	70-75	2×10^{-7}	
7	1.7×10^{-3}	60	12.4	100-118	1×10^{-7}	-1.58
8	3.5×10^{-3}	43	11.1	60-84	2×10^{-7}	-2.66

* Secondary dendrite arm spacing.

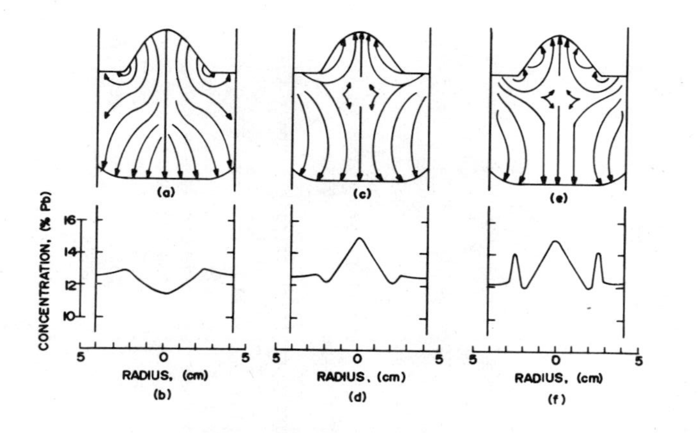

Fig. 6 - Calculated flow patterns and macrosegregation profiles using the geometry of the mushy zone in Ingot 7: (a) and (b) are for no convection in the liquid pool; (c) and (d) are for convection at 60 rpm; (e) and (f) are for convection at 60 rpm plus localized pressure reduction due to flow past a cylinder.

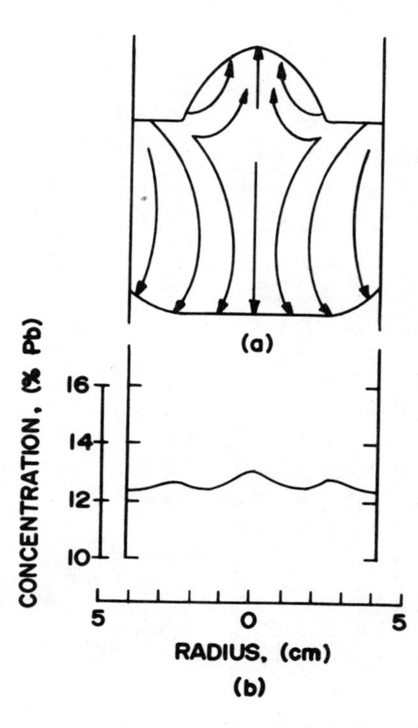

Fig. 7 - Calculated flow pattern (a) and macrosegregation profile (b) in Ingot 7 with reduced convection in the liquid pool (43 rpm).

APPARATUS AND EXPERIMENTAL PROCEDURE

The apparatus used for experiments is shown in Fig. 8; this apparatus is a modification of the system used by Kou et al[1] to simulate solidification behavior of remelted ingots. After modification for this work it consisted of four major components: (1) a mold for ingot solidification, (2) a heated container to maintain a superheated melt, (3) a driving system to control the vertical solidification rate, and (4) a rotating heater immersed in the liquid pool. The apparatus was used to make ingots of Sn-Pb alloys of compositions between 9 and 12.4 pct Pb.

Three stainless steel tubes, 3 mm in diameter, were positioned vertically inside the mold; one was along the ingot axis, another slightly beyond mid-radius, and the remaining very close to the mold wall. Each tube contained a chromel-alumel thermocouple which was free to be manually moved up and down within its tube. The mold was a stainless steel tube with an inside diameter of 8.3 cm and a length of 33 cm.

Two different electrical heaters were employed; they are shown in Fig. 9. They were comprised of cartridge-type resistance elements mounted on the stainless steel fixtures, which were 3.8 cm in diameter and 7.6 cm long. During an experiment they were immersed in the liquid metal above the mushy zone, and the power to them was controlled with a variable transformer. Convection in the liquid pool was produced by rotating the heater with the pulley-belt system shown in Fig. 8. The rotational speed was adjustable and in this work, rotational speeds from 10-60 rpm were used. The experimental procedure was the same as employed in the case of no rotation[1], except that the immersed heater was rotated during solidification.

In addition to the ingot described in Fig. 2 (Ingot 5), rotation of the liquid pool was used in six ingots. One ingot was produced with no rotation so that a total of eight experimental ingots were made. During solidification of three of the ingots, the liquid pool was not rotated for the lower portion of the ingot and then rotated as the upper portion solidified. Another ingot was solidified with different rotational speeds for the lower and upper portions. Table I summarizes experimental variables and solidification parameters associated with the eight experimental ingots.

Chemical analyses of the macrosegregation profiles across the ingots were done by x-ray fluorescence with primary white radiation from a platinum tube. The radiation impinged on an area of 2.8 mm in diameter which resulted in the local average composition since the secondary dendrite arm spacings encountered were in the range of 35-118 microns. The secondary radiation from the sample was diffracted from a lithium fluoride crystal and the intensity of the lead L_α line was compared to the intensities obtained from a set of known standards. The set of standards was analyzed before and

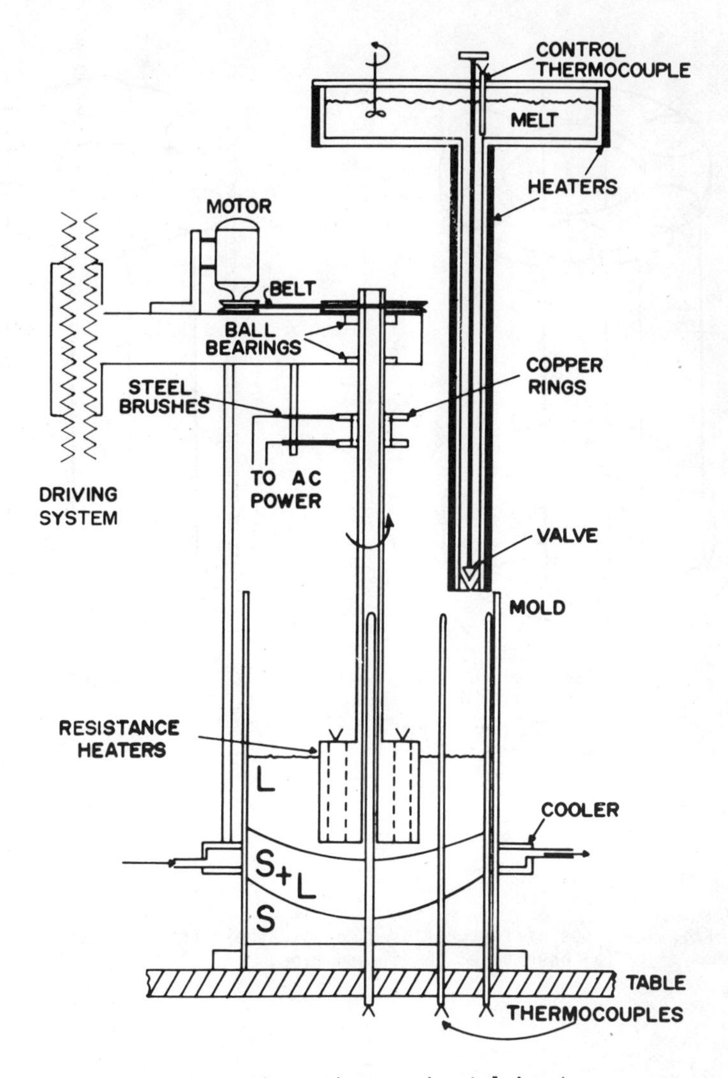

Fig. 8 - Apparatus used to make experimental ingots.

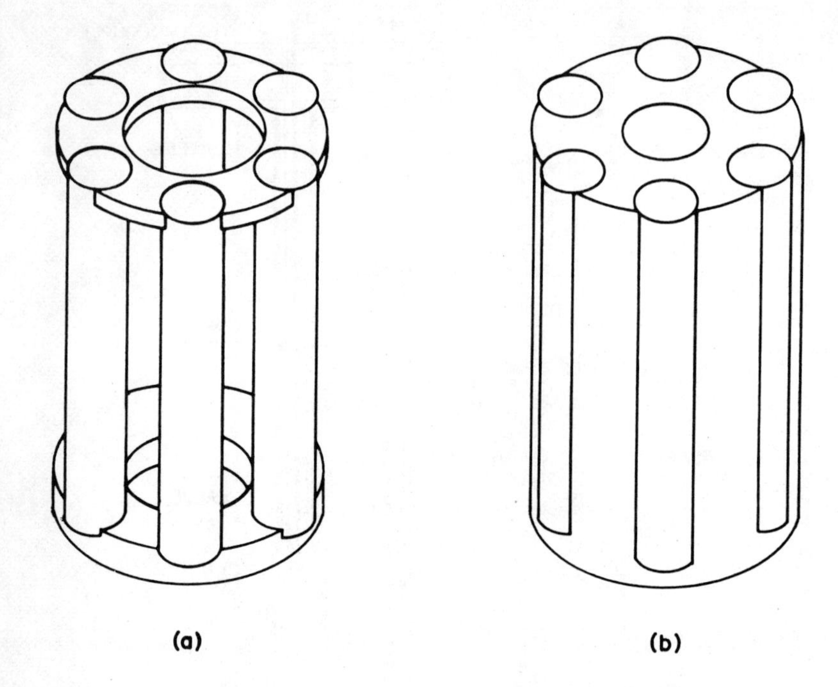

Fig. 9 - Types of fixtures used to hold the electrical heating elements: (a) open type; (b) closed type.

after each sample to insure that fluctuations in tube voltage and current did not occur during the analyses. Intensities were obtained by collecting counts for 100 seconds using a scintillation counter. Typical counts were 22,300/100 s for 5 pct. Pb and 87,500/100 s for 30 pct. Pb.

EXPERIMENTAL RESULTS

Convex Liquidus Isotherm

Among the ingots produced in this study (Table 1), the rotational speed required to cause a convex liquidus isotherm exceeded 25 rpm, with the exception of the upper portion of Ingot 4 which was rotated at 18 rpm. Figure 10 illustrates the effects of rotation of the liquid pool on the shape of the liquidus isotherm in Ingot 5. At 10 rpm, the shape of the isotherm is as in ingots solidified in the same apparatus with no rotation employed. Slightly after 15 minutes, the rotation was increased to 40 rpm which abruptly increased solidification rate until a somewhat greater steady solidification rate was achieved. In this particular ingot, solidification rate increased from 3.7×10^{-3} cm/s to 5.4×10^{-3} cm/s when the rotation was increased from 10 to 40 rpm. Reference to solidification rates for Ingots 2-5 in Table 1 show that solidification rate increases slightly with an increase in rotational speed.

Also, the increase in rotational speed of the liquid pool produced a convex liquidus isotherm similar to that reported by Edwards and Spittle[3] for a liquid pool stirred electromagnetically. In this work, convex isotherms were obtained in four ingots when rotation of the liquid pool was sufficiently strong. To supplement these observations, thermal measurements were made on an ingot with a stationary mushy zone under conditions of no rotation and of rotation of the liquid pool at 73 rpm. These results are shown in Fig. 11. The major effect of rotation is to decrease the temperature gradient in the center of the liquid pool from about 30°C/cm to 12°C/cm. Also, there are probably vortices or secondary circulations within the liquid pool which convert heat more efficiently to the outer region of the liquid pool than to the central region. When fluid flows between two concentric cylinders, with the inner rotating, the purely tangential flow becomes unstable and there appear vortices whose axes are located along a circle within the annulus and which rotate in alternately opposite directions.[14] In this case two such vortices were probably established; one was approximately 1cm above the bottom of the rotating fixture with the heaters and one was below. In the former vortex the liquid moved upward next to the mold wall and downward next to the heaters; in the latter, the liquid moved in the opposite direction contributing to the convex shape of the liquidus isotherm.

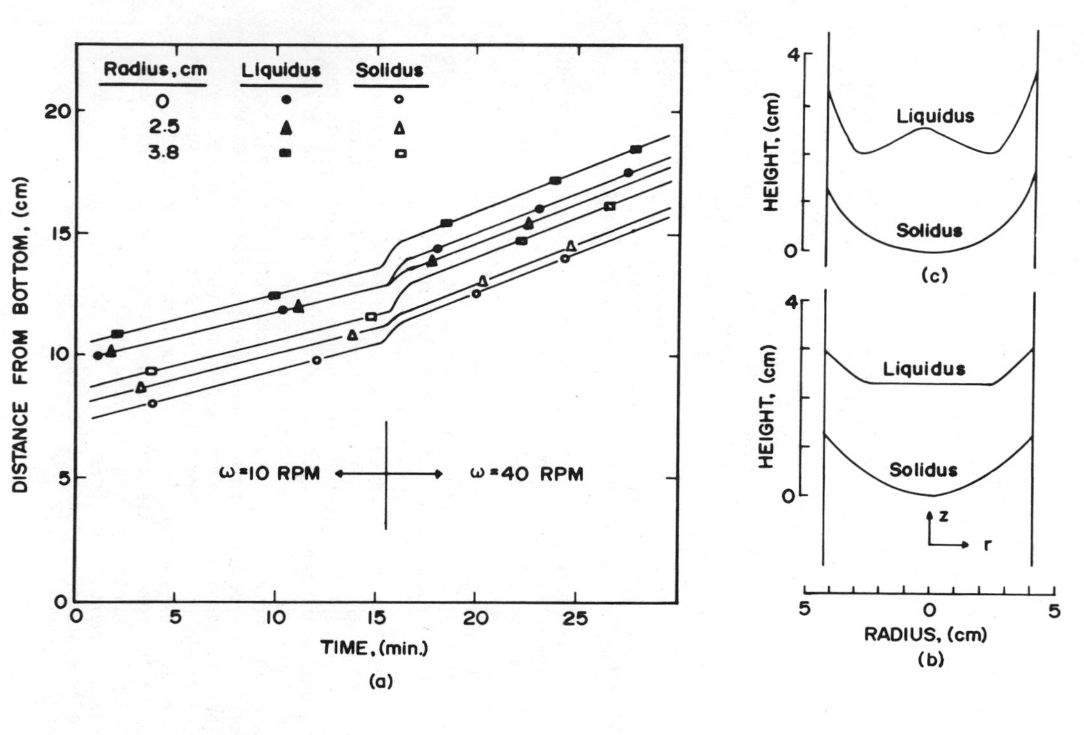

Fig. 10 - Thermal data for Ingot 5: (a) positions of solidus and liquidus isotherms; (b) shape of mushy zone in lower portion of the ingot; (c) shape of mushy zone in the upper portion.

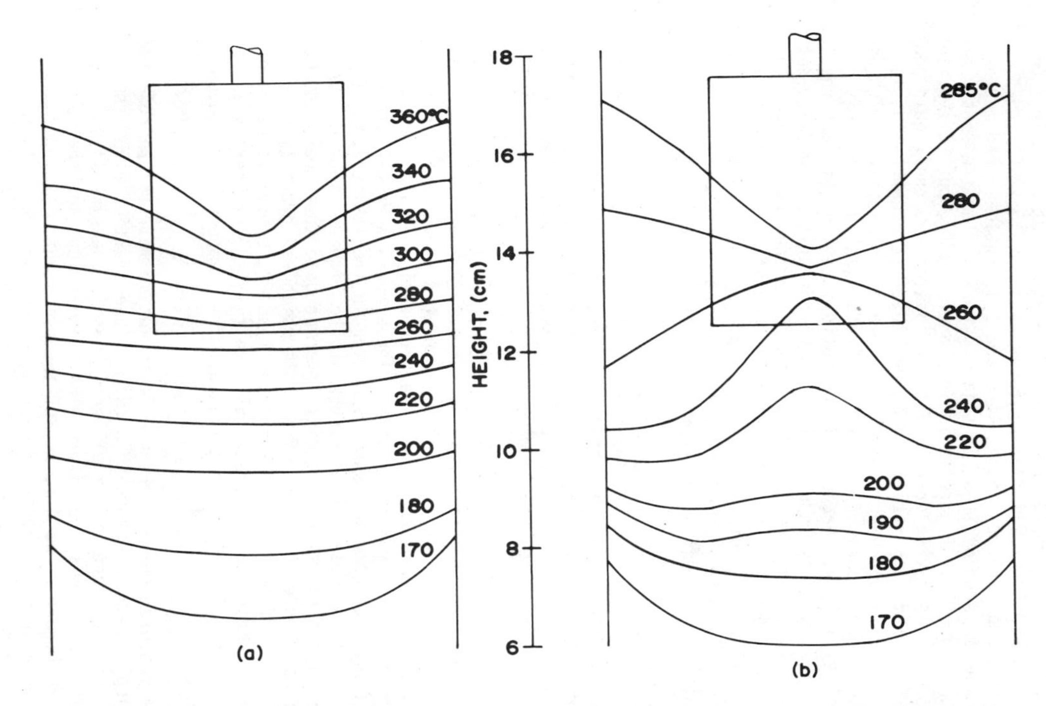

Fig. 11 - Shape of isotherms in stationary mushy zones and within the liquid pool: (a) 0 rpm; (b) 73 rpm.

Macrosegregation

It is convenient to discuss results for macrosegregation in three groups: (1) ingots which solidified at rates greater than 6×10^{-3} cm/s, (2) ingots which solidified at about 5.5×10^{-3} cm/s, and (3) those which solidified at rates less than 5×10^{-3} cm/s. The first group includes Ingots 1, 2 and 3 in Table 1, and macrosegregation in each of these is slightly positive at the center as shown in Figure 12. Figure 12 is for Ingot 2 only, but it typifies results for this entire group. With no rotation of the liquid pool (Figure 12b) and with rotation (Figure 12d), macrosegregation is slight primarily because solidification rate is relatively high and the dendrite arm spacings are relatively low (Table 1) so that permeability is low. Because of the solidification rate and the low permeability, gravity induced flow of interdendritic liquid is not strong. Rotation of the liquid pool at 23 rpm neither changes the liquidus isotherm to a convex shape nor does it alter the pressure along the liquidus isotherm enough to have a significant effect on the flow of interdendritic liquid and the macrosegregation. These results are consistent with Figure 10 in reference 2.

Ingot 4 was solidified at the intermediate rate between 5×10^{-3} and 6×10^{-3} cm/s. Results for this ingot (Figure 13) are similar to those of Figure 12 except that the effect of rotation at 18 rpm is somewhat apparent. The liquidus isotherm is convex at the center (Figure 13c) and there is more positive segregation at the center, although it too is relatively minor.

The effect of a greater rotation of the liquid pool at the intermediate solidification rate is shown in Figures 2c and 2d for the upper portion of Ingot 5. The calculated curve in Figure 2d is axisymmetric and shows some positive segregation at the center with strong peaks just beyond midradius. These peaks were calculated by assuming the localized pressure reduction in the last column of Table 1. In this case, the pressure at the liquidus isotherm and at the position of the intermediate thermocouple tube, was reduced to the point where the value of the dimensionless group ($\vec{v} \cdot \nabla T/\varepsilon$) equals -1. As discussed in References 1 and 2, calculations do not accurately predict macrosegregation when ($\vec{v} \cdot \nabla T/\varepsilon$)<-1 where $\vec{v}$ is velocity of the interdendritic liquid, ∇T is temperature gradient and ε is cooling rate.

The measured curve in Figure 2d is not symmetrical because the thermocouple tube, of course, is only at one position along the liquidus isotherm and not at all angles as in the simplified calculation for two-dimensional flow. The plane of cut for the measured results in Figure 2d is just behind the plane which included the thermocouple tube at the position of the maximum peak. There is also a peak opposite the maximum peak where no tube exists. This peak is also attributed to the tube and, indeed as discussed below,

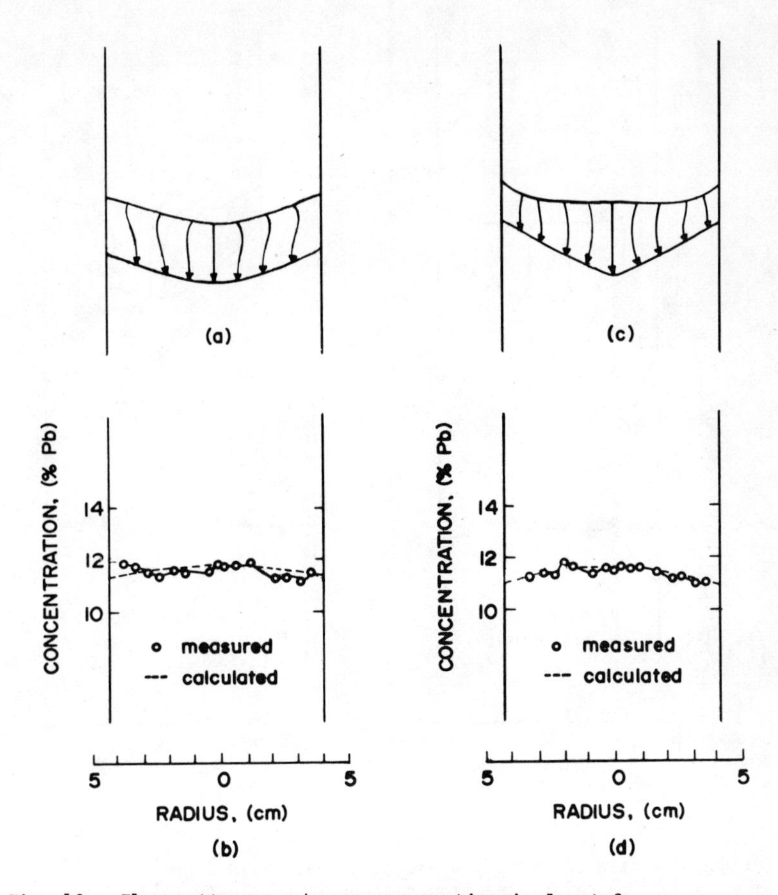

Fig. 12 - Flow patterns and macrosegregation in Ingot 2:
(a) calculated flow pattern in lower half (0 rpm);
(b) measured and calculated macrosegregation of the lower half;
(c) calculated flow pattern in the upper half (23 rpm);
(d) measured and calculated macrosegregation of the upper half.

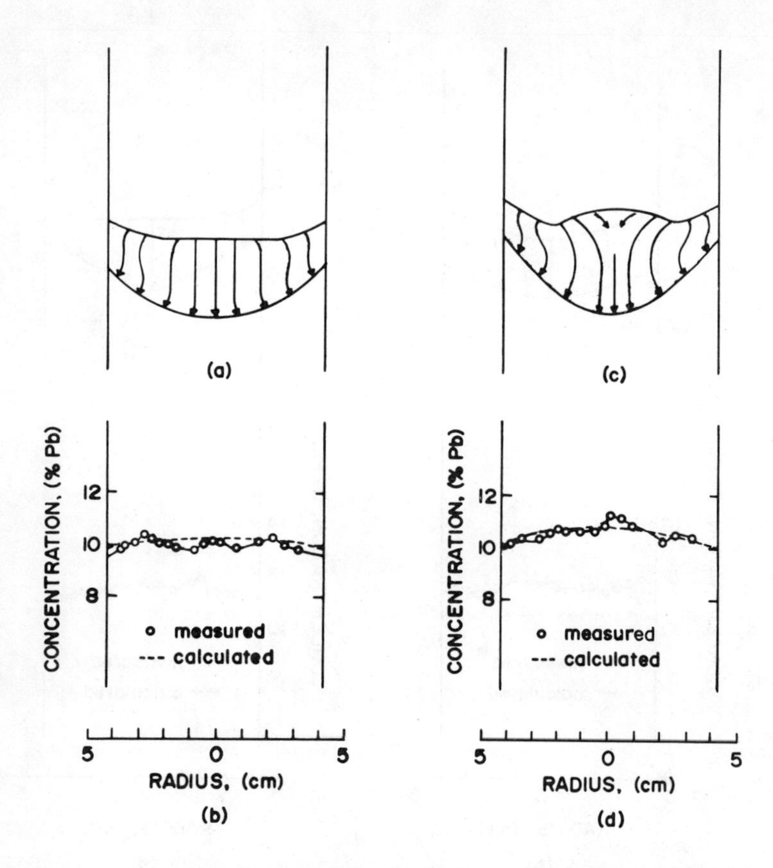

Fig. 13 - Flow patterns and macrosegregation in Ingot 4:
(a) calculated flow pattern in the lower half (0 rpm);
(b) measured and calculated macrosegregation in the lower half;
(c) calculated flow pattern in the upper half (18 rpm);
(d) measured and calculated macrosegregation of the upper half.

the tube causes a ring of segregation whose peak value diminishes with an increasing angle from the tube location.

For the group of ingots solidified at rates less than 5×10^{-3} cm/s, segregation is strongly positive at the center when the rate of rotation of the liquid pool is weak. This is shown in Figure 2b for Ingot 5 with rotation of 10 rpm and in Figure 14 for Ingot 6 also with a rotation at 10 rpm. That there is a strong segregation at the center of these two ingots is attributed to the relatively slow solidification rates of 3.7×10^{-3} and 4.5×10^{-3} cm/s. Rotation of the liquid pool at only 10 rpm is weak and only slightly affects flow of interdendritic liquid; thus the results are predictable from the work of Kou et al[1] in which no rotation was employed.

Ingots 7 and 8 were solidified at rates of 1.7×10^{-3} and 3.5×10^{-3} cm/s, respectively, and their liquid pools were strongly rotated at 60 and 43 rpm, respectively. Figure 15 shows results obtained for Ingot 7. At 60 rpm, the convection is sufficiently strong to cause (1) segregation at the center and (2) segregation associated with flow past the thermocouple tube. Both of these effects are predictable by calculation as shown by Figure 6f. Results for Ingot 8 are shown in Figure 16. Here the rotation of the liquid pool has produced segregation at the radial position of the thermocouple tube. Without the tube, segregation would be minimal as illustrated by Figure 7 in which results of calculations are shown for an ingot (Ingot 7) with a mushy zone of almost the same shape as that of Ingot 8.

Finally, several sections were cut from Ingots 7 and 8 to determine macrosegregation caused by the flow of liquid past the intermediate thermocouple tube; results for Ingot 7 are shown in Fig. 17. Maximum segregation occurs just behind the tube, Fig. 17b, and in this ingot a freckle exists at this location. The magnitude of the peak decreases as the angle from the tube increases as shown by Figures 17c-17f. Presumably, this decrease is associated with the attenuation of the pressure reduction at the position of the tube itself.

CONCLUSIONS

1. Calculations which predict macrosegregation in remelted ingots under conditions of a rotating liquid pool compare well with experiments.

2. When the liquid pool is rotated during solidification, the shape of the liquidus isotherm is atypical in that it assumes a convex shape. In this work a convex shape was achieved when the rotation exceeded 25 rpm, although in one instance a convex shape was observed at 18 rpm.

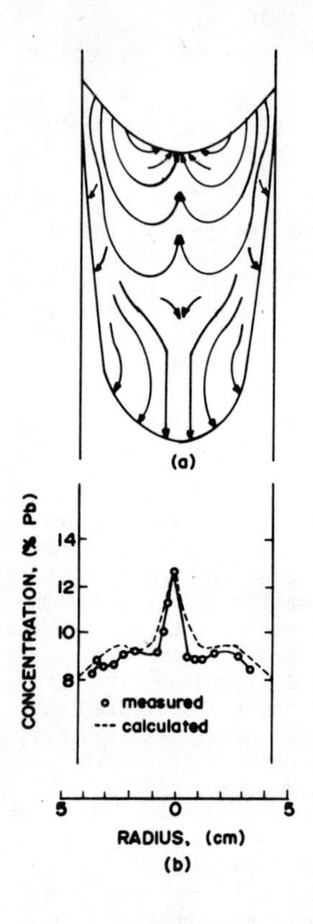

Fig. 14 - Flow pattern and macro-segregation in Ingot 6:
(a) calculated flow pattern;
(b) calculated and measured macro-segregation.

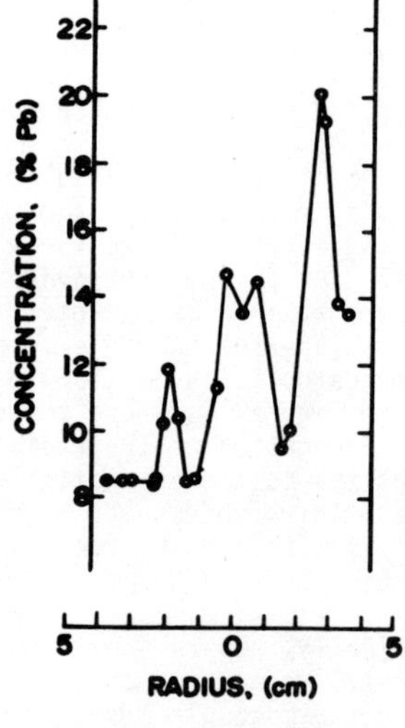

Fig. 15 - Measured macrosegregation in Ingot 7.

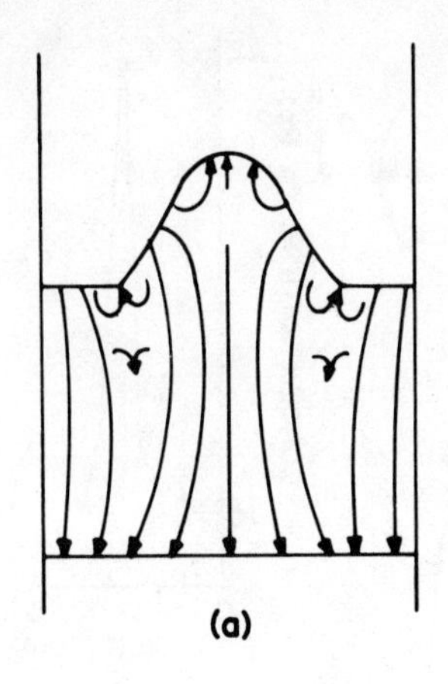

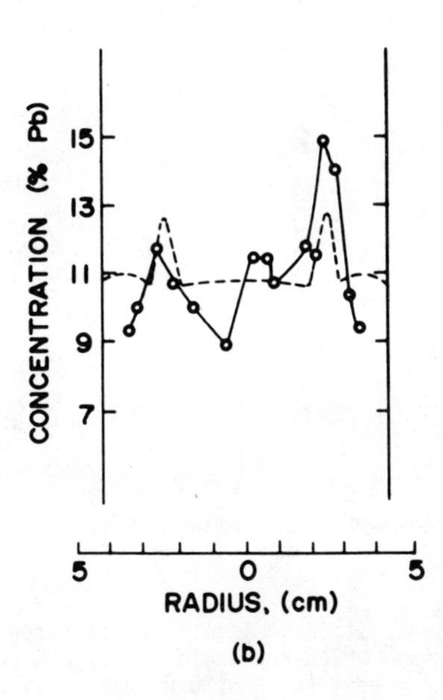

Fig. 16 - Flow pattern and macrosegregation in Ingot 8:
(a) calculated flow pattern;
(b) calculated and measured macrosegregation.

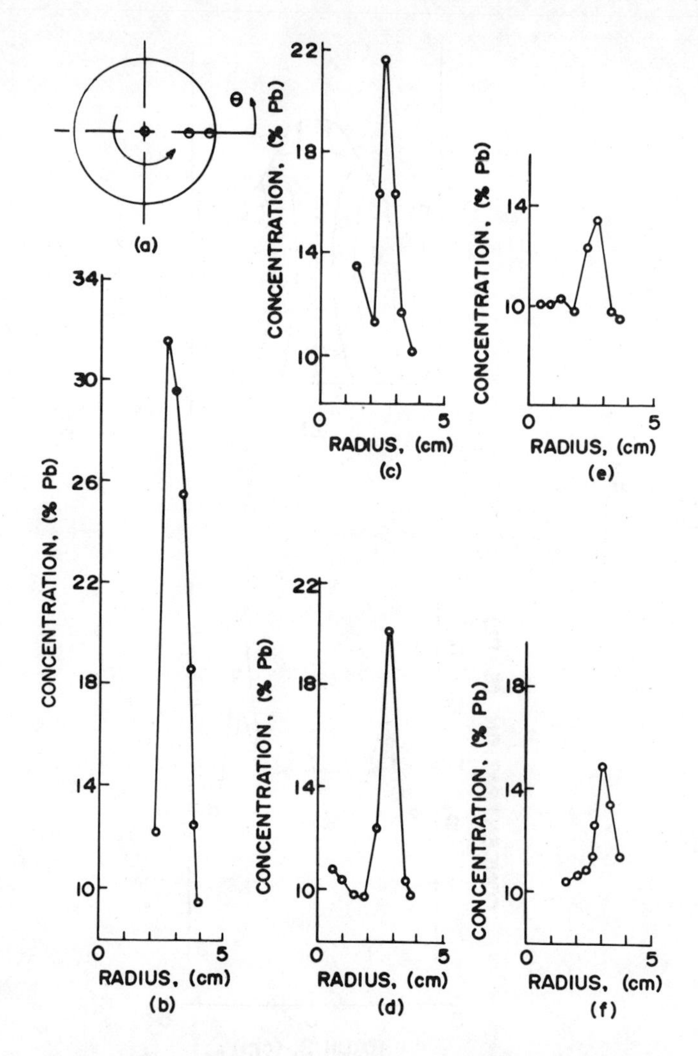

Fig. 17 - Measured segregation in Ingot 7 at different angles: (a) plane from which the angle Θ is measured; (b) Θ = 2 degrees; (c) Θ = 20 degrees; (d) Θ = 40 degrees; (e) Θ = 60 degrees; (f) Θ = 358 degrees.

3. By rotating the liquid pool, the pressure at the center of the ingot and on the liquidus isotherm is reduced and causes flow of solute rich interdendritic liquid towards the center of an ingot. Conversely, the convex liquidus isotherm has the effect of causing flow of interdendritic liquid away from the center. When the latter effect exceeds the former, the macrosegregation in a non-rotated ingot, which normally exhibits strong positive segregation at the center, can be significantly reduced by rotating the liquid pool. The effectiveness of employing a rotating liquid pool to reduce segregation depends upon the solidification rate and the rate of rotation. In this study, a desirable reduction in macrosegregation is demonstrated for solidification rates less than 5×10^{-3} cm/s and for rotation of approximately 40-60 rpm.

4. In this work, vertical thermocouple tubes extended through the solidified and partially solid zone and into the liquid pool. The presence of thermocouple tubes strongly influenced the segregation profiles such that in the case of rates of rotation greater than 40 rpm, localized segregation with a composition as rich as 32% Pb behind the thermocouple tube was found. The effect of the thermocouple tubes in causing positive segregation is decreased with decreased convection in the liquid pool (less than 25 rpm), and increased solidification rates (greater than 6×10^{-3} cm/sec). The localized segregation associated with a thermocouple tube attenuates as angular distance, at a given radius, from the tube increases. The effect is due to a decrease of the dynamic pressure at the position of the thermocouple tube at the liquidus isotherm and is predictable by calculation.

ACKNOWLEDGEMENTS

One of the authors (D.N.P.) gratefully acknowledges a Domestic Mining and Minerals and Mineral Fuel Conservation Fellowship from the U.S. Office of Education. This Fellowship was administered by Dean W. Cosart of the College of Mines, The University of Arizona. The computer work was funded by the Graduate College, The University of Arizona under Dean L. Jones.

REFERENCES

1. S. Kou, D. R. Poirier, and M. C. Flemings: Proceedings of the Electric Furnace Conference, Iron and Steel Society of AIME, December, 1977, pp. 221-228.

2. S. Kou, D. R. Poirier, and M. C. Flemings: Met. Trans. B., 1978, vol. 9B, pp. 711-719.

3. K. P. Edwards and J. A. Spittle, J. Inst. Met., v. 100, 1972, pp. 244-248.

4. S. D. Ridder, F. C. Reyes, S. Chakravorty, R. Mehrabian, J. D. Nauman, J. H. Chen and H. J. Klein, Met. Trans. B, 1978, vol. 9B, pp. 415-425.

5. S. D. Ridder, S. Kou and R. Mehrabian, "Effect of Fluid Flow on Macrosegregation in Axi-Symmetric Ingots", Conference on Modelling of Casting and Welding Processes, Franklin Pierce College, Rindge, New Hampshire, August 4-8, 1980. See p. 261 of this volume.

6. G. S. Cole, "Transport Processes and Fluid Flow in Solidification", Chapter 7 in Solidification, American Society for Metals, Metals Park, Ohio, 1971.

7. A. A. Tzavaras and J. F. Wallace, J. of Crystal Growth, v. 13-14, 1972, pp. 782-786.

8. J. Campbell, J. of Metals, v. 23, 1970, pp. 23-25.

9. Y. Nakamura, N. Tokumitsu and K. Harashima, Trans. ISIJ, v. 14, 1974, pp. 170-175.

10. A. H. Dilawari and J. Szekely, Met. Trans. B, v. 8B, 1977, pp. 227-236.

11. A. H. Dilawari and J. Szekely, Met. Trans. B, v. 9B, 1978, pp. 77-87.

12. T. Takahashi, I. Hagiwara and K. Ichikawa, Trans. ISIJ, v. 12, 1972, pp. 412-421.

13. T. Takahashi, K. Ichikawa, M. Kudon and K. Shimahara, Trans. ISIJ, v. 16, 1976, pp. 283-291.

14. H. Schlichting, Boundary Layer Theory, McGraw-Hill, NY, 1968.

MODELING SOLIDIFICATION

IN AN ELECTROSLAG REMELTED INGOT

C. L. Jeanfils, J. H. Chen, H. J. Klein

Cabot Corporation
Kokomo, Indiana

A transient heat transfer model and a steady state macrosegregation model are presented for an ESR ingot.

In the heat transfer model, the instantaneous electrode melt rate is considered known. Empirical values of the heat transfer coefficients are used. The release of latent heat is accounted for through an effective heat capacity. The latter can be calculated using a Scheil solidification model. The finite difference equations are solved using the alternating direction implicit method. The model accurately predicts the pool profiles, but underestimates somewhat the local solidification time. The model shows that, during transient conditions, the response of a given isotherm is the slowest at the ingot axis and also the lower the value of the isotherm temperature, the more sluggish is the response.

In the steady state macrosegregation model, the shape of the mushy zone and its translation rate are considered known. The model treats the case of a multicomponent alloy for which only one solid phase forms upon solidification. Limited experimental testing of the model indicates that the trend of macrosegregation is correctly predicted for the various solutes while the absolute degree of macrosegregation could only be predicted for a few solutes.

Introduction

Electroslag remelting provides a means to produce high quality ingots under controlled chemical and melting conditions. The melt rate has a determining effect on the local solidification time (LST) (1), and therefore on the characteristic size of the as-cast microstructure. A model is useful in the determination of the range of remelting conditions for an appropriate LST.

The extent of macrosegregation in an electroslag remelted ingot is generally very small (2,3). Yet in some alloys, a channel-like defect called a freckle is sometimes observed after electroslag or vacuum arc remelting (4). The operators have resorted to trial and error to determine remelting conditions that produce freckle-free material. The range of operating parameters that can be practically explored in this way is limited. A mathematical model that can predict freckle formation is therefore a useful tool for the selection of optimal remelting conditions.

The approach taken in the development of an ESR solidification model is as follows. A heat transfer model for the solidifying ingot was developed first. The model treats the transient case because the remelting operation is a batch process in nature and because the determination of an optimal melt rate profile is of interest. Aside from its usefulness in calculating pool profiles and local solidification times, the heat transfer model can be used as input to the macrosegregation model. For a steady translation of the mushy zone, this model calculates the local composition of the solidified alloy. An intermediate step in this calculation is the computation of the velocity of the interdentritic liquid. A criterion for freckle formation (5) is that the liquid velocity component which is perpendicular to the local isotherm be larger than the component of the translation velocity taken along this same direction.

Heat Transfer in the Ingot

Background

Under normal ESR operation, only a small part (≈20%) of the heat generated in the slag is used to bring the electrode temperature to just above the liquidus. This sensible heat constitutes the major contribution (≈85%) to the total heat flux at the moving slag-metal interface. The balance of the heat flux is made up of the superheat of the metallic droplets and of the heat transfer by slag-metal contact. In the present model, the melt rate is considered to be a known function of time. The superheat and the slag-metal heat transfer coefficient are estimated. This approach has been generally followed in modeling the heat transfer in the ingot, as is shown in the comprehensive review on the subject by Ballantyne et al.(6) During the operation, the instantaneous melt rate can be calculated and controlled from load cell or electrode travel data (7).

An improved characterization of the heat flux at the slag line can be obtained by modeling the heat transfer in the slag (8,9) and by treating jointly the heat transfer problems in the slag and in the metal (9). It has been pointed out (10) that the radiative heat transfer in the slag can be an important contributor to the transport of heat. There is, however, little data on the absorption coefficients for the slags of interest. Thus, heat transfer in the slag is not treated in the present model.

Some portion of the heat flux passing through the slag-metal interface is accumulated in the ingot. For example, with a 450 kg/hr melt rate and a 40 cm diameter mold, 67% of this heat flux is being accumulated when the ingot height is 20 cm (11). This rate of accumulation decreases to 57% at a height of 40 cm, and to 41% at 120 cm. The balance of the heat input is extracted by the cooling water. The heat flux at the mold wall is almost always modeled by Newton's heat transfer law with a heat transfer coefficient that applies between the ingot surface and the cooling water. The two barriers to heat transfer are the slag skin and the air gap. Since the latter increases as cooling proceeds, the heat transfer coefficient at a given location decreases with time. This is reflected in the model by letting the heat transfer coefficient vary as a decreasing function of the distance measured from the slag line. Approaches to calculate the size of the air gap have been published (12), but have not been incorporated in the model at this time.

Mathematical Model

This model was presented recently (11) and is summarized here. The conservation of heat for an elementary volume element is expressed by the following equation:

$$\frac{\partial H}{\partial t} = \nabla(k \nabla T) + v \frac{\partial H}{\partial z} \tag{1}$$

where,

- H = the enthalpy per unit volume
- t = time
- k = thermal conductivity
- T = temperature
- z = spatial coordinate, increasing in value when going from the base to the top of the ingot

The velocity term $\underline{v}$ is zero if Equation 1 is expressed in a coordinate system fixed at the base of the ingot. It becomes equal to the growth rate of the ingot if the coordinate system moves with the slag line.

The enthalpy per unit volume is related to the temperature and to the volume fraction of liquid g_L by:

$$dH = \rho c_p \, dT + \rho \, L \, dg_L$$

where,

- ρ = density
- c_p = heat capacity
- L = heat of fusion (2)

For an alloy, the release of latent heat takes place over a range of temperatures and the variation of g_L with temperature is a function of the solidification conditions. A good approximation to the variation of g_L with temperature is obtained through the Scheil's model (13) for which numerical integration is necessary if the solute partition ratio is not a constant. Figure 1 shows the results of such an integration for the alloy Ni-27% Mo.

If g_L depends only on temperature, an effective heat capacity per unit volume c_{ve} can be used (14):

$$c_{ve} = \rho c_p + \rho L \frac{dg_L}{dT} \tag{3}$$

This quantity is shown in Figure 2 for the Ni-27% Mo alloy. The use of an effective heat capacity allows Equation 1 to be written:

$$c_{ve} \frac{\partial T}{\partial t} = \nabla (k \nabla T) + v c_{ve} \frac{\partial T}{\partial z} \tag{4}$$

The same approach can be followed for a multicomponent alloy. The latent heat used in the calculation was 256 kJ/kg for the Ni-27% Mo binary alloy and for the "Waspaloy"* alloy.

The convective flow in the liquid metal pool is not calculated. An effective thermal conductivity equal to three times the thermal conductivity is used in this region.

The boundary conditions to the heat transfer problem are:

at the base: $$-k\frac{\partial T}{\partial z} = -h_b (T - T_W) \tag{5}$$

at the mold wall: $$-k\frac{\partial T}{\partial r} = +h_W (T - T_W) \tag{6}$$

at the slag-metal interface:

$$-k\frac{\partial T}{\partial z} + h_{sm} (T_{slag} - T) + \frac{M.R.}{\rho \times area} \times [H (T_{add}) - H (T)] = 0 \tag{7}$$

where:
- T_W is the cooling water temperature
- h_b and h_W are the heat transfer coefficients at the base and at the mold wall
- h_{sm} is the heat transfer coefficient between slag and metal
- M.R. is the instantaneous melt rate
- T_{add} is the temperature of the liquid metal droplets

The heat transfer coefficient at the mold wall is a function of the distance below the liquidus. The results of Paton et al. (15) are used in the present model. The value of the heat transfer coefficient at the slag-metal interface is taken to be 3 kw/m^2,K, and a superheat of 110°C is assumed for the liquid metal droplets. The slag temperature at the slag-

*Registered trademark, United Technologies Corporation

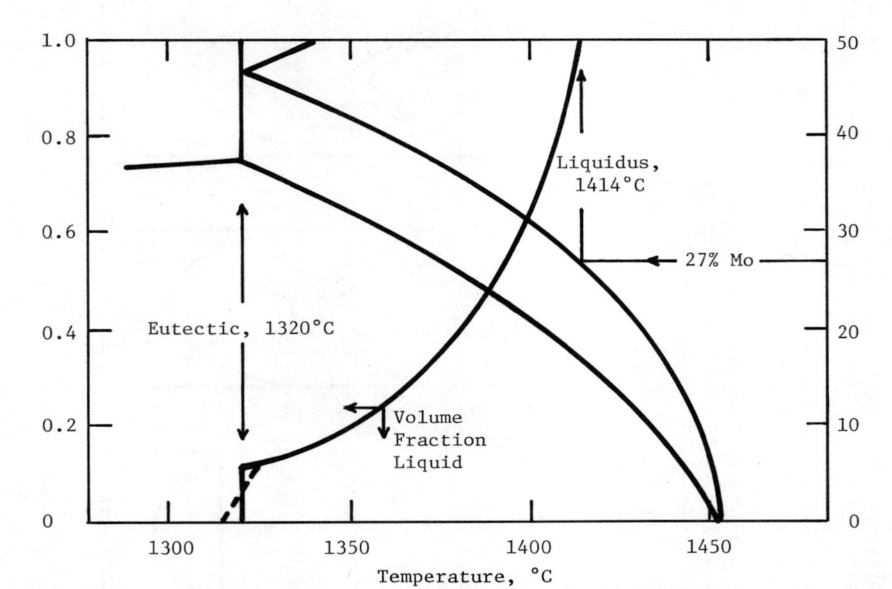

Fig. 1 - Prediction of the volume fraction of liquid in the mushy zone according to the Scheil's model in Ni-27% Mo and Ni-rich side of Ni-Mo phase diagram.

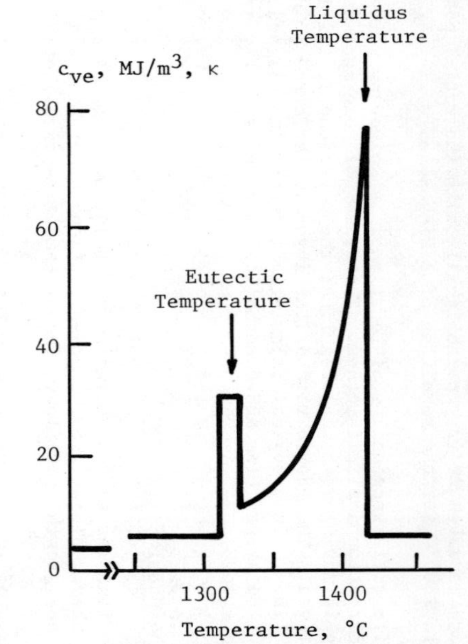

Fig. 2 - Effective heat capacity per unit volume for the Ni-27% Mo alloy according to the Scheil's model of solidification.

metal interface is taken to vary parabolically with the radial position from 1700°C to 1600°C for the remelting of a Ni-base alloy.

The finite difference analog of Equation 1 is solved by using the alternating direction implicit method. Two frames of reference are considered. One is moving with the slag line; the other one is fixed at the base of the ingot. The growth rate of the ingot is reflected by the addition of mesh lines. In a typical calculation, a mesh line is added every 20 to 40 ADI time steps. The added mesh line is moved slightly every half-ADI time step.

Testing of the Model

The algorithm was tested against exact solutions for the hypothetical cases of one-phase, linear growth, one-directional heat flow and of two-dimensional, no - growth heat flow. The agreement is excellent.

The pool profile predictions were tested experimentally with 20 cm and 40 cm diameter ingots. The experimental pool profile is determined by the addition of tungsten powder to the liquid pool. Figure 3 shows a comparison of the experimental pool profiles and of the predicted liquidus profiles before and after a 40% increase in the growth rate of a 40 cm diameter "Waspaloy" alloy experimental ingot. The agreement between prediction and experiment is seen to be good.

Dendrite arm spacing (DAS) measurements were performed for this ingot and local solidification times and mushy zone depths were calculated from the DAS with the input from solidification experiments by Ridder (16). The

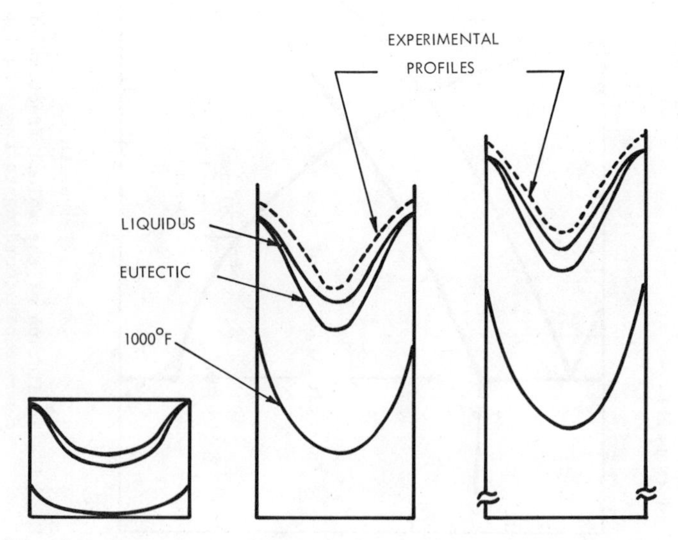

Fig. 3 - Comparison of pool profile prediction and experimental pool profiles in a 40 cm diameter "Waspaloy" alloy ingot. The eutectic and the 540°C (1000°F) isotherms are also shown.

predicted solidification times were generally 30% to 40% shorter than the experimental values. Also, the measured dendrite arm spacing reached a maximum at mid-radius, a phenomenon also observed by Poirier et al. in a maraging steel ESR ingot (3). The predictions become quite reasonable if a value of 350 kJ/kg is taken to be the latent heat of fusion for "Waspaloy."

Parametric Study and Discussion of the Heat Transfer Model

A few results from a parametric study presented in (11) are reproduced here. Figure 4 shows the calculated change in pool shape and mushy zone depth for the initial stages of remelting. If the electrode melt rate is held constant, the steady pool profile is reached at an ingot height equal to 1.5 to 2 ingot diameters. The lower temperature isotherms reach steady profiles at a slower rate. The electrode melt rate affects the steady state pool characteristics as shown in Figure 5. The higher the melt rate, the deeper the pool and the mushy zone, and the steeper the slope of the isotherms. It has been shown (1) that the local solidification time at steady state is not a monotonic function of the melt rate. There is an intermediate melt rate under which the LST is at a minimum. This entails that two ingots of the same alloy can be produced at different melt rates but experience the same local solidification time at comparable locations. Because the pool characteristics in the two ingots would not be the same, the intensity of the mechanism driving the interdendritic fluid flow would differ, and the extent of macrosegregation would not be expected to be the same in the two ingots.

The relationships between the pool characteristics and the melt rate illustrated in Figure 5 cease to hold if a varying instantaneous melt rate is considered, as seen in Figure 6. The melt rate is the predominant factor affecting the pool characteristics, but the response of these to changes in melt rate is slow. As can be seen in the calculated example shown in Figure 6, the pool depth reaches a maximum ten to fifteen minutes after the melt rate has reached a maximum. The mushy zone depth reflects the depths of the liquidus and solidus lines and lags behind the melt rate by another 30 minutes.

For a given upper limit to the acceptable characteristic size (DAS,

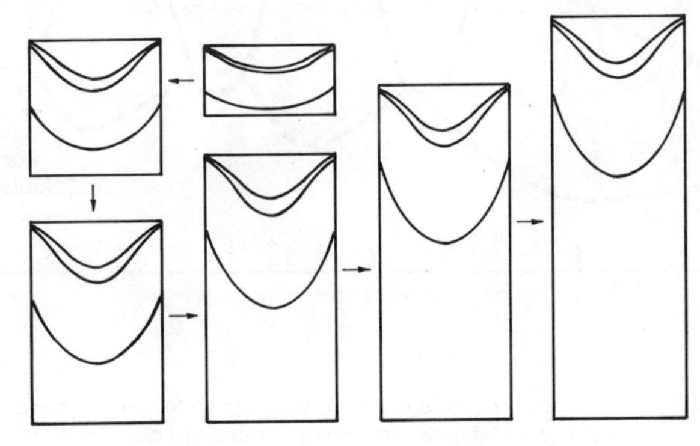

Fig. 4 - Computer prediction of liquidus, eutectic, and 540°C isotherms in a 40 cm diameter ingot; Ni-27% Mo; melt rate = 360 kg/hr.

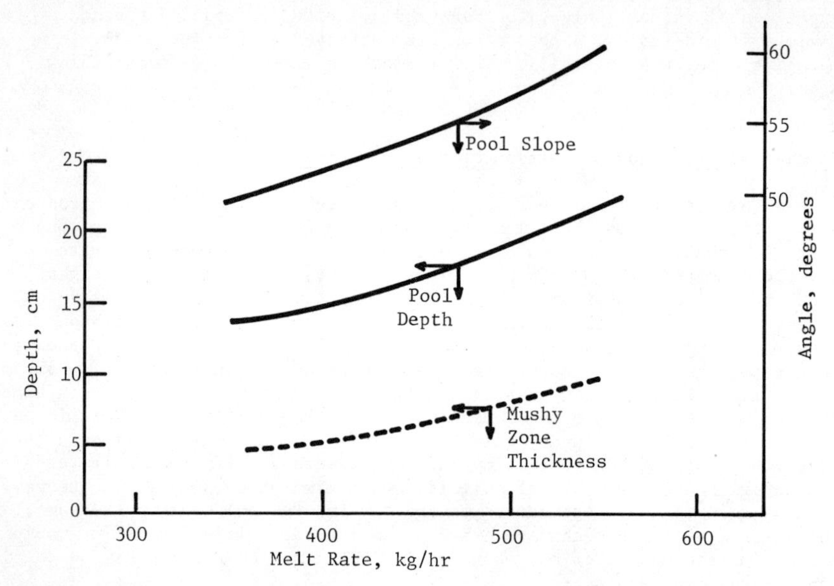

Fig. 5 - Effect of melt rate on pool characteristics at steady state for a 40 cm diameter ingot.

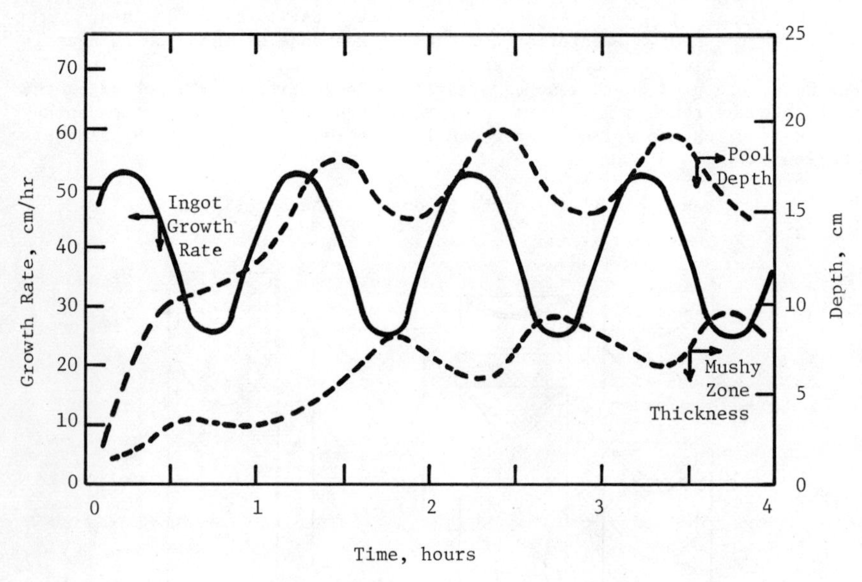

Fig. 6 - Effect of sinusoidal variation of melt rate on pool depth and mushy zone thickness for a 40 cm diameter ingot. Computer predictions for a base melt rate of 450 kg/hr with a variation amplitude of 180 kg/hr; period one hour.

primary precipitates) of the as-cast microstructure, the transient heat transfer model can be used to search for an optimum melt rate profile, as indicated by Mitchell et al. (1). Melt rate and ingot size are thus limited by the consideration of the length of the local solidification time as well as by the occurrence of macrosegregation, as is discussed below.

Macrosegregation Model

Background

The extent of macrosegregation in ESR ingots is generally very small which is one of the major benefits of the remelting operation. The liquid droplets falling from an originally segregated electrode are homogenized in the liquid metal pool which acts as a mixer. The surface to axis macrosegregation present in the electrode is minimized and end-to-end macrosegregation is smoothed out. A change in the chemical activity of the slag constituents with time can lead to a nonuniformity of composition from the ingot bottom to the hot top. Slag chemistries are therefore selected so that this factor will be minimized.

Macrosegregation, however, does occur occasionally in an ESR ingot. A nonuniformity of the flow of the interdendritic liquid has been demonstrated by Flemings, Mehrabian, and Nereo (17-19) to be the most common mechanism responsible for macrosegregation in castings. This theory is applied to ESR, and one of the objectives of this study is to calculate the magnitude of the two dimensional (r, z) flow that takes place in the mushy zone of a cylindrical ESR ingot and to examine whether this flow is stable. Instabilities in this flow can result in channel segregation, as shown by Mehrabian et al. (5). If the two dimensional flow is stable, the average local composition of the solidified ingot can be calculated. The remelting parameters should be selected to minimize the two dimensional flow in the mushy zone and to avoid flow instabilities. Which of these two requirements is the more stringent depends on the alloy composition and on the ingot size.

Mathematical Model for Macrosegregation

In the present model, the nonuniformity of the interdendritic fluid flow is considered to be the only cause of macrosegregation. The equations that quantitatively describe the phenomenon can be found in the original articles on the subject by Flemings et al. (17-19) and by Meharabian et al. (5). These equations have been applied to macrosegregation in ESR by Kou et al. (20), Ridder et al. (21,22), Poirier et al. (3), and Petrakis et al. (23). The extension of the macrosegregation equations from binary to multicomponent alloys can be found in papers by Mehrabian et al. (24), Poirier et al. (3), and Fujii et al. (25).

Because many of these references are readily available, only a brief description of the method followed in the derivation of the basic equations is given here. The alloy is considered to solidify dendritically without undercooling. The assumptions of constant solid density and of equilibrium of composition at the solid-liquid interface are made. Differential volume elements are considered, and local quantities such as volume fraction of liquid are defined by integration over these elementary volume elements and therefore can be differentiated. The conservation of the mass of each constituent is formulated. The equation of motion for the liquid is replaced by Darcy's law:

$$\underline{w} = \frac{K}{\mu g_L} (-\nabla P + \rho \underline{g}) \tag{8}$$

where:

$\underline{w}$ is the interdendritic liquid velocity measured in a frame fixed in the solidified ingot.

K is the permeability, of dimension length squared.

μ is the dynamic viscosity of the liquid.

P is the total pressure.

$\underline{g}$ is the acceleration of gravity.

This results in a set of partial differential equations in which the dependent variables are scalars: the pressure, the volume fraction liquid and the temperature. The task of solving this set of partial differential equations is simplified if the temperature problem is uncoupled from the fluid flow problem. Only the approach followed to treat multicomponent systems is given here in some detail. In the case of a binary alloy, the phase diagram gives the depression of the liquidus due to solute enrichment as well as the partition ratio k for the solute where k is defined as the ratio of the mass concentration in the solid to that in the liquid in equilibrium with the solid. In a multicomponent alloy, if only one phase is solidifying and if elements 1 and 2 do not diffuse in the solidified phase, then Mehrabian et al. (24) show that:

$$\frac{D \ln C_{L2}}{D \ln C_{L1}} = \frac{1 - k_2}{1 - k_1} \tag{9}$$

where:

D is the differential following the fluid motion.

C_{Li} is the weight fraction of component i in the liquid.

k_i is the partition ratio for component i.

If, on the other hand, Solute 1 does not diffuse in the solid while Solute 2 diffuses readily in the solid, Equation 9 must be modified. The local solute redistribution equations for 1 and 2 are given by (17, 25):

$$\frac{D \ln C_{L1}}{Dt} = -(1 - k_1) \frac{\rho_S}{\rho_L} \frac{1}{g_L} \frac{\partial g_L}{\partial t} \tag{10}$$

$$\frac{D \ln C_{L2}}{Dt} = -(1 - k_2) \frac{\rho_S}{\rho_S k_2 (1 - g_L) + \rho_L g_L} \frac{\partial g_L}{\partial t} \tag{11}$$

where:

$$\frac{D}{Dt} = \frac{\partial}{\partial t} + \underline{w} \cdot \nabla \text{ is the substantive derivative} \tag{12}$$

Taking the ratio of (10) and (11) yields:

$$\frac{D \ln C_{L2}}{D \ln C_{L1}} = \frac{(1 - k_2)}{(1 - k_1)} \frac{\rho_L g_L}{[\rho_L \, g_L + (1 - g_L) \, \rho_S \, k_2]} \tag{13}$$

An expression giving the depression of the liquidus due to solute enrichment is also required, i.e.

$$T_L = T_L \, (C_{Li}) \ , \ i = 1, 2, \ldots \tag{14}$$

as well as the variation of the density with composition and temperature, i.e.

$$\rho_L = \rho_L \, (T, C_{Li}) \ , \ i = 1, 2, \ldots \tag{15}$$

Parameters of the Model

For multicomponent alloys, the partition ratios and Expressions 14 and 15 must be determined or estimated. For a dilute multicomponent alloy, estimates derived from binary phase diagrams provide good approximations (26). For a concentrated solution, it is desirable to determine experimentally as many parameters as practical. The procedure followed to derive the model parameters for the solidification of a "Waspaloy" alloy ingot is given as one of the possible approaches.

Partition Ratios. A simple method is followed based on the following argument. If the partition ratios k_1 and k_2 are constant, then Equation 9 may be written as:

$$\frac{D \ln (C_{s2/k2})}{D \ln (C_{s1/k1})} = \frac{1 - k_2}{1 - k_1} \tag{16}$$

If microchemistries are obtained at two random locations A and B in an as-cast sample, and if the assumptions (constant k's, no diffusion in solid, one solid phase) are valid, then Equation 16 yields:

$$\frac{\ln C_{S2}\big|_A - \ln C_{S2}\big|_B}{\ln C_{S1}\big|_A - \ln C_{S1}\big|_B} = \frac{1 - k_2}{1 - k_1} \tag{17}$$

The nominal composition of "Waspaloy" alloy is 19.5% Cr, 13.5% Co, 4.3% Mo, 3.0% Ti, 1.5% Al, 0.8% C, bal. Ni. Microchemistries were obtained at random locations on as-cast samples, and the logarithm of the concentrations of chromium, cobalt, molybdenum, and aluminum were plotted against the logarithm of a concentration of titanium. Locations corresponding to primary carbides were disregarded. A linear correlation was obtained in all the four cases, namely chromium, cobalt, molybdenum, and aluminum. If the partition ratio of one element is known, Equation 17 and these correlations provide a rapid estimation of the partition ratios for the other elements. The partition ratio for titanium was estimated to be 0.80 and that for carbon to be 0.25. The following partition ratios were then used in the model:

Titanium:	0.80
Aluminum:	0.80
Molybdenum:	0.90
Cobalt:	1.10
Chromium:	1.00
Carbon:	0.25

Depression of the Liquidus. The liquidus temperature was measured by differential thermal analysis for eight alloys enriched in solute content and the range of composition covered was: Ti, 3% to 4.5%; Al, 1.3% to 3%; Mo, 4.3% to 7.6%; Cr, 19% to 22.4%; C, 0.09% to 0.14%. The following equation predicts the liquidus of the eight samples to within 1°C.

$$T_L\ (°C) = 1364 - 15.2\ (\%\ Ti - 3.00) + 0.4\ (\%\ Co - 13.50) - 9.4\ (\%\ Al - 1.50) - 2.8\ (\%\ Cr - 19.50) - 4.6\ (\%\ Mo - 4.30) - 104\ (\%\ C - 0.08) \quad (18)$$

The coefficient for cobalt was estimated from the binary cobalt-nickel diagram. The coefficients in this linear relation are of the same order of magnitude as the values that would be derived from binary phase diagrams. The experimentally measured depression of the liquidus by the solutes has, however, a consistently larger effect than would be expected from binary phase diagrams. The coefficient for carbon should be viewed with caution since the composition range covered experimentally is narrow.

Composition of the Liquid as a Function of Temperature. It is now desired to invert Equation 18 to obtain the titanium concentration in the liquid as a function of temperature. Because of Equations 9 and 13, the concentrations of the other solutes are not independent from the concentration of titanium. Since Equation 13 also requires the knowledge of g_L, Scheil's model was used as an approximation. An auxiliary computer program is used to calculate the composition of the liquid as a function of temperature. The results are supplied to the macrosegregation model in the form of a numerical function. The enrichment in the various solutes as a function of temperature is shown in Figure 7.

Density as a Function of Composition and Temperature. The solid and liquid densities were calculated using literature data and assuming that the specific volume of the alloy is the weighted sum of the specific volumes of its constituents (27-29). However, for the alloying elements that do not solidify to form a close-packed structure, the specific volume of that element in a solid nickel-based superalloy was taken to be 5% lower than its specific volume in the liquid. Due to the lack of data for carbon, it was assumed that it had the same specific volume in the solid and in the liquid. Figure 8 shows the result of the density calculation for the interdendritic liquid and for the average solid. The latter density is obtained by integrating the specific volume of the last fraction solidified from $g_s = 0$ to its current value. A constant solid density of 7.625 g/cm^3 was used in the model. The liquid density is supplied to the model in the form of a numerical function.

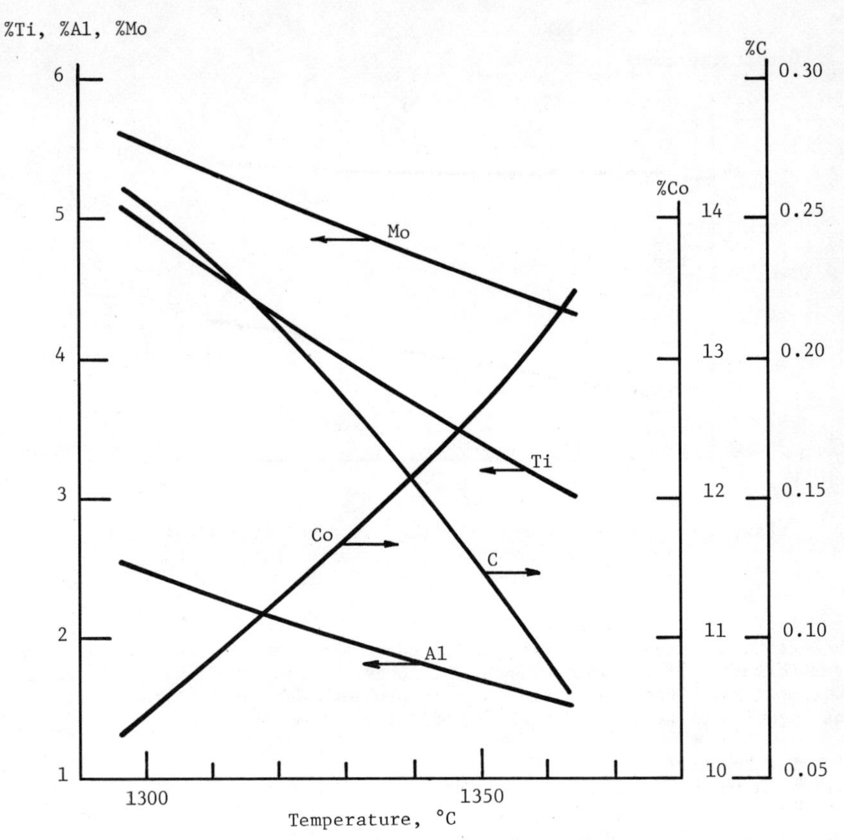

Fig. 7 - Calculated composition of the interdendritic liquid as a function of temperature for "Waspaloy."

Viscosity and Permeability. The value of the viscosity used in the model was 0.003 Pa·s. The permeability was taken to vary with the square of the volume fraction liquid, thus:

$$K = \gamma \, g_L^{\,2} \tag{19}$$

This behavior was demonstrated experimentally by Piwonka et al. (30) and by Apelian et al. (31) for $g_L \leq 0.35$. The coefficient γ was taken to be a constant and was used as an adjustable parameter. The results of the calculations which are presented here are for $\gamma = 6.10^{-5}$ mm^2.

Implementation of the Model

The partial differential equations (20-22 and 3) with pressure and g_L as dependent variables are to be solved jointly. The temperature field is assumed known. The first step taken was to transform the independent variables from the set r,z to r,T. A program to solve the finite difference

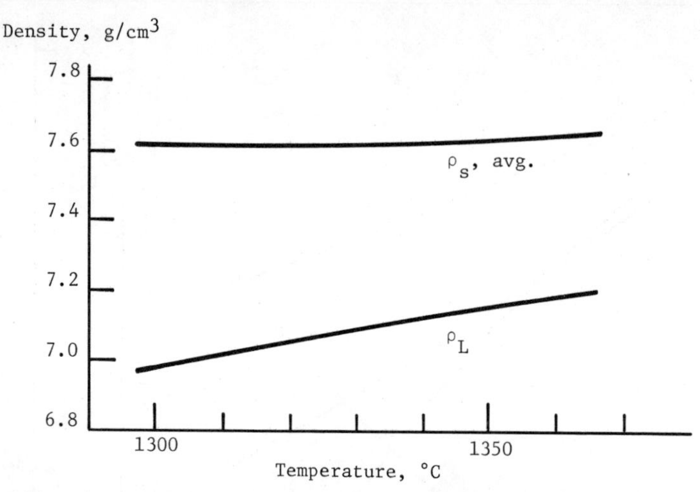

Fig. 8 - Calculated density of the interdendritic liquid and average density of the solidified fraction as a function of temperature for "Waspaloy."

analogs to the transformed PDEs was coded on the computer (PDP 11/70). The transformation of variables was done in order to keep the memory size requirement low. Computational stability problems were encountered in both the Jacobi and the Gauss-Seidel methods of solution.

The computer program was therefore modified so that it solves the finite difference analogs to the original PDEs. The procedure followed during a typical run of the program is given here in detail. First, the shapes of the liquidus, of the solidus, and of several intermediate isotherms are calculated or deduced from experimental data. The shapes of the liquidus and of the solidus are entered into a computer program that automatically generates a mesh in the mushy zone. A resulting mesh is sketched in Figure 9. The mesh line locations are adjusted so that the solidus intersects the mesh at a grid point. The pressure and the volume fraction of liquid are known at the liquidus line as boundary conditions to the problem. Therefore, the points where mesh lines intersect the liquidus line are not regular grid points at which the finite difference equations are to be solved. In a typical example, which is described here, 25 vertical and 33 horizontal lines are used. The number of mesh points is 281, which is less than 33 x 25, since there is no need for mesh points in the fully liquid and in the fully solid regions. The grid points are indexed in a manner which minimizes the computer memory requirements.

A second auxiliary program is then run to generate the numerical functions that relate the composition and density of the interdendritic liquid to the temperature. The pressure and volume fraction of liquid are initialized. The pressure is calculated by the successive overrelaxation iterative method (SOR). The SOR parameter value is 1.5. With 300 SOR iterations, convergence to within the machine precision is reached. The velocity field is then calculated, and this is followed by the calculation of the volume fraction of liquid as a function of position and by the calculation of the resulting local composition of the fully solidified alloy. The pressure is

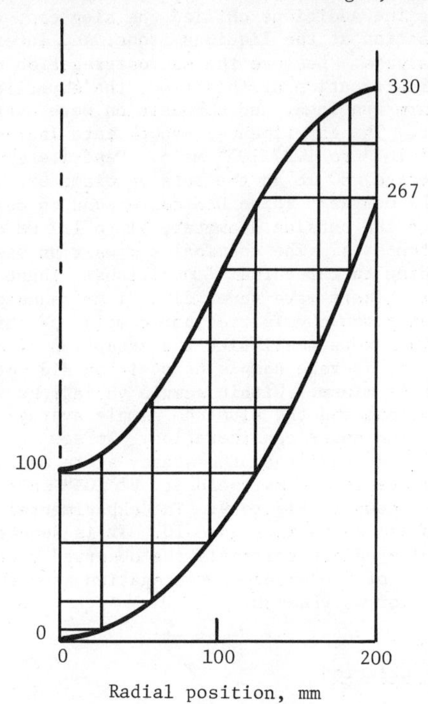

Fig. 9 - Solidus and liquidus profiles that were used in the macrosegregation calculation. The coarse grid illustrates the manner in which the finite difference mesh is constructed.

then calculated anew, and the sequence is repeated until the average local composition has converged. In this example, three of these sequences were sufficient to reach convergence.

Testing of the Model

A test of the internal consistency of the program is to examine whether the average composition of the solidified ingot matches exactly the average composition of the starting liquid. For the mesh size presented in the example cited here, the largest relative discrepancy was for the element titanium. The calculated final average composition is 3.03 percent as opposed to a starting composition of 3.00 percent; a relative discrepancy of one percent.

Limited experimental data are available at this time to test the model. Local average compositions were determined in a 40-cm diameter experimental "Waspaloy" alloy ingot. The region for which the measurements were made corresponds to a transient melt rate condition. The purpose of the transient was to test the heat transfer model and to deliberately induce freckle formation. Elongated channel segregates were formed at mid-radius. The tungsten powder used for marking the pool shape was found to interfere with the experiment. The additions chilled the slag cap, caused nucleation and rapid heat extraction at the liquidus front, and interfered with sampling for chemical analysis. Because the macrosegregation model treats only the steady state solidification at this time, the experimental values of melt rate, dendrite arm spacings, and composition were averaged over the duration of the transient. The experimental growth rate increased during this period from 0.85 10^{-4} to 1.27 10^{-4} ms^{-1}. Dendrite arm spacing measurements were, on the average, 80 μm at the outside diameter, 120 μm at mid-radius, and 100 μm at the center. These DAS correspond to mushy zone thicknesses of 30 to 46 mm at the outside diameter, 91 to 137 mm at mid-radius, and 55 to 83 mm at the axis. The chemical composition was determined for locations corresponding to three radial positions: ingot axis, mid-radius, and outside diameter. There were seven 21 x 21 mm^2 samples per radial position and five electron probe chemistries per sample. Within a sample, the chemical variations were small with the exception of aluminum and of the freckled area. The average sample composition did not change as the ingot grew, except for aluminum. Within sample variations of 0.5 percent were observed for aluminum, and the aluminum sample average decreased by 0.5 percent over the period under consideration. To see if the correct composition trends could be predicted, the steady state macrosegregation model was used with an average isothermal velocity of 10^{-4} ms^{-1} and the liquidus and solidus profiles shown in Figure 9. The experimental averages and the calculated profiles are shown in Figure 10. It is seen that the steady state segregation model predicts correctly the observed macrosegregation trend but that the magnitude of the observed segregation is much larger than predicted for aluminum and for molybdenum.

Discussion and Conclusion

The addition of tungsten powder and the transient melt rate conditions were experimental conditions not taken into account in the model. X-ray emission and x-ray fluorescence chemical analysis data from a production ingot indeed show that the model gives reasonable predictions of both the trend and the extent of macrosegregation in this alloy.

Several simplifying assumptions in the macrosegregation model deserve much futher attention:

(a) In some alloys, the primary carbide phase forms close to the liquidus temperature, and the calculation of the rate of enrichment of the interdendritic liquid is affected by the relative rates of growth of the two solid phases.

(b) The magnetic body force is not calculated in this model; yet, like gravity, it drives the interdendritic fluid flow. It is expected to play a significant role for combinations of ingot diameter and current frequency for which the electrical skin depth is less than the ingot radius.

(c) In the model, the permeability is assumed to vary as the square of the volume fraction of liquid. This has only been well established experimentally up to $g_L \simeq 0.35$.

(d) Finally, it has been implicitly assumed that there is a stable solution to the macrosegregation problem and that this solution is unique. This assumption certainly deserves further analysis.

It is believed that a refined treatment that considers these effects

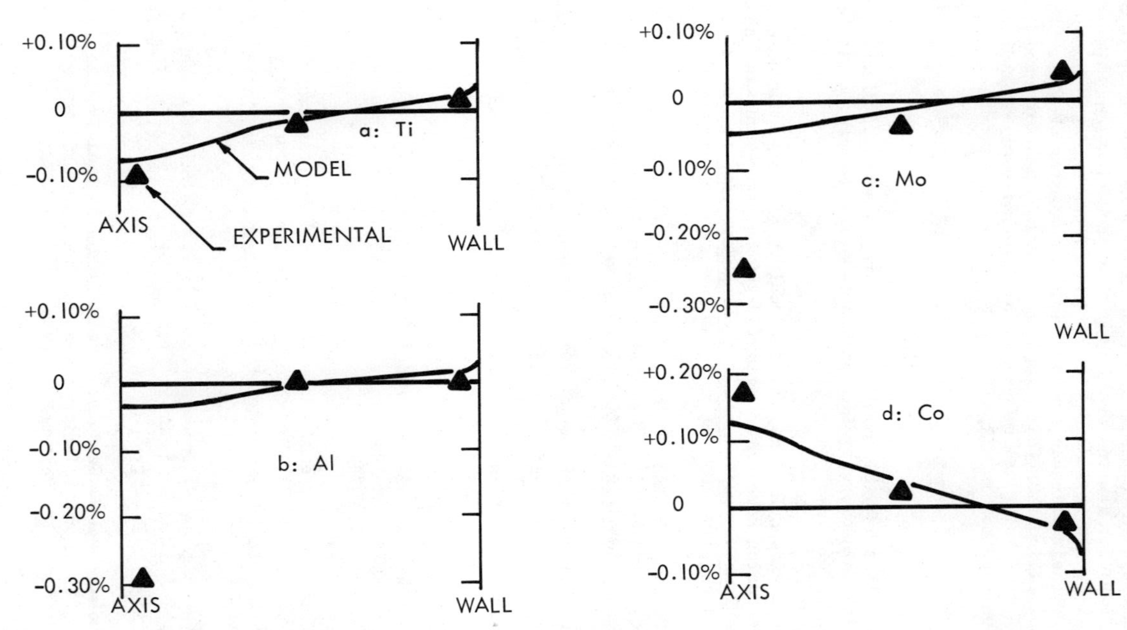

Fig. 10 - Difference between local ingot composition and average composition as a function of radial position.

would no doubt improve the validity of the present model and that more work needs to be done on this important subject.

Acknowledgements

The authors wish to thank the National Science Foundation (Grant Number DMR07819699) for its support during this investigation. They also wish to thank Drs. R. Mehrabian and S. D. Ridder of the National Bureau of Standards for their valuable suggestions and comments. While at the University of Illinois at Champaign-Urbana, Drs. R. Mehrabian and S. D. Ridder participated in a joint Cabot Corporation-University of Illinois research program on segregation in ESR.

References

1. A. Mitchell and R. M. Smailer, "Practical Aspects of Electroslag Remelting Technology," International Metals Reviews, 24 (5-6) (1979) pp. 231-264.

2. T. Mochizuki and E. Ohga, "On the Quality of Forged Hardened Rolls Made by Electroslag Remelting," pp. 253-265 in Proceedings of the Fourth International Symposium on Electroslag Remelting Processes. ISIJ, Tokyo, Japan, 1973.

3. D. R. Poirier, S. Kou, T. Fujii, and M. C. Flemings, "Segregation in an Electroslag Remelted Ingot of Maraging Steel," pp. 91-118 in Electroslag Remelting, AMMRC TR 78-28 Report, June 1978.

4. A. F. Giamei and B. H. Kear, "On the Nature of Freckles in Nickel-Base Superalloys," Metallurgical Transactions, 1 (8) (1970) pp. 2185-2192.

5. R. Mehrabian, M. Keane, and M. C. Flemings, "Interdendritic Fluid Flow and Macrosegregation; Influence of Gravity," Metallurgical Transactions, 1 (5) (1970) pp. 1209-1220.

6. A. S. Ballantyne and A. Mitchell, "Modeling of Ingot Thermal Fields in Consumable Electrode Remelting Processes," Ironmaking and Steelmaking, (4) (1977) pp. 222-239.

7. J. H. Chen, R. C. Myers, and D. R. Engel, "Computer Control of the ESR Process," pp. 831-847 in Proceedings of the 6th International Vacuum Metallurgy Conference, G. K. Bhat and R. Schlatter, eds. San Diego, California, 1979.

8. J. Kreyenberg and K. Schwerdtfeger, "Stirring Velocities and Temperature Field in the Slag During Electroslag Remelting," Arch. Eisenh., 50 (1) (1979) pp. 1-6.

9. M. Choudhary and J. Szekely, "The Modeling of Pool Profiles, Temperature Profiles, and Velocity Fields in ESR Systems," Metallurgical Transactions, 11B (September) (1980) pp. 439-453.

10. A. Mitchell, Questions and Answers Session at the 6th International Vacuum Metallurgy Conference, April 23-27, 1979, San Diego, California.

11. C. L. Jeanfils, J. H. Chen, and H. J. Klein, "Temperature Distribution in an Electroslag Remelted Ingot During Transient Conditions,"

pp. 543-555 in Proceedings of the 6th International Vacuum Metallurgy Conference, G. K. Bhat and R. Schlatter, eds., San Diego, California, 1979.

12. A Grill, K. Sorimachi, and J. K. Brimacombe, "Heat Flow, Gap Formation, and Break-Outs in the Continuous Casting of Steel Slabs," Metallurgical Transactions, 7B (June) (1976) pp. 177-189.

13. W. E. Pfann, "Principles of Zone-Melting," Transactions AIME, 199 (7) (1952) pp. 747-751.

14. A. I. Veinik, Thermodynamics for the Foundryman, Maclaren and Sons, London, 1968.

15. B. E. Paton, B. I. Medovar, V. L. Shevtzov, C. S. Marinsky, and V. I. Sagan, "Study of Heat Exchange in Electroslag Remelting According to Different Routes," pp. 410-420 in Proceedings of the Fifth International Symposium on Electroslag Remelting and Other Special Melting Technologies, G. K. Bhat and R. Simkovich, eds., Pittsburgh, PA, 1974.

16. Private Comminication, S. D. Ridder , University of Illinois, Dec., 1979.

17. M. C. Flemings and G. E. Nereo, "Macrosegregation, Part I," Transactions TMS-AIME, 239 (9) (1967) pp. 1449-1461.

18. M. C. Flemings, R. Mehrabian, and G. E. Nereo, "Macrosegregation, Part II," Transactions TMS-AIME, 242 (1) (1968) pp. 41-49.

19. M. C. Flemings and G. E. Nereo, "Macrosegregation, Part III," Transactions TMS-AIME, 242 (1) (1968) pp. 50-55.

20. S. Kou, D. R. Poirer, and M. C. Flemings, "Macrosegregation in Rotated Remelted Ingots," Metallurgical Transactions, 9B (December) (1978) pp. 711-719.

21. S. D. Ridder, F. C. Reyes, S. Chakravorty, R. Mehrabian, J. D. Nauman, J. H. Chen, and H. J. Klein, "Steady State Segregation and Heat Flow in ESR," Metallurgical Transactions, 9B (September) (1978) pp. 415-425.

22. S. D. Ridder, S. Kou, and R. Mehrabian, "Effect of Fluid Flow on Macrosegregation in Axi-Symmetric Ingots," Proceedings of this Conference.

23. D. N. Petrakis, D. R. Poirer, and M. C. Flemings, "Some Effects of Forced Convection on Macrosegregation," Proceedings of this Conference.

24. R. Mehrabian and M. C. Flemings, "Macrosegregation in Ternary Alloys," Metallurgical Transactions, 1 (2) (1970) pp. 455-464.

25. T. Fujii, D. R. Poirier, and M. C. Flemings, "Macrosegregation in a Multicomponent Low Alloy Steel," Metallurgical Transactions, 9B (September) (1979), pp. 331-339.

26. M. C. Flemings, "Principles of Control of Soundness and Homogeneity of Large Ingots," Scandinavian Journal of Metallurgy, 5 (1) (1976) pp. 1-15.

27. L. D. Lucas, "Liquid Density Measurements," pp. 268-284 in Physiochemical Measurements in Metals Research, Vol. IV, Part 2, R. A. Rapp, ed.; Interscholastic Publications, New York, 1970.

28. A. Goldsmith et al., eds., Thermophysical Properties of Solid Materials, Vol. 12, McMillan Co., New York, 1969.

29. Y. S. Touloukian et al., eds., Thermophysical Properties of Matter, Vol. 12: Thermal Expansion-Metallic Elements and Alloys, Plenum Publishing Co., New York, 1975.

30. T. S. Piwonka and M. C. Flemings, "Pore Formation in Solidification," Transactions TMS-AIME, 236 (8) (1966) pp. 1157-1165.

31. D. Apelian, M. C. Flemings, and R. Mehrabian, "Specific Permeability of Partially Solidified Dendritic Networks of Al-Si Alloys," Metallurgical Transactions, 5 (12) (1974) pp. 2533-2537.

INTERACTION BETWEEN COMPUTATIONAL MODELING AND EXPERIMENTS FOR VACUUM CONSUMABLE ARC REMELTING*

L. A. Bertram and F. J. Zanner
Sandia National Laboratories
Albuquerque, New Mexico 87185

A combined computational-experimental modeling effort is currently being conducted at Sandia National Laboratories to characterize the vacuum consumable arc remelt process. This effort involves the coupling of experimental results with a magnetohydrodynamic flow model which is capable of producing time accurate solutions of the interdependent fluid flow-solidification processes in the ingot. At this time, the modeling and experimental efforts are in the development stage. Therefore, the results presented in this work are intended to provide only a qualitative picture of the Lorentz and buoyancy-induced flows.

Models such as this are driven by boundary conditions. Considerable data have been compiled from direct observations of the electrode tip and molten pool surface by means of high-speed photography, producing a better understanding of the processes at the pool surface and the appropriate corresponding boundary conditions. The crucible wall/molten metal meniscus conditions are not yet as well understood.

Pool volumes computed at different melting currents are in reasonable agreement with experimentally determined values. Current flow through the ingot is evaluated numerically, and the results indicate that a significant portion of the melt current does not reach the interior of the ingot.

*This work supported by the U.S. Department of Energy.

I. Introduction

Vacuum consumable arc remelting is a casting process carried out in vacuum with the aim of remelting the consumable electrode in such a way that the new ingot has improved chemical and physical homogeneity. The energy which causes the remelting is supplied by a vacuum arc between the bottom surface of the eletrode (cathode) and the ingot top surface/crucible wall (anode), as shown in Figure 1. Casting rate ("melt rate"), $\dot{m}$, is controlled by varying the melt current, I_m.

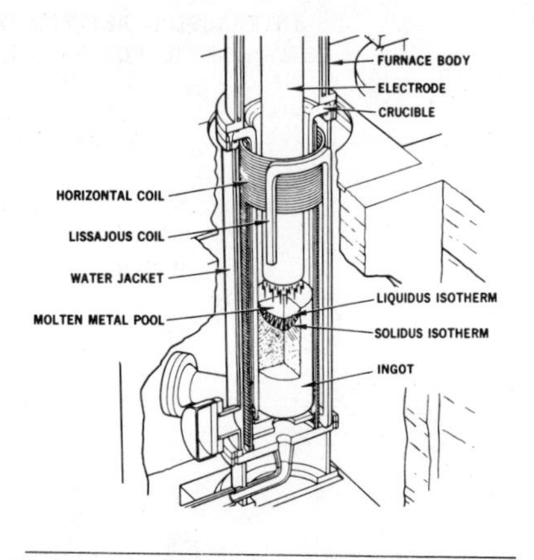

Figure 1. Vacuum consumable arc remelt furnace cut-away view.

The primary reason for using this process is to increase the homogeneity of the resulting ingot. However, its application is limited by the onset of significant macrosegregation in segregation sensitive alloy systems. Since increasing the ingot diameter increases the propensity for macrosegregation, segregation sensitive alloys are limited to small ingot diameters. This is believed to be caused by adverse fluid flow and solute transport within the melt pool and the adjacent interdendritic region. The goal of the research program described below is to understand how melt parameters such as I_m relate to fluid flow, solute transport, and the resulting ingot homogeneity.

The U-6w/oNb alloy used in this research was chosen because of the following features:

1. Intense macrosegregation is observed in small diameter ingots (0.208 m).

2. Niobium is completely soluble in both liquid and solid uranium near the solidification temperature range.

3. Niobium diffusion in solid uranium is low.

In particular, the macrosegregation observed in U-6w/oNb vacuum arc remelted ingots can be subdivided into three types. First, and most conspicuously (Figure 2), bands of varying concentration can be seen. Second, the composition of the alloy can vary with distance from the axis (Figure 3). Finally, the ends of the ingot may have different solute contents than the interior which was formed under quasi-steady solidification conditions.

The actual development of macrosegregation clearly must be a consequence of the conditions on the moving solid-liquid interface in the ingot. As shown by Flemings, Mehrabian, and others (1,2,3) over the past several years, a quantitative connection can be made between the fluid flow and the thermal environment in the interdendritic spaces of the "mushy zone." However, this theory requires that these conditions be determined with great precision and detail in order that predictions of macrosegregation become possible. It is the requirement of great detail and the accompanying wish to maximize reliability and relevance of conclusions which have motivated the combined numerical (detailed) and experimental (realistic) analysis of vacuum consumable arc remelt casting.

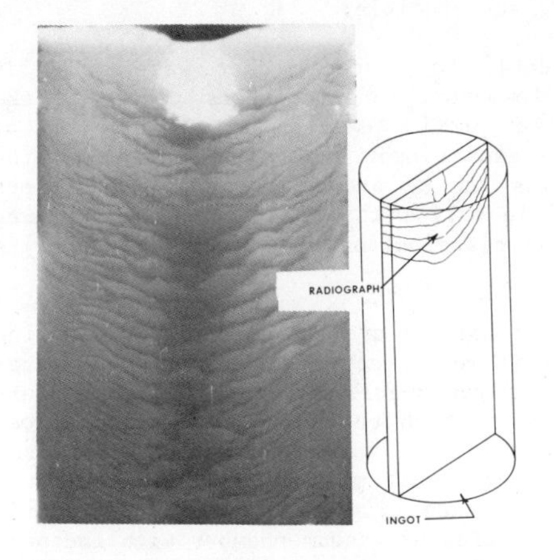

Figure 2. Positive radiograph of VAR melted 8-in (0.208 m) ingot. Nb content varies from 4.5 w/o inside bands to 8.5 w/o between bands.

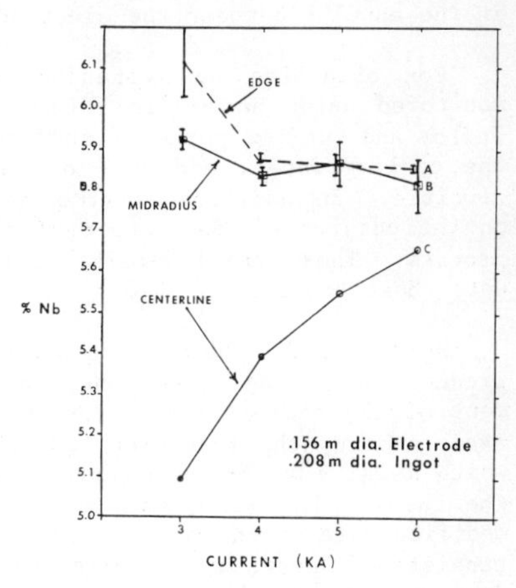

Figure 3. Centerline to edge macrosegregation in the ingot of Figure 2 as a function of melt current.

At this time, the modeling-experimental efforts are still in the developmental stage so the results presented here are intended to provide only a qualitative picture of the Lorentz and buoyancy-induced flows in the region ahead of the dendrites and in the part of the mushy zone where they are very coarse (volume fraction solid $g_s \lesssim 0.5$).

Basically, the Sandia research program is driven by the data obtained from an instrumented 10-in (0.254 m), highly coaxial (axisymmetric) furnace. The instrumentation is briefly described in Section II, followed in Section III by the formulation of a magnetohydrodynamic flow model. The model is capable of time-accurate simulation of the coupled convection-solidification processes in the ingot. Section IV describes experiments from which "first-generation" boundary conditions are constructed and presents the resulting boundary conditions. Finally, in Section V, a parameter study using the model produces predicted pool volumes as a function of melting current and measured melt rate. These calculated volumes are compared to experimentally determined values in order to infer the fraction of current passing through the melt pool.

II. Experimental Facilities

The basic measurements made on the furnace when it operates in a normal production mode (as depicted in Figure 1), consist of melt current I_m, voltage across the bus bars V_m, furnace pressure p_m, and furnace gas composition. Melt current is measured with a shunt in series with the negative bus bar and simultaneously by a Hall-effect transducer. Both I_m and V_m are recorded at a 5 kHz resolution on magnetic tape and are subsequently analyzed on a minicomputer. Furnace gas composition is obtained from a mass spectrometer which obtains its samples through a differentially pumped quartz tube with a 0.015-in (0.00038 m) orifice placed approximately 3 in (0.076 m) above the finished ingot's top surface in the annulus between the electrode and crucible.

For obtaining heat transfer data, the water jacket water flow is monitored using an orifice plate flowmeter. Thermocouples placed at the inflow and outflow ports of the jacket provide data on temperature rise of the cooling water, and thus on the global heat removal rate through the crucible. In addition, thermocouples can be placed in various positions on the outside of the copper crucible wall without affecting the melting process. These yield details on the distribution of the crucible-cooling water heat transfer.

While all the above-described measurements are conveniently made during production melting conditions, none allows direct observation or measurement of the active zone of the vacuum consumable arc remelt casting furnace; namely, the interelectrode gap in which the arc operates and across which metal must be transferred from the electrode to the melt pool atop the ingot. To obtain such data, experimental melting is performed in a modified arrangement of the standard furnace (Figure 4). The modification consists of inserting an additional section of vacuum chamber with independent cooling, sight ports, and instrument ports, between the crucible flange and the furnace body. Inside the crucible, a movable stool ("retractable hearth") is mounted on a hydraulically driven screw mechanism which can withdraw it at the rate new ingot forms. This retractable hearth arrangement allows long-term observation of the interelectrode gap since electrode feed and ingot withdrawal can be matched so that the gap remains in the field of view. The sight ports of the chamber are arranged for both direct viewing across the gap and for oblique viewing of a large fraction of the pool surface (the electrode blocks a full view of the surface). Both high-speed color movies (~10,000 frames/ses) and color still photographs have been made in this way.

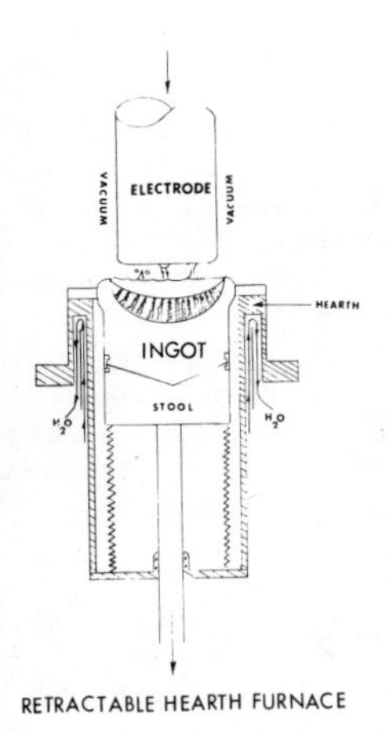

a. Retractable hearth schematic b. Melting chamber and observation port

Figure 4. The Sandia retractable hearth apparatus

Besides these measurements made at Sandia National Laboratories, additional data has been obtained at the Union Carbide Nuclear Division's Y-12 plant. This support consisted of melting an 8-in (0.208 m) uranium-6w/o niobium ingot, sectioning the resulting material, and doing radiography on the sections. (Figure 2 is a positive taken from these radiographs.) For the reasons noted in the Introduction, the U-Nb system is ideal for segregation studies, so the majority of the analyses done to date have involved this ingot.

For the radiographic information to be reduced to quantitative data, it is necessary to combine specimen thickness information, calibration of the photographic emulsions (provided by step wedges of pure U and of U-6w/oNb which were radiographed along with the specimens), and the raw radiographs to calculate pointwise densities. This is currently being done by computer-based image processing techniques (4). More detailed descriptions of experimental apparatus and technique is found in the references (5,6,7).

Fortuitously, the strong banding of this alloy provides an additional datum; namely, if it is true that the bands are created during the latter stages of solidification, then they serve as reliable markers of the outline of the melt pool at the time of their formation. To be precise, what the bands represent is contours of some fraction solid (not, of course, $g_s = 1$, the solidus). In this report, "measured pool volume" will refer to the volume enclosed by the solid of revolution generated by a band and the plane through the intersections of the band with the edge of the ingot.

III. Mathematical Model

In order that the numerical model have as simple a form as possible while retaining the relevant physics for the determination of the macroscopic heat and fluid transfer, the following forms of the governing equations are applied:

$$\frac{\partial \zeta}{\partial t} + \frac{\partial (u\zeta)}{\partial x} + \frac{\partial (v\zeta)}{\partial r} = \Pr\left[\frac{\partial^2 (\nu\zeta)}{\partial x^2} + \frac{\partial}{\partial r}\left(\frac{\partial (r\nu\zeta)}{r\partial r}\right) + S\right] + \Pr^2 Gr\,\frac{\partial \theta}{\partial r} - \frac{1}{A_o^2}\frac{\partial B^2}{r\partial x}$$

$$\text{with } S = 2\left(\frac{\partial v}{\partial r} - \frac{\partial u}{\partial x}\right)\frac{\partial^2 \nu}{\partial x \partial r} + 2\left(\frac{\partial u}{\partial r}\frac{\partial^2 \nu}{\partial x^2} - \frac{\partial v}{\partial x}\frac{\partial^2 \nu}{\partial r^2}\right) \quad . \tag{1}$$

$$\frac{\partial h}{\partial t} + \frac{\partial (uh)}{\partial x} + \frac{\partial (rvh)}{r\partial r} = \frac{\partial}{\partial x}\left(\kappa \frac{\partial h}{\partial x}\right) + \frac{\partial}{r\partial r}\left(r\kappa \frac{\partial h}{\partial r}\right) + \frac{j^2}{\Sigma} \quad . \tag{2}$$

$$\frac{\partial^2 \psi}{\partial x^2} + r\frac{\partial}{\partial r}\left(\frac{\partial \psi}{r\partial r}\right) = -r\zeta \quad . \tag{3}$$

$$u = \frac{\partial \psi}{r\partial r} \qquad v = -\frac{\partial \psi}{r\partial x}. \tag{4}$$

where (x,r) are the axial and radial cylindrical coordinates scaled by ingot radius R having origin on the melt pool axis at the surface; t is time, scaled by R^2/κ_o, κ_o being the reference value of the thermal diffusivity; ζ is vorticity, $= \partial v/\partial x - \partial u/\partial r$ in terms of the velocity components (u,v); A_o^2 is the square of an Alfvén number based on thermal diffusion speed, $A_o^2 = (\kappa_o/R)^2/(B_o^2/\mu\rho_m)$; B is the magnetic induction azimuthal component scaled by $B_o = \mu I_m/2\pi R$; θ is temperature difference from the mid-mushy zone temperature $T_o = \frac{1}{2}(T_L + T_s)$, scaled by $\Delta T_o = \frac{1}{2}(T_L - T_s)$ with $T_{L,s}$ being respectively liquidus and nonequilibrium solidus temperatures; Pr is Prandtl number, ν_o/κ_o with ν_o being a reference kinematic viscosity of the liquid metal; Gr is Grashof number $= g\alpha\Delta T_o R^3/\nu_o^2$ with g being acceleration due to gravity and α being the volumetric coefficient of thermal expansion of the liquid; h is the enthalpy, scaled by $\frac{1}{2}L$ with L being the latent heat of fusion; j is current density, scaled by $I_m/\pi R^2$; and Σ is a Joule heating parameter, $\Sigma = \rho_m L\sigma\kappa_o R^2/2I_m^2$ with ρ_m being mass density and σ the electrical conductivity of the liquid. The only variation in mass density considered is contained in the term of (1) multiplied by Gr, which assumes that $\hat{\rho}_m = \rho_m(1 - \alpha[T - T_r])$, T_r being some reference temperature in the liquid ("Boussinesq approximation"). The function $\psi(x,r,t)$ is the stream function, introduced to obtain exact conservation of volume. When the flow is steady, contours of constant ψ ("streamlines") are also particle paths.

The essential physics of the problem can be seen in Equation (1), where the rightmost pair of terms represent the sources of vorticity due to radial temperature gradient $\partial\theta/\partial r$ and axial magnetic pressure gradient $\partial B^2/\partial x$, respectively. The two terms are of opposite sign, and are almost exactly the same magnitude for the ingot sizes considered here. Thus the net motion is determined as the result of near-total cancellation of these two opposing driving forces (buoyancy and Lorentz forces).

Accompanying Equations (1)-(4) are the following equations of state:

$$\kappa(\theta) = c_L k(\theta)/(|\nabla h|/|\nabla T|) \tag{5}$$

with the $|\nabla h|/|\nabla T|$ value being calculated from the points in the neighborhood rather than specified as a function of θ, and with k being the dimensionless thermal conductivity, with $c_L = 2C_{po}\Delta T_o/L$, and C_{po} being a reference specific heat (the same value used in defining κ_o);

$$\nu = \nu(\theta) \tag{6}$$

being given as a closed-form function (curve fit); and finally,

$$h = h(\theta) \tag{7}$$

being specified as an "average" nonequilibrium cooling history by assigning the regular-mixture enthalpy of the components (liquid and solid) which would be present according to the Scheil microsegregation equation (8). Rather than calculating the Scheil prediction of fraction liquid and the solid composition every time an enthalpy value is desired, the curve $h(\theta)$ is computed once and a piecewise linear curve fit is made to it. This linear function is then used to assign $h(\theta)$ (7).

Equations (1)-(7) are solved numerically in a rectangular region representing a cross section of the ingot top (Figure 5). The actual numerical operations consist of solving the two parabolic Equations (1), (2) by an explicit, conservative, upwind-differenced scheme, and then performing a symmetric successive over-relaxation on the elliptic Equation (3). For efficiency, these steps are carried out in analytically strained cylindrical coordinates (7).

Explicit field equations for the electromagnetic variables B and j are not included because these have been treated by a fully decoupled magnetostatic approximation. This is possible because of the small size of the ingots and the low fluid velocities which together imply small magnetic Reynolds number $Rm = U_o R \mu \sigma$, and thus small induced fields (10). In particular the dimensionless azimuthal magnetic induction vector component is taken to be

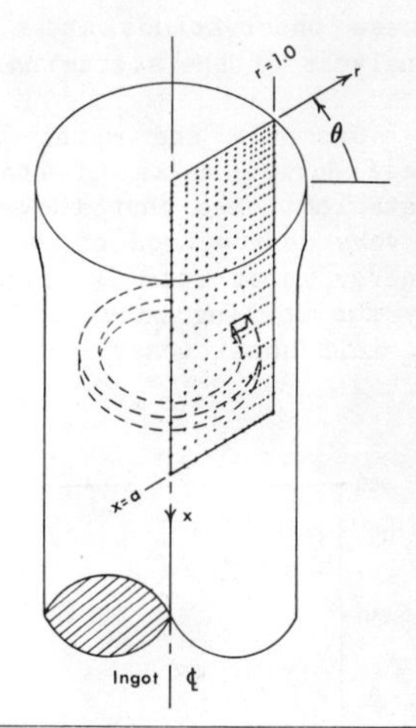

Figure 5. Coordinates and finite difference grid for numerical solutions.

$$B(x,r) = \frac{J_1(\beta_o r)\ \cosh\beta_o(a - x)}{J_1(\beta_o)\ \cosh\beta_o a}\ C_B \tag{8}$$

where a is the aspect ratio of the computational zone (Figure 5), $J_1(\cdot)$ is the Bessel function of order one, and β_o is the first zero of the zero-order Bessel function. Once B(x,r) is given, Ampere's equation determines j(x,r). Note that a parameter $0 \leq C_B \leq 1$, the fraction of melt current passing through the pool, has been inserted because the detailed current budget in a vacuum consumable arc remelt casting furnace has not been measured to the authors' knowledge.

IV. Boundary Conditions

A. Initial Experiments

The facilities described in Section II have yielded a mixture of qualitative and quantitative information concerning the vacuum consumable arc remelt casting process. From the retractable hearth experiments (6,9), which were all done with stainless steel or superalloy ingots, direct observation has determined that the mode of metal transfer is chiefly bulk liquid transport; that individual vacuum arcs behave as they do in vacuum switches, etc. (i.e., individual cathode spots were created near the centerline of the electrode face, do a high-speed random walk with a mean radial speed of about 10^2 m/s, and are extinguished at the edge of the electrode or in some cases, on the lateral surface of the electrode); that the bridging of the electrode gap by a liquid droplet causes immediate (on time scales shorter than 10^{-4} s) extinguishing of all cathode spots so that all melt current must pass through the resulting "molten wire" between the electrodes; and that both pool surface and electrode surface are at low uniform superheats, probably < 200 K (7). An immediate consequence of these observations is a strategy for controlling electode gap based on analysis of the electrical waveforms $V_m(t)$ (9).

Crucible and water jacket thermocouple measurements have also been made during melts of stainless steel (see Figure 6). This quantitative data indicates that the significant heat removal from the ingot occurs at a very narrow band of contact on the crucible wall and that the electrical energy input rate is nearly matched by the rate of thermal energy removal by the cooling water (after the startup transient, under constant current I_m melt conditions).

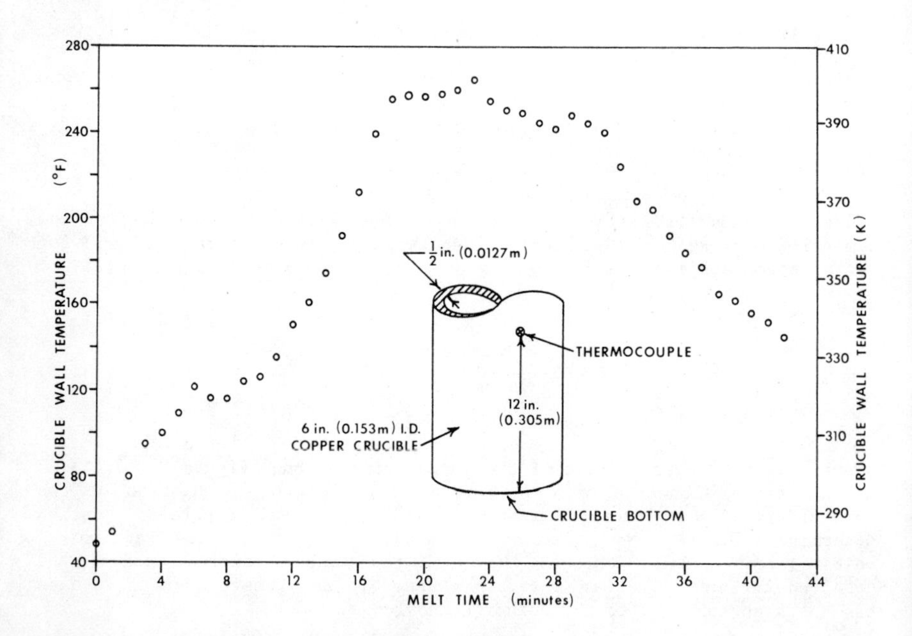

Figure 6. Measured temperatures on crucible wall during maraging steel melting. Ingot growth rate 0.51 in/min (2.1×10^{-4} m/s).

B. Initial Numerical Boundary Conditions

Taken together, these observations and measurements suggest the following boundary conditions. First, the low observed superheats suggest

$$T^*(x = 0,r,t) = T_L + \Delta T_S^* H(R_e - r) \tag{9}$$

where $H(R_e - r) = \begin{cases} 1 & r \leq R_e \\ 0 & r > R_e \end{cases}$ is the Heaviside unit step function and ΔT_S^* is the superheat of the pool surface directly under the electrode (the value used in the calculations given below was $\Delta T_S^* = 122$ K). The surface also requires a stream function boundary condition (uniform inflow was used since pool velocities so greatly exceed inflow velocity) and a vorticity boundary condition (which was taken to be zero vorticity assuming that no net shear stress is applied to the pool surface). These conditions are

$$\psi(x = 0, r, t) = \psi_1 r^2 \tag{10}$$

and

$$\zeta(x = 0, r, t) = 0 \tag{11}$$

where $\psi_1 = u_o^*/(2\kappa_o/R)$ is a Peclet number based on average inflow speed $u_o^* = \dot{m}/\rho_m \pi R^2$.

On the lateral surface of the ingot (crucible wall), the flow conditions are the usual

$$\psi(x, r = 1, t) = \psi_1 \tag{12}$$

ζ is derived from $\partial u/\partial r)_{r=1}$ unless solidification has occurred at $r < 1$. (13)

The thermal boundary conditions at the wall are not so easily determined. The data measured on the outside of the crucible has a good deal of the detail smeared out by diffusion, and this detail cannot be restored by analytical means since this restoration would be a classical ill-posed boundary value problem. The situation is further complicated by the fact that the crucible wall temperatures indicate that heat transfer is probably a combination of film cooling and nucleate boiling. Thus exterior crucible wall data alone are not sufficient to provide a reliable accurate boundary condition. Furthermore, the detailed analysis discussed by Richmond and Wray earlier in this meeting (11), indicates that the behavior of the ingot-crucible contact is at least as complicated as the crucible-cooling water contact. A crude parameterization of the expected temperature dependent ability of the ingot to reject heat to the cooling water has been used with parameter values estimated rather than measured. Specifically, the contact zone is assumed to have a thermal resistance which varies with local temperature (see Figure 7):

$$-k^* \frac{\partial T^*}{\partial r^*} = \epsilon \sigma^* T_w^{*4} + \frac{T_w^* - T_{H_2O}^*}{R_T} \tag{14}$$

where the "w" subscript denotes "wall" values, and H_2O refers to the cooling water. The thermal resistance is defined in Figure 7. This form of R_T is chosen so that the minimum resistance R_o operates when temperature is above T_{RO}; weakening contact stress causes increasing R_T for intermediate temperatures; and finally, conduction ceases below T_R

when thermal contraction has caused the ingot to shrink out of contact with the wall entirely. Parameter values assigned have been: R_o = 1.0 W/cm^2-K (about four times the thermal resistance of the copper crucible alone); $T_{RO} = T_s$, the nonequilibrium solidus temperature; $T_R = T_s - 10^{-3}/\alpha_s$ where α_s is the volumetric coefficient of thermal expansion of the solid. This T_R means that the solid has undergone a 0.1 percent volumetric contraction at the break of wall contact; for U-6w/oNb alloy T_R was 200 K below T_s.

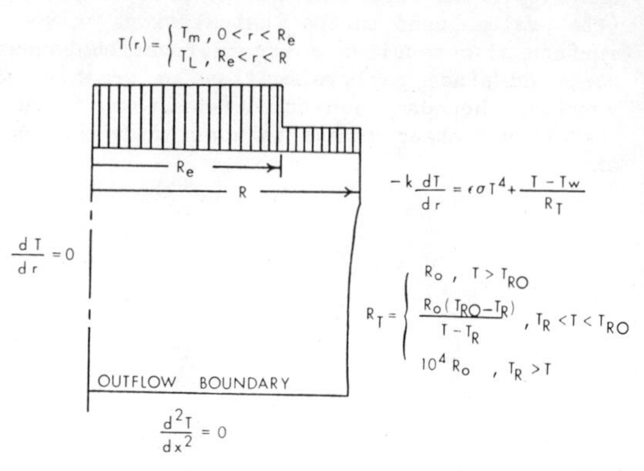

Figure 7. Thermal boundary condition summary (see text for notation).

On the axis of the ingot, symmetry provides boundary conditions unambiguously:

$$\psi(x,r = 0,t) = 0 \tag{15}$$

$$\zeta(x,r = 0,t) = 0 \tag{16}$$

$$\frac{\partial\theta}{\partial r}(x,r = 0,t) = 0 \quad . \tag{17}$$

At the bottom of a computational mesh, typical "outflow" boundary conditions are given on stream function and enthalpy (12):

$$\frac{\partial^2 h}{\partial x^2}(x = a,r,t) = 0 \tag{18}$$

$$\frac{\partial^2 \psi}{\partial x^2}(x = a,r,t) = 0 \tag{19}$$

and vorticity is zero because the code forbids melting on the outflow boundary. At interior solid surfaces (taken here to be points for which θ first drops below 0.333, i.e., where the Scheil equation first indicates that fraction solid exceeds 0.55), the stream function is specified to be rigid body motion

$$\psi = \psi_1 r^2 \tag{20}$$

and vorticity is computed from (12)

$$\zeta = \frac{\partial v_t}{\partial n} \tag{21}$$

where n is the normal to the solid surface (immobilization isotherm) pointing into the melt, and v_t is the velocity tangential to the surface at the adjacent liquid mesh point.

V. Limited Experimental and Computational Results

To briefly recapitulate the above, then, direct observation of the electrode gap region has produced several qualitative and quantitative improvements in the understanding of the vacuum consumable arc remelt casting process. The most important of these have perhaps been the observations of the transfer of metal from the consumable electrode to the pool by the bridging of the gap with liquid metal droplets, and the synchronization of these events with the characteristic millisecond lifetime waveforms of the furnace electrical signal (5,6,9). These observations have clearly established that the bulk of metal transfer is by liquid flow, and that when a droplet shorts across the gap, all arcs are extinguished.

These observations have also suggested the boundary conditions detailed above in Equations (9)-(21) for the ingot top. With these conditions as input, along with the melting rates $\dot{m}$ obtained simply by dividing the total ingot weight by melt time, it is possible to produce numerical models corresponding to the experimental melts. The utility of these models consists mainly in providing a base for a parameter sensitivity study as well as some qualitative insight into the phenomena which may occur in the system when it is driven by realistic inputs. Quantitatively, the only meaningful comparison which can now be made between the predictions and the observations is computed vs. measured pool volumes (see Figure 8). The computed volume is that enclosed by the immobilization isotherm, and thus represents a lower bound on the volume which is enclosed by a band formed at a temperature between immobilization (θ_I = 0.333) and nonequilibrium solidus (θ_s = -1.0).

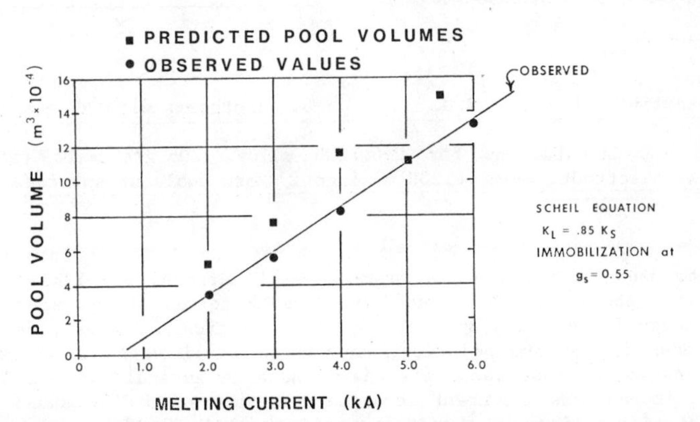

Figure 8. Computed vs. measured pool volumes as functions of melt current. All current passing through pool (C_B = 1.0 in Eq. (8)).

While the numerical model produces reasonable volumes at each current level of the experiment for $I_m \leq 5$ kA, it cannot produce a bounded pool volume at $I_m = 6$ kA because the Lorentz force is so dominant that heat transfer to the crucible by convection is sufficiently impeded that the melt pool depth exceeds the length of the computational mesh (about 1.33 radii). Clearly, either $C_B \neq 1.0$, or the current path through the ingot is not accurately given by Equation (8). Present evidence indicates that both assumptions need to be examined carefully.

A simple numerical experiment can serve to obtain an estimate of C_B which would be appropriate to the current model in (8). Suppose first that a continuous caster is to be run at the same rate as the U-6w/oNb ingot at 6 kA (205 g/s; 3.25×10^{-4} m/s); there would be no Lorentz force in the absence of current passing through the pool. The resulting motion in the melt pool would, of course, be upward along the axis and downward along the cold outer edge of the pool. This is the usual thermal convection circulation due to buoyancy forces acting alone (see Figure 9).

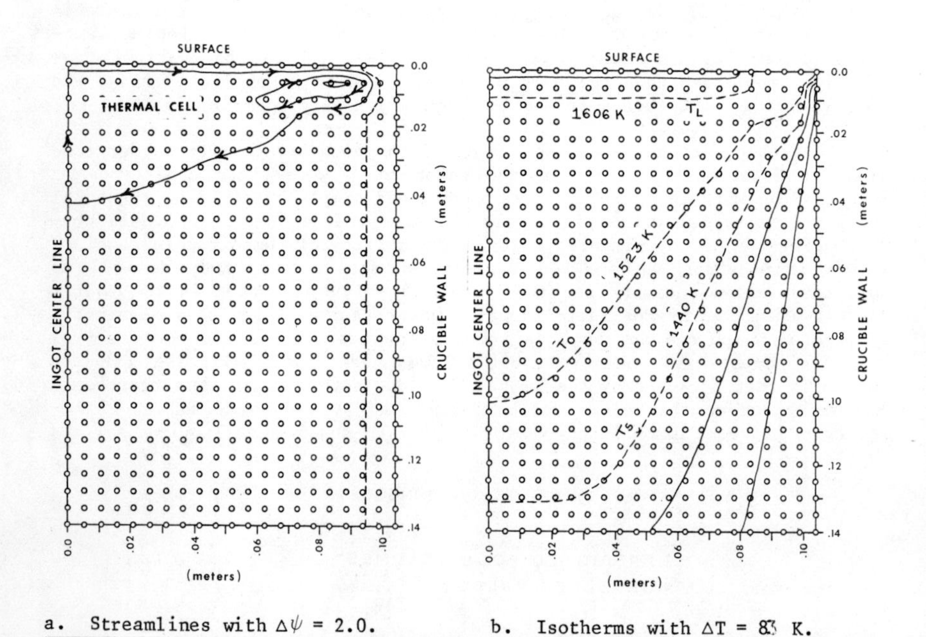

a. Streamlines with $\Delta\psi = 2.0$.   b. Isotherms with $\Delta T = 83$ K.

Figure 9. Calculations for U-6w/oNb alloy, 205 g/s melt rate, 6-in (0.156 m) electrode, 8-in (0.208 m) ingot. Zero pool current case.

Next, consider the case in which a small current, say 1 kA, is passed through the pool. In this case, the Lorentz force, though weak, opposes the buoyancy force and slows the convective motion. However, despite the slight drop in maximum velocity, convection still totally dominates the heat transfer and pool shape (isotherm shape in general) is not affected. In fact, it requires a current strong enough to essentially cancel buoyancy force before a noticeable change is produced in the isotherms.

For 1.75 kA $\lesssim I_m \lesssim$ 3.25 kA, the buoyancy and Lorentz forces virtually cancel one another, and the net motion is weaker by nearly an order of magnitude than the motion in the unopposed thermal convection case. (Note that the spacing of stream function contours, which is a rough measure of velocity magnitude according to Equation (4), has dropped from 2.0 in Figure 9 to 0.8 in Figure 10.) Throughout this curent range, the weaker flows result in deeper pools. At 2.5 kA pool current, the motions consist of separate zones of magnetic and thermal circulations (Figure 10a). Furthermore, there is evidence of long transient times for disturbances to these flows, so their stability appears to be just about marginal.

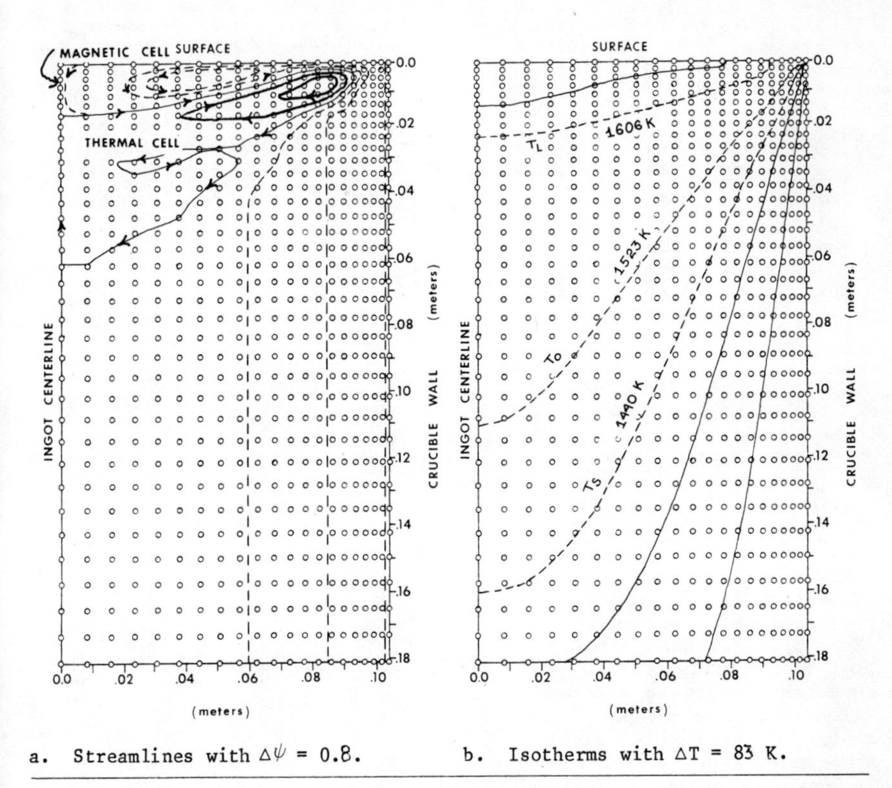

a. Streamlines with $\Delta\psi$ = 0.8. b. Isotherms with ΔT = 83 K.

Figure 10. Calculated flow under same conditions as Fig. 9, except that pool current is 2.5 kA.

When the current level becomes high enough, the Lorentz forces dominate the pool dynamics, and the principal circulation counters the thermal convection circulation direction. This must result in enhanced gradients of temperature at the edge of the dominant cell so that a thermal cell always appears at the pool edge. Maximum velocities must again increase, but with a reverse sense from the buoyancy-dominated case. This has the effect of reducing convective heat transfer, and thus increases the depth of the pool dramatically. In fact, 3.5 kA is sufficiently high current to cause even an extended (1.75 radii) computational zone to become mushy to its bottom (Figure 11).

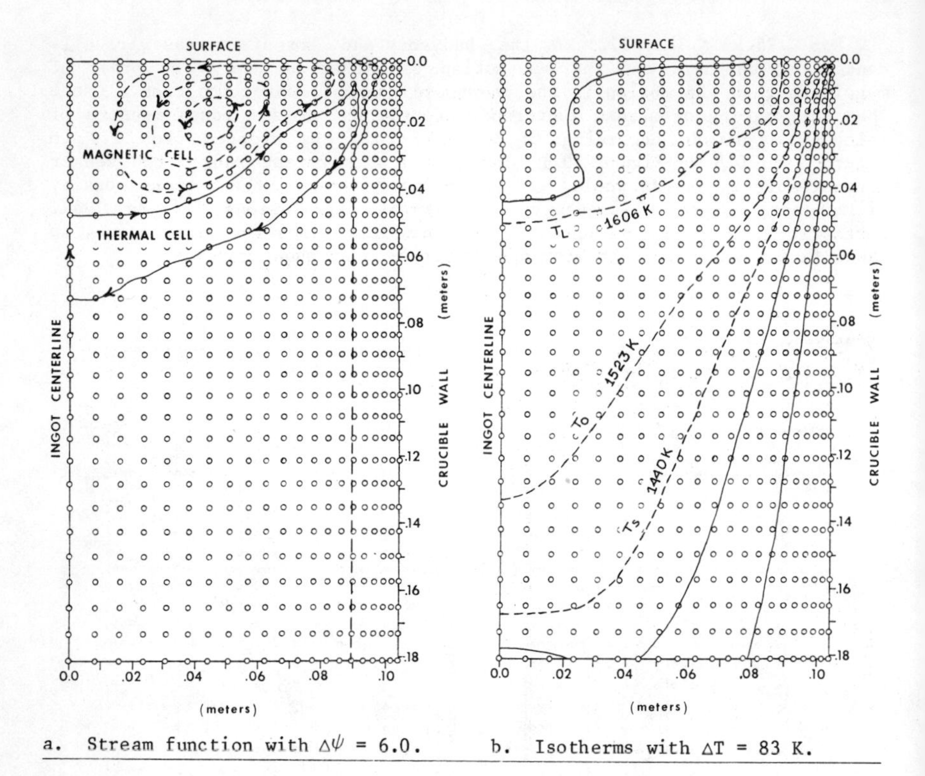

a. Stream function with $\Delta\psi = 6.0$. b. Isotherms with $\Delta T = 83$ K.

Figure 11. Calculated flow under same conditions as Fig. 9, except that pool current is 3.5 kA.

Now, these calculations each give a pool volume prediction which may be plotted vs. pool current (Figure 12). The curve appears to be monotonically increasing in pool current, as one would expect, since increased current means increasingly strong inhibition of heat transfer from the pool. The observed pool volume at 6 kA (1.33×10^{-3} m^3) cuts this curve at the right-hand end, indicating that a current of $I_p = 3.3$ kA would reproduce the observed volume. Thus for current distribution given by Equation (8), $C_B = 0.55$ would seem to be appropriate at the $I_m = 6$ kA melt current level.

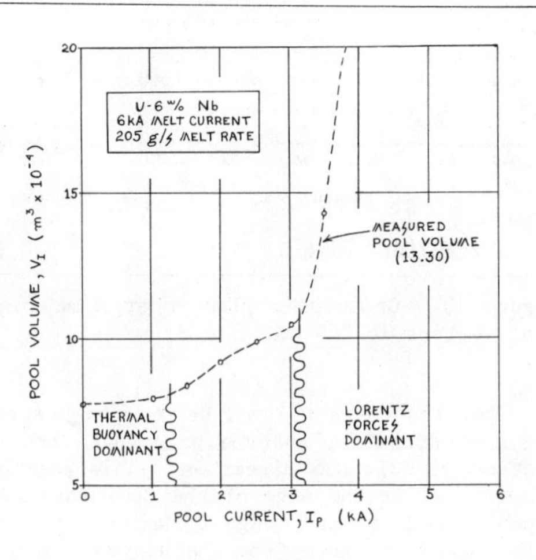

Figure 12. Calculated pool volume as a function of pool current at fixed melt rate as a means of estimating the fraction of current passing through melt pool.

From this and similar parameter studies using the

mathematical model, those parameters which significantly affect the behavior of the ingot can be isolated much more economically than by experimentation. The above example indicates the need for accurate inputs for the boundary conditions at the ingot meniscus-crucible wall contact, for accurate determination of current paths, and for accurate determination of the thermal loading of the pool surface by plasma and radiation as well as by the conventional liquid transfer. At the same time, any changes made in estimated values of the parameters must be consistent with the observations already made and with such global constraints as overall energy balance.

Specific experiments suggested by this first-generation model consist of:

1. Measurement of surface temperatures in the retractable hearth.

2. Measurements of surface velocities in the retractable hearth.

3. Measurements of crucible wall heat fluxes in the production configuration.

4. Measurements of anode current partitions between crucible wall and pool surface to improve Equation (8) and to fix C_B.

This experimentation must be accompanied by the corresponding arc physics analysis, fluid dynamics analysis, electromagnetic analysis, etc., to assure consistency, see Figure 13.

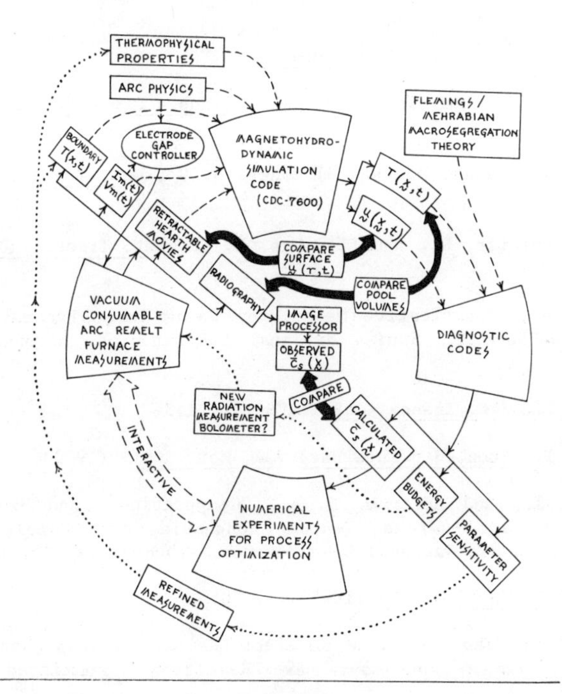

Figure 13. Sandia vacuum arc remelting research program.

V. Summary

At the outset, the goal of this program was stated to be application of the Flemings-Mehrabian macrosegregation theory to the modeling of macrosegregation in vacuum consumable arc remelted ingots. Because the interaction between the experiments and the model has not yet yielded a well-focused picture of the motions in the ingot or of the boundary conditions which produce them, the attempt has not yet been made to incorporate the equations of the macrosegregation theory into the model. When that step can be carried out with some confidence, process optimization can be undertaken (Figure 13). However, the model is adequate for coarsely delineating the relationships of the important operational parameters of the furnace, as suggested by the agreement between observed and calculated pool volumes (Figure 8). At present, it is clear that the immense complexity of the physics of the process requires continuation of the intimate step-by-step interdependence of the modeling and experimental results to keep the former realistic and the latter comprehensible.

Acknowledgement

The authors are indebted to Drs. R. Mehrabian (National Bureau of Standards), G. W. McClure, and B. Marder (both of Sandia National Laboratories) for helpful discussions involving macrosegregation and arc physics as related to the vacuum consumable arc remelt process.

Dave Beck at Union Carbide's Y-12 facility provided valuable experimental support, as did Robert Fisher, F. M. Hosking, Alan Netz, and James Maroone at Sandia National Laboratories.

References

1. Flemings, M. C. and Nereo, G. C., Trans. TMS-AIME, 212 (1967), p. 1449.

2. Mehrabian, R., Keane, M., and Flemings, M. C., Met. Trans., 1 (1970), p. 3238.

3. Fujii, T., Poirier, D., and Flemings, M. C., Met. Trans., 10B (1979), p. 331.

4. Jones, H. D., "Quantitative Metallurgy Using Image Processing Techniques," SAND80-8229, Sandia National Laboratories, Albuquerque, NM 1980.

5. Zanner, F. J., Met. Trans., 10B (1979), p. 133.

6. Zanner, F. J., Proc. 6th Internat. Vac. Met. Conf. (1979), p. 417.

7. Zanner, F. J., and Bertram, L. A., "Computational and Experimental Analysis of a U-6w/oNb Vacuum Consumble Arc Remelt Ingot," SAND80-1156, Sandia National Laboratories, Albuquerque, NM, 1980.

8. Scheil, E., Z. Metallk., 34 (1942), p. 70.

9. Zanner, F. J., "The Influence of Electrode Gap on Drop Short Properties During Vacuum Consumable Arc Remelting," submitted to Met. Trans. B.

10. Bertram, L. A., Proc. 1st Internat. Conf. Math. Model., Vol. III (1977), p. 1173.

11. Richmond, O. and Wray, P., at Engineering Foundation Conference on Modeling of Casting and Welding Processes, Rindge, NH, 1980.

12. Roache, P. J., Computational Fluid Dynamics, Hermosa Press, Albuquerque, NM, 1976.

CONVECTION IN MOLD CAVITIES*

P. V. Desai
F. Rastegar
School of Mechanical Engineering
Georgia Institute of Technology
Atlanta, Georgia 30332

Simulations of casting solidification in the past have been based on the assumption of an instantly filled mold cavity wherein the metal concerned possessed a degree of superheat selected to fall somewhere between the temperature of pouring and that of the commencement of solidification. To account for convection effects following filling, adjustments have been made to the values of thermal conductivity of the liquid between these two temperatures. Such assumptions have been typically validated via thermocouple measurements.

The present work quantitatively examines the importance of buoyancy generated convective currents in the liquid metal pool prior to the initiation of solidification. A mathematical model describing the fluid flow and heat transfer phenomena during the loss of superheat of the melt is constructed for a two-dimensional rectangular mold cavity in terms of its temperature and its vorticity/stream function. Time dependent heat flux data at the mold/metal interface available in the literature have been incorporated as boundary conditions on the problem. The model is discretized and computations of temperature and stream function distributions within the cavity are obtained via a combination of an alternating direction implicity scheme and a Gauss-Siedel overrelaxation procedure.

* This work is supported by the National Science Foundation under Grant No. DAR78-24301.

Introduction

In a vertical fluid layer confined between two walls held at a constant differential temperature, a buoyancy generated motion evolves under the influence of three basic parameters identified as the cavity aspect ratio, representing the ratio of the height to the width of the slot; the fluid Prandtl number, expressing the ratio of momentum and thermal diffusivities; and the system Grashof number, signifying the influence of buoyancy forces relative to the viscous forces. The influences of the Grashof and Prandtl numbers are often combined in their product which is referred to as the Rayleigh number, that itself representing a ratio of factors aiding buoyant convection to those opposing it.

The subject of free convection in liquid metals pools has not received much scrutiny in the classical heat transfer literature. Available work on convection in vertical cavities has been mainly for large Prandtl number fluids such as oils - Elder [1965, 1966], or with fluids of Prandtl number of order one such as gases (air) - Yin, Wung and Chen [1978]. Although the range of Rayleigh numbers examined in the literature for the geometry of present interest may be thought of as that encompassing the low Prandtl number liquid metals, such works do not prove to be adequate unless these deal with explicitly stated low Prandtl and high Grashof number cases. This is due to peculiar physical mechanisms of relative momentum and thermal diffusion at very low prandtl numbers which do not appear for Prandtl numbers of order one and higher.

In a mold cavity filled with superheated liquid metal, even small temperature differences give rise to sufficient free convection that affects the crystal growth directly via grain multiplication. The development of an equiaxed zone at the core of a casting has been associated by Jackson, Hunt, et al. [1966] with grain multiplication occuring by melting off of the arms of growing columnar dendrites in the columnar zone of the casting. A solute rich layer built up at the dendrite tip during the growth leads to the formation of a neck of lower melting point material. As noted by Davies [1973], the probability of remelting the neck into the bulk fluid, with its subsequent effect on the structural integrity of the casting, is enhanced due to the temperature gradients caused by free convection currents. Since it is usually desirable to obtain fine grained equiaxed structures in castings and since convection currents help develop equiaxed zones, quantitative prediction of free convection currents is a necessary preamble to controlling grain structure in cast metals.

Analysis

With reference to the planar rectangular mold cavity shown in Figure 1, the normalized continuity, x & y momentum, the energy and the vorticity transport equations may be written for the constant property Boussinesq fluid following Wilkes and Churchill [1966] as

$$\frac{\partial u}{\partial x} + \frac{\partial v}{\partial y} = 0 \quad , \tag{1}$$

$$\frac{\partial u}{\partial t} + u\frac{\partial u}{\partial x} + v\frac{\partial u}{\partial y} = -\,Gr\,\theta - \frac{\partial p}{\partial x} + \frac{\partial^2 u}{\partial x^2} + \frac{\partial^2 u}{\partial y^2} \quad , \tag{2}$$

$$\frac{\partial v}{\partial t} + u\frac{\partial v}{\partial x} + v\frac{\partial v}{\partial y} = -\frac{\partial p}{\partial y} + \frac{\partial^2 v}{\partial x^2} + \frac{\partial^2 v}{\partial y^2} \quad , \tag{3}$$

$$\frac{\partial \theta}{\partial t} + u\frac{\partial \theta}{\partial x} + v\frac{\partial \theta}{\partial y} = \frac{1}{Pr}\left(\frac{\partial^2 \theta}{\partial x^2} + \frac{\partial^2 \theta}{\partial y^2}\right) \tag{4}$$

and

$$\frac{\partial \zeta}{\partial t} + u\frac{\partial \zeta}{\partial x} + v\frac{\partial \zeta}{\partial y} = Gr\frac{\partial \theta}{\partial y} + \nabla^2\zeta \quad . \tag{5}$$

In these equations

$x = X/d$; $p = \dfrac{P}{\rho\nu^2/d^2}$; $u = \dfrac{Ud}{\nu}$; $v = \dfrac{Vd}{\nu}$

$y = Y/d$; $t = \dfrac{\tau}{d^2/\nu}$; $\theta = \dfrac{T - T_c}{T_o - T_c}$; Pr = Prandtl number = $\dfrac{c_p\mu}{k}$

$L = \ell/d$; Gr = Grashof number = $\dfrac{g\beta(T_0 - T_c)d^3}{\nu^2}$,

X,Y = Distances along coordinate axes,

U,V = Velocity components along coordinate axes,

$\underset{\sim}{V}$ = Velocity Vector,

ℓ,d = Height and Width of the cavity,

T = Temperature, with subscripts 0 and c representing the initial and solidification values,

τ = Time

ψ = Normalized stream function; $u = \frac{\partial \psi}{\partial y}$ and $v = -\frac{\partial \psi}{\partial x}$,

$\zeta = (\underset{\sim}{\nabla} \times \underset{\sim}{\nabla})_z = -\nabla^2 \psi$ = Normalized component of vorticity along z - axis

ν, μ = Kinematic and dynamic fluid viscosities,

k, c_p = Thermal conductivity and specific heat of the fluid,

∇^2 = Normalized Laplacian operator, and

β = Volumetric coefficient of thermal expansion of fluid.

The heat flux boundary condition is obtained from the measurements by Ruddle [1957], whose results for the cooling rates at the interface for copper and aluminum sand mold castings for the first five minutes may be respectively expressed in normalized forms as

$$q'' = 0.067de^{-0.8d^2t} \tag{6}$$

and

$$q'' = 0.054de^{-0.36d^2t}, \tag{7}$$

where $\frac{k(T_0 - T_c)}{d}$ has been used as the reference heat flux quantity, and d is in centimeters. The other boundary conditions involve the no slip velocity at the interface, where the stream function also vanishes. The boundary values of vorticity at the interface may not be computed by evaluating the vorticity transport equation at the interface, since the local generation of vorticity on the surface is not described by the kinetic process of vorticity transport. Just the same, the total vorticity must be conserved at all times within the fluid. The no slip condition at the boundary continues to produce vorticity at the boundary. The diffusion and subsequent advection of this wall vorticity is a major driving factor of the problem.

The parabolic temperature and vorticity transport equations are solved by an alternating-direction implicit method and a successive over-relaxation method is applied to the Poisson's equation for the stream function to obtain the numerical solution. Details of the computation precedure, as well as the discrete and the differential models, all are available in a thesis by Rastegar [1980].

Discussion of Results

Natural convection within a two-dimensional vertical cavity of semi-width d/2 and height ℓ has been examined via numerical solutions of the applicable continuum equations. Calculations have been made for Rayleigh numbers (based on cavity width) up to 24,500 for copper and up to 5,600 for an alloy of aluminum-30 percent copper. Temperature, velocity, vorticity and stream function profiles have been obtained for cavity aspect ratios of 1/2, 1, 2, 3 and 4. A total of over 30 numerical experiments were

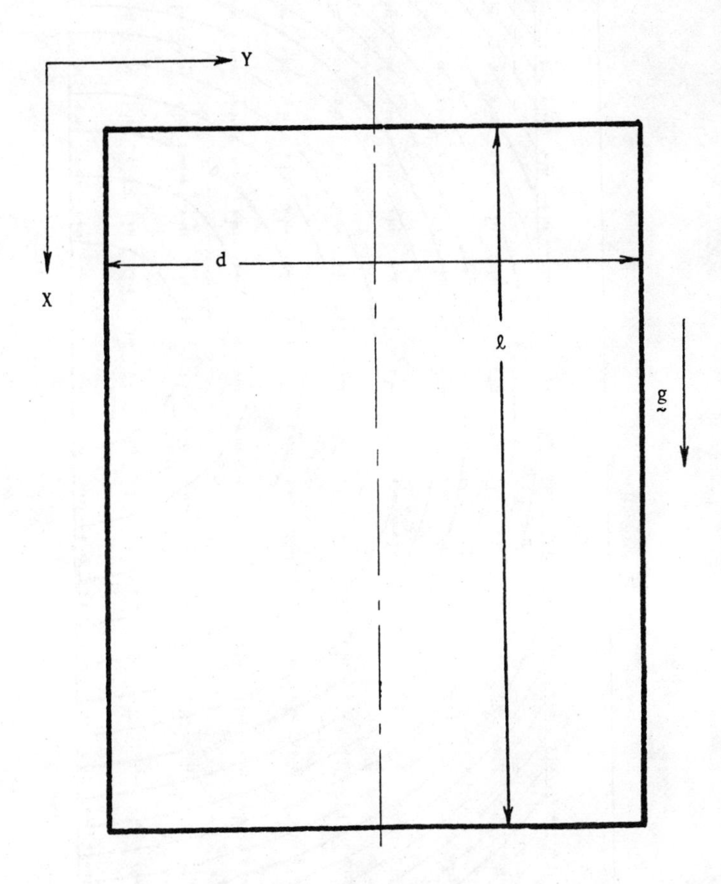

Figure 1. The Rectangular Cavity of Analysis

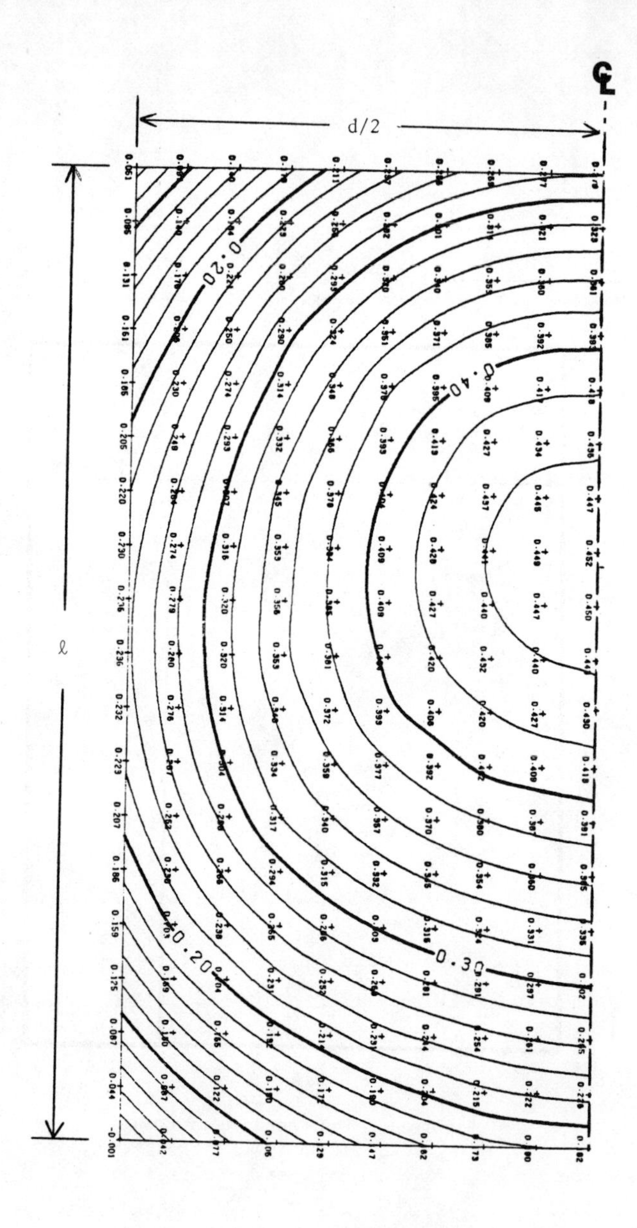

Figure 2. Profiles of Dimensionless Temperature, θ. (Aspect Ration = 1, Grashof Number = 10^6, Prandtl Number = 7×10^{-3})

performed for these two fluids. Illustrative results are presented and analyzed in this discussion. The numerical solutions obtained in the present work demonstrate excellent signs of convergence for low and moderate Rayleigh numbers ($1400 \leq Ra \leq 20,000$). For Rayleigh numbers in excess of 5,600 stable solutions are obtained only for copper.

It may be noted that the Rayleigh number represents the ratio of factors assisting the free convection to those retarding it and that the actual numerical values of the Rayleigh number as a threshold to a physical flow regime can be used only with special precautions in liquid metals vis-a-vis that for air layers and for fluids with even higher Prandtl numbers. The lowest Rayleigh number examined in this work was 1400, which corresponded to a Grashof number of 70,000 for aluminum. The lowest Rayleigh number for copper was kept at 2100, (Gr = 300,000), the results for which did not show any significant convective currents. Even at a Rayleigh number of 700 for copper the results show a flow field which is predominantly diffusive during the first thirty-six (dimensional) seconds. However, a continued heat extraction tends to develop a non-symmetric profile with reference to a horizontal axis through the cavity center, as shown in Figure 2, indicating that changes in local temperature are significant enough to vary the densities locally. The heat losses are not only by conduction, but also by free convection at this Rayleigh number. Even though the local velocities in the melt are of the order of 10^{-3} ft/sec. only, these seem to be sufficient to carry the lighter fluid lumps, which generally are at higher temperatures, upwards while the heavier ones move downwards under the action of gravity. The velocity boundary layer for Prandtl numbers less than one (0.007 for copper and 0.02 for aluminum) is quite a bit smaller than the thermal boundary layer. As a result, the local velocities are expected to be several orders of magnitude lower than one.

The stream function profiles for this case are shown in Figure 3. The direction of flow is counterclockwise. The reason for this motion is that the lighter particles are in the higher temperature zone at the bottom right corner of the figure. These particles will rise upward and travel to the colder region which is at the top left corner. Entering the colder zone at the top left results in an increase of the local density of the particle. Since there exists a boundary layer on the left wall, no matter how thin, the particle follows the velocity boundary layer, downward. However, since the thermal boundary layer is much greater than the velocity boundary layer,

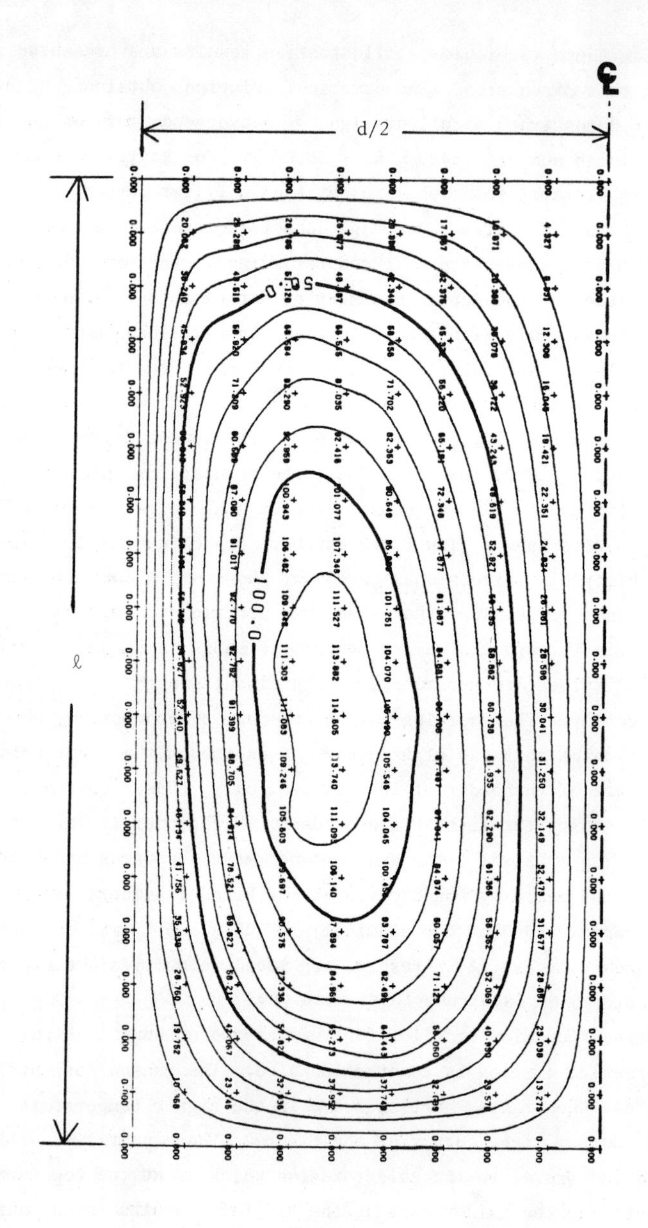

Figure 3. Final Stream Function Distribution (Aspect Ratio = 1, Grashof Number = 10^6, Prandtl Number = 7×10^{-3})

it may be concluded that the colder particles will fall from the top left corner under the action of gravity.

The bottom left corner of the cavity loses heat much faster than the rest of the mold. This is attributed to the fact that the motion from lower right to top left is against gravity, and along an adverse fluid pressure gradient. In fact, the influence of buoyancy becomes increasingly lower as the particle reaches from lower right to upper left. The mean density at top left is higher. As explained previously, the particles fall mainly under gravity near the left wall from the top left corner to the lower left corner. It may be pointed out that as the aspect ratio was increased from 1 to 2, the temperature profiles showed a greater non-symmetry and hence higher free convection effects. However, for values greater than 3, the influence of the aspect ratio is not significant.

References

1. Ruddle, R. W., The Solidification of Casting, 2nd ed., London, Richard Clay Co., [1957].

2. Elder, J. W., "Laminar Free Convection in a Vertical Slot", J. Fluid Mech., 23 [1965], pp. 77-98.

3. Elder, J. W., "Numerical Experiments with Free Convection in a Vertical Slot", J. Fluid Mech., 24 [1966], p. 149.

4. Yin, S. H., T. Y. Wung, and K. Shen, "Natural Convection in an Air Layer Enclosed within Rectangular Cavities", Int. J. Heat Mass Transfer, 71 [1978], pp. 307-315.

5. Jackson, K. A., hunt, J. D., Uhlmann, D. R. and Seward, T. P., III, Trans. Met. Soc. AIME, 236 [1966], p. 149.

6. Davies, G. J., Solidification and Casting, London, Appl. Sci. Publ. [1973].

7. Wilkes, J. O., and Churchill, W. S., "The Finite Difference Computation of Natural Convection in a Rectangular Enclosure", A.I.Ch.E.J., 12 [1966], pp. 161-166.

8. Rastegar, Freidoon, "Thermal Convection within Superheated Liquid Metal Cavities", M.S. Thesis, Georgia Institute of Technolgy, School of Mechanical Engineering, Atlanta, Georgia 30332, 1980.

Assessment of Tundish Nozzle Blockage

Mechanisms - Mathematical Modeling Approach

P. Geleta, D. Apelian, R. Mutharasan

Singh's (1,2) work on nozzle blockage during continuous casting is critically assessed. A different approach is followed here where we describe inclusion trajectories within the nozzle and subsequently determine the limiting inclusion trajectory which terminates at the nozzle wall. A mathematical model is developed which expresses the resultant force acting on an inclusion. Upon double integrating the force balance one obtains the inclusion trajectory. In practice after ≈21 minutes the nozzle begins to block and melt flow is drastically reduced. The model presented here predicts that flow reduction and blockage will occur after 21 minutes of operation when 53% of the nozzle's inner volume is occupied by the deposited inclusions. A discussion is presented on the pitfalls of mathematical modeling. Lastly, this work offers support to the hypothesis that inclusions collide with each other in the steel melt and that transport of the inclusions to the nozzle wall is due to local eddy transport phenomena.

P. Geleta formerly graduate student at Drexel University.

D. Apelian and R. Mutharasan are Associate Professors of Materials Engineering and Chemical Engineering, respectively, at Drexel University, Philadelphia, PA 19104, U.S.A.

Introduction

Continuous casting of aluminum killed steels has required special attention because of tundish nozzle blockage during pouring. Nozzle constriction and subsequent decrease of the casting rate is caused by agglomeration of alumina particles, a product of aluminum deoxidation of the steel melt. A typical speed chart (1) from U.S. Steel's Fairless Works' bloom caster is shown in Figure 1. The casting speed shows decreases after a short time of usage and after about 22 minutes of casting time the speed drops to 35 ipm requiring a nozzle change. There are two major aspects to this problem: (i) the source and origin of the depositing particles or inclusions, and (ii) the mechanism(s) by which the particles deposit.

Singh (2) has addressed both of these issues. His detailed examination showed that the major source of the depositing particles is the deoxidation and reoxidation products present in the melt. Furthermore, Singh (2) proposed a model to explain the deposition of alumina particles and subsequent nozzle constriction. His model is based upon the following two assumptions:

(i) A thin boundary layer exists adjacent to the nozzle wall. Inclusions within the boundary layer have a minimal velocity and thus make contact with the tundish nozzle wall. A schematic diagram illustrating this concept of the boundary layer is shown in Figure 2.

(ii) Once an inclusion makes contact with the nozzle wall it sinters and becomes an integral part of the refractory wall. The high temperature of the stream facilitates the sintering process.

The second assumption is a reasonable one; however, if the first assumption is true, deposition of inclusions would occur all along the wall with minimal deposition in the convergent section of the nozzle. However, Singh experimentally observed that the inclusion deposition was maximum at the convergent section of the nozzle where the boundary layer thickness is a minimum. We have attempted to better understand the mechanism of tundish nozzle blockage by modeling the trajectory of inclusions within such a convergent nozzle.

Model

To obtain the position of an inclusion in the tundish nozzle as a function of time, we examine the resultant force acting on an inclusion as it flows with the steel melt through the nozzle. From the resultant force on an inclusion, the inclusion acceleration is obtained and integrated twice to find the position of the inclusion in the tundish nozzle as a function of time. In this manner inclusion trajectories through the nozzle are obtained. Of interest is the limiting trajectory or the trajectory which terminates exactly at the end of the nozzle wall. Inclusion trajectories which lie between this limiting trajectory and the nozzle refractory wall will contact and deposit on the wall. Knowledge of the inclusion limiting trajectory allows one to theoretically calculate the fraction of inclusions which contact the nozzle wall and thus estimate the nozzle constriction rate.

A converging channel having the dimensions shown in Figure 3 is considered. This is similar to the nozzle geometry and dimensions used in Singh's studies (1,2). The assumptions made in formulating the model are:

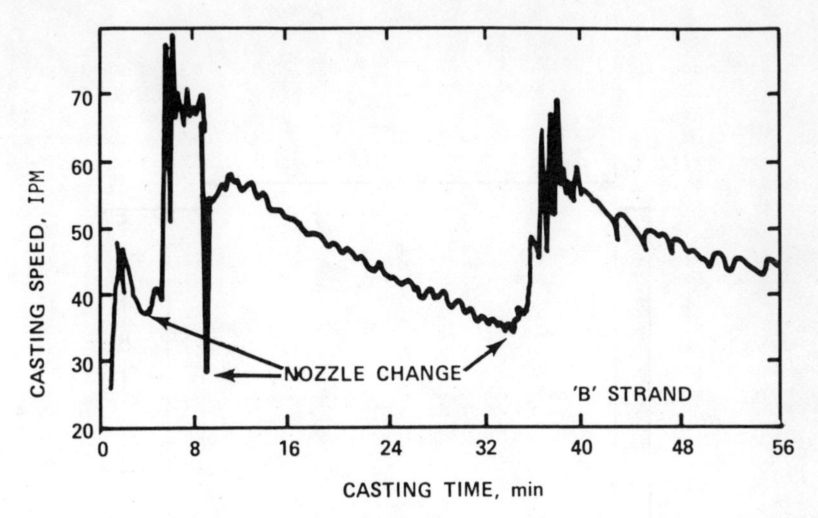

Figure 1: A typical industrial chart showing reduction of casting speed as nozzle blockage occurs (1).

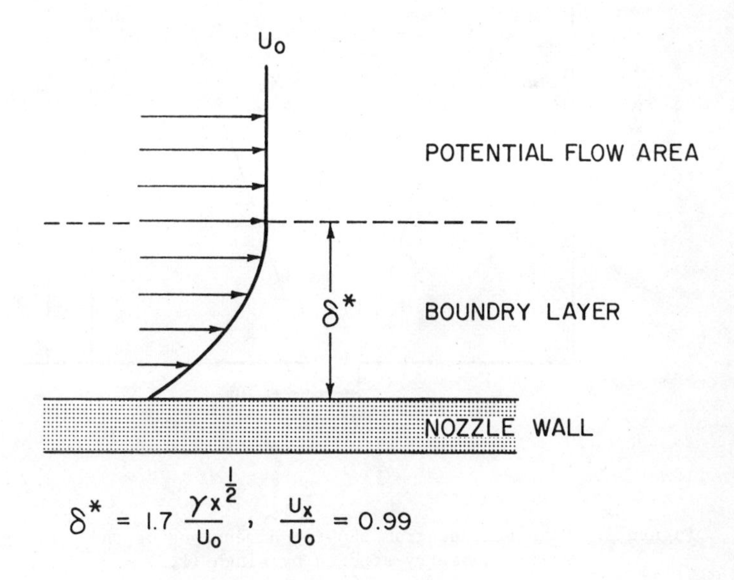

Figure 2: Concept of boundary layer formed against nozzle wall. δ^* is the boundary layer thickness; U_x is the velocity at a distance x from the approaching point; U_o is the velocity of the free stream and γ is melt kinematic viscosity.

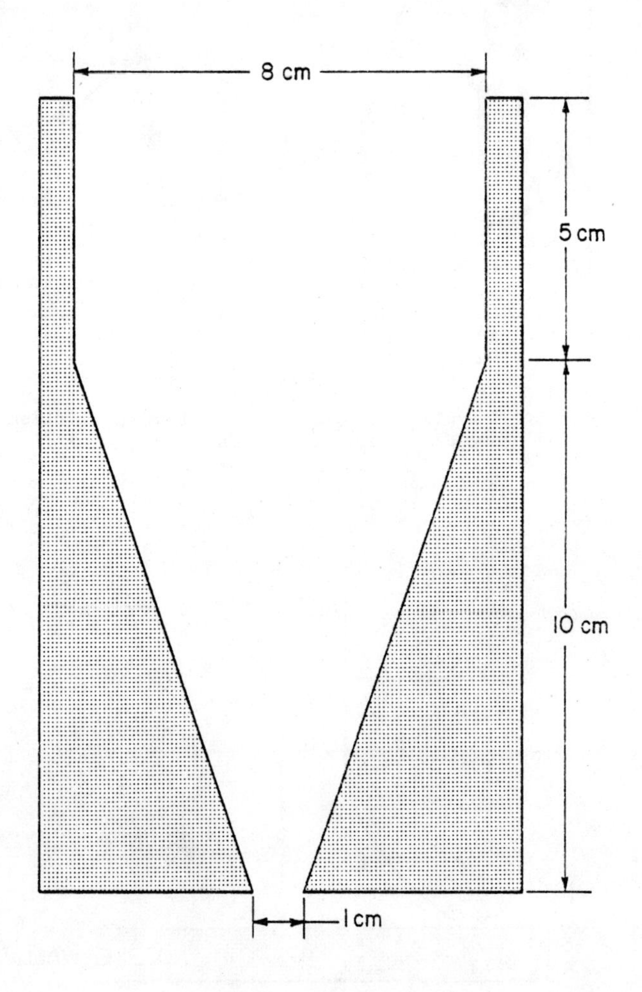

Figure 3: Schematic diagram showing dimensions of the converging nozzle geometry studied by Singh (1,2).

(i) the particles or inclusions which deposit and constrict the nozzle are spherical

(ii) these particles do not experience rotational translation

(iii) the steel melt velocity profile is unaffected by the presence of the particles, and

(iv) once the particles contact the nozzle wall, sintering unto the wall or permanent deposition takes place.

Considering next the forces acting on an inclusion flowing in the steel melt through the converging channel (Figure 3), we have:

$$\begin{Bmatrix}\text{Resultant}\\\text{Force on}\\\text{Inclusion}\end{Bmatrix} = \begin{Bmatrix}\text{Buoyant}\\\text{Force}\end{Bmatrix} - \begin{Bmatrix}\text{Gravity}\\\text{Force}\end{Bmatrix} - \begin{Bmatrix}\text{Added}\\\text{Mass}\\\text{Effect}\end{Bmatrix} - \begin{Bmatrix}\text{History}\\\text{Term}\end{Bmatrix} + \begin{Bmatrix}\text{Drag}\\\text{Force}\end{Bmatrix} \quad (1)$$

where:

- The resultant force is the net force acting on the inclusion.

  $$(\text{resultant force}) = m \cdot \frac{dU}{dt} \quad (2)$$

  m is the inclusion mass

  $\frac{dU}{dt}$ is the inclusion acceleration; U is the relative velocity of the inclusion and t is time.

- The buoyant force results from the density difference between the alumina inclusion and the steel melt and is in turn expressed by the following

  $$(\text{buoyant force}) = m_\ell \cdot g \quad (3)$$

  m_ℓ is the mass of liquid steel displaced by the inclusion

  g is the acceleration due to gravity

- The gravity force on the inclusion is simply given by

  $$(\text{gravity force}) = m \cdot g \quad (4)$$

- The added mass term allows for the fact that not only the inclusion has to be accelerated but also a portion of the fluid which adheres to the particle. In considering the acceleration of the fluid adjacent to the inclusion, the inclusion's acceleration towards the nozzle center is decreased. This is given by

  $$(\text{added mass term}) = C_A \cdot m_\ell \cdot \frac{dU}{dt} \quad (5)$$

  C_A is the added mass coefficient, a constant

- The history term takes into account the dependence of the instantaneous drag force on the state of development of the boundary layer around the inclusion. The history term is given by:

$$C_H R_p^{\,2} (\pi \rho_L \mu)^{1/2} \int_o^t \frac{dU}{d\tau} \cdot \frac{d\tau}{\sqrt{t-\tau}} \tag{6}$$

C_H is the history term coefficient, a constant

R_p is the inclusion radius

ρ_L is the steel density

μ is the melt viscosity

τ is a dummy variable of integration

- The drag force term depends on the distance of the inclusion from the tundish nozzle wall. At an infinite distance from the nozzle wall, the drag force is:

$$F_D = -\frac{\pi}{2} C_D R_p^{\,2} \rho_\ell \, U|U| \tag{7}$$

C_D is the drag force coefficient which is a function of particle Reynolds number.

However, at a distance ℓ from the nozzle wall, the drag force is increased due to inclusion motion close to a solid boundary. Inclusion motion towards the nozzle wall is impeded by the added energy required to drain fluid from between the inclusion and the nozzle wall. This drag force is given by

$$F_D = \frac{-\pi/2 \; C_D R_p^{\,2} \; \rho_\ell \; U|U|}{1 - \left(\frac{k R_p}{\ell}\right)} \tag{8}$$

k is a function of geometry.

Combining Eqs. (2)-(6) and (8) into the principal force balance, Eq. (1), we obtain:

$$m \frac{dU}{dt} = (m_\ell g) - (mg) - \left\{C_A \, m_\ell \frac{dU}{dt}\right\} - \left\{C_H R_p^{\,2} \; (\pi \rho_L \mu)^{1/2} \int_o^t \frac{dU}{d\tau} \cdot \frac{d\tau}{\sqrt{t-\tau}}\right\}$$
$$+ \left\{\frac{-\pi/2 \; C_D R_p^{\,2} \rho_\ell \quad U|U|}{1 - \left(\frac{k R_p}{\ell}\right)}\right\} \tag{9}$$

In Eq. (9), the parameter U is the relative velocity of the inclusion; U is the difference between the velocity of the inclusion with respect to stationary coordinates and the velocity of the fluid with respect to those same coordinates. This is expressed by

$$U = U_p - U_f \tag{10}$$

where

U_p is the inclusion velocity

U_f is the fluid velocity

The fluid velocity, U_f, is obtained by considering turbulent flow in a conical tundish nozzle. Applying potential flow theory for the nozzle core and boundary layer theory for flow near the nozzle wall (3), we obtain the following expressions:

$$U_f = - \frac{Q}{\pi(R^2 + Z^2) \sin^2\theta} \quad \text{for } \eta \geq 1 \tag{11}$$

$$U_f = - \frac{Q(2\eta-\eta^2)}{\pi(R^2 + Z^2) \sin^2\theta} \quad \text{for } \eta < 1 \tag{12}$$

Where $\eta = \frac{(R^2 + Z^2)^{0.5}}{\delta} \cdot (\theta - \alpha)$, and

$$\delta = \left[\frac{30}{13} (R^2 + Z^2)^{1.5} \left(\frac{\pi\mu}{\rho_\ell} \frac{\sin^2\theta}{Q}\right)\right]^{0.5}$$

In the above expression, Q is the volumetric flow rate of the steel melt; R and Z are the radial and longitudinal coordinates in the converging channel, respectively; θ is the angle of nozzle convergence in radians; and α is the angle between the nozzle wall and the location of the center of the inclusion taken at the apex of the cone. In such a tundish nozzle as shown in Figure 3, by the use of Eqs. (10) and (11) we obtain the flow velocity profile. A typical velocity profile for steel casting conditions is shown in Figure 4.

Results

The force balance given in Eq. (9) was numerically integrated twice (Runga-Kutta method) to obtain inclusion trajectories through the tundish nozzle. Trajectories were calculated for various different initial inclusion positions in the nozzle. The trajectory which terminates exactly at the end of the nozzle is called the limiting trajectory and is shown in Figure 5. All inclusion trajectories to the wall side of the limiting trajectory will contact the nozzle wall and result in inclusion collection. Whereas trajectories lying on the other side of the limiting trajectory (i.e., towards the nozzle center) pass through unimpeded. Thus, knowledge of the limiting trajectory allows one to calculate the fraction of inclusions which contact the nozzle wall - the collection efficiency. In this manner the theoretical volume of inclusions collected in the tundish nozzle over a period of time can be calculated and compared with experimental determinations of deposited volumes.

Singh (2) experimentally determined that in a converging channel (nozzle) having an inner volume of 40 cm^3 flowrate reduction and constriction occurred after 21 minutes of cast duration. The proposed inclusion trajectory model, Eq. (9), predicts that at Singh's (2) casting conditions 1.5% of the inclusions passing through the nozzle will be collected in 21 minutes. This value of 1.5% of collected Al_2O_3 inclusions translates to a total volume of collected inclusions of 21 cm^3. A reduction of the inner

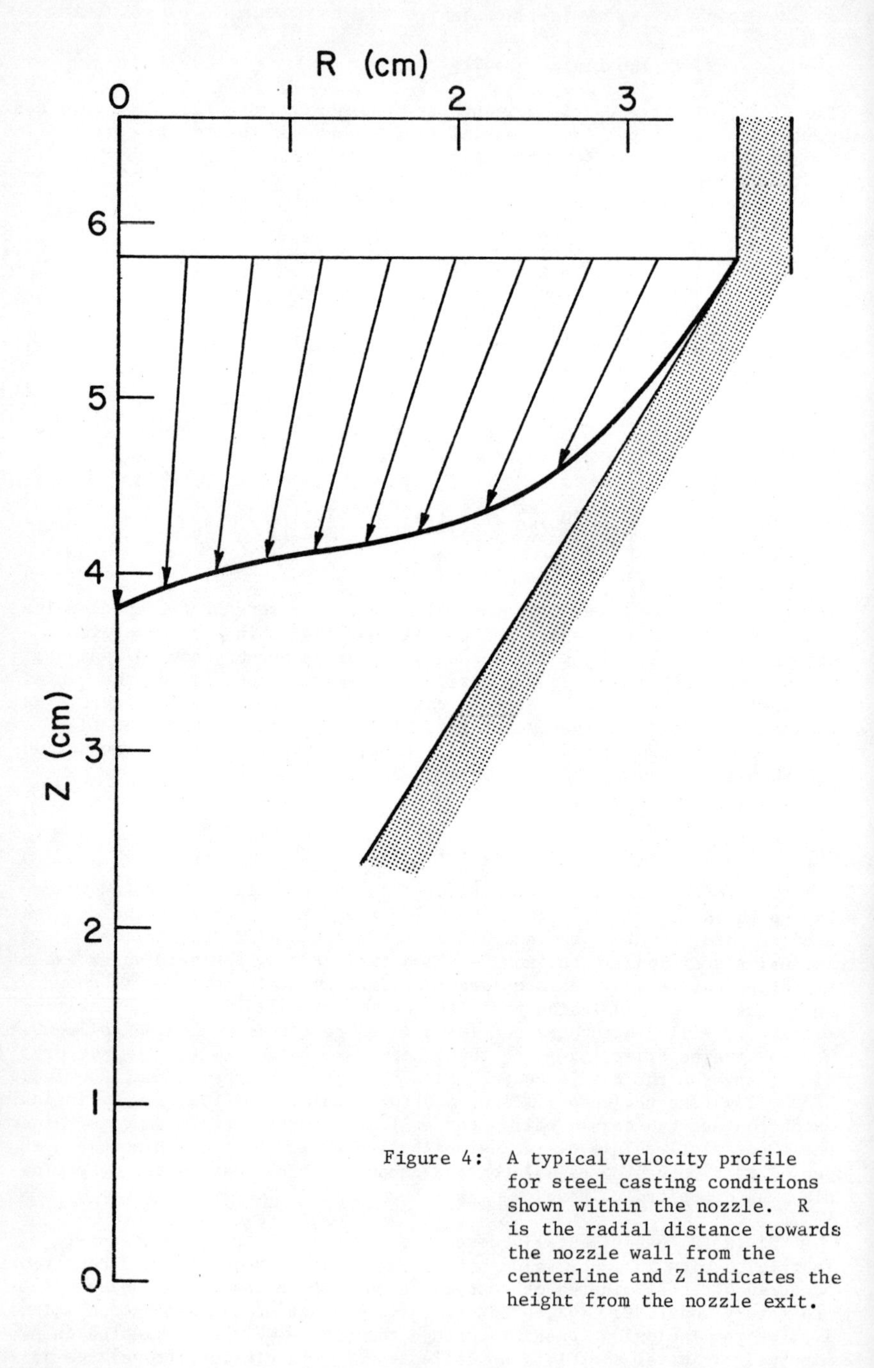

Figure 4: A typical velocity profile for steel casting conditions shown within the nozzle. R is the radial distance towards the nozzle wall from the centerline and Z indicates the height from the nozzle exit.

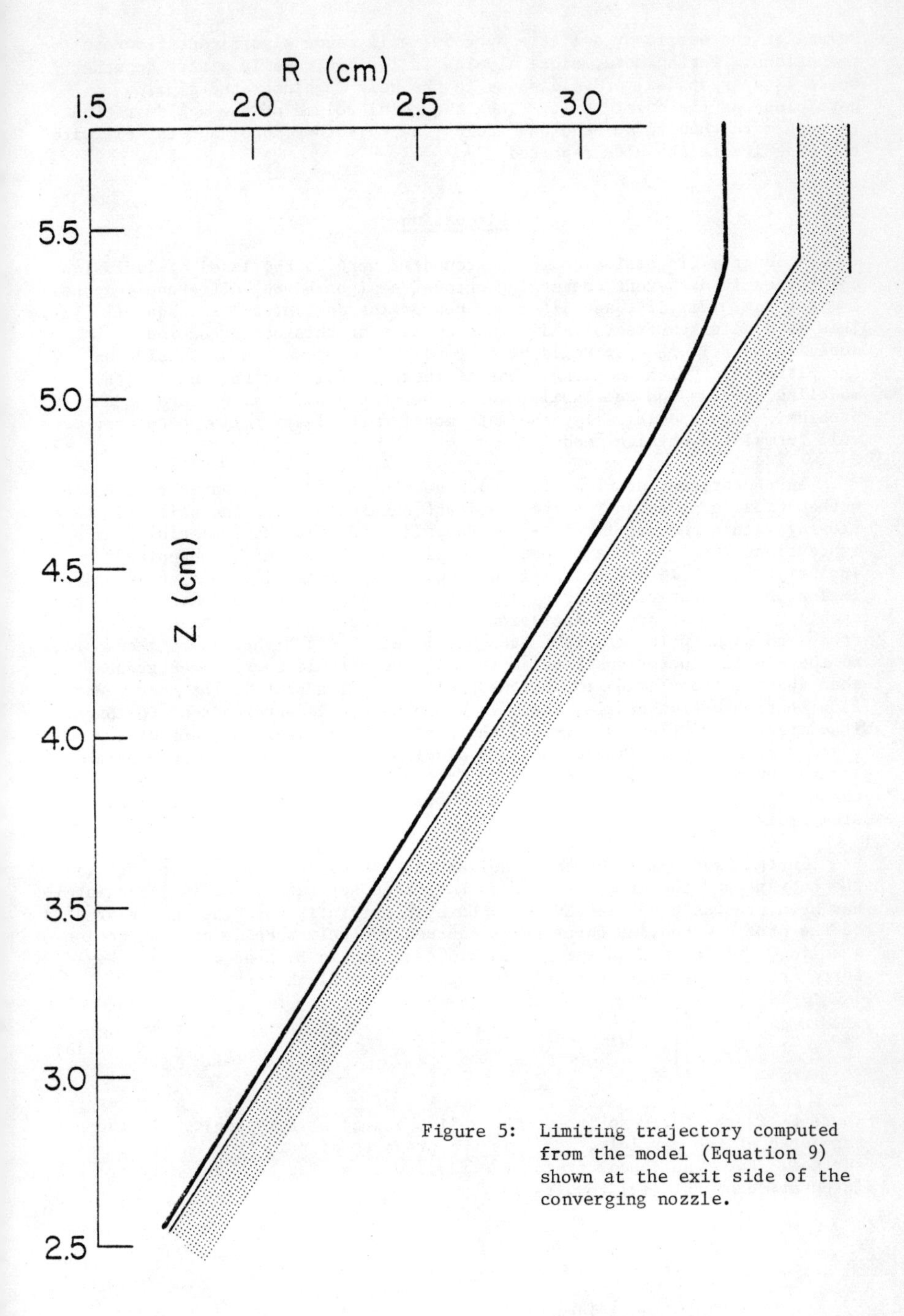

Figure 5: Limiting trajectory computed from the model (Equation 9) shown at the exit side of the converging nozzle.

volume of the nozzle of a little over 50% will cause significant flowrate reductions. Furthermore, since alumina inclusions readily sinter together there is a great deal of void space in the bulk (agglomerated alumina particles) at the nozzle exit. Thus the total volume of the agglomeration is much more than 21 cm^3 and the inclusion trajectory model agrees well with the experimental results reported (2).

Discussion

A lengthy discussion could be presented here on the level of inclusion collection in different converging channel designs having different lengths, L - see Eqs. (11, 12), and different converging angles, θ - see Eqs. (11,12). This is done elsewhere (3) and in the context of this conference on modeling we felt that it would be appropriate to present a discussion on the pitfalls of math modeling. One of these pitfalls is that during the modeling process one can easily lose sight of the physical picture of the problem. More specifically, the math model must always follow a concrete and well formulated physical model.

In the present problem of tundish nozzle blockage, we have presented a mathematical expression for the resultant net force on an inclusion as it is flowing within the steel melt - see Eqs. (1) and (9). Manipulation of these expressions resulted in a theoretical value of the volume of deposited inclusions which agreed extremely well with experimental values of deposited inclusions. However, the irony is that a physical model indicates that there should be no transport of inclusions to the nozzle wall. Consider the system from a physical point of view: the steel melt has a higher inertia compared to the alumina inclusions because the melt density is three times greater than that of the alumina inclusions. This implies that in the convergent flow section of the nozzle, the melt would have a lesser tendency to change its direction of flow as compared to that of the inclusions. Assuming viscous flow is visualized, then the alumina inclusions, being considerably lighter than the melt, would tend to "float away" from the nozzle wall and be transported to the center line of the nozzle. Where does this discrepancy stem from?

In the development of the resultant net force on an inclusion, Eq. (1), the buoyant and the drag force terms have been separated. A similar treatment has been presented by Szekely (4). However, strictly speaking such a breakdown of the drag and buoyant force terms can be made only when describing motion of a solid particle in a quiescent fluid. As shown in Bird et al. (5), the force acting on a solid particle moving within a fluid is

$$F_r = \iint_A \delta \, dA \tag{13}$$

where F_r is the force acting on the solid particle (inclusion), A is the surface area of the particle, and δ is the pressure distribution on the particle surface. For a quiescent fluid or a fluid with a plug flow velocity profile, integration of Eq. (13) results in (5):

$$F_r = m_\ell g + F_D \tag{14}$$

$$\begin{pmatrix}\text{buoyant}\\ \text{force}\end{pmatrix} + \begin{pmatrix}\text{drag}\\ \text{force}\end{pmatrix}$$

If the fluid has a velocity profile which varies as a function of position, then the pressure distribution across the surface of the inclusion is much more complicated and the separated expression for F_r as given in Eq. (14) cannot be used. If a tundish nozzle velocity profile (Figure 4) is considered where the velocity of the fluid varies with position, then the proposed model, Eqs. (1) and (9), is only applicable _if and only if_ it is assumed that the pressure distribution across the inclusion surface is an invariant. In the present problem of flow through continuous casting tundish nozzles using the expressions for buoyant and drag forces for particle motion in a quiescent fluid may lead to an erroneous conclusion even though the theoretical predictions agree extremely well with the experimental results.

The discussions presented here lead to one conclusion. The proposed inclusion trajectory model for nozzle blockage, even though it agrees well with the experimental results, is based on a mechanism which is not physically present. Generally, in math modeling we assume certain hypotheses, then mathematically develop the model and test/compare the latter with experimental results. If good agreement exists, we are then led to believe and accept the initial hypotheses. In the present problem, excellent agreement exists between the two routes even though the initial hypothesis is incorrect. Obviously, the chain of events were such that we were at first unaware of our initial incorrect hypothesis (separating the buoyant and drag forces in Eq. (1)). It only became apparent when the developed model was applied to describe the inclusion trajectory in centrifuging a similar system of steel melt with alumina inclusions. The proposed inclusion trajectory model predicted that the lighter particles will settle at the outer perimeter and the heavier fluid will settle at the core of the centrifuge. This is opposite to what occurs during centrifuging of such a two-phase system.

In terms of assessing tundish nozzle blockage mechanisms, this work has shown that an inclusion trajectory model where certain trajectories beyond a limiting one result in inclusion deposition at the nozzle wall is not a likely mechanism. The proposed math model does not agree with the physical model even though it agrees with the experimental results. Furthermore, the developed model gives erroneous results when applied to the same two phase system during centrifuging. The results of the proposed inclusion trajectory model has further reinforced our original criticism of Singh's boundary layer theory. This work offers support to the hypothesis that the alumina inclusions collide with each other in the steel melt and that transport of the inclusions to the nozzle wall is due to local eddy transport phenomena.

References

1. Singh, S. N., "A Practical Solution to the Problems of Alumina Buildup in Nozzles During Continuous Casting of Aluminum-Containing Steels", _Iron and Steelmaker_, _6_ (6), (1979), pp. 40-46.

2. Singh, S. N., "Mechanism of Alumina Buildup in Tundish Nozzle During Continuous Casting of Aluminum Killed Steels", _Met. Trans._, _5_, (1974), pp. 2156-2178.

3. Geleta, P., M.S. Thesis, Drexel University, Philadelphia, PA 19104, 1981.

4. Szekely, J., Fluid Flow Phenomena in Metals Processing, Academic Press, New York, N.Y., 1979.

5. Bird, R. B., Stewart, W. E., Lightfoot, E. N., Transport Phenomena, John Wiley & Sons, Inc., New York, N.Y., 1960.

Growth Kinetics and Morphology

QUANTITATIVE KINETIC AND MORPHOLOGICAL STUDIES USING MODEL SYSTEMS

Robert J. Schaefer
National Bureau of Standards
Washington, D.C.

Martin E. Glicksman
Rensselaer Polytechnic Institute
Troy, N.Y.

The usefulness for metallurgists of solidification studies using transparent model systems depends to a large extent on quantitative correlation to detailed theories of specific solidification phenomena. By designing experiments in which the thermal and geometrical conditions considered by the thoery can be attained as closely as possible, and by making detailed kinetic and morphological measurements of the resulting solidification behavior, one can carry out incisive tests of the theory. Thus detailed study of dendritic growth in pure succinonitrile, together with auxiliary experiments which measured relevant thermodynamic properties, led to the important conclusion that the maximum velocity hypothesis for dendrite growth was not correct. This result has stimulated further theoretical work, which now appears to relate dendrite growth velocities to morphological stability considerations. Moreover, additional experimental and theoretical work is now revealing the regimes in which convection and solute effects are significant.

Introduction

In this paper we discuss how transparent materials can serve as useful analog model systems for the study of metallic solidification. Referring primarily to the experiments which the authors have carried out in collaboration with J. D. Ayers and S. -C. Huang, we show that experiments with model systems had conclusive implications on the validity of theories of metallic solidification only when all of the relevant physical properties of the transparent materials had been carefully measured, the thermal and solutal environment of the growing crystals were under adequate experimental control, and the theories themselves had been developed sufficiently to describe in detail an experimentally attainable situation.

Before the crystalline nature of most solids was fully understood, the solidification of metals was a mysterious process. C. S. Smith (1) described some of the ideas which developed slowly in the 18th and 19th centuries, including the concept that there was some analogy between metal solidification and the crystallization of salts. However, once the crystalline nature of solid metals had been conclusively demonstrated the analogy to salts or other materials with an externally obvious crystalline form had served its purpose. Other than its intrinsic crystallinity, the dendritic structure with which metals solidify has little overt resemblance to the faceted growth forms commonly observed with salts. As a result, the extensive observations of crystal growth in transparent materials contributed very little for a long time to the understanding of metallic solidification.

During the 1950's, interest grew in the atomic kinetic process at the solid/liquid interface and several important experiments were carried out attempting to detect or measure those processes (2). Many of these experiments used ice, which under some conditions grows in the form of dendrites reminiscent of those found in metals. Ice dendrites, however, differ from those of most metals in that they form sheet-like "two-dimensional" dendritic arrays.

In 1965, Jackson and Hunt (3) pointed out that a small group of organic materials - termed plastic crystals - have low entropies of fusion similar to those of metals and provide useful transparent analogs in the study of metallic solidification. The low entropy of fusion is a result of hindered molecular rotations in the solid state and is usually associated with an unfaceted solid/liquid interface, a morphology characteristic of many solidifying metals but one not commonly observed in transparent materials. These materials solidify with fcc, bcc, or cph crystal structures, and in the presence of thermal or constitutional supercooling will solidify with a dendritic morphology.

Low-entropy-of-fusion transparent materials have already been used successfully to study several solidification processes of key importance to metallurgists, including; dendrite growth, fragmentation and multiplication, eutectic solidification, morphological stability, several convective flow phenomena, and zone melting. The usefulness of these studies has on occasion been questioned because although these materials have some characteristics in common with metals, they also differ greatly with respect to other important properties such as the thermal conductivity and kinematic viscosity. Particularly when considering relatively complex solidification processes such as convective redistribution of solute, where no general detailed theory is available to show how the process explicitly depends on these properties, one cannot assume that certain phenomena observed in transparent analogs occur at all in metals, and vice versa.

Qualitative Studies

In our earliest studies with transparent analog materials, the purity and the thermal fields were poorly known. We observed a dramatic influence of crystal imperfections, especially grain boundaries and sub-boundaries, in initiating morphological instabilities on the solid/liquid interface (4). Figures 1a-1d show the progressive development of interfacial instabilities adjacent to three intersecting grain boundaries emerging as a tri-junction, Figure 1a. The triple point initiates three small "knobs", Figure 1b,

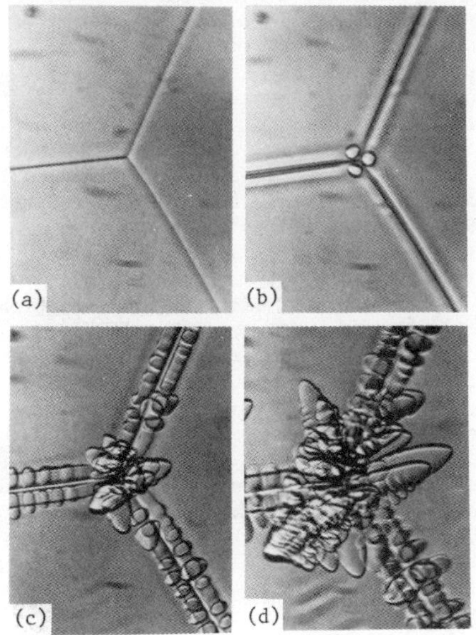

Fig. 1 - Progressive development of instabilities in a solid/liquid interface in the vicinity of a grain boundary trijunction.

which enlongate into dendrites, as periodic instabilities develop along the grain boundaries, Figure 1c. Figure 1d shows a later stage in which the grain boundary regions are undergoing dendritic growth, whereas the grain faces are still relatively featureless. A similar mechanism has been proposed by Morris and Winegard (5) to explain the development of segregation patterns near grain boundaries in a dilute Pb-Sb alloy. It appears that in the metallic system the grain-boundary induced instability can propagate across the solid/liquid interface to form a periodic cellular structure, but in the transparent materials we never observed this. Instead, we found that the instabilities localized near grain boundaries grew to ever larger amplitudes and, finally, became dendrites, while the interfacial regions without imperfections developed at most an aperiodic rumpled structure. We still do not know whether there is a real difference between the development of grain boundary induced morphological instabilities in metals and organic alloys, or if the observed differences would disappear with proper control of temperatures and impurities.

In 1971, a laser holography system was constructed at the Naval Research Laboratory making it possible to gather extensive quantitative data on crystal growth forms, shape changes, and velocities. With this system we could observe several phenomena which are probably important in metallic solidification, but we could not make any comparisons with theoretical models. Again, the heat flow was not adequately characterized, the material properties remained mostly unknown, the purity was not controlled, and the phenomena which occurred were generally too complicated for theoretical analysis. Therefore at that time, there was little point in extracting quantitative measurements from the holograms. Thus, the holographic studies added some significant detail but contributed relatively little that was new to our knowledge of the solidification process in metals.

Quantitative Studies

A large step forward occurred when we undertook a quantitative study of dendrite growth. The experimental situation considered can be idealized as follows: a single dendrite advances at steady state into an infinite bath of uniformly supercooled pure liquid, with no convection occurring in the liquid phase. We were able to demonstrate that in our experimental determination of dendritic growth kinetics these idealized conditions were approximated quite closely.

Theoretical models of dendrite growth generally predict growth velocity, V, as some function of supercooling, $\Delta\theta$, generally expressed as the power law $V = \beta G \Delta\theta^n$, where G, a lumped material parameter, is the "characteristic" velocity given by $\alpha \Delta S L / C\gamma$, in which α is the thermal diffusivity, ΔS is the entropy of fusion per unit volume; L is the latent heat of solidification; C is the specific heat; and γ is the solid-liquid surface energy. β and n are numerical constants supplied by and specific to each theory. Unfortunately, the data on the growth of metal dendrites has too much scatter for a meaningful quantitative comparison to theory, and the data on ice, which is of better quality, is of uncertain significance because of the flattened "two-dimensional" form of ice dendrites. Our experiments were carried out on succinonitrile, $CN(CH_2)_2NC$, which has a bcc crystal structure and which exhibited a greater tendency to supercool than did the other low-entropy transparent materials which we have studied.

The first experiments on succinonitrile were carried out by containing the material in a long straight tube, supercooling it to a uniform temperature, nucleating the solidification at one end, and then measuring the rate at which crystals propagated along the tube. These crystals grew along the glass walls rather than in the bulk liquid, which was not an unexpected development, but resulted in an experimental situation which has not been analyzed by theory. The main value of these early kinetic experiments was to indicate that even at the highest growth velocities observed (about 40 cm/sec), interfacial molecular attachment kinetics did not limit the growth rate to any detectable degree. In fact, recent work by Alfinstsev et al. (6) indicates that growth velocities as large as 5 m/sec were achieved in highly supercooled succinonitrile.

The study of dendritic growth into the bulk liquid required the use of a crystallization chamber which avoided "wall effects." This was accomplished by arranging that the dendrites emerged from a capillary at the center of a spherical bulb and then grew radially outward toward the chamber walls. Figure 2 shows an example of this type of dendritic growth at a supercooling of 3K. Except at extremely small growth rates, each dendrite tip then acted as if it were growing into a uniformly supercooled melt of infinite extent. Dendrite growth rates were measured in two separate succinonitrile specimens, with impurity levels differing by a factor of more than 100, with essentially identical results (7). We therefore concluded that our results represented the velocity of isolated dendrites growing in pure supercooled melts. At supercoolings of less than about 1.5 K, downward-growing dendrites clearly grew more rapidly than upward growing dendrites, indicating the effect of convection. Figure 3 demonstrates that

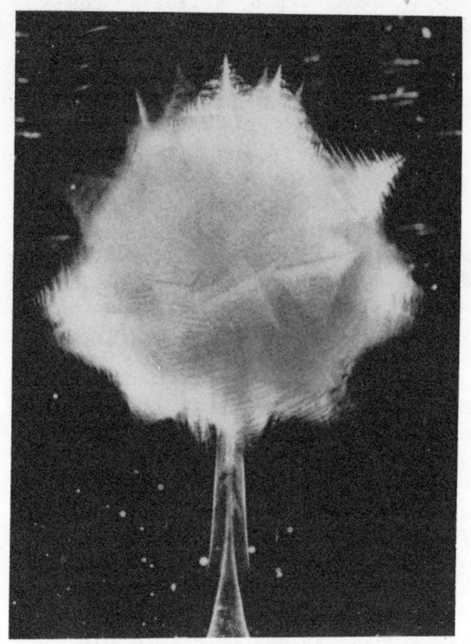

Fig. 2 - An array of dendrites growing from the end of a glass capillary into a melt with a supercooling of 3 K.

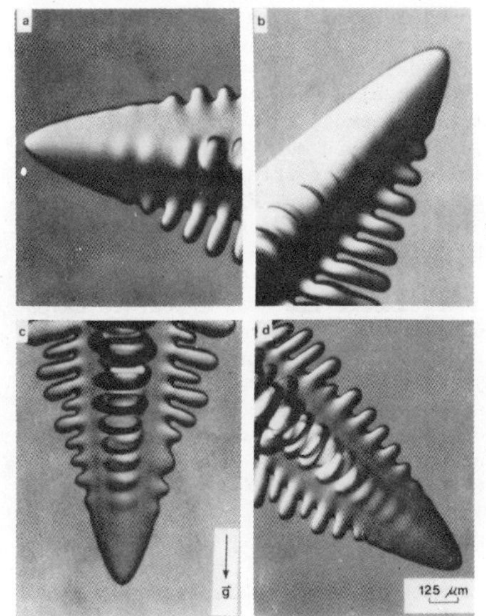

Fig. 3 - Dendrites growing with different orientations with respect to the gravity vector, showing that branching is suppressed on the upper surface due to the influence of thermal convection in the melt.

convection influences the micromorphology of dendrites by affecting the growth of side-branches. The effect of convection depends upon the relative orientation of gravity and the dendritic growth axis. At supercoolings greater than about 2 K, the growth of dendrites was spatially isotropic and there was no evidence of convection; therefore, these data were valid for comparison to the diffusion theory of dendritic growth.

Although the experimental growth rate data gave close agreement with all of the more advanced theories for the exponent n, there was not agreement with any theory regarding the constant β. There was one serious remaining uncertainty, however, because the "characteristic" material growth parameter G contains the solid/liquid surface energy γ, for which we could only estimate a value. To eliminate this uncertainty, we undertook a separate study for the measurement of the solid-liquid surface energy.

Solid-liquid surface energies can in principle be determined by measuring the groove shape which develops where a solid-liquid interface is intersected by a grain boundary or other planar "defect" such as a wall. The surface energy, if isotropic, is a simple function of the groove depth, h, the entropy of fusion, ΔS, and the temperature gradient ∇T, specifically, $\gamma = h^2 \Delta S \nabla T/2$. By using a cylindrical apparatus with an axial heater wire, and succinonitrile of sufficient purity that the temperature of the solid-liquid interface was known to within one millikelvin, it was possible to determine temperature gradients as small as 0.1 K/cm with uncertainties no larger than 1%. The result was a series of surface energy measurements with little scatter and virtually no conceivable systematic errors (8). The measured value of γ confirmed the discrepancy between the measured dendrite growth velocities and the predictions of the theoretical models.

At this point it became evident that the phenomenological dendrite growth theories should be seriously re-examined. Each of them invariably used the so-called "maximum velocity principle", although this "principle" had no rigorous physical or mathematical justification and was widely questioned. A theoretical treatment by Langer and Müller-Krumbhaar (9), which replaces the maximum velocity criterion with a morphological stability condition, now appears to give much better agreement with experiment. Measurements of the dendrite tip radius, which were made over a limited range of supercooling, provided an additional set of reference data which are also consistent with the new theoretical analysis. Figure 4 shows the method used to determine dendritic tip radii by fitting the observed profile to a parabola with a known curvature at the tip.

The studies of dendritic growth in succinonitrile thus appear to have generated experimental data which could be used to discriminate among several theories of dendrite growth. Because all relevant properties (Table I) of the system were available from the literature or were measured with high precision, and because the sample purity and thermal environment could be kept under close control, random and systematic uncertainties were reduced to a level below that which would account for the discrepancies between the experiments and the older theories.

Other Applications

Dendrite growth in pure supercooled materials is not important in commercial metallurgical processes, and the experiments on succinonitrile have been expanded recently to cover more complex processes including convection (14,15). Simultaneously, theoretical analyses have been developed to describe these processes. A major area of interest has been the influence of liquid convection on the solidification process. Extensive data on

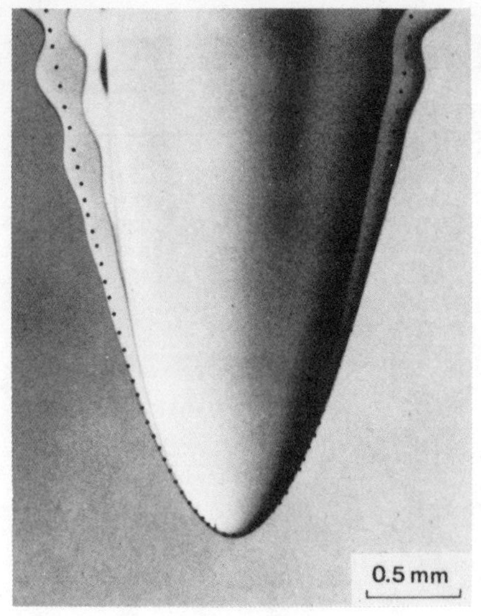

Fig. 4 - Fitting of a dendrite tip to a parabola (dots). The photograph is a double exposure showing the slight difference in profile as viewed from the [100] and [110] direction.

dendrite growth velocity as a function of orientation with respect to the gravity vector have been obtained in the small supercooling regime, where convective heat and mass transport dominate over diffusive transport. A boundary layer theory successfully predicts the supercooling level at which a sharp transition from convection to diffusion occurs, but calculation of the actual magnitude of the effect appears to be a formidable problem, and has not yet been attempted.

The effects of solutes on dendritic growth in this system are also under study. In this case theory of Langer and Müller-Krumbhaar has been extended to include the influence of solutes on the morphological stability of a dendritic crystal (16).

Turning away from dendritic growth, Coriell, Cordes, Boettinger and Sekerka (17) have analyzed convective and interfacial instabilities during unidirectional solidification of a binary alloy. They predict that in the lead-tin system, convective instabilities can arise even at extremely low solute concentrations. Experimental confirmation of this prediction in this opaque metallic system would be extremely difficult, mostly because suitable methods are not avaiable for measuring convective flow in such systems. Therefore a program is now in progress to study convection during unidirectional solidification of succinonitrile with and without small additions of a lower density solute (ethanol).

Conclusion

We have seen that transparent analog systems can be used to model complex processes of metallic solidification, but that such experiments are often qualitative and their precise relevance to the solidification of metals unclear. In contrast, by careful measurement of relatively very simple phenomena in well-characterized analog systems, it has been possible to draw the first firm conclusions concerning the theory of dendritic

Table I. Properties of Succinonitrile

Property	Symbol	Value	Ref.*
Molecular Weight	W	80.092	10
Density of Solid	ρ_s	$1016 kg/m^3$**	10*
Density of Liquid	ρ_ℓ	$988 kg/m^3$**	11
Thermal Expansivity of Solid	β_s	$-5.6x10^{-4}/K$	13
Thermal Expansivity of Liquid	β_ℓ	$-8.1x10^{-4}/K$	11*
Shear Viscosity	η	$2.6x10^{-3}$ pascal-sec.	11
Kinematic Viscosity	ν	$2.6x10^{-6} m^2/sec.$	***
Surface Tension (Liquid-Vapor)	$\gamma_{\ell V}$	$46.78 mJ/m^2$	11
Surface Tension (Solid-Liquid)	$\gamma_{s\ell}$	$8.9 mJ/m^2$	8
Refractive Index (Solid)	η_s	1.4340**	12
Refractive Index (Liquid)	η_ℓ	1.4150**	12
Equilibrium Temperature (Triple Point)	T_E	331.23K	7
Latent Heat of Fusion	L	$4.78x10^7 J/m^3$	10
Entropy of Fusion	ΔS	$1.45x10^5 J/m^3K$	10
Heat Capacity of Solid	C_{Ps}	1913J/kgK**	10
Heat Capacity of Liquid	$C_{P\ell}$	2000J/kgK**	10
Thermal Conductivity of Solid	k_s	0.225J/mKs**	8
Thermal Conductivity of Liquid	k_ℓ	0.223J/mKs**	8
Thermal Diffusivity of Solid	α_s	$1.16x10^{-7} m^2/sec$**	***
Thermal Diffusivity of Liquid	α_ℓ	$1.12x10^{-7} m^2/sec$**	***
Unit Supercooling	$L/C_{P\ell}$	24.2K	***

* Indicates value derived from properties given in cited reference
** Indicates value of property at melting point, 331.23K
*** Indicates property derived from other properties in the table

growth, one of the most basic transformation processes in metallic systems. More complex processes are now being studied, as both theoretical models and experimental techniques evolve. The future of this type of physical modeling remains quite exciting as the scope is broadened and more complex phenomena are included.

References

1. C. S. Smith, p. 3 in Metal Transformations, W. W. Mullins and M. C. Shaw, ed.; Gordon and Breach, New York, 1966.

2. Growth and Perfection of Crystals, R. H. Doremus, B. W. Roberts, and David Turnbull, ed.; John Wiley and Sons, New York, 1958.

3. K. A. Jackson and J. D. Hunt, Acta Met., 12 (1965) 1212.

4. R. J. Schaefer and M. E. Glicksman, Met. Trans., 1 (1970) 1973.

5. L. R. Morris and W. C. Winegard, J. Crystal Growth, 5 (1969) 361.

6. G. A. Alfintsev, G. P. Chemerinsky, O. P. Fedorov, Kristall und Technik, 15 (1980) 643.

7. M. E. Glicksman, R. J. Schaefer and J. D. Ayers, Met. Trans., 7A (1976) 1747.

8. R. J. Schaefer, M. E. Glicksman and J. D. Ayers, Phil. Mag., 32 (1975) 725.

9. J. S. Langer and H. Müller-Krumbhaar, J. Crystal Growth, 42 (1977) 11.

10. C. A. Wulff and E. F. Westrum, J. Phys. Chem., 67 (1963) 2376.

11. M. J. Timmermans and Mme. Hennaut-Roland, Journal de Chimie Physique, 34 (1937) 693.

12. R. M. MacFarlane, E. Courtens, and T. Bischofberger, Mol. Cryst. Liq. Cryst., 35 (1976) 27.

13. H. Fontaine and M. Bee, Bull. Soc. Franc. Mineral. Crystallogr., 95 (1972) 441.

14. M. E. Glicksman and S. -C. Huang, "Convection Heat Transfer During Dendritic Solidifcation," paper presented at AIAA 16th Aerospace Sciences Meeting, Huntsville, Alabama (1978).

15. M. E. Glicksman and S. -C. Huang, Proceedings of the 3rd European Symposium on Materials Sciences in Space, Grenoble, France (1979).

16. J. S. Langer and H. Müller-Krumbhaar, Physico Chemical Hydrodynamics, 1 (1980) 41.

17. S. R. Coriell, M. R. Cordes, W. J. Boettinger and R. F. Sekerka, J. Crystal Growth, 49 (1980) 13.

Ising Model Simulations of Crystal Growth

G. H. Gilmer
Bell Laboratories
Murray Hill, New Jersey 07974

I. Summary

The Ising model of a crystal-melt interface is described. Kinetic equations are derived for a two-component system. They apply to a model crystal simulated by the Monte Carlo method, and a fluid that is treated by the mean-field approximation. This model is equivalent to the spin-1 Ising model of a crystal-vapor system. Simulation studies provide information on the mechanism of growth, kinetics, and the distribution of the two species.

II. Introduction

The rate of crystallization is sensitive to the atomic structure of the crystal-fluid interface, and atomistic models are necessary for a study of the mechanisms of growth. When the surface of a perfect crystal is stabilized by a network of strong bonds, new layers can be generated only by the nucleation mechanism. However, the slow nucleation process is circumvented when screw dislocations intersect the surface, since they provide sites with a large binding energy where the fluid atoms can attach to the growing crystal. Certain impurities segregate at the interface and facilitate the crystallization process. Also, if the strength of the bonds in the surface layer is below a specific value, thermal roughening produces an abundance of sites with large binding energies. In this article I discuss the application of kinetic Ising model simulations to the motion of the crystal-fluid interface. Simulations of this model have provided a definitive test of theories of crystal growth, and have also produced data for more complicated surface structures that are perturbed by thermal fluctuations and impurities [1]. Computer-generated color movies are also available that illustrate the behavior of the model under various conditions [2].

III. Ising Model of the Crystal-Melt Interface

The Ising model and related systems have a long history in the theory of crystal growth [3]. Although vapor deposition systems are the most obvious applications for the model, it has also been used to represent the interface between a crystal and its melt [4]. It exhibits most of the crystal growth phenomena, and yet it is simple enough to permit efficient computation. It does not include the local rearrangements that occur when atoms in the fluid become aligned with the growing crystal. Instead, atoms are assumed to make direct transitions from the fluid to the crystalline state (and vice versa) at rates that are determined by the temperature and the activation energies. The one-component system is equivalent to the spin-½ Ising model; each site of a space lattice corresponds to an atom in one of the two states.

The model can be applied to a particular crystal-melt system by the appropriate selection of parameters describing the internal energies and entropies of the phases. Much of this information can be obtained from the equilibrium properties of the systems [5]. In terms of an Ising model with nearest-neighbor interactions, the free energy of the crystal is approximated by the mean-field expression

$$F^C = -\frac{Z}{2}\phi^C_{AA}N^C_A - \frac{Z}{2}\phi^C_{BB}N^C_B + \frac{\omega^C N^C_A N^C_B}{(N^C_A + N^C_B)} + \qquad (1)$$

$$kT\left[N^C_A \ln\left(\frac{N^C_A}{N^C_A + N^C_B}\right) + N^C_B \ln\left(\frac{N^C_B}{N^C_A + N^C_B}\right) - N^C_A S^C_A/k - N^C_B S^C_B/k\right],$$

where Z is the lattice coordination number, N_A^C and N_B^C are the total numbers of A and B atoms in the crystal, ϕ_{AA}^C, ϕ_{AB}^C, and ϕ_{BB}^C are the bond energies between the indicated atoms, ω^C is the regular solution parameter

$$\omega^C = \frac{Z}{2}(\phi_{AA}^C + \phi_{BB}^C) - Z\phi_{AB}^C\,, \tag{2}$$

kT is the product of Boltzmann's constant and the temperature, and S_A^C and S_B^C are the vibrational entropies of the A and B atoms in the crystal (the entropy of mixing is given by the previous two terms). The free energy of the liquid F^C is also represented by an equation completely analogous to (1); but for brevity we include only the expressions for the A component. The difference in the bonding in the two phases is related to the heat of fusion L_A of the pure material, i.e.,

$$L_A = \frac{Z}{2}(\phi_{AA}^C - \phi_{AA}^L)\,, \tag{3}$$

and similarly the entropy of fusion is the entropy difference

$$\Delta S_A = S_A^L - S_A^C\,. \tag{4}$$

Equilibrium between the crystal and liquid phases requires the equality of the chemical potentials of the A and B species; μ_A^C is given by

$$\mu_A^C = \frac{\partial F^C}{\partial N_A^C} = -\frac{Z}{2}\phi_{AA}^C + (1 - C_A)^2\omega^C - S_A^C\,, \tag{5}$$

where eq. (1) has been used to obtain the derivative. Equating μ_A^C and μ_A^L, we obtain

$$\ln\frac{C_A^L}{C_A^C} = \frac{Z}{2kT}(\phi_{AA}^L - \phi_{AA}^C) + S_A^L - S_A^C + (1 - C_A^C)\frac{\omega^C}{kT} - (1 - C_A^L)^2\frac{\omega^L}{kT} \tag{6}$$

relating the liquid and crystal compositions. An analogous equation relates C_B^L to C_B^C. Thurmond and Struthers [5] have shown that these relations can be accurately fit to the phase diagrams for many systems by treating ω^L and ω^C as fitting parameters. (Often it is sufficient to use the ideal alloy liquid with $\omega^L = 0$.

Kinetic analogues of these equations have been presented by Jackson [4]. The liquid-crystal transition rates of the A species (R_A^+) and vice versa (R_A^-) can be taken directly from the chemical potentials, i.e.,

$$R_A^+ = fe^{\mu_A^L/kT} = fC_A^L e^{-\frac{Z}{2}\phi_{AA}^L/kT - S_A^L + (1-C_A^L)^2 \omega^L} \tag{7}$$

$$R_A^- = fe^{\mu_A^C/kT} = fC_A^C e^{-\frac{Z}{2}\phi_{AA}^C/kT - S_A^C + (1-C_A^C)^2 \omega^C} , \tag{8}$$

and similar equations apply for R_B^+ and R_B^-. Here f is the product of an attempt frequency and a Boltzmann factor for an activation energy. These equations are obviously consistent with the equilibrium condition (6). Note that only the bulk properties are included, and effects caused by the interface cannot be treated. Still eqs. (7) and (8) are not unique, and in the absence of detailed information about the potential energy surfaces for the transition, somewhat simpler expressions are usually postulated. The change in potential energy when an atom makes a crystal-fluid transition is

$$\Delta E_A = L_A - (1 - C_A^C)^2 \omega^C + (1 - C_A^L)^2 \omega^L . \tag{9}$$

Since the potential energy of an atom is usually greater in the liquid than in the crystal, $\Delta E_A > 0$ and the simplest form of the rate equations is

$$k_A^+ = \nu C_A^L e^{-(S_A^L - S_A^C)} \tag{9}$$

$$k_A^- = \nu C_A^S e^{-\Delta E_A/kT} , \tag{10}$$

where ν is the new frequency factor. Similar equations apply to the B species, although it should be noted that $\Delta E_B < 0$ in systems with retrograde solid solubility, and ΔE_B should then appear in the expression for k_B^+ [6]. The consequences of these and related equations for the distribution of the species between crystal and liquid have been discussed by Jackson [4,6]. We treat this aspect of crystal growth below, but will now derive the rate equations for the simulations.

The kinetic mean-field model treats crystallization and melting transitions as if they always occur at a kink site. The simulation model represents in detail the evolution of the crystalline phase, and the surface is free to assume whatever structure is dictated by the kinetics. The liquid phase is treated as a continuum; the chemical potentials of its components and the interactions with the crystal phase are approximated by the mean field expressions. The two-component system is equivalent to the spin-1 Ising model; each site of the space lattice is occupied by a crystalline A or B atom, or by the liquid.

The kinetics of crystal growth are simulated by performing transitions from crystalline to liquid, and the reverse, at the sites on the surface. The transition rate at a particular site depends, in general, on the number and types of neighboring atoms in the *crystal*. The change in the energy during a liquid-crystal transition is

$$\Delta E_A(n_A,n_B) = \Delta E_A^L + Z(\phi_{LL} - \phi_{AL}) + n_A(2\phi_{AL} - \phi_{LL} - \phi_{AA}^C) + n_B(\phi_{BL} + \phi_{AL} - \phi_{LL} - \phi_{AB}^C) \,. \tag{11}$$

Here ΔE_A^L is the change in the energy of the liquid phase

$$\Delta E_A^L = Z[C_A^L C_B^L \phi_{AA}^L + C_B^L(C_B^L - C_A^L)\phi_{AB}^L - (C_B^L)^2 \phi_{BB}^L] \,. \tag{12}$$

This accounts only for the bonds internal to the original liquid; their energy is altered by the absence of the additional A atom. The change in the energy of the Z bonds surrounding the site of the transition is included in the remaining terms in (11). The average interaction energy between the contents of two neighboring sites in the liquid is ϕ_{LL} in (11), where

$$\phi_{LL} = (C_A^L)^2 \phi_{AA}^L + 2C_A^L C_B^L \phi_{AB}^L + (C_B^L)^2 \phi_{BB}^2 \,, \tag{13}$$

and the corresponding interaction between an A or B atom of the crystal and the liquid is ϕ_{AL} or ϕ_{BL}, respectively. In the absence of detailed information on these interactions, we assume that they are equal to the equivalent energy in the liquid, i.e.,

$$\phi_{AL} = C_A^L \phi_{AA}^L + C_B^L \phi_{AB}^L \,. \tag{14}$$

The rate of crystallization on sites at the interface is assumed to be independent of the surface structure, and is again given by eq. (9). The melting rate then depends on n_A and n_B, i.e.,

$$k_A^-(n_A,n_B) = \nu e^{\Delta E_A(n_A,n_B)/kT}$$

$$= \nu k e^{(n_A \Phi_{AA} + n_B \Phi_{AB})/kT} , \quad (15)$$

where $k = e^{[\Delta E_A^L + Z(\phi_{LL} - \phi_{AL})]/kT}$,

$$\Phi_{AA} = 2\phi_{AL} - \phi_{LL} - \phi_{AA}^C \quad (16)$$

$$\Phi_{AB} = \phi_{BL} + \phi_{AL} - \phi_{LL} - \phi_{AB}^C \quad (17)$$

according to eq. (11). The "effective" bond energies Φ_{AA} and Φ_{AB} largely determine the surface structure and the kinetics. The value of Φ_{AA} in a one-component system is proportional to the heat of fusion,

$$\Phi_{AA} = \phi_{LL} - \phi_{AA}^C = \frac{2}{Z} L . \quad (18)$$

This formulation permits the direct application of existing systems of crystal-vapor simulations to crystal melt systems. Rate equations similar to (9) and (15) have been simulated for a variety of lattice structures and temperatures [1,7,8]. The equivalent crystal-vapor bond energies are given by (16) and (17). However, crystal-vapor simulations performed at large driving force may not be directly applicable. This driving force is obtained by an increase in the vapor pressure (k^+), whereas the crystal-melt system is supercooled with a consequent reduction in k^-. Although the resulting differences in the rate constants are small near equilibrium, significant changes may occur at large driving force. There is a greater disparity between the melting rates of atoms at the various sites when the system is supercooled.

A complete description of the simulation procedure has been given elsewhere [1]. Briefly, a small section of the crystal surface is represented, typically this is a square area of sixty atoms on a side. The configuration of the interfacial region is stored as a list of sites occupied by atoms in the crystalline state, together with the identity of these atoms.

Periodic boundary conditions eliminate edge effects; atoms at an edge interact with a replica of the segment translated to a position adjacent to that edge. These interactions stabilize the surface orientation, and the various close-packed surfaces can be selected by the appropriate choice of the boundary conditions.

The transitions are performed with probabilities that are proportional to the rate constants k_A^+, k_A^-, k_B^+ and k_B^-. This is accomplished by standard Monte Carlo techniques using computer-generated random numbers [9]. Unfortunately, the equilibrium compositions of the crystal and liquid may differ from those given by eq. (6), which is based on a mean-field approximation. Therefore, the equilibrium concentrations must be obtained by simulation, although (6) provides an approximate value that may be useful as an initial guess. It is quite accurate when $C_B^C << 1$, and when ω^C is small.

IV. Surface Roughening and Crytal Morphology

The rate of solidification on different segments of a crystal surface depends on their orientation relative to the crystal lattice. Facets at low-index directions are evidence of anisotropy in the kinetics. Facets occur at orientations where the growth rate has a sharp minimum as a function of the orientation angle.

The arrangement and lateral interactions between atoms in the surface layer are important factors that affect the kinetics. Three faces of the FCC lattice are illustrated in Fig. 1. The (111) orientation has the strongest lateral bond network, since the coordination number for nearest neighbors determines the energy in this model. The fraction ξ of nearest neighbors in the surface layers was introduced by Jackson [4] as a quantitative measure of this factor, and in this case, $\xi_{(111)} = 1/2$, $\xi_{(100)} = 1/3$, and $\xi_{(110)} = 1/6$.

The simulated growth rates of these faces are illustrated in Fig. 2 for two values of L/kT. The rates are plotted as a function of the driving force

$$\Delta\mu = \frac{L(T_M - T)}{T_M}, \tag{10}$$

where $T_M = L/\Delta S$ is the melting point and T is the actual temperature of the interface. The shapes of these curves imply that two distinct mechanisms are operating on close-packed faces of perfect crystals.

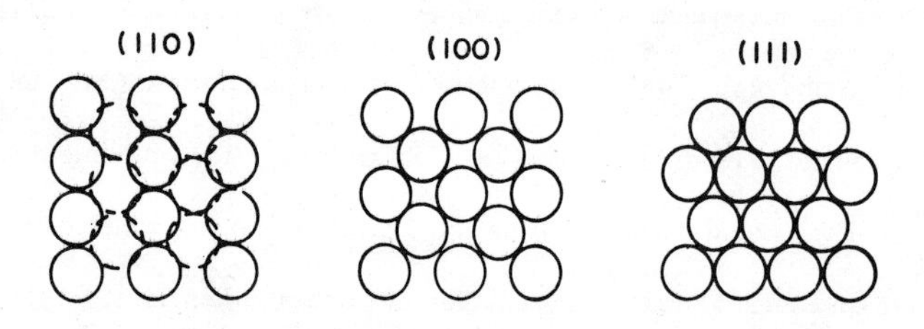

Fig. 1 Arrangement of atoms in layers parallel to the FCC (111), (100), and (110) faces. Atoms underlying the (110) layer are illustrated as dashed circles.

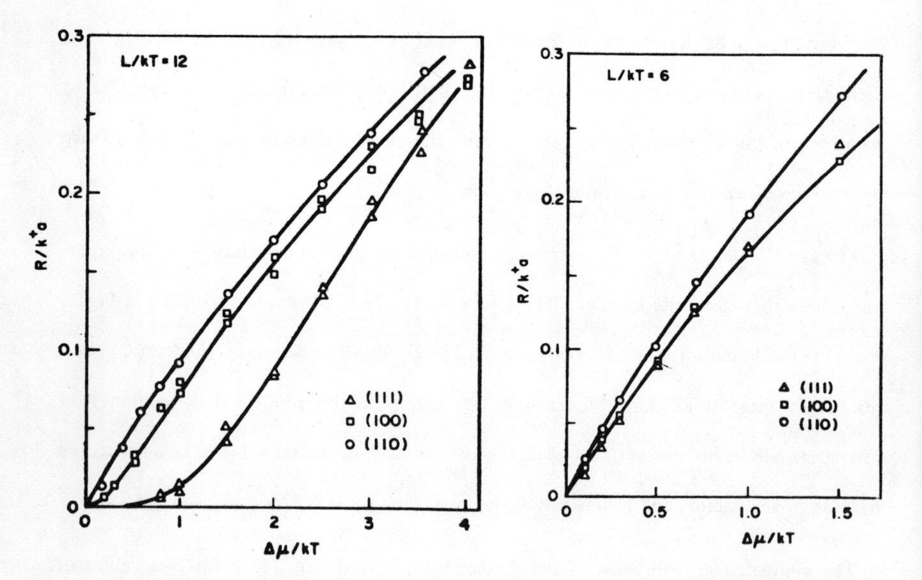

Fig. 2 Simulated growth rates on three FCC surfaces. (a) $L/kT = 12$ (b) $L/kT = 6$. The rate of motion of the surface is normalized by the product of the crystallization rate k^+ and the lattice constant a.

The kinetics at $L/kT = 12$ have considerable anisotropy, especially at the smaller values of $\Delta\mu$. The (111) face is essentially immobile when $\Delta\mu \lesssim 0.5kT$. A perfect single crystal growing with $0.1 < \frac{\Delta\mu}{kT} < 0.5$ would expand quickly until it assumed an octahedral shape, since the other orientations have appreciable growth rates. Then measurable growth would cease on this highly faceted crystal. At somewhat higher values of $\Delta\mu$ the disparity between the growth rates is smaller, and additional facets or rounded corners would appear on the growth form. At very large driving forces the growth rates on the faces become approximately equal. Thus, a gradual change from very anisotropic to nearly isotropic growth can be achieved by manipulation of the driving force. This change may not be observable during solidification, however, because of the smaller range of $\Delta\mu$ that is accessible. Rapid interface kinetics and the accompanying latent heat often prevent steady-state growth at large undercoolings.

Systems with smaller values of L/kT_M have nearly isotropic growth at all values of $\Delta\mu$, as indicated by the kinetic data of Fig. 2b. These results are in qualitative agreement with Jackson's classification of crystal morphology with L/kT_M [10]. Systems with $L/kT_M \gtrsim 3$ produce faceted crystals, whereas those with smaller values usually generate a rounded growth form. In the latter case, the smooth crystal-melt interface is located close to the melting point isotherm. The critical value of L/kT_M is apparently about a factor of two larger in the case of the Ising model, a difference that is not surprising in view of the simplified treatment of the crystal-fluid interactions used in the model system.

The interface kinetics are affected by the atomic-scale roughness of the crystal surface. The motion of a smooth, low-index surface is inhibited, since atoms in the adjacent liquid layer have few neighbors in the crystal to stabilize and position them in the lattice. It may be necessary to nucleate two-dimensional ($2d$) clusters of crystalline material in this layer, and indeed this is the mechanism of growth on the (111) face at $L/kT = 12$. Nucleation kinetics can be recognized by the characteristic region of metastable states at small values of $\Delta\mu$, as observed in Fig. 2a. The critical cluster is very large in this region, and the probability of producing a large cluster by random fluctuations is minute.

Growth on the (100) and (110) faces is much faster. These faces have weaker lateral bond networks, and at $L/kT = 12$ they have more adatoms and clusters than does the (111) face. large $2d$ holes in the surface layer of the crystal are also present as a result of thermal disordering. When a driving force is applied, equilibrium clusters serve as nuclei that facilitate the crystallization of new layers.

The (100) growth rate of Fig. 2a is non-linear near the origin, and careful examination reveals a small metastable region. At large values of L/kT_M this face has an extensive metastable region. The growth rate is linear in $\Delta\mu$ only when the bond energy is less than a critical value. The critical bond energy in the Ising model interface is determined approximately by the relation

$$\frac{\xi L}{kT_M} = 3.3 \tag{11}$$

Thus, when $L/kT_M = 12$, both (111) and (100) faces are smooth, but when $L/kT_M = 6$ they are both rough and move by the *continuous* mechanism [11]. The (110) face is an exception to (11) and does not have a finite critical bond energy. As shown in Fig. 1, atoms in the surface layer are connected only in one-dimensional chains. Thermal disordering of these chains is expected for all values of the bond energy, and nucleation should not be required. (Inclusion of second neighbor interactions does create a $2d$ bond network, and nucleated growth would occur with sufficiently large bond energies.)

A transition from nucleation kinetics to continuous growth can be observed in the case of a crystal-vapor interface. An increase in the temperature is equivalent to a reduction in the bond energy in this case. Thus, a low-index plane may grow by nucleation at low temperatures and by a continuous mechanism at high temperatures. The changeover from a faceted to a rounded growth form has been demonstrated with several plastic crystals including hexachloroethane, ammonium chloride, and adamantane [12]. The facets of a given set of low-index planes disappear at a critical roughening temperature. Since temperature is not a free variable, in the crystal-melt equilibrium, a specific low index plane in a pure system cannot exhibit the transition. Some planes may be above their critical roughening temperatures and others below.

A distinctive change in the surface structure takes place at the roughening point. Below this temperature, only small clusters of adatoms and vacancies appear on the surface. These cause deviations in the position of the crystal surface, but the average coordinate remains localized and does not change more than a fraction of a monolayer. At the transition temperature the clusters percolate, and the presence of a cluster of unlimited extent permits the surface to become delocalized. Nucleation is not required, and even equilibrium fluctuations cause the interface to move in a one-dimensional random walk [13].

V. Impurities and Defects

Thermal roughening of the crystal surface can explain the rapid growth of crystals in many cases. Dislocation-free silicon, for example, has only a small (111) facet [14] when grown from the melt, and the undercooling at the center of this facet has been estimated to be only a few degrees Centigrade. Thus, the close-packed (111) face is smooth, but is quite close to the roughening point. Often crystals with extensive facets also grow at a rapid rate. In this case dislocations or impurities probably are instrumental in the growth process.

The spiral growth mechanism permits growth on low-index facets. A dislocation that terminates with a component of its Burger's vector perpendicular to the surface will produce one or more steps. These steps are pinned at the point where the dislocation terminates. In the presence of a driving force the free edges advance and wind up into a spiral pattern, which rotates about the pinning point during steady-state growth [3].

Growth spirals have been observed in numerous experiments, and the parabolic relation, $R\ \alpha(\Delta\mu)^2$, predicted by theory [3] has been confirmed in simulations [8,15]. A somewhat larger exponent on the driving force is observed when $2d$ nucleation and spiral growth operate concurrently.

Small quantities of certain impurities may also increase the growth rate. They can influence the kinetics in one of two ways. (1) An impurity that forms strong bonds to the host species will adhere to the crystal surface and afford favorable sites for crystallization. We refer to this as an ordering impurity, since it tends to form an ordered alloy with the host when present in large quantities. (2) An impurity that segregates at the interface may stabilize host atoms in this region. Interfacial segregation often occurs when the impurity bonding energy is intermediate between that of the crystal and the melt. This type of impurity layer provides a transition region between the crystal and the melt. Crystalline clusters of host atoms are favored here because of the strong bonding. If the layer retains sufficient mobility to permit diffusional and rotational motion, the layer may enhance crystal growth. Segregation is most pronounced for clustering impurities, i.e., systems with relatively weak AB bonds. Materials with a solid phase miscibility gap at low temperatures are likely candidates.

The kinetics of a close-packed face with ordering impurities are shown in Fig. 3 (triangular symbols). The simple cubic (SC) (100) face was simulated at $L/kT_M = 12$; a large nucleation barrier inhibits growth of the pure crystal, as shown by the square symbols. In this case, $\Phi_{AB} = 2\Phi_{AA}$, and $\Phi_{AA} = \Phi_{BB}$, and the concentration of B atoms in the liquid is $C_B^L = 2.5 \times 10^{-3} C_A^L$. The small quantity of impurity clearly has a dramatic effect on the growth rate.

As might be expected, these impurities are most effective at small values of $\Delta\mu$. Here the slow nucleation rate suppresses the growth of the pure crystal. Even a minute concentration of impurity can cause a large relative increase in the nucleation rate. A single impurity atom can make the difference between an unstable cluster, and one that will persist and expand until the entire layer has crystallized. At large values of $\Delta\mu$ the nucleation rate exceeds the rate of crystallization of the B atoms. In this case, the impurities have little effect on the kinetics.

A finite growth rate is observed even when $\Delta\mu \to 0$. Here $\Delta\mu$ is the driving force measured for the pure A system, and the addition of this type of impurity to the crystal will raise the melting point. Thus, as some of the impurities in the liquid phase diffuse into sites on the crystal surface, they produce a layer that is, in effect, supercooled. Thus, the gradual transformation of the surface layers produces slow growth, even under conditions where the pure crystal would be in equilibrium.

Clustering impurities influence the kinetics by a very different mechanism. A layer of B impurities may condense on the surface of an A-rich crystal, even when the temperature is above the melting point of a crystal of B atoms. This can occur when $\Phi_{AB} > \Phi_{BB}$, and the A-rich crystal has a greater affinity for the impurities than the B crystal.

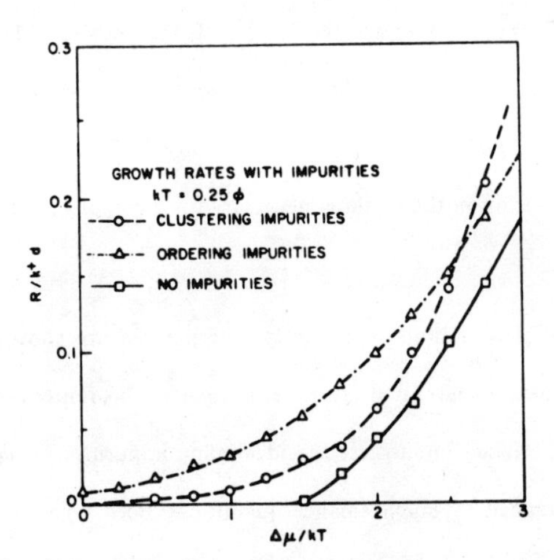

Fig. 3 Simulated growth rates in the presence of impurities are compared with the growth rates of the pure crystal.

The kinetics with clustering impurities are also shown in Fig. 3. Here $\Phi_{AB} = 0.533\Phi_{AA}$, and $\Phi_{BB} = 0.4\Phi_{AA}$. Again, the impurities enhance the growth rate, especially at the smaller values of $\Delta\mu$. The linear dependence of R on $\Delta\mu$ near the origin implies that nucleation is not required. These data suggest that spiral growth is slower than the impurity mechanisms at small $\Delta\mu$, since a parabolic dependence on $\Delta\mu$ implies a smaller growth rate as $\Delta\mu \rightarrow 0$.

The composition of the crystal may differ substantially from the value corresponding to the equilibrium phase diagram, and crystals grown with large $\Delta\mu$ values are most likely to have significant deviations. The more volatile component is trapped in excess of its equilibrium concentration. The liquid is enriched in this species, and during slow growth the impurity is rejected by the crystal-melt interface. Rapid motion of the interface permits insufficient time for this species to leave the interfacial region, and a larger concentration of impurity is trapped in the crystal. This effect may be enhanced when a segregated layer is present, since rapid growth may trap portions of this layer in the crystal [7].

This model applies only to interface trapping, and does not include diffusion in the liquid phase. The concentration in the liquid layer adjacent to the crystal surface determines the crystal composition. Since this concentration may differ from that in the bulk liquid, a diffusion calculation in the liquid is necessary for a complete determination of the crystal composition.

Simulation data for the amounts of impurity incorporated are shown in Fig. 4. The clustering impurities form a segregated layer. This layer is apparently responsible for the large concentration of trapped impurities; the ideal alloy impurities do not form a segregated layer and are trapped in much smaller quantities. Both types of impurities have approximately the same equilibrium concentration.

The concentration of clustering impurities captured by the crystal can exceed the maximum solubility of the B atoms in the A-rich crystal. The data of Fig. 4 indicate that $C_B = 0.13$ at $\Delta\mu/kT = 0.5$, whereas the maximum equilibrium concentration is

$C_B^{(e)} = 0.02$. Thus, the kinetics of the process can produce a metastable alloy. Since the B-rich phase can only be created by a nucleation event, the metastable crystal may remain indefinitely in the super-saturated condition. A small solid-state diffusion coefficient also slows the process. Laser annealing techniques permit the rapid melting and solidification of very thin layers. Solidification rates of several meters per second have been measured. Metastable alloys have been produced by ion-implantation and laser annealing with substitutional impurity concentrations that exceed the equilibrium value by several orders of magnitude [16].

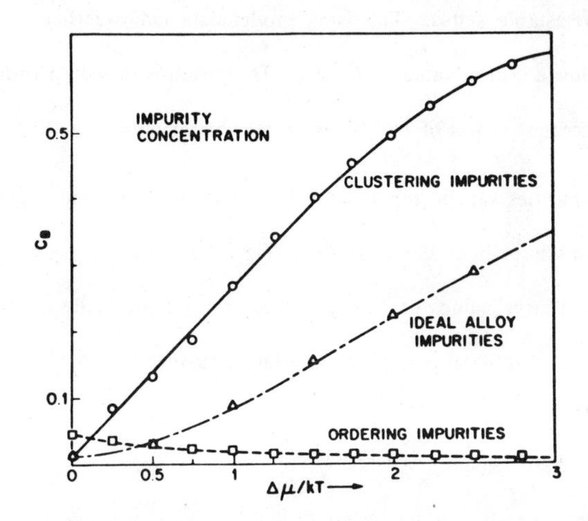

Fig. 4 The concentration of impurity trapped in the crystal. Square and circular symbols correspond to the ordering and clustering impurities, respectively, under the condition specified for Fig. 3.

The ordering impurities are incorporated in the greatest numbers at small values of $\Delta\mu$. These impurities are strongly bound to the crystal, and slow growth permits a larger number to interact with each layer. As $\Delta\mu$ is increased, fewer impurity atoms impinge on a given layer and the concentration is reduced. In general, an increase in $\Delta\mu$ causes an increase in concentration when $\Phi_{AB} < \Phi_{AA}$, and a decrease when $\Phi_{AB} > \Phi_{AA}$.

VI. Conclusions

The kinetic Ising model has been simulated under conditions that produce a number of different growth mechanisms. The two-component crystal-melt interface can be mapped onto the spin-1 model, with bond energies that are determined by the heats of fusion and the concentrations in the melt. Nucleated and continuous growth have been demonstrated on close-packed faces of a perfect crystal. Faces with a strong bond network may require nucleation, while others grow by the continuous mechanism. A large anisotropy in rate is present when L/kT_M is large, and the close-packed faces have extensive regions of metastable states. The Ising model data indicate that growth becomes nearly isotropic below a critical value of L/kT_M. This result is in accord with solidification data, although the critical values of L/kT_M are somewhat smaller.

Defects and impurities can be represented by the Ising model; spiral growth and two impurity mechanisms have been demonstrated. These mechanisms are most effective on close-packed faces at large values of L/kT_M; faces that are thermally roughened include active growth sites as an integral part of the interface structure.

REFERENCES

[1] Reviews of Monte Carlo crystal growth simulations include: J. P. van der Eerden, P. Bennema, T. A. Cherepanova, in: *Progress in Crystal Growth and Characterization,* B. R. Pamplin, Ed. (Pergamon, Oxford, 1979), vol. 3, p. 219; J. D. Weeks and G. H. Gilmer, Adv. Chem. Phys. **40,** 157 (1979); G. H. Gilmer and K. A. Jackson, in: *Crystal Growth and Materials,* E. Kaldis and H. J. Scheel, Eds. (North-Holland, Amsterdam, 1977) p. 80.

[2] A color movie depicting nucleated, continuous, and spiral growth is: *Computer Simulation of Crystal Growth* by G. H. Gilmer, K. C. Knowlton, H. J. Leamy and K. A. Jackson. This film is available for loan from the Film Library, Bell Laboratories, Murray Hill, N. J.

[3] W. K. Burton, N. Cabrera, and F. C. Frank, Phil. Trans. Roy. Soc., London, **243A,** 299 (1951).

[4] K. A. Jackson, Canadian J. Phys. **36,** 603 (1958).

[5] C. D. Thurmond and J. D. Struthers, J. Phys. Chem. **57,** 831 (1953).

[6] K. A. Jackson, G. H. Gilmer, and H.J. Leamy, in: *Laser and Electron Beam Processing of Materials,* eds. C. W. White and P. S. Peercy (Academic Press, New York, 1980) p. 104.

[7] G. H. Gilmer, Science **208,** 355 (1980).

[8] G. H. Gilmer, J. Cryst. Growth **35,** 15 (1976).

[9] K. Binder, in: *Phase Transitions and Critical Phenomena,* Vol. 5, C. Domb and M. S. Green, eds. (Academic Press, New York, 1975).

[10] K. A. Jackson, D. R. Uhlmann, and J. D. Hunt, J. Crystal Growth **1,** 1 (1967).

[11] Continuous growth is discussed by N. Cabrera and R. V. Coleman, in: *The Art and Science of Growing Crystals,* J. J. Gilman, ed. (Wiley, New York, 1963) p. 3.

[12] K. A. Jackson and C. E. Miller, J. Cryst. Growth **40,** 169 (1977) and A. Pavlovska, *ibid.,* **46,** 551 (1979).

[13] J. D. Weeks in: *Ordering in Strongly Fluctuating Condensed Matter Systems,* ed. T. Riste (Plenum, London, 1980). Also see H. Müller-Krumbhaar, in: *Current Topics in Materials Science* (North-Holland, Amsterdam, 1978) vol. 2, p. 115.

[14] T. F. Ciszek, J. Cryst. Growth **10,** 263 (1971).

[15] R. H. Swendsen, P. J. Kortman, D. P. Landau, and H. Müller-Krumbhaar, J. Crystal Growth **35,** 73 (1976).

[16] H. J. Leamy, J. C. Bean, J. M. Poate, G. K. Celler, J. Cryst. Growth **48,** 379 (1980).

THE STABILITY OF TWO-DIMENSIONAL

LAMELLAR EUTECTIC SOLIDIFICATION

Harvey E. Cline
General Electric Corporate Research and Development
P.O. Box 8, Schenectady, New York 12301

The stability of two dimensional lamellar eutectic solidification was analyzed by calculating the growth of normal modes. Below a critical spacing that corresponds to the minimum undercooling, the structure was unstable with respect to variations in the lamellar spacing, while above the critical spacing the structure was stable. Growth of Pb-Sn eutectic thin films were observed in the stable regime at spacings larger than found in bulk material. A mechanism involving the propagation of bent plates eliminated variations in the lamellar spacing.

Introduction

Eutectic alloys form a striking periodic morphology consisting of alternate plates of the two equilibrium solid phases. The spacing of the periodic structure depends on the rate of solidification, because the solid phases separate from the liquid by liquid diffusion. Solidification occurs at a temperature slightly below the equilibrium eutectic temperature because of the irreversible process of diffusion in the liquid and the free energy needed to form interphase boundaries (1). Previously, the steady state diffusion ahead of the planar solid-liquid interface was solved by the Fourier series method (2-4). There was a problem in determining the unique spacing, because solutions exist for all values of the spacing. Tiller (1) assumed that growth occurs at a minimum in the undercooling to explain the observed relationship between the spacing λ and the growth velocity V, $\lambda^2 V$ = constant. Jackson and Hunt (5) observed the growth of a transparent eutectic film and found that the spacing was controlled by the formation of faults.

Another approach to understanding what controls the spacing of a eutectic was to analyze the stability of the growth with respect to shape perturbations (6). Fluctuations in the growth conditions induce variations in the spacing of the lamellar eutectic. These variations in the geometry decrease with time in the regime of stable growth. To test the steady state solutions for stability involves solving the diffusion problem in the liquid ahead of the perturbed interface. The mathematical techniques and detailed stability calculations were described in another paper (6). A qualitative description of the stability analysis of a two-dimensional eutectic will be presented. Observations of the growth of Pb-Sn thin films were made to test the predictions of the two-dimensional stability analysis and to investigate the mechanism that controls the lamellar spacing.

Diffusion in the Liquid

The solution to the diffusion in the liquid ahead of the planar interface is found by superimposing point sources along the interface. A unit source located on the moving solid-liquid interface at position X' produces a composition in the liquid at X given by the Green's function $G(X,X')$. The influence of a point source decreases with distance along the interface (6), Fig. 1. For example, in the Pb-Sn eutectic each Sn plate is a source of Pb, and each Pb plate is a sink. If all the plates are the same width, the sources and sinks produce a diffusion field with the periodicity of the lamellar structure. However, if one Sn plate is wider than the others, a source of Pb is introduced into the liquid which influences the composition at the solid-liquid interface, the influence decreases with distance from the perturbed plate, Fig. 1.

Normal Modes

An arbitrary variation in the widths of the plates may be decomposed into a sum of normal modes. The growth of normal modes is calculated separately which greatly simplifies the stability calculation. The geometry of the structure may vary in two ways which yields the branches of normal modes. First, variations in the spacing of the structure occur along the solid-liquid interface which give a gradual change in the widths of the alpha and beta phases. The alpha-beta and beta-alpha junction points both move in the same direction. In an analogous problem of lattice vibrations of a one-dimensional diatomic chain of atoms, this mode is known as the acoustical mode, Fig. 2. A second type of variation, the optical mode, occurs if each phase varies in the opposite sense. In this case, the composition of the solid varies slowly along the interface.

The number of normal modes equals the degrees of freedom of the system. If there are N plates of each phase, then there are 2N degrees of freedom. Each mode

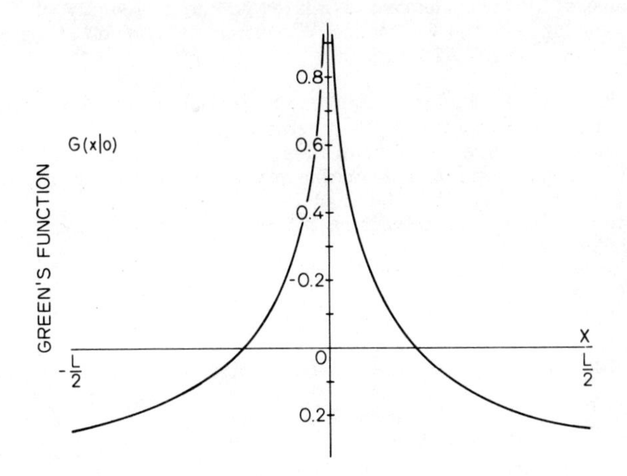

Fig. 1 Green's function for the steady-state diffusion equation for a planar interface of length L for a source located at X' = 0. This function is the composition in the liquid due to a point source.

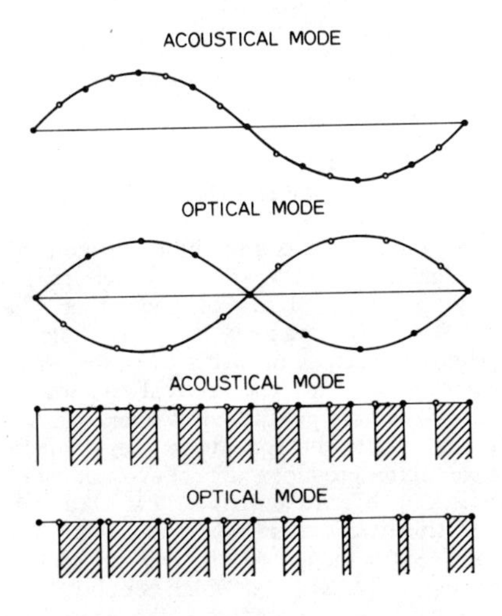

Fig. 2 In the acoustical mode, the amplitude of the displacements of the two types of junction points move together; (b) in the optical mode, they move oppositely; (c) the acoustical mode produces spacing variations, (d) while the optical mode produces variations in the composition of the solid.

has a discrete wavelength λ_p and a corresponding wavenumber $K_p = 2\pi/\lambda_p$. The maximum value of the wavenumber is related to the periodicity of the structure, $2\pi/\lambda$. The spacing between the modes depends on the number of plates N of each phase, there are N modes in each branch.

The diffusion in the liquid was calculated for each mode. A sinusoidal variation in the diffusion field of amplitude E_p is generated by the p-th mode. The composition variation tends to stabilize the growth of a mode. The critical mode is the least stable mode, E_p is a minimum. This minimum value occurs on the acoustical branch at long wavelengths, Fig. 3. The magnitude of the composition variation, E_p, increases with increasing wavenumber on the acoustical branch and decreasing wave number on the optical branch.

Undercooling

The solidification temperature is undercooled from the equilibrium eutectic temperature by an amount ΔT which depends on the average composition of the liquid at the interface and on the average curvature of the solid-liquid interface. The undercooling is a minimum when the terms representing the diffusion in the liquid equal those terms representing the interphase boundary energy. Both the average composition of the liquid and the curvature of each plate are influenced by the shape perturbation. The orientation of the alpha-beta boundaries are locally modified by the average curvature which influences the groove angles. Variations in the widths of the plates induce composition variations which modify the local geometry and control the growth of the variation. Each normal mode is examined separately. The main result of this calculation (6) was that below a critical spacing corresponding to the minimum undercooling all the normal modes decrease in amplitude with time. Hence, the structure is stable provided that

$$\lambda^2 V > \text{constant} \quad (1)$$

while at spacings below the critical spacing the long wavelength acoustical mode grows with time to generate an unstable structure.

In the unstable regime, variations in the plate widths increase with time under the influence of capillarity. Smaller-than-average plates decrease in size and become overgrown by adjacent plates of the other phase; the smaller plates terminate to increase the average spacing. Faults form until the spacing increases to the critical spacing where the effect of diffusion in the liquid produces stable growth. In contrast, at spacings larger than the critical spacing, variations in the spacing decrease with time to give stable growth. However, there is a mechanism described by Jackson and Hunt that limits the spacing in the stable regime. Diffusion in the liquid ahead of a large plate produces a concave solid-liquid interface which may catastrophically split into two plates with a new plate nucleated in the region of negative curvature. In thin two-dimensional films there is a stable region of growth where the plates do not split, Fig. 4. At the critical spacing, the undercooling does not depend on the spacing, $T/X = 0$, which is equivalent to Tiller's assumption of growth at an extremum (1). In contrast, the stability analysis (6) makes no assumption about the spacing; rather, it predicts a range of stable growth above the critical spacing. Details of the mechanism of control of the spacing are obtained from observations of the growth of eutectic thin films.

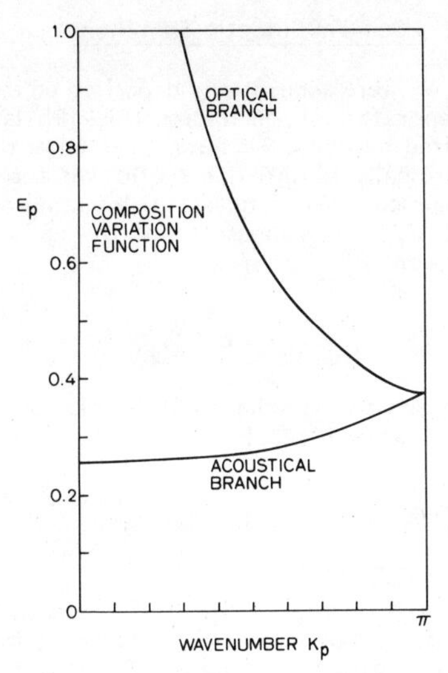

Fig. 3 The calculated value of the composition variation function, E_p, at different wavenumbers, K_p, for both the optical and acoustic modes.

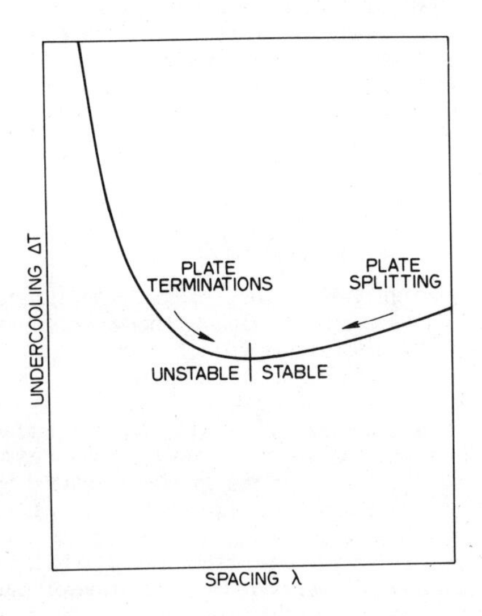

Fig. 4 The critical spacing occurs at the minimum undercooling. The spacing is maintained near the critical value by the mechanisms of plate termination and plate splitting.

Lead-Tin Eutectic Thin Films

Thin films of Pb and Sn were sequentially deposited on a pyrex microscope slide with an electron beam evaporator. A .56 micron thick Pb layer and a 1.44 micron thick Sn layer were deposited to give a 2-micron thick eutectic film after the layers melt together. The directionally solidified eutectic was formed by moving a glass slide through a heat zone produced by a quartz-iodine lamp and elliptical reflector. A thermal gradient of $200^{\circ}C/cm$ was measured at the melting point of the Pb-Sn eutectic with an applied power of 20 watts/cm.

Lamellar Spacing

The relationship between the lamellar spacing and the growth velocity in the Pb-Sn eutectic follows the relationship (7-12):

$$\lambda^2 V = 3.8 \times 10^{-11} cm^3/sec. \quad (2)$$

over a wide range of solidification rates, Fig. 5. The spacing observed in thin films is larger than that measured on bulk material. According to the stability analysis the spacing may be larger than the critical value. The structure of the eutectic thin film consists of alternate stripes of Pb and Sn phases, Fig. 6. By carefully controlling the solidification conditions and eliminating defects in the film, a fault-free area may be produced. Defects in the film such as pinholes produce faults in the structure. As shown in Fig. 7, several plates terminate at a pinhole in the film. However, as the planar solid-liquid interface passes the hole in the film, the plates bend around the hole. New plates do not nucleate in the small volume of liquid near the hole. The hole produces a variation in the lamellar spacing. Since several plates were eliminated by the pinhole, the spacing after the defect was larger than the average spacing. The variations in the lamellar spacing decrease during solidification by the propagation in the bend of the plates along the interface. According to the stability analysis, variations in the spacing above the critical spacing decrease with time in agreement with this observation.

Discussion

An analysis of the stability of two-dimensional eutectic solidification predicts a range of stable growth above a critical spacing that corresponds to the minimum in the undercooling. Observations of the growth of Pb-Sn eutectic films show fault-free structures at spacings larger than the $\lambda^2 V = 3.8 \times 10^{-11} cm^3/sec.$ value observed in bulk eutectic alloys. It is easier to nucleate new plates in bulk alloys than in thin films. A mechanism of fault motion in the solid-liquid interface may finely adjust the spacing in bulk alloys (5). In thin films stable growth exists at spacings larger than the critical spacing. Variations in the spacing produced by defects in the thin film were observed to decrease during solidification. The mechanism for producing a uniform spacing involves the propagation of bent plates along the solid-liquid interface, much like the motion of edge dislocations. Thus, it is possible to make adjustments in the spacing locally without the formation of faults. It is possible to produce a fault-free structure in directionally solidified thin films. There is considerable interest today in the possibility of fabricating such submicron periodic structures.

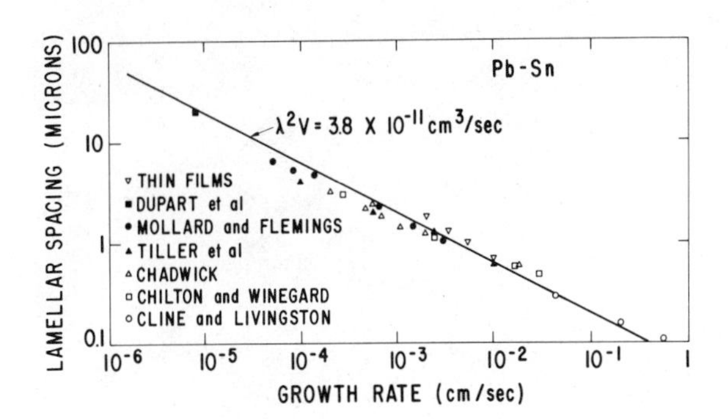

Fig. 5 The lamellar spacing vs. the growth velocity for the Pb-Sn eutectic. Thin films show a higher value of the spacing than bulk material which shows there is a range of stable growth.

Fig. 6 A scanning electron micrograph of a Pb-Sn eutectic thin film with a 1.8 micron spacing solidified at .0021 cm/sec. The Pb phase appears lighter than the Sn phase.

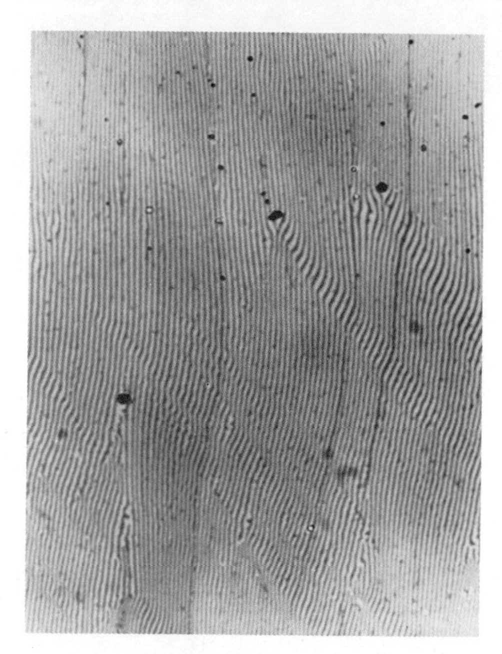

Fig. 7 Plates bend around a pin hole in a Pb-Sn eutectic thin film solidified at .0021 cm/sec. The spacing was adjusted during growth by the propagation of the bent plates along the solid-liquid interface (1000X).

References

1. W.A. Tiller, Liquid Metals and Solidification (ASM, Cleveland, 1958) pp. 276-318.
2. C. Zener, Trans. AIME 167, 550 (1946).
3. W.H. Brandt, J. Appl. Phys. 16, 139 (1945).
4. W. Hillert, Jernkantorets Ann. 141, 757 (1957).
5. K.A. Jackson and J.D. Hunt, Trans. AIME 236, 1129 (1966).
6. H.E. Cline, J. Appl. Phys. 50, 4780 (1979).
7. F.R. Mollard and M.C. Flemings, Trans. TMS-AIME 129, 1526 (1967).
8. W.A. Tiller and Mrdjenovich, J. Appl. Phys. 34, 3639 (1963).
9. G.A. Chadwick, J. Inst. Metals 92, 18 (1963).
10. J.P. Chilton and W.C. Windgaurd, J. Inst. Metals 89, 162 (1960).
11. H.E. Cline and J.D. Livingston, Trans. AIME 245, 1987 (1969).
12. J.M. Dupart, P.G. Fournet, G. Fontaine, M.G. Blanchin, M. Turpin and R. Racek, Proc. of Conf. on In Situ Composites, Lakeville, Conn. III, 85 (1972).

DIRECTIONAL GROWTH OF EUTECTICS AND DENDRITES AT THE LIMIT OF STABILITY

W. Kurz, D.J. Fisher

Department of Materials
Swiss Federal Institute of Technology (EPFL)
Lausanne - Switzerland

Microstructures of metals and alloys after solidification are generally composed of either dendrites or eutectic or both. Solidification theory applies an analogous approach to these structures: calculation of the diffusion field in the stationary state, and of curvature effects. This always leads to an ambiguous relationship between supersaturation and Péclet number. To determine a unique relationship between growth conditions and microstructure, a new criterion recently proposed for finding the 'operating point' of such systems (growth at the limit of morphological stability) is useful. It is shown that, in this way, cases where the extremum criterion has failed can be better understood.

Introduction

There are only two basic growth modes in solidifying metals and alloys; dendritic and eutectic. These forms appear in the columnar and equiaxed grains of cast products. Both can be treated theoretically by analogous approaches which consider two opposing effects; diffusion of heat or mass and capillarity.

Until now, the extremum criterion proposed by Zener [1] has generally been used in order to avoid the uncertainty of scale (dendrite tip radius or eutectic interphase spacing) which arises due to ignoring the time dependence of the corresponding growth forms. Recently, two independent attempts have been made to use a stability criterion to find the operating point of such systems:

- for dendrites, Langer and Müller-Krumbhaar [2 - 3]
- for irregular eutectics, Fisher and Kurz [4 - 6]

The aim of this paper is to point out the similarity of these two approaches for directional growth, each of which permits an improved fit between experiment and theory.

The Stability Criterion

The above criterion, which permits the determination of the operating region of the 2 growth morphologies in solidification, is growth with a characteristic size which corresponds to the wavelength of instability. This wavelength can be obtained with the aid of the Mullins-Sekerka analysis [7]. In order to obtain a more practical relationship some simplifications have been made [6]: equal conductivities of both phases, same temperature gradient in liquid and solid, very small distribution coefficient, k, and planar s/ℓ interface. This leads to

$$E = \frac{\dot{\delta}}{\delta} = \frac{V}{mG_c}\,(-\,\Gamma\omega^2 - G + mG_c)\,(\omega' - \frac{V}{D}) \qquad (1)$$

(see list of symbols).

Figure 1 visualizes this relationship. The interesting value of the wavelength for our cases is λ_i, the minimum wavelength for morphological instability. (There is only a small difference between the λ_i values for eqn. (1) and the full Mullins-Sekerka solution). λ_i can be calculated by setting the first term in the bracket of eqn. (1) equal to zero.

$$\lambda_i = 2\pi\,(\frac{\Gamma}{\Delta})^{\frac{1}{2}} \qquad (2)$$

where the gradient difference $\Delta = mG_c - G$. Because eutectics as well as dendrites grow far from the limit of constitutional supercooling of the plane front, $G << mG_c$ and

$$\lambda_i = 2\pi\,[(\frac{\Gamma}{\Delta T_o})\,(\frac{D}{V})]^{\frac{1}{2}} \qquad (3)$$

as $\Delta = mG_c = \Delta T_o V/D$ for the planar interface in steady state. From this equation it becomes evident that λ_i is the geometric mean of a capillarity

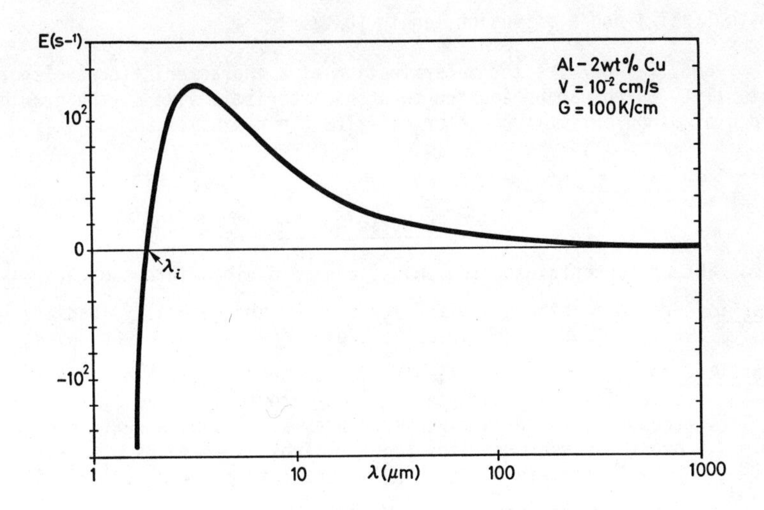

Fig. 1 Growth rate of perturbation of planar s/ℓ interface as a function of wavelength (eqn. 1)

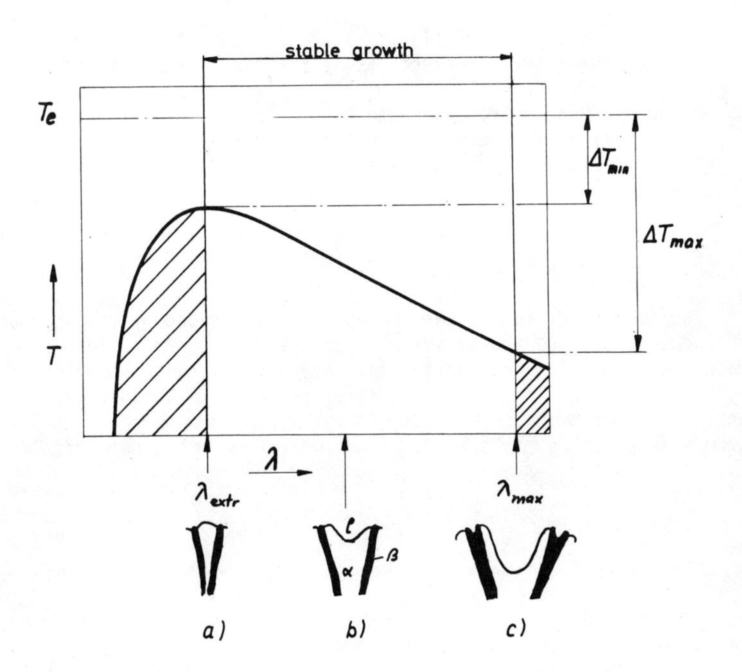

Fig. 2 Growth temperature and interface morphology of class II eutectics

length ($\Gamma/\Delta T_0$) and a diffusion length (D/V).

Eqn. (2) permits the determination of a characteristic wavelength of instability which can be equated to a characteristic size of the growth form (width of minor eutectic phase or dendrite tip radius).

Eutectic Growth

Eutectics containing two phases can be divided into two classes [8]:

Class I: Growth is strongly diffusion coupled and generally leads to regular structures. Typical examples are: Ag-Cu, Cr-Cu, Ni-W, Pb-Sn.

Class II: Growth is branching-limited due to anisotropic growth of the faceted (minor) phase (this type of growth is often called weakly coupled). The resulting structures still show a certain degree of order but are much less regular. Typical examples of this class usually contain a metal and a metalloid phase e.g. Fe-C, Al-Si.

Other observed differences between these classes are, for comparable growth conditions;

- spacings, spacing variations and undercoolings of class II eutectics are generally one order of magnitude larger than those of class I;
- spacings and undercoolings of class I eutectics do not depend on the temperature gradient, as is the case for class II.

An acceptable theory of growth has to explain all of these differences. The key lies in the very different microscopic interface morphologies which have been observed, during growth, with the aid of organic eutectics [5, 9].

Regular (class I) eutectics

The theory of steady-state growth can be treated quite rigorously for class I eutectics, due to the symmetry and planarity of the solid-liquid interface [10 - 13]. The general physical approach of these theories is the same; Calculation of the diffusion field, taking into account capillarity effects, coupling of these effects and maximizing growth rate with respect to growth temperature. This leads to the well know relationships:

$$\lambda^2 V = K_1 \quad (4)$$

$$\Delta T/\sqrt{V} = K_2 \quad (5)$$

$$\Delta T \lambda = K_3 \quad (6)$$

It has been shown experimentally that for class I eutectics growth indeed occurs very close to the extremum [14]. Therefore eqns. (4 - 6), with the constants obtained by theory [11], give quantitatively correct results [15, 16]. However it seems that systematically, slightly larger λ-values than predicted by the extremum have been found, which is consistent with the new criterion of growth at the limit of stability.

Irregular (class II) eutectics

In view of the above mentioned differences between class I and II, eqns. (4 - 6) no longer quantitatively describe the growth of the economically more important second category, which includes Al-Si and cast iron. Looking at the interface morphology, one must question the applicability of the analysis. However, use of a more realistic geometry makes the treatment extremely complex. Therefore, it has been important to show by careful temperature measurements in a single organic system which can grow either as a class I or class II eutectic that the solution of the diffusion problem for a planar interface can also be applied, to a first approximation, to the complicated class II eutectics [6]. The complex morphology is the result of changing local spacings which are the result of branching difficulties (creation of branches and terminations).

For the stable part of the growth curve (the range where $dT/d\lambda < 0$*)) there are three characteristic interface morphologies (fig. 2):

a) close to the extremum both phases show the same positive curvature.

b) at larger λ the higher volume phase (the metal in Al-Si or Fe-C) forms a depression (changing from a positive curvature at the three-phase junction to a negative curvature in the centre of the phase).

c) finally, at still larger λ, the faceted phase also forms a depression after having reached the limit of stability, this representing the point of branching, λ_{max}.

In systems where the anisotropy of growth of the faceted phase prevents easy adaption of λ to the local growth conditions, the eutectic spacing has to reach a spacing which activates branching. This is obtained at ΔT_{max}, the minimum undercooling required to cause a depression (instability) in the faceted phase. Fig. 2 shows the branching point at the limit of morphological stability of the minor phase in the schematic fashion in which it was used to simplify calculation**). Growth proceeds between branches (λ_{max}) and terminations (λ_{extr}) leading to large spacings, λ, and large spacing variations. The maximum spacing needed by class II eutectics to form branches can be obtained by simple geometric reasoning (fig. 8b):

$$\lambda_{max} \cong 1,5\ \lambda_i / f_\beta \qquad (7)$$

The growth of these eutectics should not therefore be described by one operating point, but rather by an operating (growth) range between λ_{extr} and λ_{max}. As in these cases the lamellae are mostly straight, the mean spacing of class II eutectics is

$$\bar{\lambda} = 0,5\ (\lambda_{max} + \lambda_{extr}) \qquad (8)$$

---

*) To the left of the extremum, terminations occur [11]

**) In reality, the instability of the faceted phase would develop first along the width, rather than across the thickness of the plate, because of the longer interface available to accommodate the minimum wavelength for instability (fig. 3).

Fig. 3 Branching of graphite flakes in eutectic Ni-C alloy by lateral instability and bending [24]

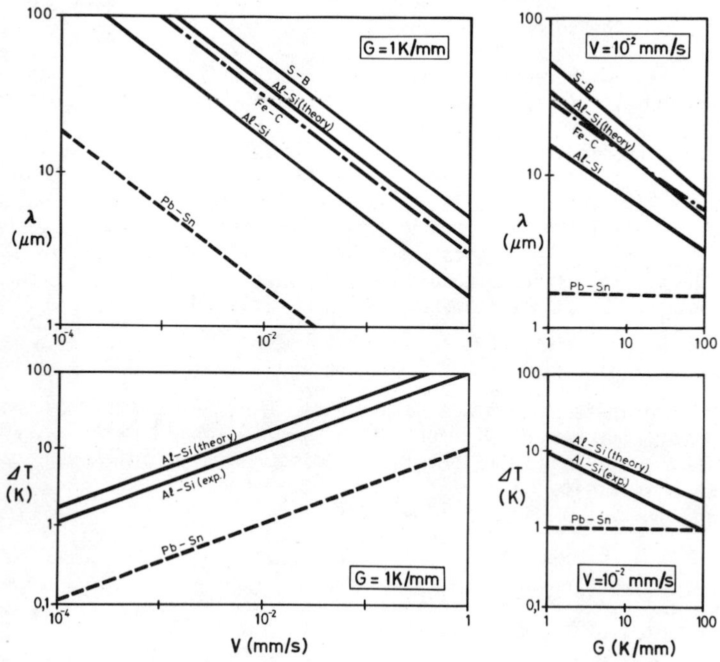

Fig. 4 Comparison of experimental results, on nf-nf (Pb-Sn) and f-nf (Al-Si, Fe-C, succinonitrile-borneol = S-B) eutectics, with calculations [6]

Fitting the results of numerical calculations to equations 4 and 5, and expressing the temperature gradient dependence of the constants in the same way as Toloui and Hellawell [17],

$$\lambda^2 V = K_4 G^{-x} \quad (9)$$

$$\Delta T/\sqrt{V} = K_5 G^{-y} \quad (10)$$

one obtains the results in fig. 4. Despite the many approximations introduced in the theory, the results are in reasonable agreement with experiment.

Dendrite Growth

Similarly to eutectic growth, single-phase, dendrite growth can be analysed in terms of diffusion (heat, mass) and interface phenomena. Theories have been summarized and tested against experiment by Glicksman et al [18]. All of these models, however, use the extremum criterion to arrive at a unique solution for the tip radius, R, as a function of undercooling or growth rate. For all models, higher theoretical growth rates for a given undercooling are predicted than are found by experiment (fig. 5). As in the case of class II eutectics, it becomes evident that the basic solution is reasonable, but not the optimization criterion.

Recently, Langer and Müller-Krumbhaar [2, 3] have applied a stability analysis to the tip of a free dendrite. Their calculations indicate that the tip grows with a radius which is at the limit of stability. They showed that this new criterion permits a very good fit between experiment and theory (much better than is possible using the extremum criterion).

Applying the simplified Mullins-Sekerka stability criterion (eqn. 2) to the hemispherical dendrite tip growing in a positive temperature gradient, the following equation can be derived [19];

$$V = \frac{2D\,(GR^2 + 4\pi^2\Gamma)}{R^3 pG - 2R^2 pC_0 m + 4\pi^2 \Gamma R p} \quad (11)$$

This relationship between R, V and G is in satisfactory agreement when compared with a recent more complete analysis by Trivedi [20]. Equation 11 can be divided into two growth regimes, one at low and one at high growth rate, V. This leads to the derivation of a series of simple functions between the important properties of the system [19].

It is interesting to note that the treatment by Burden and Hunt [21] of dendrite growth in a positive temperature gradient also proposes, for low rates, growth at the limit of stability. This can easily be seen in their simpler model which gives:

$$\Delta T = \frac{GD}{V} + K_6 V^{\frac{1}{2}} \quad (12)$$

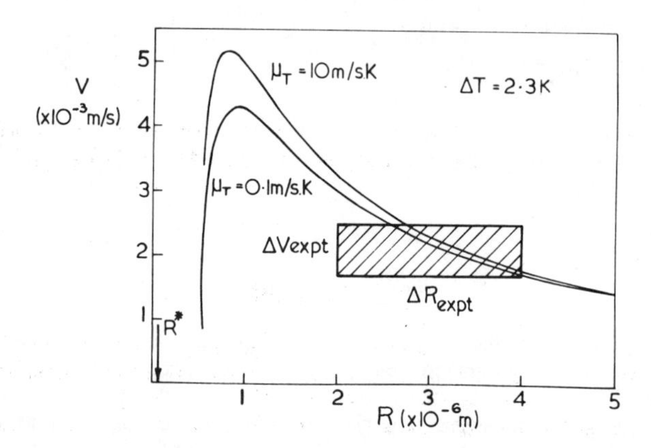

Fig. 5 Growth rate of Succinonitril dendrites as a function of tip radius. Full lines: Trivedi's theory for 2 interfacial mobilities; hatched region: experiment [23]

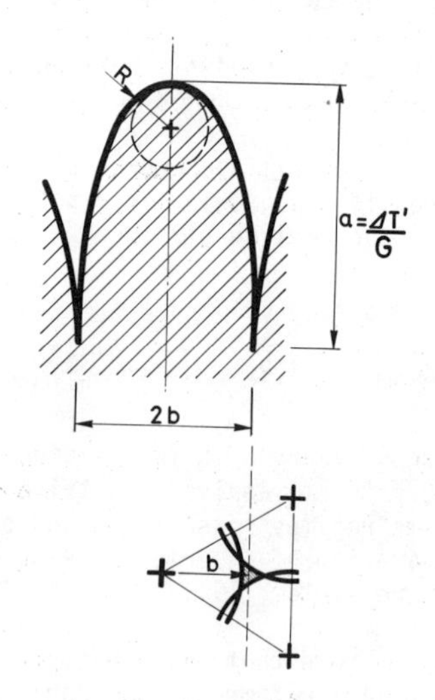

Fig. 6 Geometry of cell rsp. dendrite (neglecting details of side branching)

The first term on the RHS of eqn. 12 is nothing more than the constitutional undercooling criterion where the solidus/liquidus range, ΔT_0, for steady state plane front growth is replaced by the dendrite tip undercooling, ΔT.

Admitting that the mean envelope of the dendrite can be approximated by an ellipsoid, the primary spacing of columnar dendrites can be calculated. As defined in fig. 6, $R = b^2/a$, leading to:

$$\lambda_1 = (3R\Delta T'/G)^{\frac{1}{2}} \tag{13}$$

where $R = \lambda_i$ (eqn. 2). An ellipsoid was used because of the similarity in form compared to the volume fraction - temperature curve corresponding to the Scheil equation, permitting a simple description of planar, cellular or dendritic growth (fig. 7).

Eqns (11) and (13) permit the calculation of λ_1 as a function of V and G. Again the results can be separated into low and high velocity regimes. The transition velocity, V_{tr} (fig. 7), is given by

$$V_{tr} = \frac{GD}{\Delta T_0 k} = V_{cs}/k \tag{14}$$

At high V, the most important range for castings, and putting $\Delta T' = \Delta T_0$, the present model gives:

$$\lambda_1 \approx 4{,}3 \left(\frac{\Delta T_0 D\Gamma}{k}\right)^{\frac{1}{4}} V^{-\frac{1}{4}} G^{-\frac{1}{2}} \tag{15}$$

It is interesting to compare this simplified equation with that of Hunt [22] which has been obtained by using an extremum approach:

$$\lambda_1 = 2{,}83 \, (k\Delta T_0 D\Gamma)^{\frac{1}{4}} V^{-\frac{1}{4}} G^{-\frac{1}{2}} \tag{16}$$

The similarity of both results is striking (eqn. (15) leads generally to higher λ_1-values). In Fig. 7 R and λ_1 from eqn. (11) and (13) are plotted against V for 2 G-values. It can be seen that close to the planar interface at low V, R increases strongly with decreasing V*) and λ_1 goes to zero. Cellular morphologies develop between the limits of constitutional supercooling, V_{cs}, and the transition velocity, V_{tr}, whilst in the high velocity regime, well developed dendrites prevail. There appears to be another limit of stability at very high growth rates (which could be the limit of absolute stability).

---

*) Due to the approximations involved in the hemispherical dendrite model, R only reaches large values (centimeters) at the limit of planar interface stability, instead of infinite values as predicted by Trivedi [20].

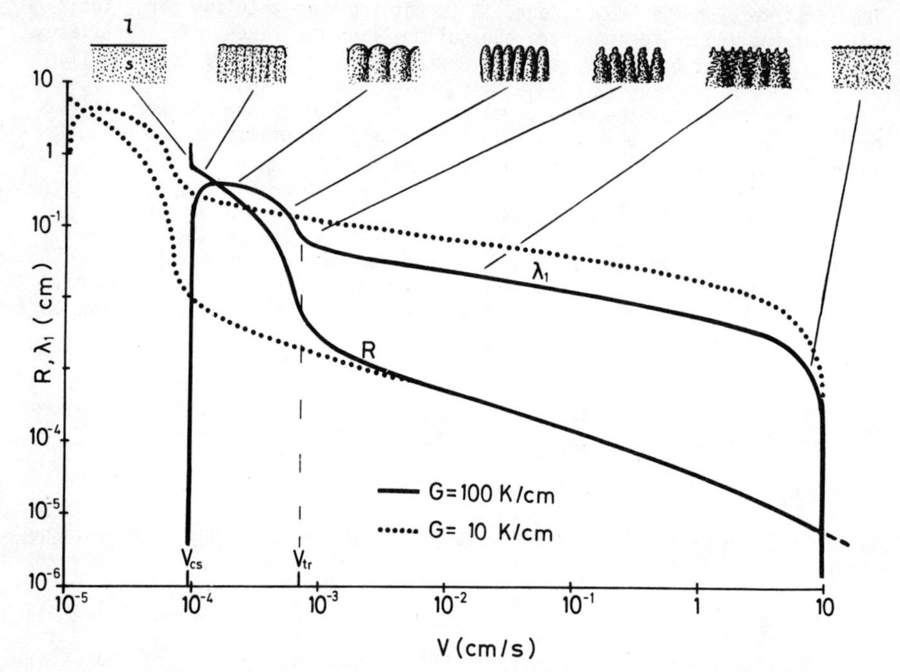

Fig. 7 Tip radius, R, and primary arm spacing, λ_1 as a function of V and G for Al-2wt%Cu. The diagram indicates the interface morphology to be expected in the different growth regimes for G = 100 K/cm. The transition from low to high V regime (eqn. 11) happens at $V_{tr} = V_{Cs}/k$

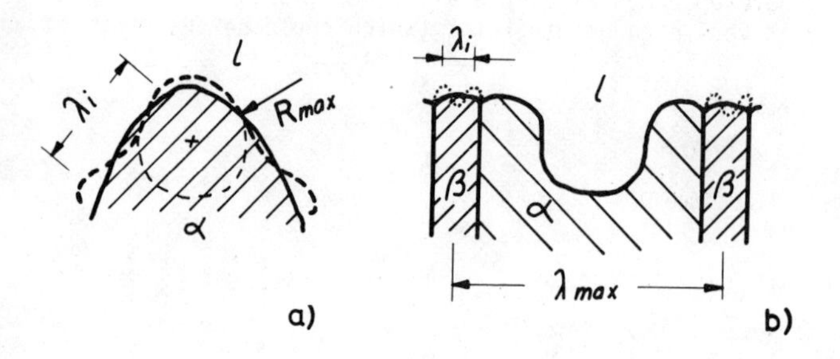

Fig. 8 Growth at limit of stability: a) dendrite tip growth at $R_{max} \cong \lambda_i$, b) class II eutectic growth between λ_{extr} and $\lambda_{max} \cong 1,5\ \lambda_i/f_\beta$

Conclusions

There is a marked similarity between theoretical treatments of dendrite and eutectic growth. Both morphologies show a maximum in the curve of V or T versus R (λ). Smaller scale structures lead to easy diffusion but large capillarity effects, whilst larger scale microstructures decrease curvature effects at the expense of increased diffusion difficulties. The extremum criterion does not seem to be the one which nature chooses generally, but rather, growth at the limit of stability, giving coarser structures. Figure 8 summarises the interface structure of both growth forms, and the corresponding criteria:

- The dendrite tips grow at their limit of stability, i.e. $R = \lambda_i$
- The (class II) eutectics grow between the limits of branching (λ_{max}, Fig. 8b) and termination (λ_{extr})

Therefore, as the scale of the structures becomes small, capillarity has not (as in the case of growth at the extremum) a direct influence on the interface morphology chosen by the system i.e. capillarity does not cut off the fine scale of both growth morphologies. Instead, the interface energy seems to act indirectly, via its influence on the wavelength of instability, on the observed (maximum) scale of growth structures in solidification.

List of Symbols

a, b	half axis of ellipsoid
C_o	alloy composition
D	diffusion coefficient in liquid
E	relative growth rate of amplitude of perturbation
f_β	volume fraction of β-phase
G	temperature gradient (dT/dz)
G_c	concentration gradient (dC/dz)
H	latent heat of melting
K	constant
k	distribution coefficient
m	liquidus slope
p	(1 - k)
R	dendrite tip radius
S	melting entopy (H/T_m)
T	temperature
T_e	equilibrium temperature
T_m	melting temperature

ΔT_o solidus/liquidus (equilibrium) intervall

$\Delta T'$ non equilibrium temperature difference between tip and root of dendrite

ΔT undercooling

t time

V growth rate

V_{cs} limit of constitutional supercooling

V_{tr} transition between low and high velocity regime of eqn. (11)

z distance from s/ℓ interface

Γ Gibbs-Thompson parameter (σ/S)

Δ gradient difference (mG_c - G)

δ amplitude of perturbation

$\dot{\delta}$ growth rate of perturbation ($d\delta/dt$)

λ eutectic spacing

λ_1 primary dendrite spacing

λ_i critical wavelength of instability

σ solid/liquid interface energy

ω frequency ($2\pi/\lambda$)

ω' $V/2D + [(V/2D)^2 + \omega^2]^{\frac{1}{2}}$

References

[1] C. Zener, AIME Trans. 167 (1946) 550

[2] J.S. Langer and H. Müller-Krumbhaar, J. Cryst. Growth 42 (1977) 11

[3] J.S. Langer and H. Müller-Krumbhaar, Acta Metall. 26 (1978) 1681 + 1689 + 1697

[4] D.J. Fisher and W. Kurz, "Proc. Quality Control of Engineering Alloys and the Role of Metals Science" (edited by H. Nieswaag and J.W. Schut) p. 59, University of Delft (1977)

[5] D.J. Fisher, Sc. D. Thesis, Ecole Polytechnique Fédérale de Lausanne (1978)

[6] D.J. Fisher and W. Kurz, Acta Metall. 28 (1980) 777

[7] W.W. Mullins, R.F. Sekerka, J. Apply. Phys. 35 (1964) 444

[8] W. Kurz and D.J. Fisher, International Metall. Rev. 24 (1979) 177

[9] J.D. Hunt and K.A. Jackson, Trans. Met. Soc. AIME 236 (1966) 843

[10] M. Hillert, Jernkont. Ann. 141 (1957) 757

[11] K.A. Jackson and J.D. Hunt, Trans. Met. Soc. AIME 236 (1966) 1129

[12] G.E. Nash and M.E. Glicksman, Acta Metall. 22 (1974) 1283, 1291

[13] S. Strässler and W.K. Schneider, Phys. Cond. Matter 17 (1974) 153

[14] R.M. Jordan and J.D. Hunt, Met. Trans. 3 (1972) 1305

[15] M. Tassa and J.D. Hunt, J. Cryst. Growth 34 (1976) 38

[16] J.N. Clark and R. Elliott, Metal Science 10 (1976) 101

[17] V. Toloui and A. Hellawell, Acta Metall. 24 (1976) 565

[18] M.E. Glicksman, R.J. Schaefer and J.D. Ayers, Met. Trans 7A (1976) 1747

[19] W. Kurz, D.J. Fisher "Dendrite Growth at the Limit of Stability: Tip Radius and Spacing", Acta Metall., in press

[20] R. Trivedi, J. Cryst. Growth 49 (1980) 219

[21] M.H. Burden and J.D. Hunt, J. Cryst. Growth 22 (1974) 109

[22] J.D. Hunt, Solidification and Casting of Metals, p. 1, Metals Society, London, 1979

[23] R.D. Doherty, B. Cantor, S.J.M. Fairs, Met. Trans. 9A (1978) 621

[24] B. Lux, M. Grages, W. Kurz, Praktische Metallographie 5 (1968) 567

EFFECT OF DENDRITE MIGRATION ON SOLUTE REDISTRIBUTION

Lawrence A. Lalli
Alcoa Laboratories
Alcoa Center, Pa.

Summary

Solidification is a process that typically occurs in the presence of a temperature gradient. The mathematical model of dendritic solidification that will be developed considers the effect of a temperature gradient on both the liquid and solid solute diffusion in the secondary dendrite arms. The diffusion equations are solved by finite difference methods using a moving boundary formulation in a planar coordinate system. Calculations are made which indicate that the secondary dendrite arms will migrate along the primary stalk through a process of temperature gradient zone melting. The migration distance is in agreement with Allen & Hunt's analytical treatment. Additional calculations are also made concerning the final amount of eutectic, the amount of coarsening of the secondary arms, and the concentration distributions during solidification. It will be shown that the final as-cast eutectic decreases as the migration distance divided by the secondary arm spacing increases. Also, secondary arm coarsening is predicted to increase with the temperature gradient divided by the isotherm velocity.

Introduction

Solute redistribution during solidification has been quantified theoretically and verified experimentally in terms of the non-equilibrium Scheil equation[1] and solute diffusion in the solidifying dendrite.[2,3] Experimentally, it has also been observed that secondary dendrite arms coarsen[4] during solidification and can migrate relative to the primary arm.[5,6] Also, non-symmetrical "saw-tooth" concentration distributions[7,8] have measured across secondary arms. It has been hypothesized by Allen and Hunt that the above experimental observations may be partly due to temperature gradient zone melting (TGZM)[9] during solidification, and they have derived an expression for the migration distance of a solidifying secondary dendrite arm. The present work confirms that of Allen and Hunt, but goes on to make detailed numerical calculations of concentration distributions, amount of eutectic, and amount of coarsening, as well as the migration distance. These calculations have been summarized in terms of solidification variables such as temperature gradient, isotherm velocity, secondary dendrite arm spacing, cooling rate, and diffusion coefficient.

Description of Model

Figure 1 depicts the concepts embodied in Allen and Hunt's model and in the computer simulation presented here. A constant linear temperature gradient, G(^{o}C/mm), is assumed to exist across adjacent secondary arms. The cooling rate, ε (^{o}C/sec), at any point in space is also constant. It follows that ε=GxR where R is the isotherm velocity in mm/sec. However, since it will be shown that this gradient causes the arms to migrate up the temperature gradient, the cooling rate experienced by a given arm will not necessarily be constant or equal to ε. The equilibrium liquidius and solidus lines corresponding to this temperature gradient are also shown in Figure 1. The computer model assumes these equilibrium concentrations are valid locally at the four solid-liquid interfaces and uses these as the boundary conditions for solving Fick's second law within both the solid and the two surrounding liquid regions. At the two interfaces of the solid dendrite arm, a solute mass balance was used to solve for the velocity of the interface. Since the colder liquid pool causes solute to diffuse into the colder interface of the solid dendrite at a rate in excess of what the dendrite can diffuse away, the cold side of the solid dendrite will tend to melt. Simultaneously, the dendrite is cooling which causes the liquidius concentration to increase resulting in solute diffusing away from the interface into the liquid. These two competing mechanisms can result in either net melting or solidifying at this interface. Conversely, on the hot side of the solid dendrite, both dendrite cooling and liquid diffusion tend to cause a solidifying interface. If the colder interface melts while the hotter interface solidifies, both interfaces are migrating from left to right at different velocities. Because material that originally solidified at a high temperature then melts, and later re-solidifies at a colder temperature, the dendrite arms will contain more solute, thus resulting in less eutectic.

Similar conditions were used at the two exterior boundaries, except that the calculation of the solute flux in the outlying solid was not modeled, but was taken as being equal to the corresponding boundary of the interior solid. Since the solid solute flux is relatively small, little error is expected to be introduced here. Since the difference

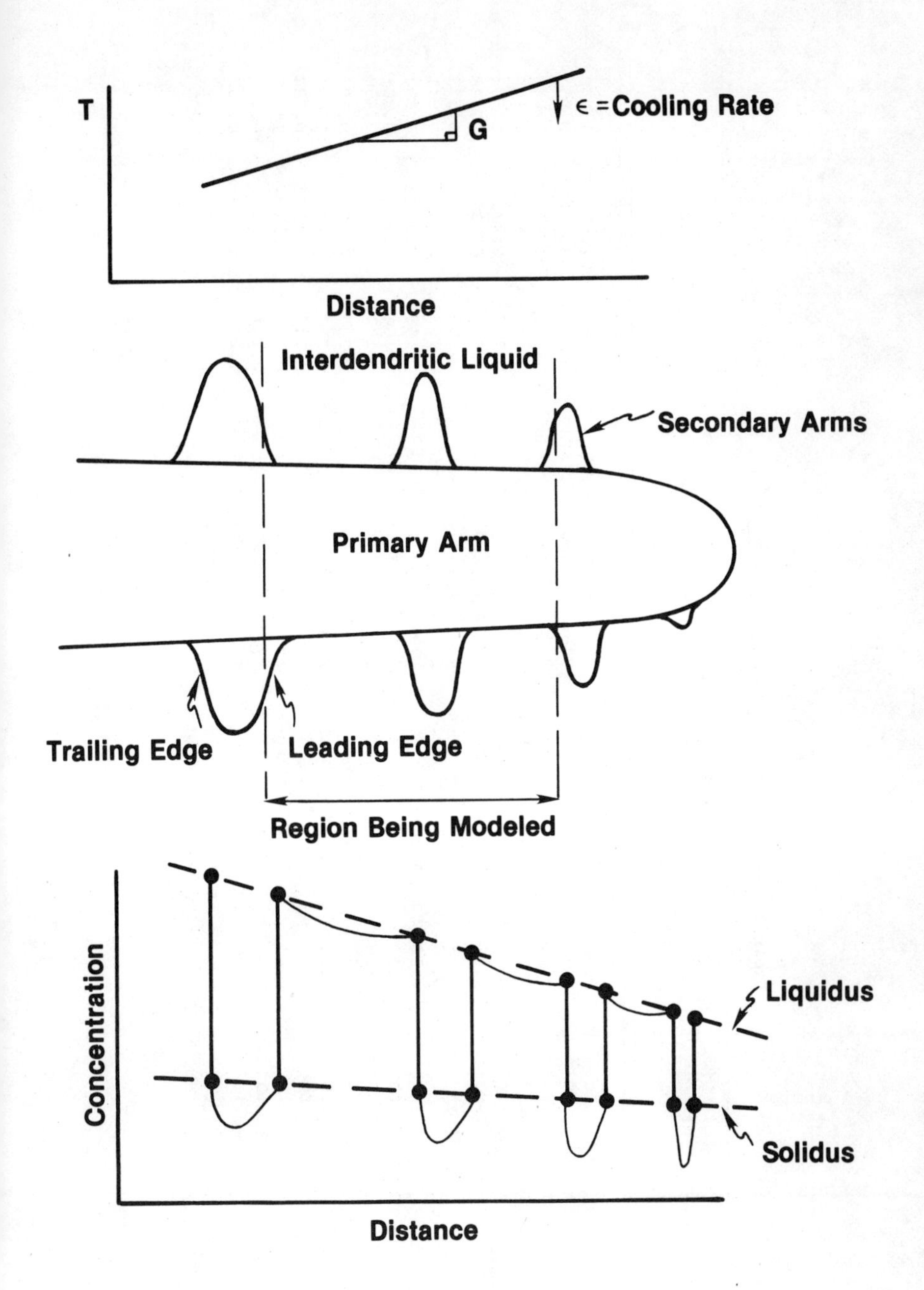

Figure 1 - Secondary Dendrite Arm Migration during Solidification.

Table I - Results of Computer Calculations

ε(°C/sec)	G(°C/mm)	R(mm/sec)	d_{2_i} (mm)	d_{2_f} (mm)	d_m (mm)	t_f(sec)		Co_f(at%)	$Co_f^{1/k-1}$	f_E(vol.%)	C_{min}(at%)
0.028	0	∞	0.10	0.10	0	3492.	349,200	4.70	0.155	2.55	1.280
0.056	0	∞	0.10	0.10	0	1746.	174,600	4.75	0.153	3.81	0.953
0.056	0.28	0.20	0.10	0.10	0.064	1747.	174,700	4.67	0.156	3.38	1.016
0.056	0.56	0.10	0.10	0.10	0.128	1748.	174,800	4.60	0.159	2.68	1.160
0.11	0	∞	0.10	0.10	0	873.	87,300	4.66	0.157	5.27	0.693
0.56	0	∞	0.10	0.10	0	174.6	17,460	4.47	0.165	6.39	0.428
0.56	2.8	0.20	0.10	0.101	0.064	174.9	17,490	4.48	0.164	5.00	0.532
0.56	5.6	0.10	0.10	0.103	0.128	175.9	17,590	4.50	0.163	3.27	0.811
0.56	11.	0.05	0.10	0.109	0.264	179.9	17,990	4.40	0.168	1.02	1.252
5.6	0	∞	0.10	0.10	0	17.46	1746	4.49	0.164	7.08	0.361
5.6	28.	0.20	0.10	0.106	0.067	17.79	1779	4.31	0.172	5.12	0.450
5.6	56.	0.10	0.10	0.125	0.142	18.88	1888	3.69	0.207	2.44	0.603
5.6	111.	0.05	0.10	0.182	0.309	23.64	2364	3.16	0.250	0.53	0.811
5.6	0	∞	0.05	0.05	0	17.46	6984	4.44	0.166	6.73	0.384
5.6	56.	0.10	0.05	0.062	0.140	18.86	7544	3.87	0.196	0.20	1.102
5.6	0	∞	0.20	0.20	0	17.46	437	4.41	0.167	6.88	0.352
5.6	56.	0.10	0.20	0.268	0.162	19.08	477	3.24	0.243	3.45	0.457

between liquidus and solidus concentrations is smaller at the hottest interface, this tends to move at a faster rate than the coldest interface. Consequently, a differential migration results which will cause the final spacing of the dendrite arm to be larger than the initial spacing. As this coarsening occurs, the total amount of solute is conserved within the region being modeled. Since the total solute is now distributed over a larger distance, a lower overall concentration results at the end of solidification.

The differential equations were solved using a finite difference method with a Murray-Landis[10,11] moving grid formulation. A Crank-Nicholson technique was used for the time derivative and an upwinding technique was required for the mass transport term in order to avoid numerical instabilities. The finite difference equations are contained in the appendix.

Presentation of Results

The results of the computer runs are summarized in Table I. All calculations have been made for an Al-4-1/2 wt % Cu binary alloy with a constant value for the partition ratio of 0.17. The simulation was terminated at the eutectic temperature (548°C). Also, a constant value for the liquid diffusivity of 0.005 mm^2/sec was used; for the solid, a value of 330 mm^2/sec was used multiplied by an Arhennius term with an activation energy of 146 kJ/mole.[12] The cooling rate, temperature gradient, and initial arm spacing were all varied. The freezing time was calculated including the effects of migration. Brody and Fleming's solidification parameter, $\eta = t_f/d^2_{2_i}$, where t_f=freezing time (sec) and d_{2_i}=initial secondary arm spacing (mm), was also calculated and used to describe the effects of solute diffusion in the solid.

Discussion of Results

Figure 2 is a plot of the migration distance versus the inverse of the isotherm velocity, R. It agrees very well with the expression developed by Allen and Hunt, although higher cooling rates result in slightly higher migration distances than are calculated by their expression. This agreement and the fact that this model agrees with Brody & Fleming's finite difference model for the case of zero temperature gradient, serve as useful checks for the finite difference formulation used here. The coarsening behavior predicted by the model is shown in Figure 3. The final arm spacing divided by the initially assumed value is plotted versus the temperature gradient divided by the isotherm velocity, resulting in a fairly good correlation. Spittle & Lloyd[13] have experimentally shown a similar dependence of secondary arm spacing with the inverse of the isotherm velocity. This coarsening mechanism is one that has not been previously quantified in the literature.

Figure 4 depicts the dependence of volume % eutectic (final eutectic liquid divided by final arm spacing) on Brody & Fleming's η parameter defined above. The maximum eutectic occurs at small η and for a infinite isotherm velocity (or zero gradient since $R=\varepsilon/G$). Since the exterior boundary conditions that conserved solute, also caused the coarsening behavior, a lower overall final composition can occur (shown in parenthesis in Figure 4). This is the reason why the curves in Figure 4 have a maximum instead of becoming horizontal at low η. As is also

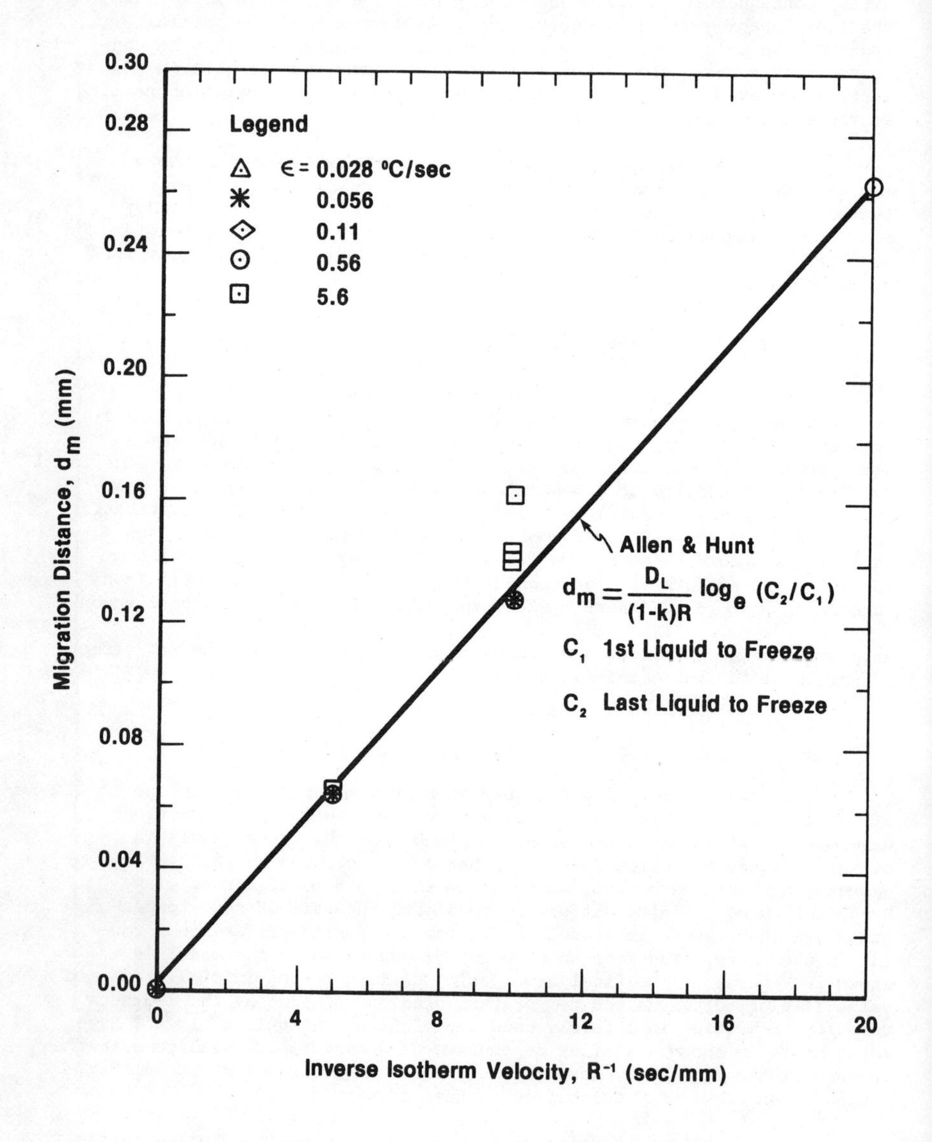

Figure 2 - Migration Distance vs. Isotherm Velocity.

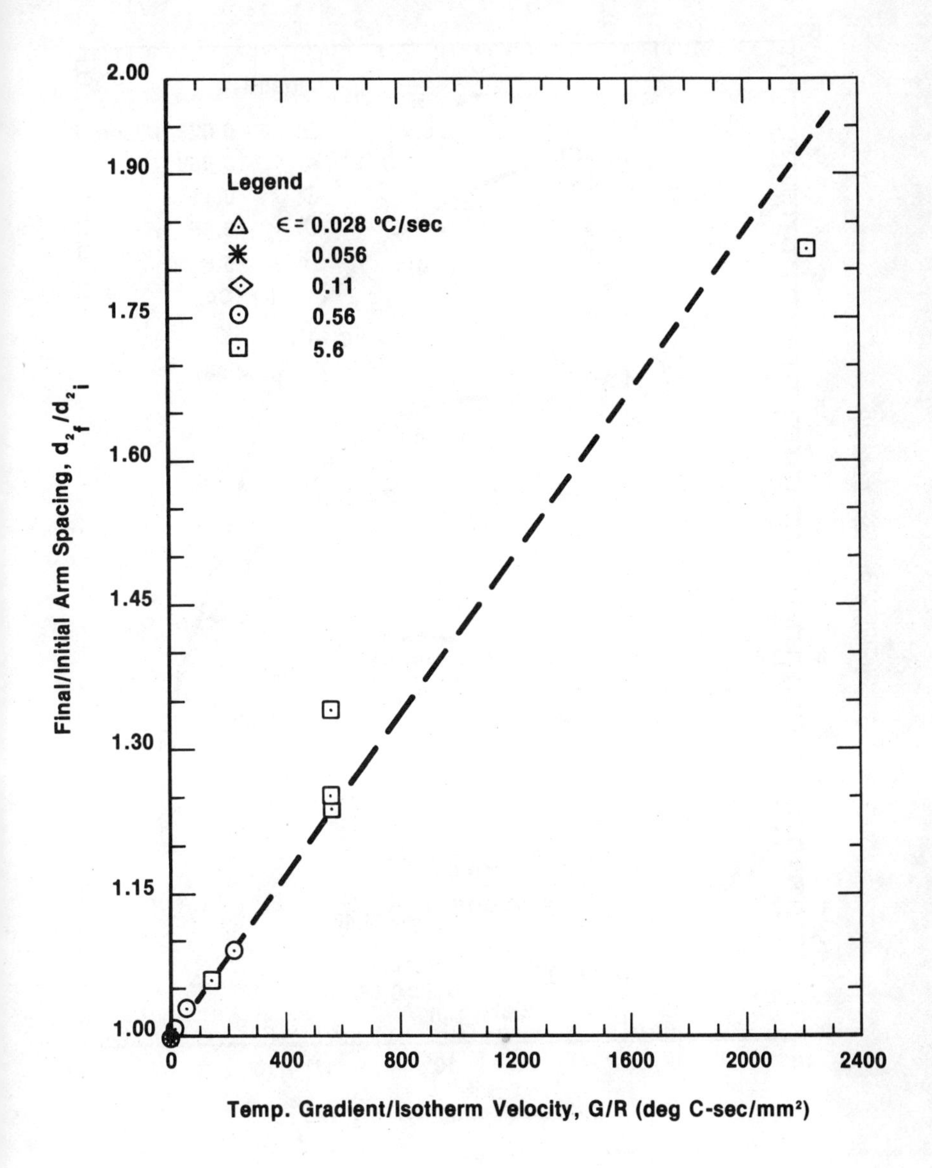

Figure 3 - Dendrite Coarsening vs. G/R.

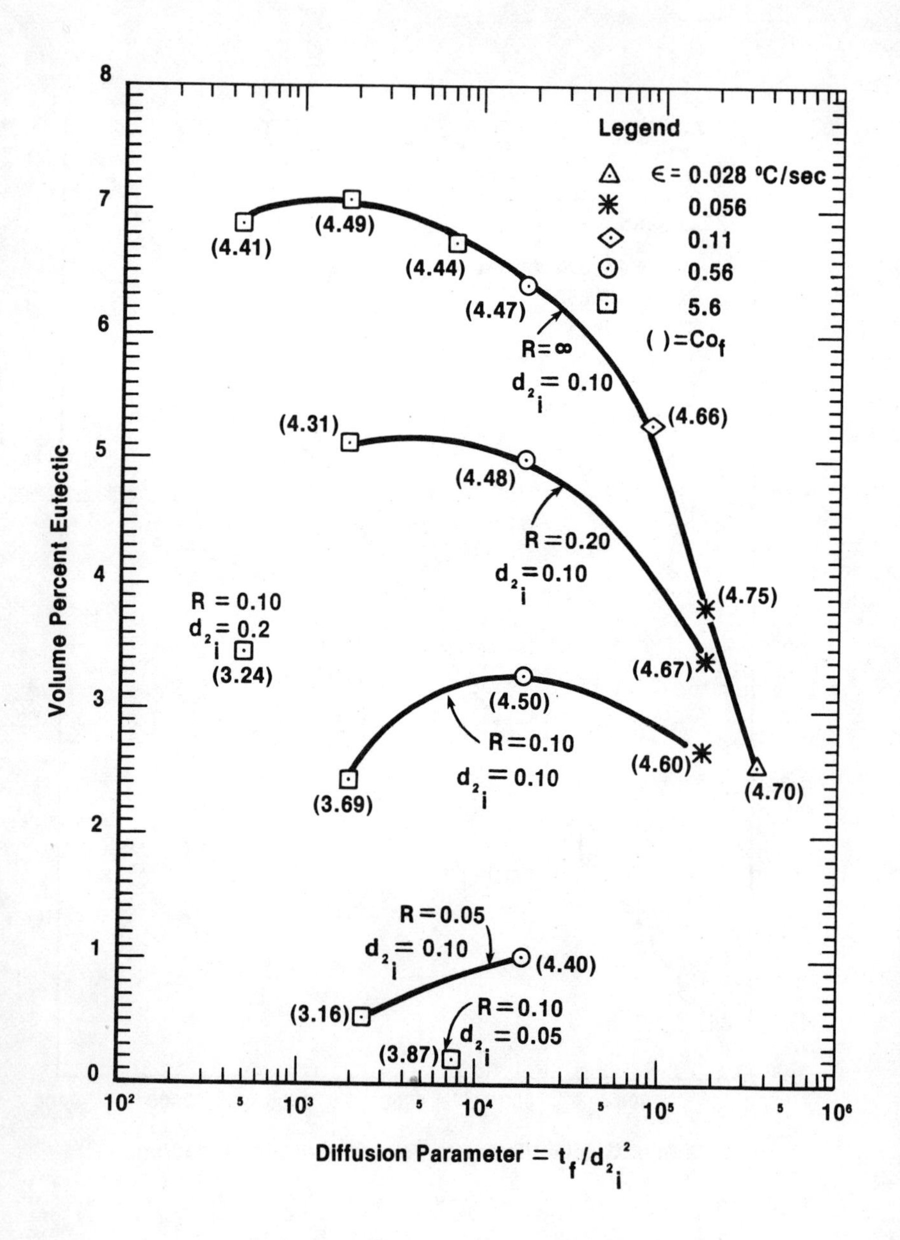

Figure 4 - Effects of Diffusion, Migration, and Coarsening.

apparent in Figure 4, as the isotherm velocity decreases, the amount of eutectic decreases for the same value of η. Only a small part of this decrease is due to a longer freezing time due to migration. Rather, the migration distance itself and the resulting melting and resolidification are responsible for this decrease in eutectic. There is not, however, a good correlation with the isotherm velocity for the two data points at 0.05 & 0.20 mm initial spacing as compared with the 0.10 mm spacing data. The reason for this is that the amount of decrease in eutectic is apparently better related to the distance migrated divided by the initial arm spacing. This seems to make sense physically in that it says that the amount of remelting is significant only relative to the total original volume of metal. Figure 5 demonstrates this relationship where the vertical axis is the eutectic calculated for a given gradient divided by that for zero gradient at the same value of η (thus eliminating the effects of solid diffusion). A slightly improved correlation is shown in Figure 6 where the vertical axis has been further normalized to account for the decrease in average solute concentration that is calculated due to the coarsening behavior. This normalization is based solely on the behavior that the Scheil equation predicts for percent eutectic at differing overall concentration levels.

Figure 7 is a plot of minimum solute concentration in the solidifying dendrites as a function of η. Again, the effects of isotherm velocity and the differing arm spacings are apparent here as in Figure 4.

Figure 8 demonstrates the final concentration distributions and migration distances in the final solid secondary arms (the eutectic has been omitted on this plot). These distinctly different distributions resulted from freezing the same dendrite at the same cooling rate, but under different temperature gradients. Here, the non-symmetrical "sawtooth" pattern is evident.

Conclusions

Solidification inherently involves temperature gradients. It has been shown that an assumed temperature gradient across the secondary arms can substantially affect solute redistribution, resulting in less as-cast eutectic formation and increased minimum solute content. Furthermore, this has been shown to be directly related to the migration distance of the secondary arms relative to their initial spacing, where the migration distance is inversely related to the isotherm velocity. These phenomena are due to melting and resolidification of the trailing edge of the secondary arms through the mechanism of temperature gradient zone melting. In addition, this temperature gradient can contribute to coarsening behavior of the secondary arms during solidification, as well as resulting in non-symmetrical "saw-tooth" concentration distributions across the secondary arms. The amount of coarsening increases directly with the temperature gradient divided by the isotherm velocity. All of these phenomena have been experimentally observed and reported in the literature. The finite difference formulation using a moving boundary coordinate system has proven to be a convenient method for treating this problem. Extensions of this technique to consider an expanded number of secondary arms, the nucleation of additional arms, and two dimensional diffusion should be possible.

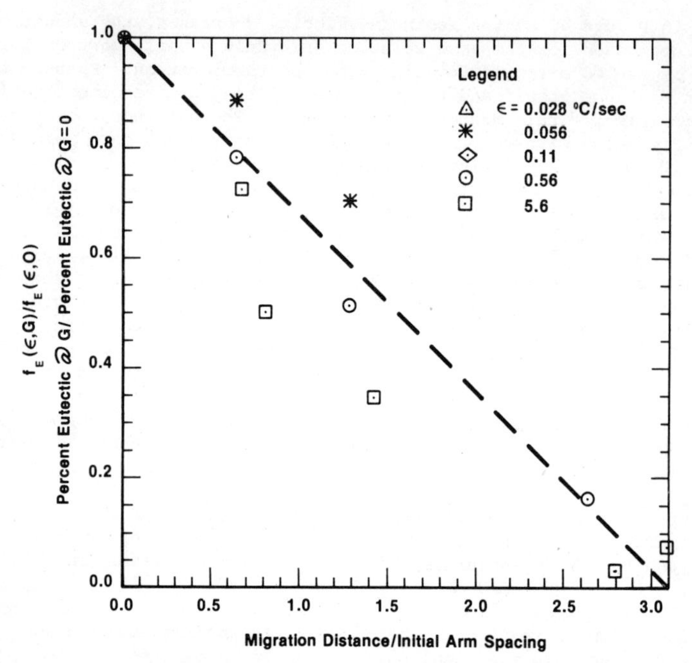

Figure 5 - Normalized Percent Eutectic vs. Normalized Migration Distance.

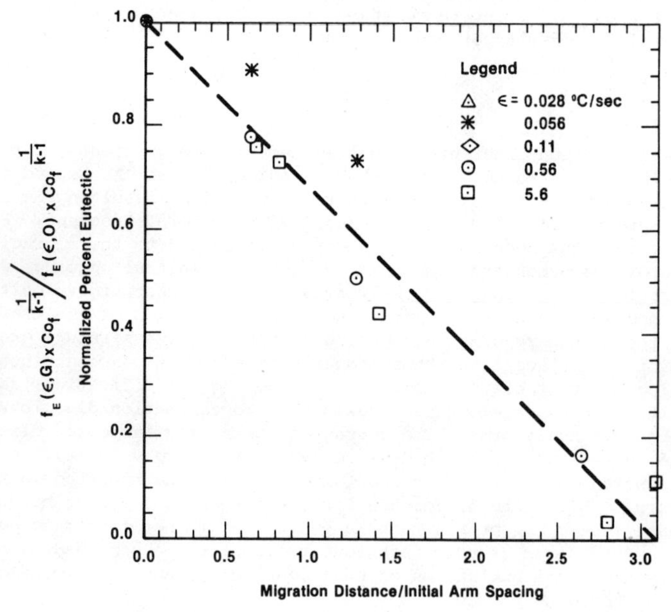

Figure 6 - Normalized Percent Eutectic vs. Normalized Migration Distance.

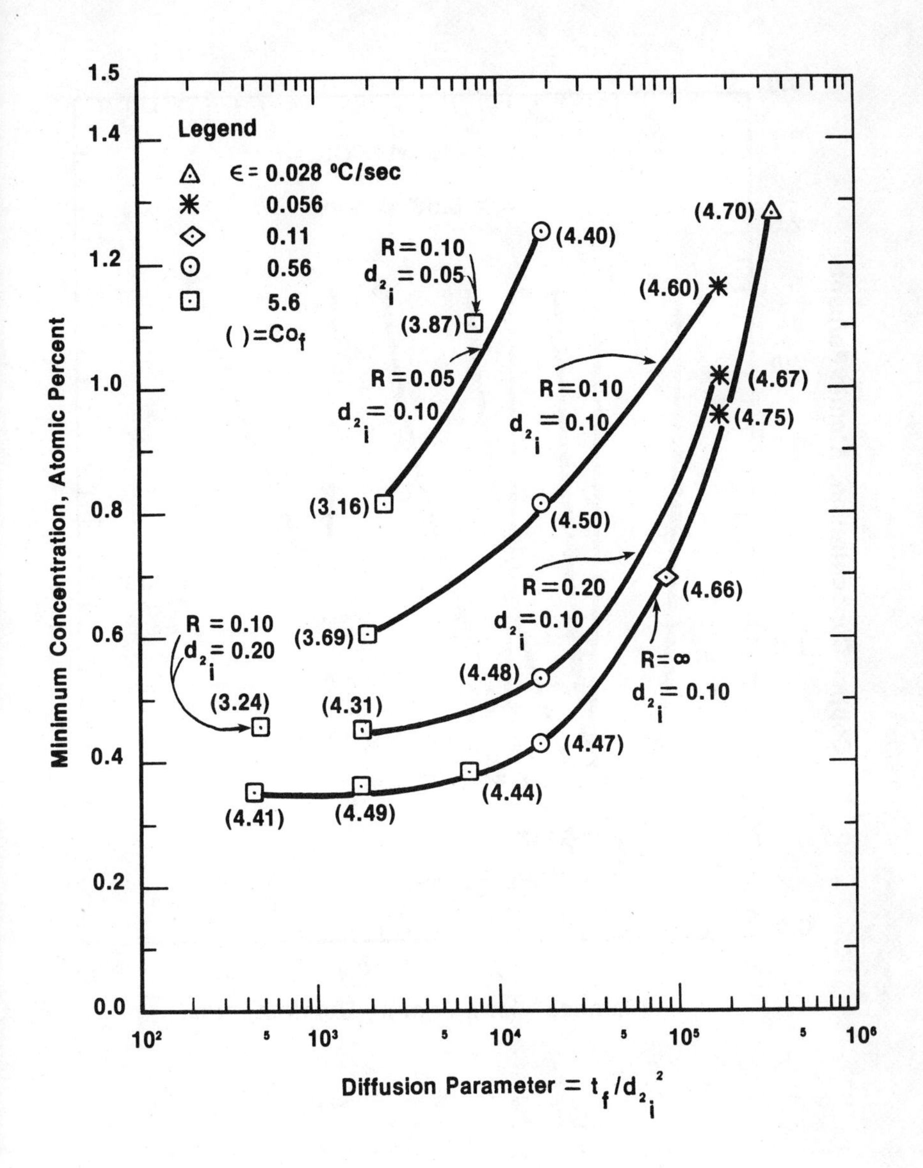

Figure 7 - Minimum Solute Concentration.

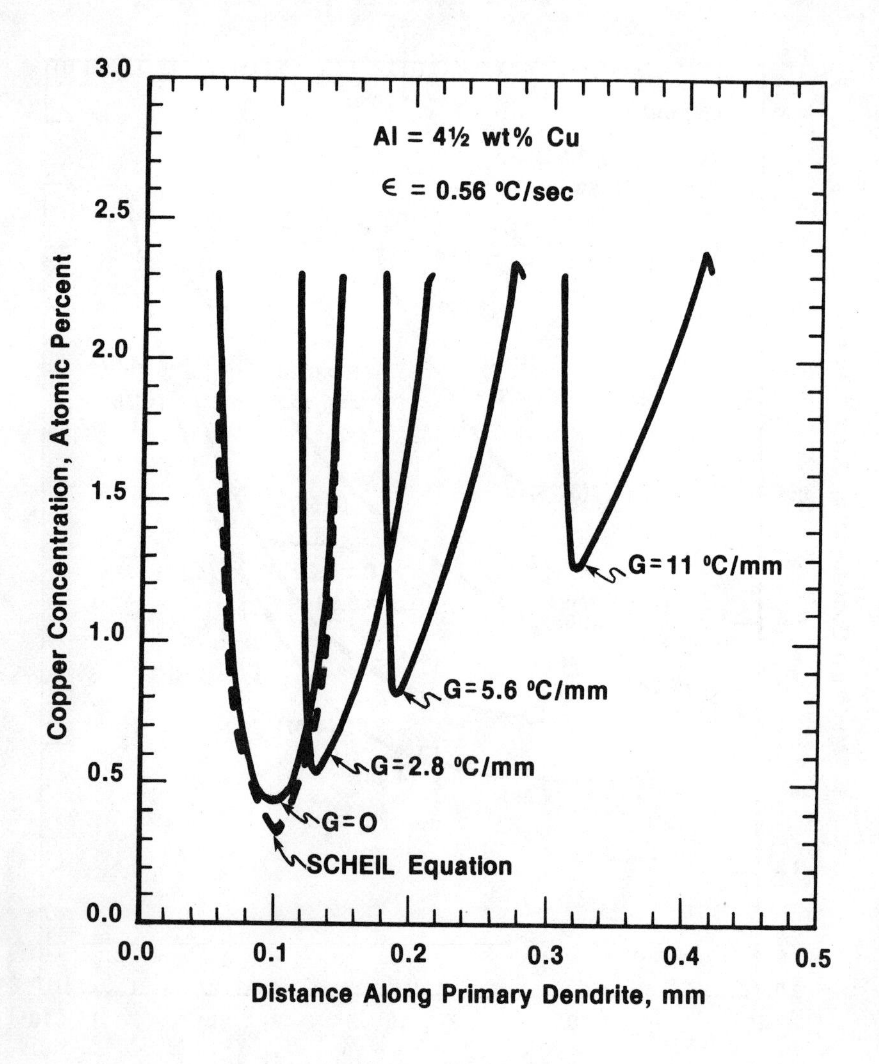

Figure 8 - Final Concentration Distributions Effect of Temperature Gradients.

NOMENCLATURE

ε	=	Cooling Rate, ^{o}C/Sec
G	=	Temperature Gradient, ^{o}C/mm
R	=	Isotherm velocity, mm/sec
d_{2_i}	=	Initial secondary arm spacing, mm
d_{2_f}	=	Final secondary arm spacing, mm
t_f	=	Freezing time, sec
η	=	$t_f/d_{2_i}^2$, diffusion parameter
C_{o_f}	=	Final overall composition, atomic %
k	=	Partion Ratio
f_E	=	Volume percent eutectic
C_{min}	=	Minimum solute concentration, atomic %
D_L	=	Diffusivity in liquid, mm^2/sec
f_E	=	Volume % Eutectic
d_m	=	Migration distance, mm
C_L^*	=	Liquidus Concentration, atomic fraction
C_s^*	=	Solidus Concentration, atomic fraction

References

1. Flemings, M. C., Solidification Processing, McGraw-Hill, 1974
2. Brody, H.D. and Flemings, M.C., TMS-AIME, 1966, 236, 615.
3. Bower, T.F., Brody, H.D., and Flemings, M.C., TMS-AIME, 1966, 236, 624
4. Kattamis, T.Z., Coughlin, J.D., and Flemings, M.C., TMS-AIME,1967, 239, 1504.
5. Allen, D.J., and Hunt, J.D., Met. Trans., 1976, 7A, 767.
6. Allen, D.J., and Hunt, J.D., Sheffield International Conference on Solidification and Casting, 1976, 1, p.8/1
7. Weinberg, F. and Teghtsoonian, E., Met. Trans., 1972, 3, 93.
8. Rickinsoon, B.A. and Kirkwood, D.H., Sheffield International Conference on Solidification and Casting, 1976, 1, p9/1.
9. Pfann, W.G., Trans. AIME, 1955, 961.
10. Murray, D. and Landis, F., Trans. ASME, 1959, 81, 106.
11. Moren, A., Randich, E., and Goldstein, J.I., Proc. of International Conference on Computer Simulation for Materials Application in Nuclear Metallurgy, 1976, 20, part. 1.
12. Fricke, W. G., Scripta Met., 1972, 6, 1139.
13. Spittle, J.A. and Lloyd, D.M., Sheffield International Conference on Solidification and Casting, 1976, 1, p3/1.

Acknowledgment

This paper was initially written as a term paper for Professor H.D. Brody at the University of Pittsburgh.

APPENDIX

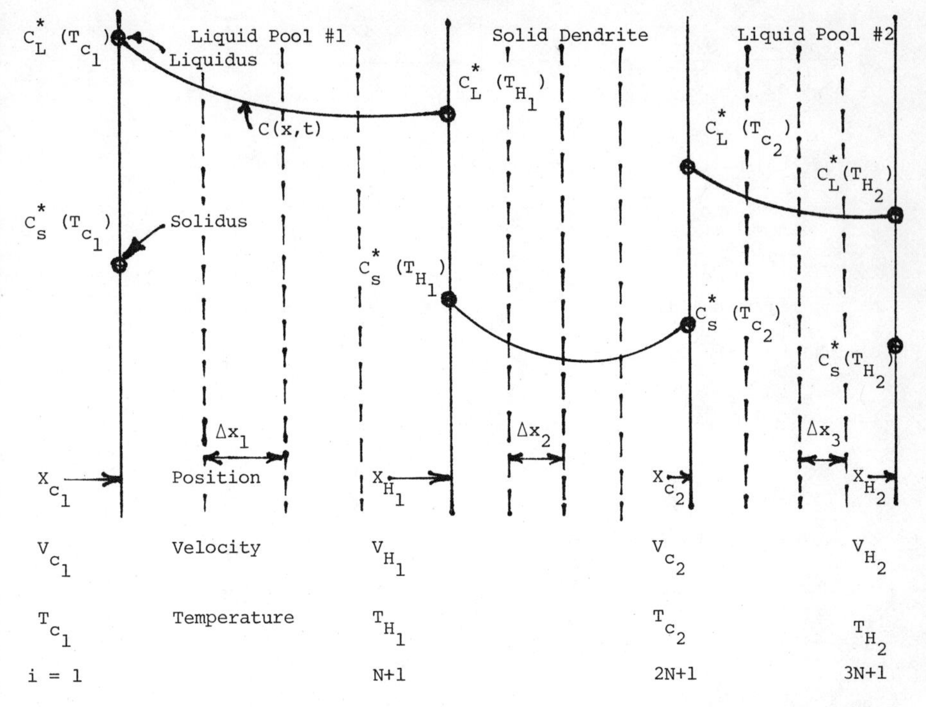

V_{c_1}	Velocity	V_{H_1}	V_{c_2}	V_{H_2}
T_{c_1}	Temperature	T_{H_1}	T_{c_2}	T_{H_2}
i = 1		N+1	2N+1	3N+1

Position, Velocity, Temperature and Concentration are all functions of time.

Let $\xi_1 = \dfrac{x - x_{c_1}}{x_{H_1} - x_{c_1}}$, $\xi_2 = \dfrac{x - x_{H_1}}{x_{c_2} - x_{H_1}}$, $\xi_3 = \dfrac{x - x_{c_2}}{x_{H_2} - x_{c_2}}$

The dashed lines represent the finite difference grid and are lines of constant ξ. Solving for x in the above expressions and differentiating,

$$v = \frac{dx}{dt} = \frac{dx_{c_1}}{dt} + \xi_1 \left[\frac{dx_{H_1}}{dt} - \frac{dx_{c_1}}{dt}\right] = v_{c_1} + \xi_1 (v_{H_1} - v_{c_1}),$$

where $\frac{dx}{dt}$ is the rate of movement of the finite difference grid system. The diffusion equation in this moving grid system becomes

$$D(T) \frac{\partial^2 C(x,t)}{\partial x^2} = \frac{\partial C(x,t)}{\partial t} - v(x,t) \frac{\partial C(x,t)}{\partial x}$$

where the minus sign preceding the velocity indicates that there is an apparent mass transport term in the negative x direction due to a positive velocity, v, of the grid system. The solute concentration, C, is in units of atomic fraction.

In terms of ξ_1, we have for liquid pool #1,

$$\frac{1}{(x_{H_1} - x_{c_1})^2} \frac{\partial^2 C}{\partial \xi_1^2} = \frac{1}{D_L} \frac{\partial C}{\partial t} - \left[\frac{v_{c_1} + \xi_1 (v_{H_1} - v_{c_1})}{(x_{H_1} - x_{c_1}) D_L} \right] \frac{\partial C}{\partial \xi_1}$$

subject to the the boundary conditions

$$@ \xi_1 = 0, \qquad C = C_L^* \ (T_{c_1})$$

$$@ \xi_1 = 1, \qquad C = C_L^* \ (T_{H_1})$$

At the liquid-solid interfaces, x_{H_1} for example, a solute balance can be derived from the differential equation, assuming $\frac{\partial C}{\partial t}$ is small.

$$\frac{D_s}{(x_{c_2} - x_{H_1})} \left. \frac{\partial C}{\partial \xi_2} \right|_{\xi_2 = 0} - \frac{D_L}{(x_{H_1} - x_{c_1})} \left. \frac{\partial C}{\partial \xi_1} \right|_{\xi_1 = 1} = -v_{H_1} \left(C_s^* - C_L^* + \frac{\partial C}{\partial \xi} \overline{\Delta x} \right)$$

where $\overline{\Delta x}$ is the thickness of the interface element $= \frac{1}{2} (\Delta x_1 + \Delta x_2)$.

The above expression is used to determine a value of the interface velocity, v_{H_1}.

The following finite difference expressions have been used:

$$\frac{\partial^2 C}{\partial \xi^2} = \frac{1}{2} \left[\frac{C'_{i-1} - 2C'_i + C'_{i+1}}{\Delta \xi^2} \right] + \frac{1}{2} \left[\frac{C_{i-1} - 2C_i + C_{i+1}}{\Delta \xi^2} \right]$$

$$\frac{\partial C}{\partial \xi} = \frac{1}{2} \left[\frac{(\lambda - 2) C'_{i-1} + (2 - 2\lambda) C'_i + \lambda C'_{i+1}}{2 \Delta \xi} \right] + \frac{1}{2} \left[\frac{(\lambda - 2) C_{i-1} + (2 - 2\lambda) C_i + \lambda C_{i+1}}{2 \Delta \xi} \right]$$

and $$\frac{\partial C}{\partial t} = \frac{C'_i - C_i}{\Delta t}$$

where the primes indicate values at the next time step, $t + \Delta t$. Evaluating the derivatives at both the present and future time, and averaging, is the Crank-Nicholson technique. The factor, λ, is determined based on upwinding considerations. When the velocity, v, for the current time step is positive, λ is set equal to 2, and

$$\frac{\partial C}{\partial \xi} = \frac{1}{2} \left[\frac{C'_{i+1} - C'_i}{\Delta \xi} \right] + \frac{1}{2} \left[\frac{C_{i+1} - C_i}{\Delta \xi} \right]$$

and when v is negative, $\lambda = 0$ and

$$\frac{\partial C}{\partial \xi} = \frac{1}{2} \left[\frac{C'_i - C'_{i-1}}{\Delta \xi} \right] + \frac{1}{2} \left[\frac{C_i - C_{i-1}}{\Delta \xi} \right]$$

Without this upwinding technique, the concentration calculations was found to be unstable.

The simultaneous equations for N-1 uniformly spaced interior nodes in the 1st liquid pool, i = 2,N, thus become

$$A_i C'_{i-1} + B_i C'_i + D_i C'_{i+1} = R_i$$

where

$$A_i = \frac{1}{2} \frac{1}{(x_{H_1} - x_{c_1})^2 \Delta\xi_1^2} + (\lambda-2) E$$

$$B_i = - \frac{1}{D_L \Delta t} - \frac{1}{(x_{H_1} - x_{c_1})^2 \Delta\xi_1^2} + (2-2\lambda)E$$

$$D_i = \frac{1}{2} \frac{1}{(x_{H_1} - x_{c_1})^2 \Delta\xi_1^2} + \lambda E$$

$$R_i = - \frac{C_i}{D_L \Delta t} - \frac{1}{2} \frac{(C_{i-1} - 2C_i + C_{i+1})}{(x_{H_1} - x_{c_1})^2 \Delta\xi_1^2}$$

$$- E \; ((\lambda-2) C_{i-1} + (2-2\lambda)C_i + \lambda C_{i+1})$$

where
$$E = \frac{v_{c_1} + \xi_i (v_{H_1} - v_{c_1})}{4 D_L (x_{H_1} - x_{c_1}) \Delta\xi_1}$$

$$\Delta\xi = \frac{1}{N} = \frac{\Delta x}{(x_{H_1} - x_{c_1})}$$

$$\xi_i = (i-1) \Delta\xi$$

At $x=x_{H_1}$, i=N+1

$$C = C_L^* (T_{H_1})$$

and
$$v_{H_1} = \frac{\left[\frac{D_s}{(x_{c_2} - x_{H_1})} \frac{\partial C}{\partial \xi_2}\bigg|_{\xi_2=0} - \frac{D_L}{(x_{H_1} - x_{c_1})} \frac{\partial C}{\partial \xi_1}\bigg|_{\xi_1=1} \right]}{C_L^* - C_s^* - \left[\frac{\lambda}{2} \frac{1}{(x_{c_2} - x_{H_1})} \frac{\partial C}{\partial \xi_2}\bigg|_{\xi_2=0} + \frac{2-\lambda}{2} \frac{1}{(x_{H_1} - x_{c_1})} \frac{\partial C}{\partial \xi_1}\bigg|_{\xi_1=1} \right] \overline{\Delta x}}$$

where

$$\left.\frac{\partial C}{\partial \xi_2}\right|_{\xi_2=0} = \frac{-3\, C_s^*(T_{H_1}) + 4\, C_{N+2} - C_{N+3}}{2\,(x_{c_2} - x_{H_1})\, \Delta\xi_2}$$

$$\left.\frac{\partial C}{\partial \xi_1}\right|_{\xi_1=0} = \frac{C_{N-1} - 4C_N + 3C_L^*(T_{H_1})}{2\,(x_{H_1} - x_{c_1})\, \Delta\xi_1}$$

The time integration begins by assuming values for x_{c_1}, x_{H_1}, x_{c_2}, x_{H_2}, T_{c_1}, T_{H_1}, T_{c_2}, T_{H_2}. The interface concentrations, C_s^* and C_L^* are calculated, and values of C_i are then assumed to be equal to the nominal composition, C_o. The interface velocities are then calculated, v_{c_1}, v_{H_1}, v_{c_2}, v_{H_2}. New interface positions are calculated at t + Δt/2 (i.e., $x(t+\Delta t/2) = x(t) + \frac{v\,\Delta t}{2}$) as well as the grid spacing for the period t to t+Δt and the diffusivities at t+Δt/2. New values of concentration, C_i', can now be determined and the process repeated, using new values of x at t+Δt.

MODELING SOLUTE REDISTRIBUTION DURING SOLIDIFICATION

OF AUSTENITIC STAINLESS STEEL WELDMENTS*

J. C. Lippold
Materials Science Division
Sandia National Laboratories, Livermore

W. F. Savage
Rensselaer Polytechnic Institute
Troy, New York

The scanning transmission electron microscope has been employed to determine the concentration profiles associated with the initial stages of solidification of austenitic stainless steel weldments. When delta ferrite is the primary solidification product the initial solid is enriched in chromium and depleted in nickel but as solidification proceeds the nominal composition is rapidly approached. A model has been proposed which can predict the solute redistribution during the initial transient stage of solidification. The experimental data obtained from the as-welded samples correlate remarkably well with the predicted solute profiles.

*This work supported in part by the U.S. Department of Energy (DOE) under Contract #DE-AC04-75DP00789.

Introduction

The relationship between the hot cracking susceptibility of 300-series stainless steel weldments and the amount of retained delta ferrite in the weld metal has been the subject of numerous investigations.[1-4] In general, the presence of 5-10 volume percent ferrite ensures that the weld will be crack-free. Unfortunately, the absence of a concise model for describing the solidification sequence of these alloys has prevented investigators from gaining a full understanding of the role of the retained delta ferrite in reducing the hot-cracking susceptibility.

More recent evidence has shown that austenitic stainless steels solidify with either delta ferrite or austenite as the primary phase, and that this behavior is a function of the ratio of austenite-forming elements (Ni, Mn, C, N) to ferrite-forming elements (Cr, Si, Mo).[5-7] Alloys which form delta ferrite as the initial product of solidification have been found to be inherently more resistant to hot cracking than alloys of similar composition which solidify as austenite.[8] Thus, the criterion that requires 5-10 volume percent delta ferrite in the weld metal merely serves to ensure that solidification occurs as delta ferrite. A decrease in retained ferrite below 2-3 volume percent usually indicates a shift to solidification as austenite and is often accompanied by an increase in hot cracking susceptibility.

The present investigation attempts to describe the solidification sequence when the primary phase is delta ferrite and to rationalize the resultant as-welded microstructure using current theories of solute redistribution. The results provide a basis for understanding solute microsegregation during weld soldification in a variety of alloy systems.

Theoretical Background

The Iron-Chromium-Nickel Ternary System

The Fe-Cr-Ni Ternary system provides the basis for predicting phase equilibrium in austenitic stainless steels. Since most commercial stainless steels are composed of 16-25 wt.% chromium and 8-20 wt.% nickel, their compositions are localized in the iron-rich corner of the ternary diagram. The liquidus surface of the system can be represented by a series of isotherms which reach a minimum along a line running from the Fe-Ni peritectic reaction (Fe-4Ni) to the ternary eutectic point at 49 Cr - 43 Ni - 8 Fe, as illustrated in Fig. 1.[9] Alloy compositions lying on the Cr-rich side of this line will solidify predominantly as BCC delta ferrite while those on the Ni-rich side of this line will solidify predominantly as FCC austenite.

Phase equilibria in the iron-rich region over a range of temperatures can be represented by a pseudo-binary diagram at a constant iron content. An example of such a diagram at 70% Fe is illustrated in Fig. 2. The two-phase liquid-plus-delta and liquid-plus-austenite regions are separated by a triangular three-phase "eutectic region" where mixtures of delta ferrite, austenite, and liquid exist in equilibrium. Below this region is a two-phase austenite-plus-delta ferrite region enclosed by the austenite and ferrite solvus lines.

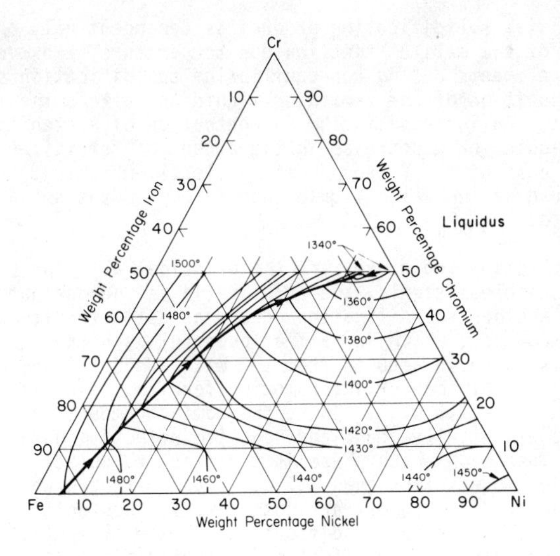

Figure 1. Fe-Cr-Ni Ternary Liquidus Surface.

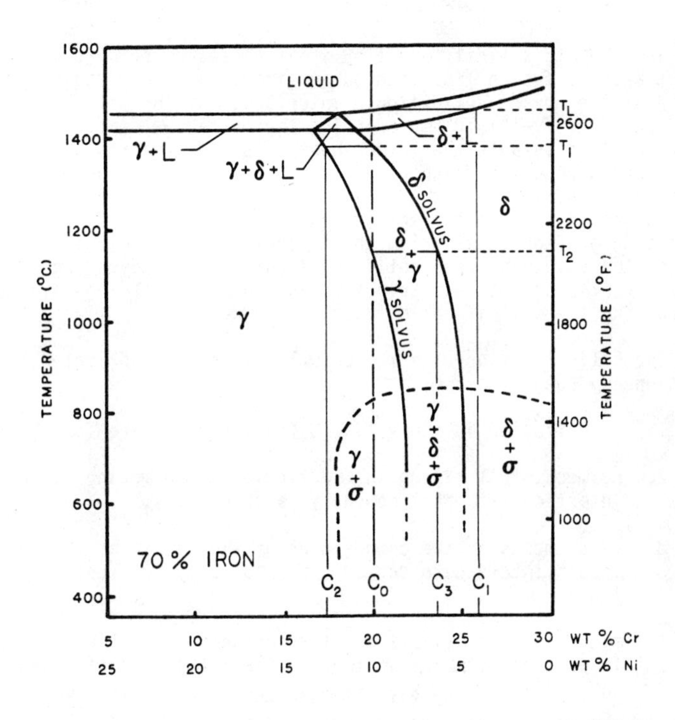

Figure 2. Fe-Cr-Ni Pseudo-Binary Equilibrium Diagram at 70% Fe.

The initial solidification product is dependent only on the nominal composition of the melt at the liquidus temperature. However, segregation of alloying elements during non-equilibrium solidification shifts the overall composition of the remaining liquid and alters the final solidification product. An increase in the concentration of austenitizers in the remaining liquid, or a decrease in the amount of ferritizers, favors solidification as austenite. Correspondingly, local enrichment in ferrite forming elements ahead of the solid-liquid interface promotes solidification as delta ferrite.

Since chromium and nickel are the principal alloying elements in austenitic stainless steels, the Cr/Ni ratio is the dominant factor in controlling whether solidification occurs as delta ferrite or austenite. In addition, manganese, which is half as powerful as nickel in stabilizing austenite, is added in amounts from 1-2 wt.%, and silicon, which is 1.5 times more powerful than Cr in promoting ferrite, is normally present in amounts from 0.5 - 1.0 wt.%. Thus, the nominal amounts of manganese and silicon have essentially equal and opposite effects and should have little combined influence on which phase is the first to solidify. Carbon, which is present in amounts less than 0.1 wt.%, and nitrogen which may be picked up during the welding process are both powerful austenitizers and tend to promote primary austenite solidification. Trace elements, especially sulfur and phosphorus, have little effect on the solidification mode, although segregation of these elements during freezing is the major cause of hot cracking.

Solidification of a Fusion Weld

In order to solidify an alloy it is necessary both to dissipate the latent heat of fusion liberated during freezing and to redistribute the solute at the moving solid-liquid interface. During welding, the base metal forms an effective "mold" surrounding the molten weld pool and serves as a heat-sink to dissipate the latent heat of solidification. The type of solute redistribution which accompanies solidification depends upon the nominal composition, C_0, the equilibrium distribution coefficient, k_0, and the growth rate at the solid-liquid interface. Since welding is a classic case of nonequilibrium solidification and requires rapid redistribution of the solute between solid and liquid at the advancing solid-liquid interface, the solute redistribution accompanying welding is a major factor controlling the solidification mode.

The solidification of a fusion weld is best approximated by the following boundary conditions:

1. no diffusion in the solid (diffusion times are too short),

2. no mechanical mixing in the liquid at the moving solid-liquid interface (stagnant boundary layer effect),

3. adjustments in the composition of the liquid at the moving solid-liquid interface occur by diffusion only (diffusion is rapid in the liquid),

4. microscopic equilibrium exists between the solid and liquid phases in contact with one another at the moving interface such that: $k_e^n = C_S^n/C_L^n$ at any given temperature for an alloy with n components

   where: k_e^n = the effective distribution coefficient for the nth component

C_S^n = concentration of the nth component in the solid at the advancing solid-liquid interface

C_L^n = concentration of the nth component in the liquid at the moving solid-liquid interface

Solidification under these conditions is often called Case III Solidification. Solidification under Case III conditions exhibits three characteristic stages:

1. the initial transient stage,

2. the steady-state stage,

3. the terminal transient stage.

Figure 3 summarizes the changes in solute distribution accompanying Case III Solidification for $k_e^n = 0.5$ and 2.0. During the initial transient, the concentration of the solid phase, C_S^n, varies from $k_E^n C_o^n$ in the first solid to form to C_o^n at the end of the initial transient, where C_o^n = the nominal concentration of the nth component in the original alloy. Meanwhile, the composition of the liquid on the opposite side of the moving interface, C_L^n, varies as C_S^n/k_E^n and a concentration gradient is established by diffusion such that the concentration corresponds to C_o^n at some point ahead of the moving interface.

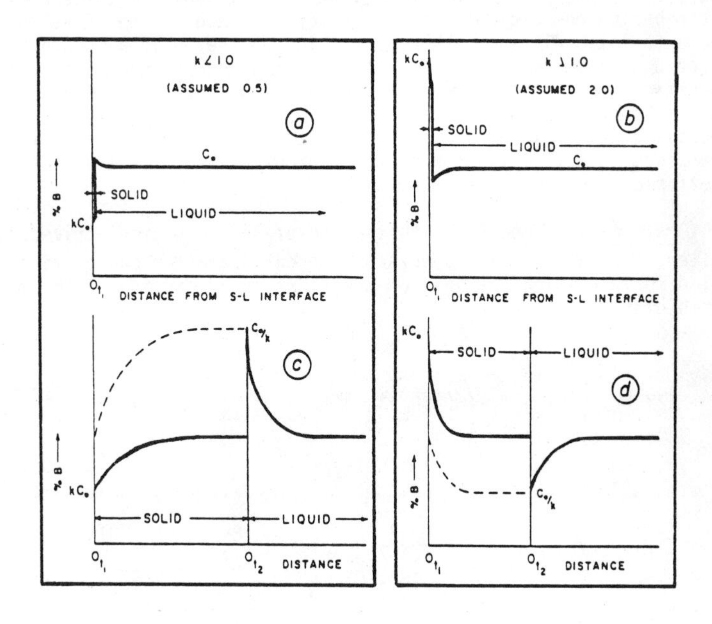

Figure 3. Solute Redistribution During the Initial Transient Stage of Case III Solidification.

Once such a diffusion gradient is established in the liquid, the steady-state stage begins and solid of concentration C_o^n is able to form continuously from the liquid of concentration $C_L^n = C_o^n/k_E^n$ on the opposite side of the moving interface as shown at position O_{t_2} in Fig. 3c and Fig. 3d. The concentration gradient advances ahead of the moving interface by diffusion and remains unchanged in shape so long as the velocity of the interface remains constant.

Ultimately, either exhaustion of the liquid phase or interaction with the diffusion gradient associated with a neighboring solidification front occurs and the terminal transient stage begins. During this stage, the concentration rapidly shifts to that of an invariant point, such as an eutectic, and the effective solidus is depressed markedly below that predicted by equilibrium.

The Initial Transient - By reference to Fig. 2 it can be seen that under the conditions of Case III Solidification the solid composition changes rapidly from the initial concentration, C_1, to C_o, the composition of the solid formed during steady state solidification. The length of this initial transient is dependent upon the characteristic distance, X_c, which may be expressed as

$$X_c = \frac{D_L}{k_o R} \tag{1}$$

where D_L is the diffusion coefficient in the liquid; k_o, the equilibrium distribution coefficient; and R, the growth rate. After five to seven characteristic distances the composition of the solid approaches to within 1% of C_o. From this point on solidification occurs under steady state conditions until the terminal transient stage begins. Since the diffusion coefficient of the liquid is assumed to be of the order of 10^{-5} - 10^{-6} cm^2/sec for substitutional solutes in molten metal, the length of the initial transient is extremely short for the growth velocities and distribution coefficients encountered in welding.

Mathematical Model - With the aid of an expression derived by Smith, Tiller, and Rutter[10] it is possible to calculate the composition of the solid formed during the initial transient as a function of the characteristic distance, where

$$C_s = 1/2\, C_o \left\{ 1 + \text{erf} \left(\frac{\sqrt{(R/D_L)X_c} +}{2} \right) + (2k_o - 1) \exp\left[-k_o(1-k_o)\frac{R}{D_L} X_c\right] \text{erfc} \left[\frac{(2k_o-1)\sqrt{(R/D)X_c}}{2} \right] \right\} \tag{2}$$

Figure 4 shows the concentration gradients determined by this relationship for chromium and nickel as functions of distance and R = 4, 10, and

25 in/min (1.7, 4.2, and 10.5 mm/sec). Since initially, $k_{o_{Cr}} = 1.3$ and $k_{o_{ni}} = 0.4$ in a 20 Cr - 10 Ni Alloy (again assuming the tie lines lie on the plane of the pseudo-binary diagram in Fig. 2), the first solid to form would be approximately 26 Cr - 4 Ni. As solidification proceeds, the solid becomes progressively richer in nickel and depleted in chromium until ultimately the solid composition approaches that of the nominal composition and the steady state stage of solidification begins. Note that the Ni gradient (with $k_o < 1.0$) approaches the nominal composition less rapidly than that of Cr (with $k_o > 1.0$) because of the fact that k_o appears in the denominator of the characteristic distance (Equation 1).

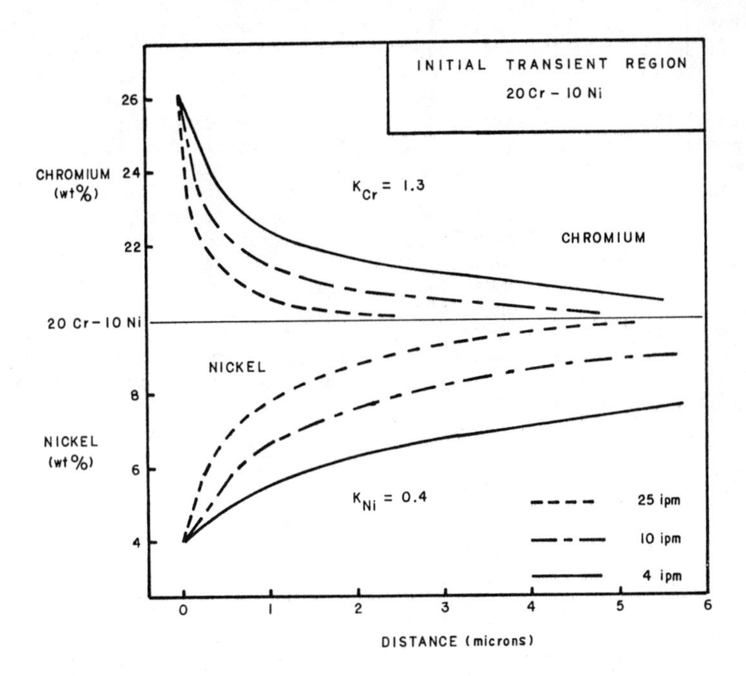

Figure 4. Effect of Travel Speed on the Solute Redistribution of Chromium and Nickel in the Initial Transient Region for $D_L = 5 \times 10^{-5}$ cm^2/sec.

Since the width of the initial transient is inversely proportional to the growth rate, R, increasing the interface velocity from 4 to 25 in/min decreases the width of the initial transient by a factor of approximately six and thus the solute gradients are steepened. Therefore, steady state solidification conditions are approached more rapidly as the welding speeds are increased and the solute enriched (or depleted) zones produced during the initial transient become narrower. This causes a steep transverse concentration gradient at the cores of the cellular or cellular dendritic subgrains.

Generalized Theory of Solidification as Primary Delta Ferrite

Based upon the above discussion, it is possible to propose a generalized mechanism for the solidification of austenitic stainless steel weldments. This theory separates alloys located on the Cr-rich side of the Fe-Cr-Ni eutectic liquidus in which the primary solid phase is delta ferrite from alloys on the Ni-rich side of the eutectic liquidus in which the primary solid phase is austenite.

Under normal conditions, solidification in alloy weldments and castings generally occurs under Case III conditions to form either cellular or cellular dendritic subgrains. If a cell or cellular dendrite is visualized to be composed of a series of small volume elements, each of which can be considered to grow into the liquid as a planar surface, the principles of Case III solidification can be applied on a microscopic scale along the entire interface.[11,12] Thus each advancing cell, or cellular dendrite, has an initial transient region at its core and a final transient region located at the dendrite interstices. Between these two transient regions lies a region where growth occurred under steady-state conditions.

Thus, the variation in composition transverse to the axis of a typical subgrain formed during solidification as primary delta ferrite approximates the concentration profiles in Fig. 4. The initial transient stage occurs at the cores of the cells, and the portion of the ferrite subgrain that solidifies under steady-state conditions forms solid of nominal composition until the advancing interface impinges upon an adjacent solidification front. Once impingement occurs the terminal transient stage begins and solidification is completed by the formation of a divorced eutectic mixture of austenite and ferrite. The distribution of chromium and nickel following solidification as primary delta ferrite is summarizing schematically in Fig. 5.

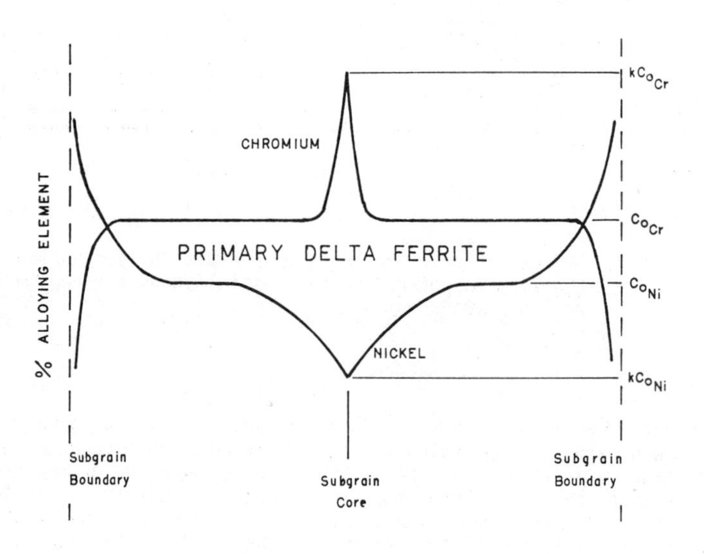

Figure 5. Schematic of Chromium and Nickel Distribution Following Solidification of a Dendritic Subgrain as Primary Delta Ferrite.

Experimental Procedure

Material

The parent material employed in this investigation was a Type 304 stainless steel with the following composition (wt%): 19.3 Cr, 9.5 Ni, 1.9 Mn, 0.56 Si, 0.37 Mo, 0.06 C, 0.044 N, 0.022 P, 0.024 S. Autogenous, partial penetration welds in 6.35 mm (0.25 in) thick plate were made using the gas tungsten - arc (GTA) welding process. The resultant weld microstructure, shown in Fig. 6, contained 7 volume percent ferrite. The vermicular morphology of the dark-etching ferrite is characteristic of austenitic stainless steel weldments which solidifiy with delta ferrite as the primary phase.

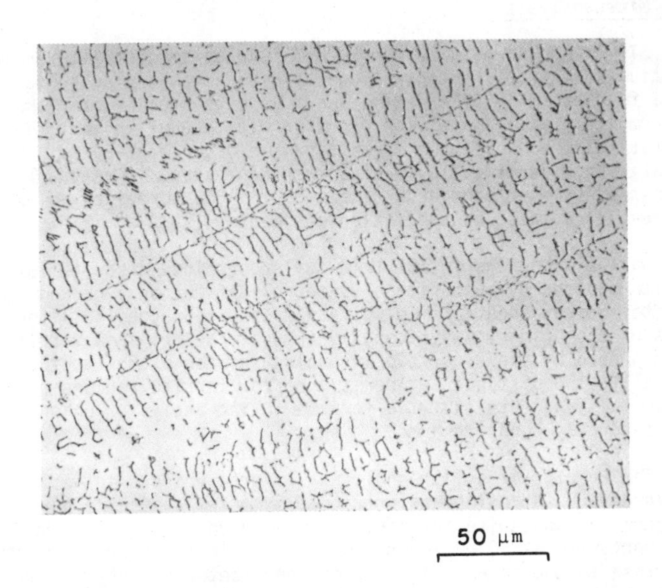

Figure 6. Microstructure of GTA Weld in Type 304, 400X.

Specimen Preparation

Sections of the fusion zone exhibiting the ferrite morphology shown in Fig. 6 were removed and lapped to 125 μm thickness. Discs were then punched from this section and chemically thinned in a dual-jet electropolishing unit using a solution containing 30 parts of perchloric acid, 175 parts of butyl alchohol, and 275 parts of methanol at -40°C (-40°F) and 20 volts.

Thin Foil Microanalysis

Microanalysis of 0.1 - 0.2 μm thick regions of the thin foils was performed with a Phillips 400 scanning transmission electron microscope (STEM) fitted with an energy dispersive X-ray spectrometer. The vacuum in the specimen chamber was approximately 10^{-7} torr (1.3×10^{-5} Pa). An electron probe size of 20 nm (200 Å) was maintained at an accelerating voltage of 120 kV.

The Fe K_{α}, Cr K_{α}, and Ni K_{α} peaks from the resultant x-ray spectrum were integrated within a window which was preselected for each element. A computer-generated background was then subtracted from the spectrum for each measurement. The number of integrated counts for chromium and nickel were then ratioed to the number of iron counts and converted to chemical compositions using the Cliff-Lorimer equations.[13]

Results

STEM Microanalysis

STEM microanalysis was performed to experimentally determine the chromium and nickel concentration profiles accompanying solidification as delta ferrite. The STEM offers the advantage over similar analytical instruments (for example, the electron microprobe) in that the spatial resolution of the electron beam is on the order of 1-20 nm. Since compositional changes during the initial transient stage of weld solidification occur rapidly over submicron distance, the STEM provides an excellent means for measuring these concentration gradients.

Analysis regions in the thin foils contained discrete islands of ferrite bounded by austenite. Since solidification occurred as delta ferrite the retained ferrite which was observed occupied sites along the cores of the primary and secondary arms of the original ferrite dendrites, as suggested by the dendritic morphology of the microstructure of Fig. 6. Typically, the distance between the narrow ferrite regions was less than 5 μm. This suggests that the ferrite in the analysis regions was coincident with secondary dendrite arms.

The results of a microanalysis traverse which originated in a thin retained ferrite region and proceeded into the austenite perpendicular to the ferrite/ austenite interface is presented in Fig. 7. Individual analyses were performed at 50 nm intervals both within the ferrite and in the austenite close to the interface. The distance between analysis points was increased in regions farther from the ferrite/austenite interface as the concentration gradient became shallower.

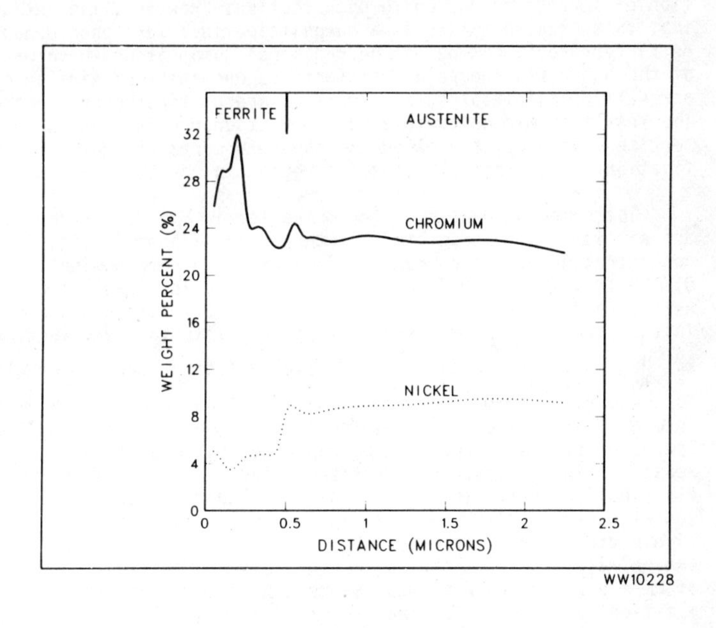

Figure 7. STEM Microanalysis Profile in Type 304 Weld Metal.

Discussion

The validity of the solidification model which has been proposed for austenitic stainless steel weldments and alloy weldments in general is contingent on the hypothesis that a plane front soldification model can be applied to the dendritic solidification of multi-component, multi-phase alloy systems. This requires that the dendritic solidification front be considered as a series of small volume elements each of which obey the rules for plane front solidification on a microscopic scale such that the initial transient behavior can be predicted by the relationship in Equation 2. Bower et al[14] have shown that such an assumption is reasonable if the boundary conditions governing Case III solidification are valid and that both the constitutional supercooling and radius of curvature effects are relatively small. An additional constraint requires that the synergistic effects among alloying elements during solute redistribution are negligible.

The use of STEM microanalysis as an investigative tool for studying the solidification process is predicated by the assumption that bulk diffusion during cooling to room temperature does not significantly alter the concentration profiles which are established during solute redistribution. Since the cooling rates normally encountered during fusion welding are on the order of 10^2 - 10^3 C/sec, diffusion of solute elements subsequent to solidification should be negligible. This is particularly true for substitutional alloying elements such as chromium and nickel. For austenitic stainless steels which solidify as delta ferrite the situation is further complicated by the transformation of ferrite to austenite upon cooling. However, the authors have shown that this transformation is a composition-invariant phenomenon whereby the delta ferrite of nominal or near-nominal composition which makes up the bulk of the dendritic subgrain transforms to austenite of similar composition. As a result, the initial solute profiles are unaffected by the transformation. The resultant microstructure is somewhat unique in that the untransformed ferrite lies precisely along the subgrain cores and facilitates the identification of the initial transient region in the thin foil samples.

The compositional profiles presented in Fig. 7 for chromium and nickel are similar to the predicted solute contours shown in Fig. 4. The chromium concentration decreases rapidly and reaches a near-nominal value less than 0.5 μm from the dendrite core. The increase in nickel concentration is more gradual and reflects the fact that the characteristic distance of the initial transient is inversely proportional to the distribution coefficient, k_E^n. Again referring to Fig. 4, the actual data corresponds closely to the profile predicted for a solidification rate of 10.6 mm/sec (25 in/min) despite the fact that the welding speed was only 1.7 mm/sec. The apparent discrepancy can be explained by considering the functional relationship expressed by Equations 1 and 2. Note that by holding both the distribution coefficient (k_0) and the solidification rate (R) constant, a set of curves similar to those in Fig. 4 can be generated by varying the liquid diffusion coefficient, D_L. A series of such curves for a solidification rate of 1.7 mm/sec (4 in/min) is shown in Fig. 8. The largest value of $D_L = 5 \times 10^{-5}$ cm^2/sec represents the profile generated at a solidification rate similar to that reported in Fig. 4.

Superimposed on the curves in Fig. 8 are the data from Fig. 7 in addition to published data reported by other investigations.[15-17] The individual data points were obtained using STEM microanalysis on a variety of 300-series stainless steel weldments which solidified as delta ferrite at nearly equivalent solidification rates. The correlation of this data is remarkable considering the variety of instruments which were used and the relative difficulty of the experimental technique.

The experimental results indicate that the diffusion of solute elements in the liquid boundary layer of the solid-liquid interface is more sluggish than previously reported[11,12] and that a value of $D_L = 2 \times 10^{-6}$ cm^2/sec is more representative for austenitic stainless steels. This assumes that the distribution coefficient remains relatively constant throughout the initial transient stage of solidification. Lyman[15] has arrived at a similar value for D_L as a result of his investigation of Type 304L weldments.

It can be concluded that the functional relationship of Equation 2 can be applied to the dendritic solidification of weldments if the boundary conditions for Case III solidification are legitimate and the appropriate values are selected for the pertinent variables affecting solute redistribution, namely, the solidification rate, the liquid diffusion coefficient, and the distribution coefficient. The experimental solute profiles for austenitic stainless steel weldments which solidified as delta ferrite were in close agreement with the theoretical profiles and reinforce the validity of the solidification model proposed for these alloys.

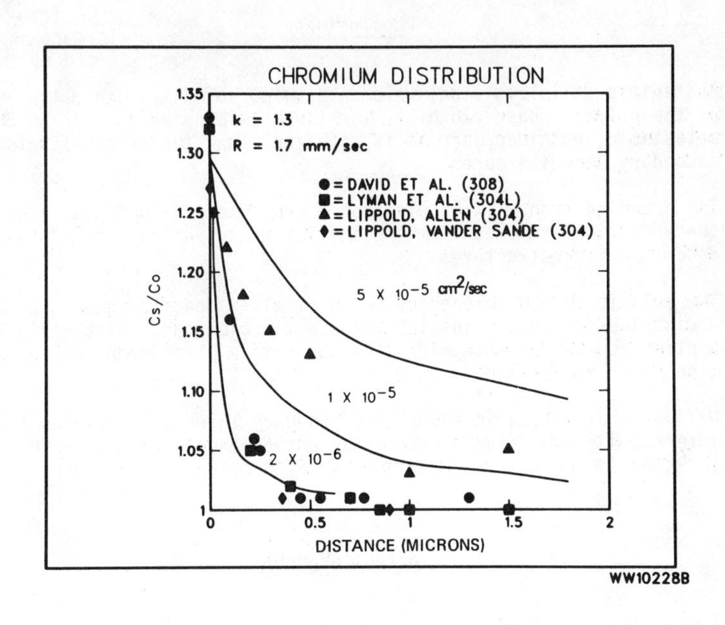

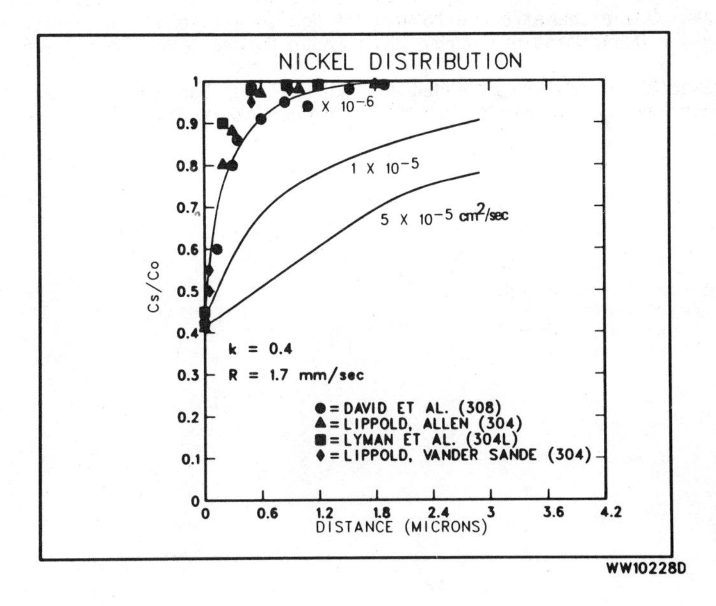

Figure 8. Effect of D_L on the Solute Redistribution of Chromium and Nickel in the Initial Transient Region for R = 1.7 mm/sec (4 ipm).

Conclusions

1. Austenitic stainless steel weldments which solidify with delta ferrite as the primary phase exhibit a vermicular microstructure in which metastable, retained ferrite is situated along the original primary and secondary dendrite cores.

2. The scanning transmission electron microscope provided high resolution concentration profiles from the initial transient regions of the dendritic microstructures.

3. The relationship developed by Smith *et al.*[10] describing solute redistribution in the initial transient region can be successfully applied to complex weld microstructures if a plane front solidification model is assumed.

4. Diffusion of solute in the liquid boundary layer at the solid-liquid interface is more sluggish than was anticipated. A value of $D_L = 2 \times 10^{-6}$ cm^2/sec was determined from the STEM microanalysis data.

Acknowledgments

This work was jointly supported by the Chemetron Corporation under the auspices of the Chemetron Fellowship at Rensselaer Polytechnic Institute and by the Department of Energy (DOE) under Contract No. De-AC04-76DP00789.

Special thanks are extended to Rich Bell and Don Lind for assisting in specimen preparation and to Rob Allen for performing the STEM microanalysis.

References

1. Borland, J. C., and Younger, R. N., "Some Aspects of Cracking in Welded Cr-Ni Austenitic Steels," British Welding Journal, January 1960, pp. 22-59.

2. Hull, F. C., "Effect of Delta Ferrite on the Hot Cracking of Stainless Steel," Welding Journal, 46 (9), (1967), pp. 399s-409s.

3. Arata, Y., Matsuda, F., and Saruwatari, S., "Varestraint Test for Solidification Crack Susceptibility in Weld Metal of Austenitic Stainless Steels," Trans. JWRI, 3(1), (1974), pp. 79-88.

4. Lundin, C. D., DeLong, W. T., and Spond, D. F., "Ferrite-Fissuring Relationship in Austenitic Stainless Steel Weld Metals," Welding Journal, 54(8), (1975) pp. 241s-246s.

5. Arata, Y., Matsuda, F., and Katayama, S., "Fundamental Investigation on Solidification Behavior of Fully Austenitic and Duplex Microstructures and Effect of Ferrite on Microsegregation," Trans. JWRI, 5(2), 1976, p. 35-51.

6. Fredrikkson, H., "Solidification Sequence in 18-8 Stainless Steel, Investigated by Directional Solidification," Met. Trans., 3(11), 1976, p. 35-51.

7. Lippold, J. C., Savage, W. F., "Solidification of Austenitic Stainless Steel Weldments: Part 1 - A Proposed Mechanism," Welding Journal, 59(12), 1979, Research Suppl., p. 362s-374s.

8. Masumoto, I., Tamaki, K., and Kutsuna, M., "Hot Cracking of Austenitic Steel Weld Metal," Trans. JWS, 41(11), 1972, p. 1306-1314.

9. Metals Handbook, American Society for Metals, 1973, vol. 8, p. 424.

10. Smith, V. G., Tiller, W. A., and Rutter, J. W., "A Mathematical Analysis of Solute Redistribution During Solidification," Canadian Journal of Physics, 33, 1955, 723-745.

11. Flemings, M. C., Solidification Processing, McGraw-Hill Book Company, New York, 1974.

12. Chalmers, B., Principles of Solidification, John Wiley and Sons, New York, 1964.

13. Cliff, G., and Lorimer, G. W., "Quantitative Analysis of Thin-Metal Foils Using EMMA-4, the Ratio Technique," Proc. 5th European Congress on Electron Microscopy, Manchester, Institute of Physics (London), 1972, pp. 140-141.

14. Bower, T. F., Brody, H. D., and Flemings, M. C., Trans. AIME, 236, (1966) p. 624.

15. Lyman, C. E., "Analytical Electron Microscopy of Stainless Steel Weld Metal," Welding Journal, 59(7), (1979), pp. 189s-194s.

16. David, S. A., Goodwin, G. M., and Braski, D. N., "Solidification Behavior of Austenitic Stainless Steel Filler Metals," Welding Journal, 59(11), (1979), pp. 330s-336s.

17. Lippold, J. C., and Vander Sande, J. B., unpublished research performed at Massachusetts Institute of Technology, 1979.

Graphics and Geometric Modeling

COMPUTER GRAPHICS IN HEAT-TRANSFER SIMULATIONS

Griffith Hamlin, Jr.
Los Alamos Scientific Laboratory
Los Alamos, New Mexico

Computer graphics can be very useful in the setup of heat transfer simulations and in the display of the results of such simulations. The potential use of recently available low-cost graphics devices in the setup of such simulations has not been fully exploited. Several types of graphics devices and their potential usefulness are discussed, and some configurations of graphics equipment are presented in the low-, medium-, and high-price ranges.

1. Introduction

Simulation of heat transfer during solidification has progressed over the years until now it is reported to have reached a point where its application to industrial problems can be considered. There are essentially three steps in performing such a computer simulation: the problem setup, the actual heat-transfer simulation, and the displaying of results of the simulation.

Computer graphics deals with the computer hardware and software needed to display and manipulate computer-drawn pictures. Where can computer graphics help in heat-transfer simulations? Computer graphics can make its largest contribution in the first step, problem setup. In this step, the user specifies to the computer a two- or three-dimensional heat-transfer mesh consisting of nodes, boundaries between various materials, thermal properties of various materials, etc. This mesh, being a model of a physical object, has an inherent graphical representation. Therefore it seems natural to use computer graphics input devices to specify it. However, computer graphics is not yet in general use in this step due to a lack of readily available software. This may be due, in part, to the availability only recently of inexpensive graphics devices, and in part to the different style of software design needed for interactive applications as opposed to applications designed for batch execution. The next section of this paper will consider this problem setup in detail.

In the second step, the thermal analysis calculations, computer graphics cannot provide much help. This step is traditionally performed on large mainframe computers. There now exist several well-known computer programs for performing this step.

The third step, the display of results, can and is using computer graphics. Output in the form of charts, graphs, (some with color showing temperature variations, some with temperature contours, etc.) are in moderate use now. New, inexpensive color-display devices will surely expand this usage. Section 3 will consider this step in more detail.

In Section 4 we configure a few examples of computer graphics systems, using readily available components suitable for use in heat-transfer simulations.

2. Computer Graphics In Problem Setup

Setting up a moderately complex problem for any of the well-known heat-transfer analysis programs can be tedious and costly in terms of the time required to perform the initial setup and modify it until all errors are eliminated. Mesh generator programs have helped some in this area. However, these programs still require the user to specify an inherently graphical object (a mesh) in a nongraphical way: supplying a data file of coordinates of nodes, numbering of nodes and edges for later reference, etc. It is difficult to detect errors in this textual representation of the mesh. Recently available low-cost graphics input devices of all kinds have the potential to greatly simplify the problem setup phase of computer simulations, thus eliminating one of the remaining obstacles to its more widespread use. In this section, we will consider how each of several such input devices could be used in the setup of heat transfer problems. We hope that this may help motivate the production of the required software.

Locator Devices

Locator devices are graphical input devices that are used to specify an x,y position or series of positions. The data tablet is such a device, and it is a natural device for entering the positions of mesh nodes, material boundaries, etc. A data tablet consists of a flat rectangular pad with electronic sensing devices underneath. An electronic pen or stylus is moved over the rectangular tablet area, and the sensors can detect the position of the stylus on the tablet, usually with an accuracy of about 1/10 millimeter. Software that samples this position several times per second can be used to enter freehand sketches or tracings of parts into the computer. The stylus typically contains a small switch that closes when the tip is pressed down on the tablet surface. This can be used to indicate specific positions of nodes or other objects in the mesh. Data tablets come in various sizes, from about a foot square to four or five feet on a side. Smaller sizes can be purchased for about $700, with larger sizes ranging to several thousand dollars.

Other locator devices include joysticks, thumbwheels, trackballs, and various combinations of buttons and switches that control the position of a small indicator, called a cursor, upon a display screen. The user moves the cursor to the desired position with the device and then indicates this to the computer by some action. Although these devices are relatively inexpensive, they typically cannot provide the accuracy and overall versatility of a tablet. With tablets recently available in the $700 range, I think their use will become widespread.

Pick Devices

A light pen or any of a class of devices known as "pick" devices are very useful for correcting errors or making changes in heat transfer meshes. A light pen can be used to point at an area of the display screen to indicate to the software the particular point or edge of the mesh the user wants to move or delete or modify in some way. Usually, nodes and sides of mesh areas are numbered in input data files. The numbers are used to specify nodes upon which the user wants to perform some operation. With the ability to point at various nodes, there is no longer a need for the user to be aware of their numbers. This ability of graphical input devices to specify various mesh components on a display WITHOUT NAMING THE COMPONENTS is perhaps their most useful characteristic for fixing errors and modifying meshes. For less expensive graphics systems, a data tablet or other locator device can be made to simulate a pick device by a graphics software technique known as "correlation." Similarly, a pick device can be made to simulate a locator device. Thus one device can usually be used successfully to perform both types of functions.

Textual Input Devices

A keyboard or other traditional textual input device can be used for specifying boundary conditions. However, many of these parameters, such as initial temperatures, are numeric quantities that need not be specified with great accuracy (not over three significant digits). For these parameters, various types of knobs and dials can, with an appropriate visual indicator on the display screen, be used as "valuator" devices for quickly indicating to the computer an approximate value of something. However, when a very accurate or exact value is required, a keyboard is usually used.

Software

Software controlling the graphical devices should allow the user to operate them in a natural interactive man/machine "dialogue." One useful technique is to display, along one side of the graphics display, a list or "menu" of commands that the user can select. Pointing at one of these with an input device is invariably easier and less error prone than typing commands on a keyboard.

Software commands would normally include the functions available on current mesh-generating programs, such as the ability to specify lines and curves for sides of meshes and the ability to indicate how to divide mesh segments with equal or proportional spacing. In addition, well-designed software would include commands for allowing any part of the mesh to be deleted, moved, or modified in various ways. Also, the graphics system software should provide commands for quickly defining or retrieving predefined, often-used geometric shapes from disk storage, positioning them with knobs, tablet, etc., into the correct position as part of a larger mesh geometry. This facility is quite useful after the system has been in use for a while and a library of commonly used pieces has been built up. Other such libraries could contain material properties (melting points, specific heats, and various other thermal coefficients) of commonly used materials to be called up by the user. Thus the user need only specify material types and normally need not remember and enter the various properties of the material for each and every problem using that type of material.

3. Graphics in Display of Results

Display Devices

Use of any locator device should be coupled with a CRT or some other type of display that shows the effect upon the mesh of the user's actions. Recently, several low-cost raster-scan display devices have become available that use TV technology parts and low-cost memory. These displays allow the user to selectively erase and redraw specified parts of the picture, so that he can instantly view the results of correcting errors in a mesh. This is not possible with traditional storage tube displays without erasing the entire screen and redrawing the modified image. In addition, color is available on raster-scan displays. Using color to indicate different materials in a heat-transfer mesh can be very effective. Most inexpensive color displays of this type still suffer from poor resolution, but that should improve in the near future.

When purchasing such displays, resolution and the number of simultaneous displayable colors are probably the two most important parameters. Displays using TV standards can achieve resolutions of up to 640 x 480 pixels. However, experience has shown that useful work can be done with much less resolution, namely, in the 200 x 200 range. Displays in this resolution class are quite inexpensive (a few hundred dollars to about $3,000). They will have most trouble displaying meshes with many curved boundaries, and the least trouble with meshes with mostly vertical and horizontal boundaries. For finer meshes, the graphics software should provide a "zoom" capability to allow the user to quickly zoom in on an area of interest in the mesh.

Color shading techniques are well suited for displaying variations in temperature, stress, or other scalar functions over an area of the heat mesh. For less expensive displays without color or shading capabilities, contour lines of constant temperature, pressure, etc., can be used.

Hardcopy Devices

Devices for producing hardcopy pictures, especially color hardcopy, have heretofore been rather expensive or not available. However, within the last year inexpensive ($3,000 and up) color-camera attachments have become available that will attach to most raster-scan displays and produce a Polaroid picture within a minute. For better quality hardcopy, pen plotters with multiple-color pens are now available at moderate prices ($1,100 to $4,500). These devices, however, will require several minutes to plot a moderately complex mesh, and a very long time to plot color-shaded images. They are probably most useful for producing that one last hardcopy when everything is correct, to be used for presentation of results. Ink-jet plotters allow good quality with much faster plotting speeds than pen plotters, but are considerably more expensive. Impact plotters with three-color ribbons provide moderate quality and moderate speeds (2 minutes for a moderately complex shaded plot) for prices in the $12,000 range.

4. Examples of Graphics System Configurations

Small System

A very inexpensive home computer, such as the APPLE II, Ohio Scientific, or North Star, can be the basis for a small graphics system capable of setting up and displaying results for small to moderate heat transfer problems. The limiting factor in this system will be the resolution of the display and the memory available to hold the data describing the input mesh and results. Components required are the central processor, as much memory as one can afford (usually 65,000 bytes), a color TV for display, small floppy disks, small pen plotter (such as the Houston Instrument HI-PLOT), small 12-inch data tablet (such as the Summagraphics BIT-PAD-ONE), and a small printer. These components can be obtained for about $5,500 to $6,500.

Medium System

A medium-size graphics system can be configured from a small minicomputer such as the DEC LSI-11/23; a hard disk, such as recently-announced models from Advanced Electronic Design or Data Systems Design in the 10- to 20-megabyte capacity range; an inexpensive ($2,000) system terminal like the Lear Seagler ADM-3A terminal with black and white graphics option; and the same tablet, plotter, and printer as in the small system described above. Such a system will cost approximately $18,000.

Large System

A large system might employ a vector refresh black-and-white display with three-dimensional capabilities for tackling full three-dimensional simulation models (i.e., where no symmetry can be used to reduce the dimensionality). Megatek, Vector General, and other companies make such systems that interface to minicomputers, such as a DEC PDP-11/34 or similar minicomputers. With large disks, a fast electrostatic or ink-jet printer/plotter, tablet, light pen, and joystick, such a system would be in the $60,000 to $100,000 range.

5. Summary

Several types of inexpensive graphics devices have recently become available that can be used in heat transfer problem setup and display if the proper software exists to make use of them. To be maximally effective, this software should use the graphics devices in an interactive manner, coupled with a pictorial display of the mesh that is being created or modified. Inexpensive raster-scan display devices as well as inexpensive hardcopy devices are also available. All of this equipment can now be usefully employed in the problem setup phase and the result display phase of heat transfer simulations.

GEOMETRIC MODELING: A STATUS REPORT

Melvin R. Corley*
School of Mechanical Engineering
Georgia Institute of Technology
Atlanta, GA 30332

Computer-based geometric modeling systems have been in existence for over ten years and many developers of these research systems feel that the basic problems associated with the unambiguous representation of three-dimensional objects have been solved. This paper presents an overview of the capabilities of current modeling systems and how they can be used in the design process. Special attention is given to the problems that might be encountered in using a computer-based geometric modeling system in the design of castings. The paper concludes with a realistic assessment of the state-of-the-art and prospects for near-term development.

* The author is currently at the Mechanical Engineering Department, Louisiana Tech University, Ruston, LA 71272.

Introduction

Any time a physical object is to be designed and manufactured it must be geometrically modeled. This is true because every physical object can be described by a set of numbers which represent the dimensions of the exterior boundaries of the object. Such a set of dimensions (geometric model) may exist in a master craftsman's head, on the back of an envelope, on a set of shop drawings, or in a computer file. The role of a geometric model is to partition the universe into two parts--that part which lies within the exterior boundaries of the object being modeled and that part which does not; and any scheme which allows this determination to be made can be properly called a geometric modeling system. Such a broad definition of geometric modeling points to the fact that the way parts have been modeled and the presentations of these models through the use of paper drawings has changed little in the past one hundred years. With the present availability of powerful digital computer systems and rapidly developing software in support of three-dimensional geometry representation and closely allied fields such as interactive graphics, a more general notion of the role and functions of a geometric modeling system must be understood.

Voelker has offered the description of the general geometric modeling system that is depicted in Fig. 1 (1). In this geometric modeling system the block labelled "Useful Geometry Systems" consists of such things as

- definition translator
- question translator
- internal representation of geometry
- algorithms for applications
- answer translator

The object definitions may be in the form of part description languages which use various types of primitives, drawings input via interactive graphics terminals, or combinations of predefined complex objects retrieved from a library. Questions that relate to the geometry of the object may be asked of the modeling system. The questions must be answered by application algorithms that access the internal geometry representation. The kinds of questions that may be asked are limitless, but typical of these are:

- What are the mass properties of the object?
- Is there interference between moving parts?
- What does the object look like?
- Will a given numerically controlled machine tool program produce the modeled part?

The outputs of the modeling system algorithms are those that are appropriate for the type of question asked. Typical types of outputs are drawings, N/C tapes, sets of real numbers which give object properties, boolean values representing yes-no responses, etc.

Such an expansive model of the geometric modeling system may appear inappropriate to new users but is necessary as it points to the availability of the geometry of the object as central to the CAD/CAM system. This geometry system should not be confused with the representation of that geometry through drawings. Rather drawings are simply the product of one application algorithm applied to the geometry data base. The single most important feature of any geometric modeling system is that it should accurately and unambiguously represent the geometry of any three-dimensional solid object (2). Given such a fundamentally sound base,

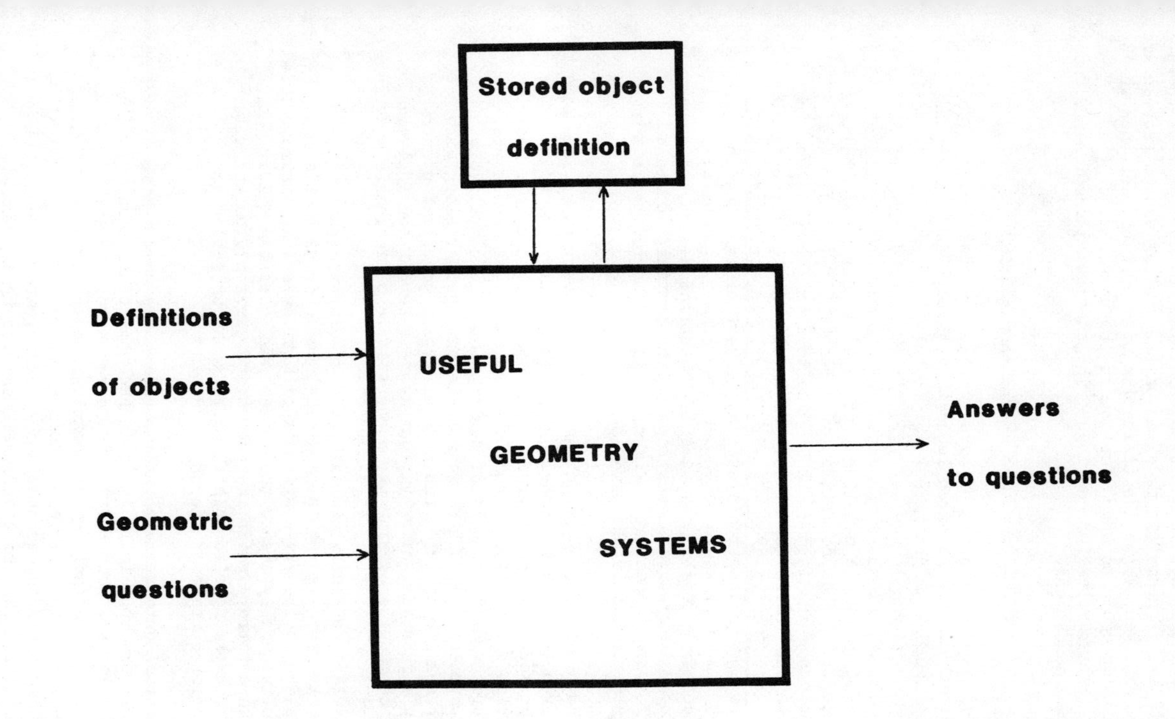

Figure 1 - Schematic Diagram of a Geometric Modelling System.

algorithms to use this base to provide answers to user questions can be developed.

The focus of this report is the role that computer-based geometric modeling systems can play in developing a computer-aided design system for castings. The report begins with a brief review of the development of the field of modern geometric modeling and an assessment of the state-of-the-art. The points of impact of geometric modeling on the casting design process are considered next followed by the identification of a criteria set upon which an adoption decision could be based for use in the development of a computer-aided design system for castings. The concluding section presents findings and recommendations.

Development of Geometric Modeling Systems

It has been observed that it appears to take a decade for new ideas or techniques to move from the research laboratory through the development stage and to be placed into practice (3). If this is true, then it would appear that the work begun in the early 1970's in using digital computers to model three-dimensional solid objects should be coming to fruition. While developers of geometric modeling systems certainly do not believe their work is complete, the proceedings of a recent CAM-I conference reveals that the consensus is that it is time to move these systems from the laboratory to the design departments of industry to better determine what the users' needs really are (4). It was at this conference that system developers and prospective system users were brought together to focus on the problems that would be encountered in moving toward incorporating geometric modeling systems in industrial design. A number of system developers presented status reports and participated in a demonstration of system capabilities in modeling a common conversation piece offered by Messerschmitt-Bolkow-Blohm.

Voelcker and Requicha (5) present an excellent overview of some of the issues involved in geometric modeling of mechanical parts and incorporating the models in procedures for automatic production. Baer, Eastman and Henrion offer a fairly complete assessment of the current status of geometric modeling and brief descriptions and comparisons of eleven existing systems (6). These reports taken together along with other less comprehensive articles and several system description/user manuals and sample runs made using various systems permit the following observations to be made concerning the present state of geometric modeling.

1. The modelers currently available are generally capable of accurately representing three-dimensional objects that are definable with the given system's primitives and are thus able to faithfully answer the fundamental question that, given an arbitrary point in three space, does the point lie inside the modeled object, outside the object or on an exterior surface of the object?

2. A variety of data representation schemes are used by the available modelers including cell decomposition, boundary representations, constructive solid geometry, and swept volumes, but no single representation appears to be clearly the best of all applications. Some systems support multiple representations to facilitate different applications. The trade-offs here are the traditional ones for computer applications--memory size versus execution time.

3. The available systems rely on the use of a prescribed part definition language for model creation. The input phase may be either interactive, semi-interactive, or batch. The systems do accept some responsibility for the well-formedness of the model by rejecting illegal constructions and providing error messages. Still open is the question of just how many and which primitive objects or construction tools are needed to model mechanical parts.

4. The availability of dimensioning and tolerancing capability is not widespread and automatically dimensioned drawings produced by such systems appear crude.

5. Computer execution time for both model generation and application processing (e.g. isometric drawing) appear to be typically on the order of tens of seconds on a medium size computer system for even fairly simple parts. The execution time goes up rapidly also as "fine print" detail such as fillets and draft are added. Including this level of detail increases the computing time to several minutes.

6. Developers of geometric modeling systems have been concentrating on developing the basic system and generally have only begun to develop applications such as N/C tape generation, finite element analysis, etc. Most system developers feel that the research involved in modeling objects has been well explored and that the development of applications and commercially viable systems are probably best left to CAD system vendors who have the necessary software support, system documentation, and marketing skills and experience.

In short it appears that the field of computer-based geometric modeling is nearing the critical juncture where it needs to move from the research to the development and early production phases. Many of the basic research issues have been resolved, at least theoretically, but much remains to be done in improving computational efficiency, extending the range of applications and providing fully supported hardware-software packages.

Geometric Modeling for Castings

Upon preliminary examination it may appear that the role of geometric modeling for casting design might be essentially the same as for any other manufacturing process, except that the shapes involved are typically more complicated. While this difference alone is significant and will be discussed in more detail, there are many other subtle aspects of casting design which present significant challenges to modern geometric modeling systems. It is useful to consider these modeling difficulties in two groups; those that will be encountered when modeling a casting for production and those that surface when modeling castings for solidification simulation. The first case arises when a manufacturer is designing the shape of a casting to fulfill geometric constraints and objectives for the part. The second case becomes important when the preliminary design of the casting has been completed and it is turned over for detailed analysis as to how to produce the desired shape while insuring the structural and dimensional integrity of the finished product. The first phase of designing for functional completeness we will call preliminary design and the second phase we will call detailed design.

In the preliminary design phase the designer is concerned only with shaping the casting to insure that it accomplishes its volumetric responsibility. This aspect of the design of parts such as pump housings, manifolds, engine blocks, turbine blades, and other complicated shapes is critical to the efficient operation of the finished casting as it performs its responsibility of directing fluid flow, providing operating clearances, providing attachment points, etc. During this phase of the design of the casting, the designer needs the following capabilities of the geometric modeling system:

- The capability of generating simple and complex shapes. The fact that the casting has its origins as a fluid encourages its use in manufacturing those parts that have complicated geometries and internal voids that would render them difficult or impossible to manufacture by other processes.

- An interactive part definition procedure that provides construction aids similar to those found in many two-dimensional drafting systems and immediate graphical feedback of the partial model. Editing capabilities allowing the designer to modify or delete previously defined segments of the part is also a desirable feature. Geometric design of castings like any other type of design is an iterative, interactive process given to frequent revisions before completion.

- Automatic calculation of the various area/volumetric properties of the part. Frequently cross-sectional areas, enclosed volumes, moments of inertia, etc. are primary design objectives and need to be checked at the earliest stage of the design process.

As the casting design moves from the preliminary to the detail phase, new aspects of the design that impact the geometric modeler begin to surface. During the detailed phase of the design more specific questions relating to the structural integrity, manufacturability, cost of production and solidification of the casting must be answered. These requirements must be satisfied while retaining the essential dimensional requirements set forth in the preliminary design. During this detailed design phase the geometric modeling system will be called upon to support several new applications in addition to those needed for the preliminary design. A partial list of these additional modeler capabilities would include the following:

- Automatic volumetric enmeshment for analysis by various finite element method programs. Of special concern will be structural analysis programs for determining the strength of the casting under various static and dynamic external loads and freezing solidification simulation programs for determining the progress of the solidification front in the casting versus time. The freezing simulation is especially important as the freezing pattern significantly affects the metallurgical integrity of the casting. In particular, for many common metals used in castings the area surrounding the last point to solidify will be filled with microscopic or larger voids due to shrinkage of the material as it freezes. Thus the casting designer must always assure that these seriously weakened regions are not located in the body of the casting but rather in risers or other locations external to the casting proper.

- Multiple part modeling and assembly verification. Complex castings are made using intricate combinations of patterns and destructible cores which must all be in place simultaneously before the metal is poured. Since each pattern piece and core must be separately manufactured, the geometric modeling system should facilitate modeling of each subpart and also be able to handle the assembled set of parts as a unit and check for mating of subpart surfaces. In a similar way, at the detailed modeling stage, it will be necessary to model the mold and all the metal feeding/gating systems, for while in the mold they form an integral part of the casting and can strongly affect solidification patterns. Also separate heat sinks, or chills, placed in the mold material to facilitate heat removal will need to be molded individually and also incorporated as a part of the mold assembly as a whole. Thus it is clear that metal casting does place on a geometric modeling system the requirement that many separate parts be modeled individually and then be assembled to form a whole.

- The ability to easily and efficiently handle complex "fine point" detail. At the detailed modeling level the casting designer is responsible for determining how the casting pattern can be made. The principle pattern pieces must have convex surfaces and even straight surfaces projecting into the mold material are given a slight taper, or draft, to facilitate pattern removal. For the same reason sharp corners are usually rounded with fillets. These features combine to present the modeling systems with considerable difficulty. At the present state of development, such details not only increase the model definition time on the part of the designer, but greatly lengthen the computer execution time when processing the model for various applications. Some have suggested that such detail be simply ignored or treated symbolically much like threaded parts are treated (7). This is a reasonable suggestion for the preliminary design phase but studies would have to be made to determine the impact of such a shallow treatment at the detailed design level. It is certain, however, that at some point before the pattern is produced, the draft and filleting will have to be applied.

There may be other aspects of particular kinds of castings that place special demands on the geometric modeling system used in their design. We have not included these or such aspects as the interface of the geometric modeler with a process planning system for production analysis of both the casting and finishing operations. The points listed above have been offered to guide in the development of geometric modeling systems for casting applications.

The Future of Geometric Modeling

It is clear that a geometric modeler will be at the heart of well integrated CAD/CAM systems as systems move from simple two-dimensional drafting tools to full three-dimensional design systems. The commercialization of present research systems has already begun (8) and the pace should begin to accelerate. Most system developers feel that the basic research into three-dimensional part representation has been essentially complete; however, much work needs to be done in developing application packages that will attract the end user. Such applications include refined drawing packages with automatic dimensioning, finite element enmeshment, N/C tape preparation or verification, etc. As these application packages begin to appear with a full geometric modeler at the core, these systems will move from the research laboratory to the design department of industries worldwide and a new generation of part modeling and manufacturing will have begun.

Acknowledgements

The preceding review was conducted under the sponsorship of NSF Award No. DAR78-24301, Division of Applied Research Project Monitor, Dr. Richard Schoen. The author wishes to thank Dr. John T. Berry (Project Co-Director) and Mr. Chin-Shing Wei of the Georgia Tech staff of the Computer-Aided Design System for Castings project for their considerable assistance in information gathering and system evaluations.

References

(1) H. B. Voelcker, "Geometric Modeling of Rigid Solids," short course notes at University of Rochester, June 11-15, 1979.

(2) P. Veenman, "ROMULUS--The Design of a Geometric Modeller," Proceedings of the Geometric Modelling Seminar, Bournemouth, England, Nov. 27-29, 1979.

(3) I. C. Braid, "Geometric Modelling--Ten Years On," Proceedings of the Geometric Modelling Seminar, Bournemouth, England, Nov. 27-29, 1979.

(4) W. A. Carter, ed., Proceedings of the Geometric Modelling Seminar, Bournemouth, England, Nov. 27-29, 1979.

(5) H. B. Voelcker and A. A. G. Requicha, "Geometric Modeling of Mechanical Parts and Processes," COMPUTER, Dec. 1977, pp. 48-57.

(6) A. Baer, C. Eastman, and M. Henrion, "Geometric Modelling: A Survey," COMPUTER-AIDED DESIGN, Vol. 11, No. 5, Nov. 1979, pp. 253-272.

(7) P. R. Wilson, "Industries Requirements for Geometric Modelling," Proceedings of the Geometric Modelling Seminar, Bournemouth, England, No. 27-29, 1979.

(8) P. Veenman, ROMULUS: User's and Programmer's Guide, Shape Data, Ltd., Cambridge, England.

CSTMS3, A FINITE ELEMENT MESH GENERATOR FOR CASTINGS

A. Badawy and K. Schreiber
Battelle's Columbus Laboratories
Columbus, Ohio
&
J. Chevalier and T. Wassel
U.S. Army Tank Automotive Research and
Development Command
Warren, Michigan

Summary

An automatic finite-element mesh generation program (CSTMS3) for casting models is described. The program is designed to be used by foundry engineers with no previous programming or finite-element experience. The output of CSTMS3 is to be used later in the analysis of mold heat-flow, fluid-flow, and stress analyses to provide for castings soundness.

Introduction

The development of the finite-element method provides engineers with a general and powerful tool for the analysis of continuum mechanics problems. A general problem as a whole may be formulated by the finite-element method through the following procedure:

- Discretization of the continuum (to finite elements)
- Selection of shape functions (within each element)
- Obtaining the element characteristic matrix
- Assembly of the algebric equations (this includes the assembly of the element matrices into the overall master matrix of the discretized continuum as a whole)
- Solution of the algebric equations for the unknown parameters.

The theory behind each of the previous steps is well explained in many publications (1, 2, 3, . . . etc.). This paper deals only with the first topic, the discretization of the continuum or what is usually called the finite-element mesh generation.

Generation of Mesh Data

As mentioned before, the first step in the finite-element analysis is to divide the continuum (by imaginary lines or surfaces) into a number of finite elements. This process is still essentially a matter of engineering judgment to decide what number, size and arrangement of finite elements that give an effective representation of a particular continuum. Ideally, a finite element computer program should generate its own mesh data from a minimum number of geometric parameters. The amount of input data required is minimal and once the relevant coding has been written and tested, the possibility of errors is largely eliminated.

"CSTMS3", A Finite Element Mesh Generator for Castings

Program CSTMS3, described herein, is an automatic mesh generator for three-dimensional finite element meshes composed of brick (8 node) and wedge (6 node) elements. CSTMS3 constructs a mesh by assembling a number of user-defined zones (blocks), each of which is automatically subdivided into finite elements. The program can handle casting models composed of parallelpipeds and cylindrical zones. Other zone shapes are currently tested (hollow cylinders, cones, hollow cones, etc.).

"CSTMS3" is designed to be used by casting engineers with no previous programming or finite-element experience. The output of CSTMS3 is to be used later in the analysis of mold heat-flow, fluid flow, and stress analysis to provide for castings soundness. Interactive design (man-machine and machine-man) of casting molds is another important output of CSTMS3. In this paper, however, we will explain only the capabilities of CSTMS3 to generate 3D mesh data.

Since the main goal is to study the heat transfer between the molten metal of the casting model and the surrounding sand (sand mold),it is clearly advantageous to have a finer mesh in the regions describing the contact between the molten metal and the sand. For that reason, Program CSTMS3 automatically enlarges and reduces the original input geometry. The rate of enlargement and reduction decreases with distance from the contact surface (between the metal and sand) inwards (metal) and outwards (sand). To assist in understanding the capabilities and advantages of using Program CSTMS3, the following detailed example is given. The test casting model of Figure 1 would be used as the basic input geometry. To generate the mesh the user should follow the following steps:

1 - Prepare the Geometrical Data

To define the geometry of Figure 1 the user should:

A - Divide the model into a certain number of "zones". A zone is the basic building block for the model. To create a model, the user must visualize a subdivision of his casting into zones in a manner similar to that used to manually create a finite-element mesh. So far, only two zone shapes are available to the user, "brick" and "cylindrical". A brick zone is a volume bounded by six planes. A cylindrical zone is a volume bounded by a cylinder. Figure 2 shows a possible way of dividing the model (five brick zones, including the sand box, and one cylindrical zone).

B - Prepare data defining the zones. Zones 1 to 5 (brick zones) should be defined using eight X, Y, Z coordinates, e.g., Zone 1 (sand mold) would be defined by:

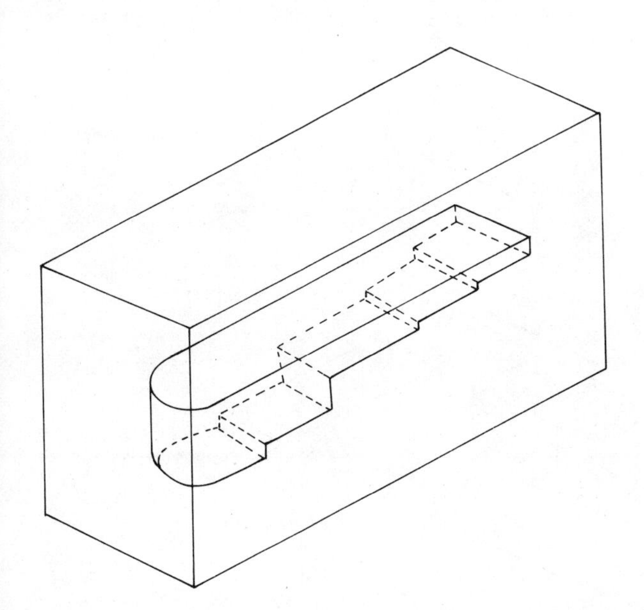

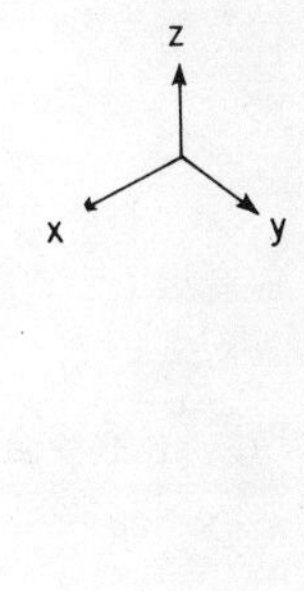

Figure 1 - Test casting model.

0.0, 0.0, 0.0
34.0, 0.0, 0.0
34.0, 14.0, 0.0
0.0, 14.0, 0.0
0.0, 0.0, 17.0
34.0, 0.0, 17.0
34.0, 14.0, 17.0
0.0, 14.0, 17.0

Zone 6, the cylindrical zone, should be defined by X, Y, Z coordinates of the lower center, radius, height, and axis of the cylinder, e.g.,

7.5, 7.5, 3.75, 3.5, 5.0, Z

C - Choose a "WORKING VIEW" (W.V.), e.g., X-Y (plan), X-Z (elevation), or Y-Z (side view), and then choose an "AIDING VIEW" (A.V.). The idea here is for the program to use the (W.V.) to generate a 2D mesh and then use the (A.V.) to generate the third dimension of the mesh by simply "slicing" the (A.V.) to a number of "layers". From Figure 2 it is most convenient to use X-Y as a (W.V.) and Y-Z as an (A.V.).

Notes:

* Steps A and B should be prepared on a piece of paper before running the program.
* The order of defining the zones to the program does not have to follow a specific sequence.
* The user does not have to define the volumes connecting the cylindrical zone to its neighboring brick zones. The program automatically defines that interconnection.

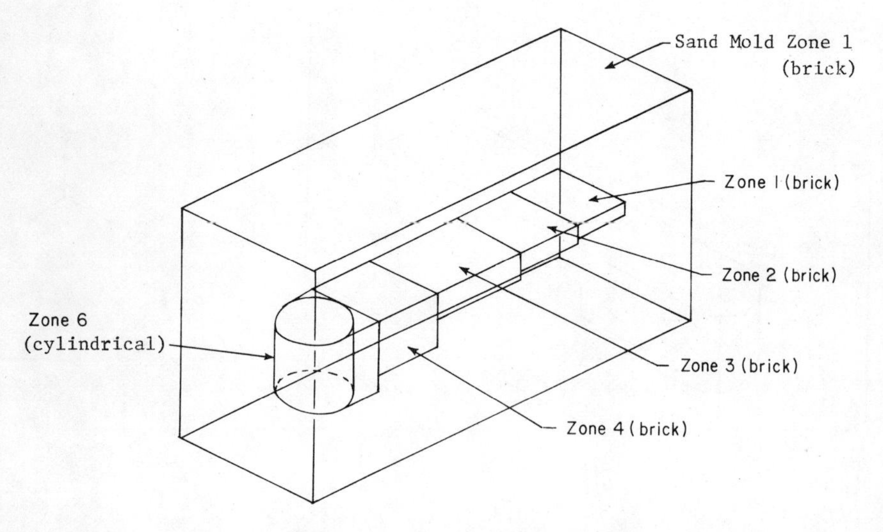

Figure 2 - Divide the model to zones.

2 - Input the Data to the Computer Program CSTMS3

Having prepared the geometrical data, the user is now ready to enter data to the program, using the following steps:

1 - Start up the computer (PDP/11 minicomputer)

2 - Run the program by entering the underlined material below. The computer will answer in the fashion indicated by the nonunderlined upper case letters.

(A) RU CSTMS3 <CR>

WELCOME TO PROGRAM CSTMS3, FOR AUTOMATIC MESH GENERATION.

(B) PLEASE DEFINE THE W.V.

W.V. IS X-Y <CR>

PLEASE DEFINE THE A.V.

A.V. IS Y-Z <CR>

(C) PROGRAM IS READY TO ACCEPT GEOMETRICAL INPUT DATA.

At this stage the program is ready to accept the input data defining the casting model. The program will proceed:

DO YOU HAVE A MASTER FILE? (Y/N)

The program is asking if you have a file containing all the geometry information defining the model. After every run the program will automatically generate a master file for subsequent use. If user enters,

Y <CR>

the program will respond

ENTER THE NAME OF MASTER FILE.

User should now enter the name, e.g.,

TEST .001 <CR>

If user does not have, or want to create a new master file he should enter

N <CR>

ENTER A NAME FOR THE MASTER FILE TO BE GENERATED

TEST .001 <CR>

READY FOR DATA DEFINITION

O-DONE, 1-BRICK, 2-CYLINDER, 3-FLASK

YOUR CHOICE

The user should choose from the number one to three, e.g. (test model of Figure 2).

1 <CR>

BRICK ZONE NUMBER [1]

(The number between brackets is a default value; if the user wants to use it he simply hits <CR>; if he does not, he can enter any other brick zone number.)

<CR>

ENTRY OF DATA FROM A FILE OR KEYBOARD [F/K]

(The X, Y, Z coordinates, defining the different zones, could be entered to the program via keyboard or through a predefined data file.)

If the user enters

F <CR>

program proceeds

ENTER NAME OF DATA FILE

e.g., ZONE1.DAT <CR>

If the user enters

K <CR>

program proceeds

ENTER X, Y, Z COORDINATES DEFINING ZONE

NUMBER 1.

The user then enters eight X, Y, Z coordinates to define Brick Zone 1 as he prepared in Step (1-B).

The user finishes defining all the zones (brick, cylinders, and sand mold) then enters

0 <CR>

to indicate DONE option.

The program will immediately display a "wire frame" diagram for the different zones on the CRT graphics terminal, as shown in Figure 3.

The program then proceeds:

ARE YOU SATISFIED WITH YOUR MODEL?

This question actually indicates the beginning of the "Design Session", i.e., updating the model, modifying, deleting, adding, rotating, scaling, . . . etc.

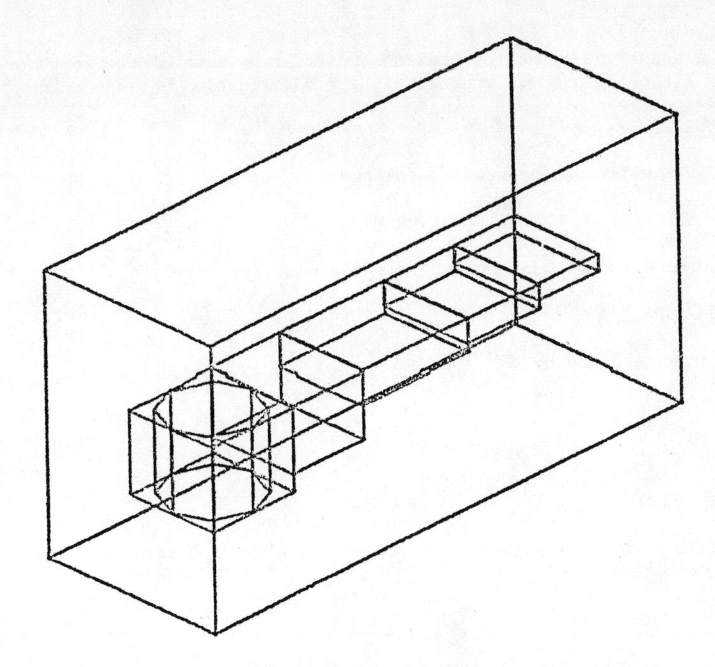

Figure 3 - Typical output on CRT for the input test casting model.

If the user is satisfied with his model, he simply enters

<u>YES</u> <CR>

If the user wants to change any geometrical data (interactively), he enters

<u>NO</u> <CR>

The program then asks for a change command

COMMAND -

User enters (in the same line) his modification command, e.g.,

COMMAND - <u>DELETE</u> <CR>

POSITION THE LIGHT PEN AT ANY PART OF THE ZONE TO BE DELETED

, the program asks. The user is required to hit any part of the zone he wants to delete. Program immediately deletes the hit zone. Other design commands are available to the user, e.g.,

ROTATE for rotating any zone or the whole model through a specified angle about a specified axis

SCALE for scaling any particulate zone or the whole model along the X, Y, or Z directions, or overall scaling.

Figure 4 shows the effect of scaling and rotating some of the zones of Figure 3. After the user is done with shape modification, he can enter

DONE <CR>

The program will then proceed as follows:

DO YOU WANT TO GENERATE THE MESH?

If the user does not want to generate the mesh, he enters

NO <CR>

The program responds by

ARE YOU SURE?

User enters

YES <CR>

Program stops at this point. Now, if the user wants to generate the mesh, he enters

YES <CR>

The program then asks for an output file name into which to load the generated mesh.

ENTER NAME OF OUTPUT FILE

User enters any name, e.g.,

OTFIL .001 <CR>

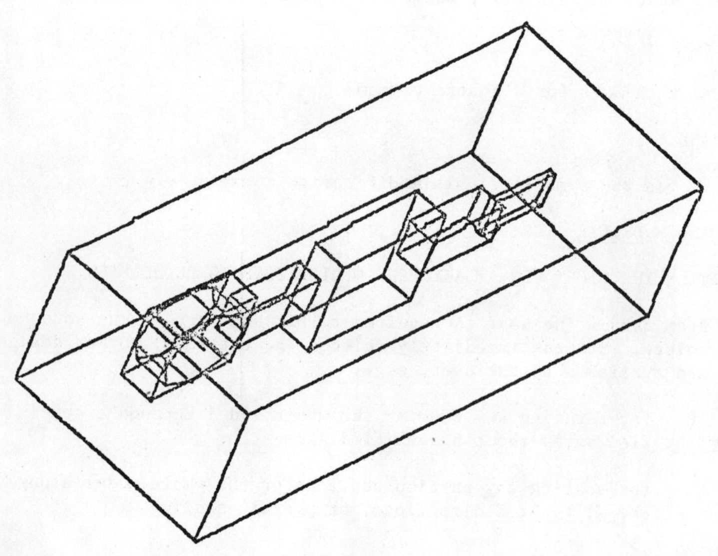

Figure 4 - Scaling and rotating different zones.

The program will automatically generate the mesh without any interaction from the user. The program will identify metallic and sand elements and wedge and brick elements, number the nodes of each element, and generate the X, Y, Z coordinates of each node for every layer. Figure 5 shows a 3-D plot for one of the layers as shown on the CRT graphics terminal.

Acknowledgment

This study, part of an overall CAD-CAM process for designing molds, is based on a project sponsored by the U.S. Army Tank Automotive Research and Development Command, Warren, Michigan, under Contract DAAK30-78-0020. The authors wish to thank the CAD-CAM program engineer, Dr. H. Brody (University of Pittsburgh) and Dr. N. Akgerman (Battelle program manager) for their review and constructive suggestions.

References

(1) Zienkiewicz, O. C., "The Finite Element Method in Engineering Science", McGraw Hill, 1971.

(2) Gallagher, R. H., "Finite Element Analysis Fundamentals", Prentice Hall, 1975.

(3) Hansteen, O. E., "Finite Element Methods as Applications of Variational Principles", published in "Finite Element Methods in Stress Analysis", by I. Holand and K. Bell, TAPIR, 1969, pp 451-473.

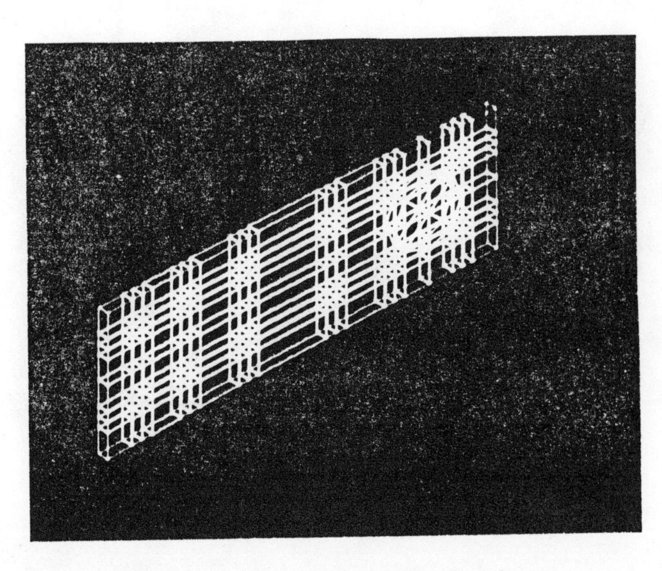

Figure 5 - 3-D plot for Layer No. 3.

Future Directions and Needs

AUTOMATED WELDING - RESEARCH NEEDS

T. W. Eagar
Department of Materials Science and Engineering
Massachusetts Institute of Technology
Cambridge, Massachusetts 02139

Increased use of automated welding in recent years has demonstrated deficiencies in our understanding of the factors controlling the geometry of the weld pool. In the present paper, all fusion welding processes are categorized in terms of the welding power density delivered to the surface of the workpiece. It is shown that there are three regimes of power density in which different factors control the weld shape. In the low heat density regime, characterized by oxyacetylene and electroslag welding, convection within the metal and slag pools controls the heat transport. In the medium heat density regime, characterized by most arc welding processes, a number of phenomena have been shown to have an influence on weld pool shape; however, no formalism has been developed which allows *a priori* estimation of the weld geometry. In the high heat density regime, characterized by electron beam and laser welding, metal evaporation is known to be important, but little has been done to quantify the effects of this evaporation. In each regime, but particularly in the high and medium heat density regimes, the shape of the weld pool is strongly dependent upon the composition of the base metal. It is argued that a better understanding of these heat source-material interactions is necessary in order to develop reliable automated welding techniques.

Introduction

During the past decade, the use of automated welding has grown dramatically. In the United States, the sales of electrodes for automatic and semi-automatic welding have grown from 103 million kilograms per year (28% of the total) to 592 million kilograms (49% of the total) (1). With continued need for improved productivity and reliability, this growth, at least in percentage terms, can be expected to continue.

In order to take full advantage of the process, any automated welding system should be reliable and reproducible. It is also desirable that the control technique be simple, inexpensive and rugged enough to operate in an industrial environment. One of the purposes of this paper is to demonstrate that many of our current welding processes do not provide adequate levels of reliability and reproducibility, and that these deficiencies are largely due to an incomplete understanding of the phenonema controlling the size and shape of the weld pool.

Many charts have been produced which categorize the various welding processes. To this list might be added the categorization given in Figure 1, which divides the processes by the heat density produced at the surface of the workpiece. It is felt that this division provides a useful starting point for discussion of heat transfer and pool geometry during welding.

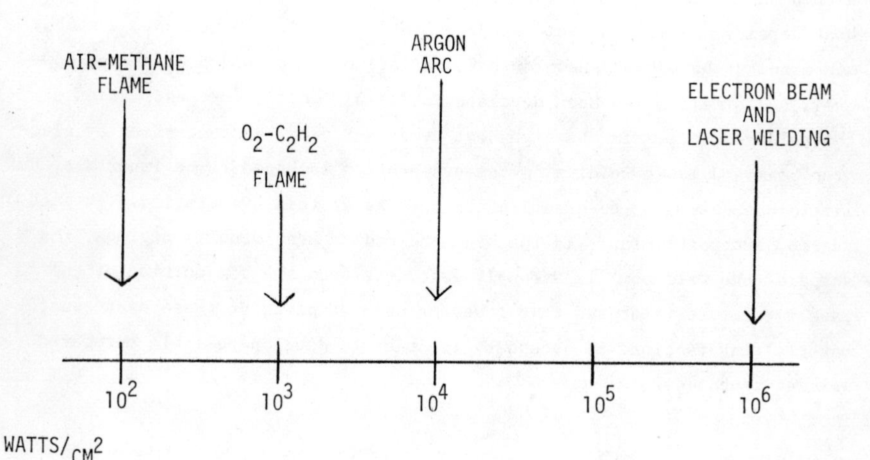

Figure 1 - Range of surface heat fluxes which are useful for fusion welding.

It should be noted that all fusion welding processes must lie between approximately 10^3 watts/cm^2 and 10^6 watts/cm^2. Below the former value, heat is conducted through the solid rapidly enough to prevent melting on the surface, while above 10^6 watts/cm^2, the metal is evaporated so violently that any molten pool is disintegrated, producing a separation rather than a fusion of metal.

On the basis of the categorization given above, three heat density regimes may be defined, viz.

Low heat density - approximately 10^3 watts/cm^2, including processes such as oxyacetylene and electroslag welding.

Medium heat density - approximately 10^4 watts/cm^2, including most arc welding processes.

High heat density - approximately 10^6 watts/cm^2, including laser and electron beam welding.

It is of interest to note that the cost of achieving high heat densities at the surface of a workpiece is in rough proportion to the heat density achieved. For example oxyacetylene welding equipment can be obtained for around 10^3 dollars, arc welding equipment for 10^4 dollars and laser and electron beam equipment for 10^6 dollars. As is also well known, the interaction time, i.e. the inverse of the speed of welding, decreases with increasing heat density; however this relationship is not linear.

In subsequent sections the formation of the weld pool in each of these heat density regimes will be examined.

<u>Low Heat Density Regime</u> (10^3 watts/cm^2)

Oxyacetylene and electroslag welding are the most important welding processes in the low heat input regime. The oxyacetylene process is generally not automated while the electroslag process is inherently automated. Both processes produce large heat affected zones, primarily due to the long interaction times required to produce melting of the base metal when using low heat density processes.

Modeling of the electroslag welding process is reasonably sophisticated, (2, 3) with current models providing good agreement with experimental data. Perhaps the greatest need today is for better sensors for control of the process. A number of defective electroslag weldments have been found in recent years, particularly in highway bridges. These defects resulted more from inadequate process control rather than insufficient scientific knowledge of the process.

The low heat density and the longer interactions times encountered in oxyacetylene and electroslag welding as compared with arc, laser or electron beam welding, make the former processes easier to control and to model theoretically. Heat transfer and melting in the low heat density regime primarily involves a knowledge of the thermal boundary conditions and fluid flow in the gas and the liquid phases. While our knowledge of these conditions remains imperfect, it is much better than it is with higher heat density welding processes.

Medium Heat Density Regime (10^4 watts/cm^2)

From a technological point of view, the medium heat density welding regime is the most important as it includes the arc processes, which include the majority of all welded parts. Unfortunately, this is the regime where heat transfer and metal pool formation are most poorly understood. At present, there is no useful technique for predicting the size or shape of the weld pool. Indeed, it has been found that minor variations in composition may produce very large changes in melting behavior. These variations have caused considerable difficulty in a number of automated welding processes.

Any study of heat transfer in welding arcs must account for the fluid flow which occurs both in the plasma and in the molten weld pool. Fluid flow within the plasma is produced by the electromagnetic pressure gradient which results from the spatially non-uniform current within the arc. (4) This pressure gradient produces a plasma jet which may reach velocities of 100 to 600 meters per second and exerts a force on the weld pool of several grams (5) as shown in Figure 2. This jet may depress the surface of the weld pool by several millimeters. Such weld pool surface depressions have been observed directly by Thompson (6). A photograph from his work for a 200 ampere gas-tungsten arc on HY-130 steel is shown in Figure 3. It is obvious from Figures 2 and 3, that depression of the weld pool surface during welding may alter the profile of the resulting weld. Friedman has also shown this analytically. (7)

The strong plasma jet developed by the electromagnetic pressure also disrupts the shielding gas flow pattern as shown by the excellent experimental work of Okada et al. (8) Their data show that a 20 ℓ/min argon gas flow which provides an inert shield out to 12 mm radius at zero weld current, provides only 4 mm of protection at 500 amperes. Others have shown

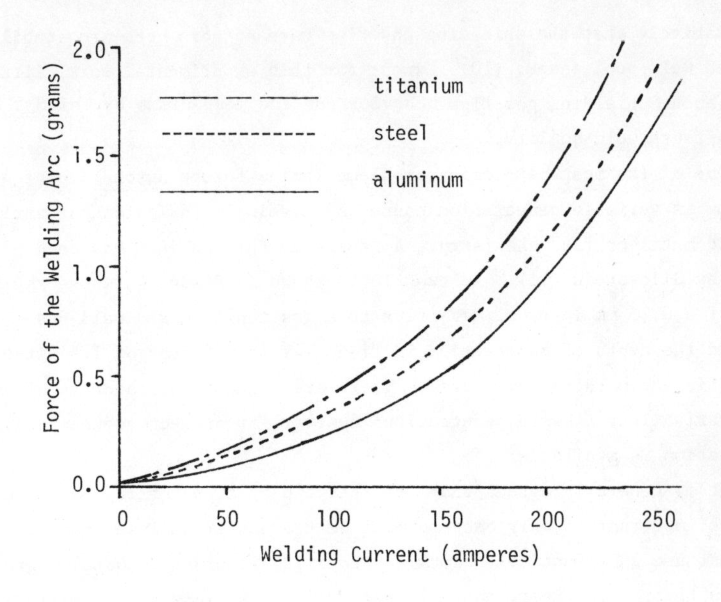

Figure 2 - Force of welding arc on surface of weld plate due to fluid flow within the plasma. After Burleigh (5)

Figure 3 - Photograph of gas tungsten arc weld on steel baseplate showing the depression of the weld pool surface due to the arc forces. The original photograph was taken with color film using a red filter to exclude much of the arc light. After Thompson (6)

qualitatively that the shielding gas flow rate affects both arc stability (9) and weld pool shape. (10) Except for this experimental work, little is known about shielding gas flow behavior and the mechanisms by which the flow affects weld quality.

One of the most perplexing problems facing future automation of arc welding is variable penetration caused by seemingly insignificant variations in weld metal or flux chemistry. A review of this subject has been provided by Glickstein (11). An example is shown in Figure 4, where the addition of a 0.25 mm layer of MgF_2 flux to a gas tungsten arc weld has increased the depth of penetration in Ti-6Al-4V by a factor of 2.5 without a change in the welding parameters. There are a number of theories given for this variability of weld penetration; however, at present no conclusive explanation is available.

It is generally agreed that convective heat transfer in the molten weld pool is important if only because most heat transfer analyses for arc welding must use an effective thermal conductivity in order to obtain agreement between theory and experiment. Indeed, Mills has claimed that variable penetration as described above, is due to changing convection patterns within the weld pool (12). Nonetheless, none of the current theories and few of the experimental observations can explain the shape of weld pools such as

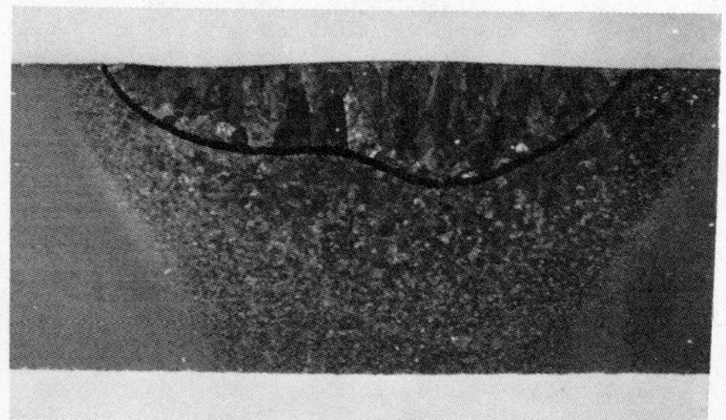

Figure 4(a)

Figure 4(b)

Figure 4 - Gas tungsten arc welds made in 4.7 mm thick Ti-6Aℓ-4V at 170 amperes and 12.7 cm/minute travel speed with a 2 mm arc length. Weld (a) was made with argon shielding gas alone, while weld (b) was made with argon plus 0.25 mm of MgF_2 spread on the surface of the plate.

Figure 4 (b) in which the depth of the weld pool is more than one half of the width. Consider a point source of heat as in Figure 5 (a). In the absence of radiation and convection, the melting isotherm will be a semicircle. In no case can the depth to width ratio exceed the value 0.5. No theory which uses a constant value of effective thermal conductivity can account for weld pools shaped such as Figure 4 (b). Any theory of heat transfer within arc weld pools should use true values of the local enhancement of thermal conductivity due to convection. Since such data is extremely difficult if not impossible to obtain experimentally, it must be calculated. This is a feasible, albeit difficult undertaking; however, the insight gained through such a study should add greatly to our understanding of the welding process.

It is known that evaporation of metal vapor occurs in both the medium and high heat density regimes of welding. (13) During arc welding this metal vapor enters the plasma causing significant changes in the temperature, thermal conductivity and current distribution in the arc. (14) Changes in these properties may have a direct influence on heat transport to the surface of the weld pool as well as the strength of the plasma jets. Klueppel et al. have shown that plasma impurities are distributed through the arc region on a timescale of microseconds. (15) Clearly, most of our previous welding research has failed to study changes which occur on such a brief timescale, even though there is ample evidence suggesting the importance of these phenomena!

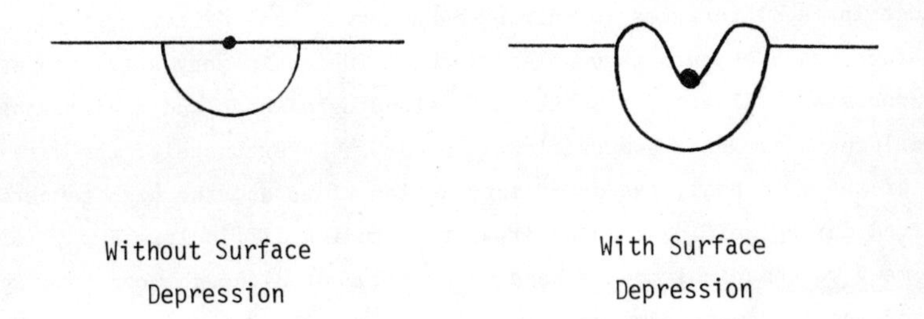

Figure 5 - Schematic of weld penetration produced by a point source of heat with and without surface depression. Welds with depth to width rations much greater than 0.5 (cf. Figure 4b) can only be rationalized in terms of weld pool surface depression and/or well directed convective heat flows in the molten metal pool.

High Heat Density Regime (10^6 watts/cm^2)

As noted previously, metal vaporization becomes important in the medium heat density regime, primarily because small variations in plasma composition can produce large differences in the properties of the welding arc. At high heat densities, as produced by lasers and electron beams, vaporization becomes intense, producing local pressures of several atmospheres. (16) These pressures displace the molten metal, producing a drilling effect which in turn produces the deep narrow welds characteristic of the high heat density regime. Even though the entire surface is vaporized, the composition of the base metal remains an important variable in both electron beam (17) and in laser welding (18). In both cases the penetration of the beam is strongly dependent on small changes in the base metal composition. For example, laser welding aluminum alloy 5083 produces a thirty percent greater penetration than alloy 5456 at equal operationg conditions. (18) Such differences cannot be attributed to the thermal properties of the materials. Shauer and Giedt have shown that composition differences may influence the spiking tendency (19) and temperature distribution (20) in electron beam welds. Most theories and models of electron beam and laser welding have not addressed the problem of how the metal composition interacts with the heat source to influence the shape of the weld pool.

Future Research Needs

In the preceeding sections it has been shown that a number of poorly characterized phenomena are known to have a significant influence on the size and shape of the weld pool. This is particularly true in the medium and high heat density regimes where metal vaporization occurs. In many cases, compositional variations within the commonly specified limits may produce large differences in melting behavior. These differences have led to defects such as burn through and lack of fusion in many automated welding processes. Clearly, a better understanding of heat and fluid transport as influenced by metal composition is needed. Unfortunately, the very small size of the weld pool, the short interaction times and the high temperatures achieved during welding make experimental studies difficult. For this reason, many mathematical models have been produced, although most have ignored many of the phenomena discussed in this paper. Progress will not be made until both better experimental boundary conditions are available for input to the model and the models are developed to include the true physical phenomena controlling the processes. Until such joining of experiment and theory is made, automated welding, while continuing to grow, will not achieve the degree of reliability and reproductability that is needed.

Acknowledgements

The author expresses his appreciation to Dr. J. E. Anderson who first introduced him to the classification scheme presented here.

References

1. T. B. Jefferson, "Another Record Year for Welding Sales", Weld. Des. & Fabr., 53 (7), 1980, p. 88.

2. T. Deb Roy, J. Szekely and T. W. Eagar, "Temperature Profiles, the Size of the Heat Affected Zone and Dilution in Electroslag Welding", submitted to Met. Trans.

3. A. L. Liby, G. P. Martins and D. L. Olsen, "Simulation of Unsteady Heat Flow in Vertical Electroslag Welding", presented at this conference.

4. C. W. Chang, T. W. Eagar and J. Szekely, "The Modelling of Gas Velocity Fields in Welding Arcs", Proceedings of the conference on Arc Physics and Weld Pool Behavior, The Welding Institute, London, May, 1979.

5. T. D. Burleigh, S. M. Thesis, Dept. of Materials Science and Engineering, Mass. Inst. of Technology, Cambridge, MA, September, 1980.

6. R. B. Thompson, S. B. Thesis, Department of Mechanical Engineering, Mass. Inst. of Technology, Cambridge, MA, June, 1980.

7. E. Friedman, "Analysis of Weld Puddle Distortion and Its Effect on Penetration", Weld J., 57 (6), 1978, p. 161-s.

8. T. Okada, H. Yamamoto and S. Harada, "Observation of the Shielding Gas Flow Pattern During Arcing by the Use of a Laser Light Source", Arc Physics and Weld Pool Behavior, op. cit. p. 203.

9. C. J. Allum, B. E. Pinford and J. H. Nixon, "Some Effects of Shielding Gas Flow on Argon-Tungsten Arcs Operating in High Pressure (1 to 14 Bars, abs.) Environments", Weld J., 59 (7), 1980, p. 199-s.

10. W. R. Reichlet, J. W. Evancho and M. G. Hoy, "Effects of Shielding Gas on Gas Metal Arc Welding Aluminum", Weld J., 59 (5), 1980, p. 147-s.

11. S. Glickstein and W. Yeniscavich", A Review of Minor Element Effects on the Welding Arc and Weld Penetration", Weld Res. Council Bulletin, No. 226, May, 1977.

12. G. S. Mills, "Fundamental Mechanisms of Penetration in GTA Welding", Weld J., 58 (1), 1979, p. 21-s.

13. J. D. Cobine and E. E. Burger, "Analysis of Electrode Phenomena in the High Current Arc", J. Appl. Phys., 26 (7), 1955, p. 895.

14. S. S. Glickstein, "Arc Modelling for Welding Analysis", Arc Phys. and Weld Pool Behavior, op. cit., p. 1.

15. R. J. Klueppel, D. M. Coleman, W. S. Eaton, S. A. Goldstein, R. D. Sacks and J. P. Walters, "A Spectrometer for Time-gated, Spatially Resolved Study of Repetitive Electrical Discharges", Spectrochimica Acta, 33B, 1978, p. 1-30.

16. J. E. Anderson and J. E. Jackson, "Theory and Application of Pulsed Laser Welding", Weld J., 44 (12), 1965, p. 1018.

17. C. M. Weber, private communication, Babcock and Wilcox Co., Alliance, Ohio, 1980.

18. D. B. Snow, M. J. Kaufman, C. M. Banas, and E. M. Breinan, "Evaluation of Basic Laser Welding Capabilities", Final Report Contract N00014-74-C-0423, United Technologies Research Center Report R79-911989-17.

19. D. A. Schauer and W. H. Giedt, "Prediction of Electron Beam Welding Spiking Tendency", Weld J., 57 (7), 1978, p. 189-s.

20. D. A. Schauer, W. H. Giedt and S. M. Shintaku, "Electron Beam Welding Cavity Temperature Distribution in Pure Metals and Alloys", Weld J., 57 (5), 1978, p. 127-s.

THE SIMULATION OF HEAT TRANSFER IN

CASTINGS AND WELDMENTS--SOME THOUGHTS OF NEEDED RESEARCH

Preben N. Hansen* and John T. Berry+

*Associate Professor,
Danish Technical University
Lyngby, Denmark
(Currently Visiting Scholar,
School of Mechanical Engineering,
Georgia Institute of Technology)

+Professor of Mechanical Engineering
Georgia Institute of Technology
Atlanta, Georgia

It is now eighteen years since the first recorded use of the digital computer in studying solidification in castings [1]. In the case of the numerical analysis of the heating and cooling of weldments using the computer, many years have also elapsed from the initial usage in 1968 [2]. Although a considerable number of publications have appeared in this field since these dates, the widespread implementation of this potentially powerful problem-solving technique in the industries concerned has been slow, being confined mainly to high value and/or high technology type products.

Although the principal rate controlling factors in implementation have often been economics or safety-related, many other factors have contributed to this aspect.

Amongst the areas of needed research upon which effective implementation hinges are such topics as:

- Availability of comprehensive, fully interactive geometric modelers or parts description schemes.
- Development of effective low-cost linkages between such modelers and existing general purpose FEM/FDM codes.
- Possession of knowledge of the extent of convective effects in molten and solidifying metal, especially during pouring and filling complex cavities prior to solidification of castings or in the weld pool during joining operations.
- Existence of convenient compact methods for describing the effective flux at the special interfaces, for example the mold-metal interface capable of describing the current generation, as well as traditional types of mold materials.
- Provision of information describing the extent of dilatational

phenomena and related thermal properties in the various media concerned with casting and welding.

° Availability of criteria functions which describe desirable solidification parameters needed to optimize microstructure, soundness, ensure freedom from hot tearing/cracking phenomena, etc.

1. Availability of Comprehensive Geometric Modeling Schemes.

This topic is discussed elsewhere in depth later in these proceedings [3], but let it suffice here to say that although much progress has been made in the last ten years in the area of modeling, much remains to be done in linking such schemes with the various computational technologies. In the present paper, the special problems of this interface will be discussed in the next section.

2. Development of a Low-Cost Linkage Between Geometric Modeling and Thermal Computation Codes.

As will be seen from the accompanying paper prepared by the authors' colleague and referred to above [3], the state of development of the art of modeling is considerably in advance of that concerned with the areas linking modeling and numerical computation. Although several two-dimensional enmeshment packages are available, as are schemes for enmeshment of three-dimensional wire-frame [4] or block models [5], a comprehensive three-dimensional enmeshment scheme in the public domain would be quite invaluable. It should be noted that such schemes have been developed in the private sector [6].

Concurrently, helpful refinements to existing FEM or FDM codes might also be undertaken. For example, in the FEM area more efficient routines for handling the liberation of latent heat might well be developed. In the case of FDM, codes which might be used with more generalized geometries are clearly needed. Some progress has been made in this area by Robertson [7] and more recently by the present authors and Wei [8]. Such approaches might eventually prove relatively inexpensive in comparison with FEM schemes. However, the difficulties to be surmounted are not inconsiderable and will not be solved overnight.

In addition to the above conventional techniques, it would appear highly desirable if alternative approximate methods could be devised, which could be utilized to give local solidification times for particular locations within a casting, somewhat along the lines of the Chvorinov rule (or casting modulus method). Knowing, for example, the geometry of the casting (established, perhaps, by a geometric modeling routine) the solidification time of a point P might be expressed as:

$$t_s = f_T \cdot f_p \qquad (1.)$$

where f_T is a constant dependent on the alloy used, the mold material, the casting temperature and the initial mold temperature. f_T can be determined experimentally, or obtained from comparison with more exact simulations. f_p would be a variable which expresses the cooling capacity of the surroundings with respect to the point P. To estimate f_p, appropriate techniques might be developed which from the known geometry of the casting and the placing of the point P relative to that geometry one could calculate how effectively the mold cools the point P. It is felt that this technique would incorporate the use of a space integral, where the distance from the point P to the free surface of the casting (the metal/mold interface) is one of the parameters to be integrated.

3. Possession of Knowledge of the Extent of Convective Effects in Molten and Solidifying Metals.

We are still some years away from a routine whereby the foundryman might use a mathematical model to solve simultaneously the equations of a) momentum, b) energy, and c) continuity to incorporate the effects of convection in the cooling and solidification characteristics of a truly three-dimensional casting. Nonetheless, work is currently in progress [9] which incorporates the effect of convection in the thermal behavior of a slab casting in a simple way in order to gain a basic understanding of the problem involved.

An alternative approach [8] has shown that reasonable simulations of the delivery of superheat may be obtained using a mathematical model which merely solves the equation of energy (Fourier's transient). In doing so it is customary to adjust the coefficient of thermal conductivity to obtain the same temperature profiles in the melt as given by the previously mentioned set of three equations. This may be approached utilizing temperature profiles given in the literature for the cases of free convection within simple mold cavity geometries and estimating the appropriate adjustment parameters.

4. Existence of Convenient Compact Methods for Describing the Effective Flux at an Interface, Such as that Between the Metal and the Mold

One of the main problems in truly three dimensional FEM/FDM simulations of the solidification and cooling of castings is the great number of volume elements needed to ensure accuracy in the result of the computations involved.

Since typically, the majority of these elements will be in the mold, it is obvious that great savings would be obtained if the cooling effect of the mold were to be described in a compact analytical way. The difference in results obtained with this proposed approach and that utilizing the mold elements would appear negligible [10]. Such a compact technique would obviously have to be extended to cover some of the newer molding media together with the many recently developed bonding agents referred to by the authors' colleagues, Hartley and Babcock in another paper in these procedures [11].

The problem of the chill-metal interface is, of course discussed elsewhere by Pehlke and Jeyarajan [12]. It will be seen on examining the literature that although some approximate theoretical models have been developed, we are still heavily dependent on empirical data. Extension of our knowledge of how the air gap forms, as well as the various modes of heat transfer during its development would seem to be of paramount importance if an exact theoretical model is to be achieved.

5. Provision of Information Describing the Extent of Dilatational Phenomena and Related Thermal Properties in the Various Media Concerned with Casting and Welding

From a thermal simulation point of view, dilatational phenomena inside solid bodies often play a relatively minor role. But the effect on the heat conductance between bodies can be quite predominant (air gap formation). When one considers crack formation (both hot and cold) this effect is seen as an important factor in both weld or casting simulation, as will be the partial contact of portions of welded structures during and immediately after the fabrication of complex joining operations.

Again, as will be noticed with the collection of data on thermal conductivity and specific heats at high temperatures, there is a considerable dearth of data on density, thermal expansion, latent heats of transformation, high temperature strength and ductility--information of considerable importance in both casting and joining.

6. Availability of Criteria Functions which Describe Desirable Solidification Parameters.

Essentially, without such criteria functions the calculation of thermal history will not be of much help to the foundryman or the fabricator and the state of the art will not exceed the level where people compare measured and calculated temperature profiles.

Some encouraging work has been done so far. Davies [13] was able by simulation to connect the solidification mode to the actual feeding lengths in a casting and has computed feeding range data for several alloys which agree well with the empirical information available. Berry has been able to relate the speed of solidification and the temperature gradient ahead of the solidification front to the mechanical properties of the cast product [14]. Hansen [15] has also shown a semiquantitative correlation between the speed and geometry of the solidification front, the speed of contraction of constrained regions in a steel casting and the hot cracking susceptibility. Ebisu [16], also using a numerical simulation, has been able to successfully apply knowledge of themal parameters to exercise microstructure control in centrifugally cast stainless steels. Even so, much more work in this area is needed before we will be able to predict from calculated temperature or solidification profiles, the process parameters which will ensure a sound and crack free product with desired mechanical properties.

In summary, it appears that although the first ten or so years since the Fursund work [1] were typified by a relatively low level of interest in simulation, there has been a considerable upswing of activity in this area, especially if the present conference is to be regarded as being indicative of the norm. Hopefully, one will witness in the next five to ten years increasingly more implementation coming about and, at the same time, the development of the various data and adjuncts to the many available computer codes which will hasten the accomplishment of this goal.

Acknowledgements

The authors would like to acknowledge the support of the NSF Division of Applied Research Grant DAR 78-24301 which is making possible work on some of the applied and implementational research referred to in the paper.

References

1. K. Fursund, 'Giesserei', (14) 1962 51.

2. O. Westby, "Temperature Distribution in the Workpiece by Welding", Inst. for Mek. Tek., Norges Tekniske Hogskole, Trondheim Norway.

3. M. R. Corley, "Geometric Modeling: A Status Report," p. 467 of this volume.

4. FLING package, Lockheed of Georgia Company, 1979.

5. A. Badaway, "CSTMS III, A Finite Element Mesh Generator for Castings," p. 475 of this volume.

6. Private Communication, Shape Data, Ltd., January 1980.

7. S. R. Robertson, "A Finite Difference Formulation of the Equation of Heat Conduction in Generalized Coordinates", Numerical Heat Transfer (2) 1979 61-80.

8. P. N. Hansen, C.-S. Wei, and J. T. Berry unpublished work, supported by NSF Grant DAR 78-24301.

9. P. V. Desai and F. Rastegar, "Convection Currents in Mold Cavities," p. 351 of this volume.

10. C.-S. Wei, P.N. Hansen and J. T. Berry, "The Q Method - A Compact Technique for Describing the Heat Flux Presented at the Mold-Metal Interface in Solidification Problems," Proceedings of Int. Conf. on Numerical Methods in Thermal Problems, Venice June 1981.

11. J. G. Hartley and D. Babcock, "Thermal Properties of Mold Materials," p. 83 of this volume.

12. K. Ho, A. J. Jeyarajan, and R. D. Pehlke, "Determination of Boundary Conditions at the Metal Mold Interface During Solidification of Castings"--presented at this conference.

13. V. de L. Davies, "Feeding Range Determination by Numerically Computed Heat Distribution", AFS Cast Metals Research Journal (11) 1975 33-44.

14. J. T. Berry, "Some Analytical Aspects of Designing Production Methods for Casting Soundness", Proceedings of 1st Cairo University Conference on Mechanical Design and Production, December 1979.

15. P. N. Hansen, Ph.D. thesis part 2, 1975, Department of Metallurgy, The Technical University of Denmark.

16. Y. Ebisu,"Computer Simulations on the Macrostructures in Centrifugal Castings", Transactions of the AFS, Vol. 85, 1977, PP 643-654.

STATUS OF MODELING FOR SHAPED CASTINGS

T. S. Piwonka

TRW Materials Technology
23555 Euclid Avenue
Cleveland, Ohio 44117

INTRODUCTION

An examination of the papers presented at this conference leads to the conclusion work will be required to bring the technique to the point where it is a production tool. It is the purpose of this paper to consider some of the areas where progress has been made as well as those where more work is needed.

Objective of Modeling

For the commercial foundry the objective of the modeling effort is to develop a computational technique which will:

1. Design gating and feeding systems for castings. This is the heart of the problem as the quality of the casting as well as the economics of its production depend directly on the efficiency of the gating and feeding system.

2. Specify optimum metal and mold temperatures. Control of metal pour and mold preheat temperatures is desirable to minimize melting costs, dross formation, misrun, and other casting defects.

3. Predict residual stresses and hot tearing. Elimination of hot tears and cold cracks is essential for reduced scrap rates. Knowledge of residual stress distribution will improve casting dimensional accuracy.

4. Predict post-casting processing and casting properties. Correlation of cooling rates with dendrite arm spacing means that heat treatment requirements can be predicted. Residual porosity which may exist in certain casting design may be closed by hot isostatic pressing. Finally, the relationship between casting structure and properties allows a prediction of the mechanical properties at each point in the casting, an invaluable aid to designers.

5. Provide the basis for plant scheduling and production planning. The exact knowledge of pour weights, pattern layout and flask design permits scheduling to be done so as to mimimize production costs.

Ideally the model will be accurate enough to optimize casting conditions so that scrap and repair costs arising from process design deficiencies are eliminated. It will also be inexpensive enough to apply that it will replace the steadily diminishing supply of "foundrymen" (those who design gating systems for castings, as opposed to "foundry engineers," who design material handling equipment and plant layout). It will be applied in an interactive mode, allowing foundrymen to alter casting parameters and gate and riser locations and sizes to try various combinations to improve casting process design.

It is clear that developing the models to this degree of sophistication will be a challenging and arduous task. Nevertheless, the work presented at this conference has begun to address the problems that must be solved.

Definitions

The normal way in which complex problems are approached is to simplify them to the point where they may be handled economically. Thus "castings" are treated as simple shapes initially, or combinations of simple shapes, despite the fact that simple shapes are rarely cast commercially. For this reason it is important to remember what really constitute the "casting" and the "mold."

The "casting" is the space filled with molten metal during pouring, and consists of the cast part itself plus the gates plus the runners plus the sprue plus the risers plus all other castings in the mold. Each of these compoennts affects the solidification of those components adjacent to it, and in turn is affected by them. These near neighbor effects may in many cases be negligible, but will often be of great importance in fine tuning a gating system.

The "mold" is all the material surrounding the molten metal. It is comprised of the mold material (or materials, if more than one are used) including mold coatings plus cores plus chills plus insulation plus exothermic compounds plus the surroundings (in the case of open risers). It may also include the flask itself when castings are made with shell molds.

The model then, to be complete, must recognize all of the components of the mold/casting system. This does not imply that models which neglect certain features may not be useful; indeed experience shows that large quantities of useful information may be obtained from simplified models, as demonstrated by Riegger (1), Grant (2) and Larson (3).

Models are constructed by mathematical techniques which describe heat flow and fluid flow for the casting. Computational methods are used to determine numerical solutions which indicate the progression of solidification, temperature distributions, and other information. Since both the disciplines of heat flow and fluid flow are involved, it is appropriate to examine each.

Heat Flow

It is apparent that the ability to model heat flow is well developed. As indicated by Patantar (4) and Jeyarajan, Jechura and Wilkes (5) there is a solid understanding of the techniques to be used for bulk calculations and at interfaces (6). As noted above, these techniques are being applied to real castings and useful results are being obtained. The models have reached the point where it is no longer necessary to ignore radiation losses and interfacial contact resistance when they materially affect the course of solidification. It also appears that both finite element and finite difference methods can be used successfully (5), very much at the discretion of the modeler, although each technique continues to have its partisans. There are, however, a number of areas where much work remains to be done:

1. Mesh Generation. The generation of the mesh used for the model is critical to the success of the model. As pointed out in the paper by Badawy et al. (7), the way the mesh is selected affects the numerical results. Furthermore, it is a task that people do well, but computers do poorly, as noted by Corley (8). It is therefore no surprise that at present, the best way of generating a mesh is to have a skilled foundryman (i.e., one who has an idea of how the casting will solidify) do the task, as recommended by Stoehr (9). Unfortunately, this requirement effectively guarantees that no commercial foundry will apply computer modeling to its castings, because the time that the foundryman spends generating the mesh could be equally well spent designing the gating system without using modeling techniques.

   However, this should not be allowed to deter attempts to develop automatic mesh generation techniques. As a first step, the use of interactive mesh generation between the comptuer and the foundryman may be a place to start. The computer program could aribitrarily generate a primitive mesh in the casting and mold which could than be corrected as necessary. Eventually it is possible that a program might be written in which the computer could recognize areas which require finer or coarser meshes.

2. Geometric Modeling. This is another area in which modeling techniques are currently insufficient. As Corley (8) noted current geometric modeling systems lack the ability to include draft angles and fillet radii - both essential to the generation of accurate models for casting. Furthermore, present geometric models do not include the mold, as such a task complicates the problem immensely. In addition, there is no way for for a geometric model to include information on the quality level desired in the final product. Thus calculations must assume that a perfect casting must be produced, whereas in actuality quality requirements of most castings vary throughout the part as a function of the local design stresses.

   While there appears to be no immediate solution to the deficiencies of current geometric modeling systems, it may be noted that there has been little attempt to combine them with Group Technology techniques. Such a marriage may prove fruitful and should be considered.

3. Data Base. Numerical computations of the models require that accurate values of physical constants for molds and alloys be available. Unfortunately, much of the data that is required is unavailable or lacks the accuracy to be useful in the models. While the present NSF effort (10) is to be applauded, the work must be extended to include mold materials other than sand, and the properties of the great variety of alloys. The national laboratories, facilitized as they are with excellent equipment, may be the logical place to make the measurements which are required.

Despite the above list of areas where more work is needed, heat flow modeling is well developed, as is reflected by its growing use in solving foundry problems.

Fluid Flow

Unfortunately, the situation regarding fluid flow is somewhat less encouraging. Modeling fluid flow is far more complex than modeling heat flow, as was demonstrated by Szekely (11) and Desai and Rastegar (12). It appears that a decade of intensive work may be required before a satisfactory model can be built. There are, however, encouraging signs.

On the dendrite level fluid effects during solidification are well understood as shown by the papers given by Ridder, Mehrabian and Kou (13), Petrakis, Poirier and

Flemings (14) and Lalli (15). This work will be valuable when it is married with the macro fluid flow models which are being developed, as calculations may then be made to predict dendrite arm spacings and the degree of segregation to be expected in each area of a casting. This information can in turn be used to develop optimal heat treatments which maximize properties and minimize energy utilization.

There is, however, one area which is receiving very little attention. It is the forced convection that arises from the pouring and feeding operation. This forced convection is crucial to the design of a good gating system in small and medium sized castings. It affects oxidation of the molten metal, dross location in the final casting, and may affect grain nucleation. Experienced foundrymen consider this forced convection carefully in their design of gating systems. No model which ignores it can hope to be wholly successful. The problem, however, is tremendously difficult to solve using current techniques, as a way must be found to account for the time-varying entrance velocity of the metal into the casting cavity (time varying because of the fluid dynamics of the gating system - location and size of gates, runners, chokes, sprues, etc. - and the varying height of the metal in the casting cavity as the cavity fills).

To obtain an adequate model it may be necessary to employ a number of novel approaches instead of attempting to use the exact solutions which arise from rigorous application of fluid flow equations and momentum and mass balances. In this regard the recently published work of St. John, Davis and Magny (16) is instructive: in modeling of fluid flow in a gating system these authors used equations derived originally for the breaking of a dam in a dry river bed. This type of ingenuity will be required if meaningful models are to be built.

After the development of models which include forced covection from pouring operations it will be necessary to develop models for solidification of "thin wall" castings, "thin wall" meaning those castings where solidification begins before pouring is complete. Because this does not normally represent a desirable situation as it indicates a high probability of casting misrun and non-fill, it must be avoided unless the gating system can be arranged to permit the establishment of strongly progressive solidification by this mechanism, thus assuring high casting soundness.

Conclusions

The progress that has been made in modeling solidificaition processes is evident. Before the models will be in general use in the foundry community, however, extensive work will be required. This is especially true in the areas of mesh generation, geometric modeling, data base establishment, fluid flow in the gating system, and the effect of forced convection during solidification. As these techniques emerge, and as the calculating and display powers of computers improve, the widespread use of models may be expected to improve the quality of castings throughout the industry.

REFERENCES

1. O. K. Riegger, "Application of a Solidification Model to the Die Casting Process," p. 39 of this volume.

2. J. W. Grant, "Thermal Modeling of a Permanent Mold Casting Cycle," p. 19 of this volume.

3. H. R. Larson, "Use of Finite Element Analysis of Cast Railroad Wheels," paper presented at the "Modeling of Casting and Welding Processes" conference held at Rindge, N. H., August 3-8, 1980.

4. S. V. Patankar, "Numerical Techniques for Heat Transfer and Fluid Flow," paper presented at the "Modeling of Casting and Welding Processes" conference held at Rindge, N. H., August 3-8, 1980.

5. A. Jeyarajan and R. D. Pehlke, "Determination of Boundary Conditions at the Metal Mold Interface During Solidification of Casting," paper presented at the "Modeling of Casting and Welding Processes" conference held at Rindge, N. H., August 3-8, 1980.

6. K. Ho, A. Jeyarajan and R. D. Pehlke, "Determination of Boundary Conditions at the Metal Mold Interface During Solidification of Casting," paper presented at the "Modeling of Casting and Welding Processes" conference held at Rindge, N. H., August 3-8, 1980.

7. A. Badawy, K. Schreiber, J. Chevalier and T. Wassel, "CST MS3, A Finite Element Mesh Generator for Casting," p. 475 of this volume.

8. M. R. Corley, "Geometric Modeling: A Status Report," p. 467 of this volume.

9. R. A. Stoehr, "Simulations in the Design of Sand Casting," p. 3 of this volume.

10. J. G. Hartley and G. Babcock, "Thermal Properties of Mold Materials," p. 83 of this volume.

11. J. Szekely, "Fluid Flow Phenomena in Metals Casting," paper presented at the "Modeling of Casting and Welding Processes" conference held at Rindge, N. H., August 3-8, 1980.

12. P. V. Desai and F. Rastegar, "On Convection in Molds," p. 351 of this volume.

13. S. D. Ridder, R. Mehrabian and S. Kou, "A Review of Our Present Understanding of Macrosegregation in Axi-Symmetric Ingots," p. 261 of this volume.

14. D. N. Petrakis, D. R. Poirier and M. C. Flemings, "Some Effects of Forced Convection on Macrosegregation," p. 285 of this volume.

15. L. A. Lalli, "Effect of Dendrite Migration on Solute Redistribution," p. 425 of this volume.

16. D. H. St. John, K. G. Davis and J. G. Magny, "Computer Modeling and Testing of Metal Flow in Gating Systems," Report MRP/PMRL-80-12 (J), Canada Centre for Mineral and Energy Technology.

SUMMARY THOUGHTS ON MODELING

FOR THE PRODUCTION FOUNDRY

Lionel J. D. Sully

Doehler-Jarvis Castings
N L Industries, Inc.
Toledo, Ohio

The technology for modeling casting processes has existed for fifteen years or more. Recent developments in software, graphics, and computers have brought numerical modeling to a cost and time regime where the tool will be useful in the production foundry. Problems not solvable economically by numerical methods should look again at the analog approach. The computational capability differences between industry and the universities is presenting a growing problem.

A magnificent week! An opportunity, especially appreciated by those of us from the industrial world, to immerse oneself in the subject of process modeling; to renew old acquaintances; and to meet new people working in the field. A grateful word of thanks to Drs. Apelian and Brody for organizing the conference and to the Engineering Foundation *for its sponsorship.*

The subject was indeed ripe for such a forum. The variety of applications of computer modeling in the casting and welding processes was impressive in the range from research at the atomic level to production foundry practice for real castings. I will confine my remarks to this cast metals aspect of the conference, but the multi-disciplinary approaches taken by many contributors reinforces the fact that the modeling is not entirely a metallurgical problem.

The progress in the development and use of computers over the past twenty years is sobering. Further, the impact of continuing hardware development on the modeling task is difficult to appreciate. However, lest we become too ready to rest on our laurels, it is perhaps useful to take a cursory glance at the status of the field 15 to 20 years ago. I would pick one paper that can put the current status of modeling in perspective "Predicting Solidification Patterns in a Steel Valve Casting by Means of a Digital Computer" by J. G. Henzel, Jr. and J. Keverian, Metals Engineering Quart, ASM, May, 1965, Pages 39 to 44. Contained within this paper are all the elements of a good solidification model. In fact, I venture to say that the paper could have been presented at this current conference and not appear dated.

Thus, we have a paradox in that the general use of the modeling tool has not grown substantially in the metal casting industry, while the computer technology on which it depends has exploded in its capability. The same observation is not true of the mechanical engineering fields that have made wide use of the general purpose finite element methods available to them. The papers presented at this conference do provide insight on this paradox and offer clear indications that the time has come for a real impact for modeling in the production foundry.

The above reference clearly indicates that the general form for numerical modeling has been available for some years. However, the time required to first access the numerical method, and second to understand the results of the computation was, I am sure, excessive. Given the short timetable for production of a casting once a drawing is received from the customer, the time to obtain the results of a model is critical. The advent of computer graphics software and systems has now clearly eliminated this critical blockage to widespread use of modeling in production casting processes. The creation of a model can be accomplished in hours where several weeks were required in the past. Graphical output, especially in color, provides a ready understanding of thermal model data.

While time was one blockage to general use of models for production castings, cost was certainly another. Several developments are working to bring the cost of modeling within the practical range of todays economical costs. First the computer hardware costs are still dropping while the capability increases. Several authors reported work from minicomputer systems and the development in low cost microcomputers will make simulation on these machines very low in cost. Second, good software is available to run casting models. Government sources currently cost software at $10 to $15 per line. Again, the time factor combined with this cost make special purpose software entirely impractical for production use. In contrast, the specific needs of research often dictate such special software. Fortunately, the general purpose codes developed, primarily by aerospace industries in the 1960's provide a firm foundation for casting models. Optimization of the broad capabilities of these codes is an open area for cost improvement: optimization for both the problem and the machine on which it must be run.

The third cost factor that is working to foster the wide use of numerical modeling methods is the economic climate in which we now must work. The pressure to reduce the cost of production and optimize quality is making the traditional "cut and try" methods of problem solving entirely inadequate. The papers presented on modeling of production casting clearly demonstrated the ability of the model to allow the metallurgical and manufacturing engineer to "see" inside the casting and mold. This new vision permits examination of design strategies at a magnitude of cost of $1,000 rather than $100,000 for in plant experimentation with extensive die fixes, in my own experience. The link between cost savings in the plant to increased sophistication (and thus cost) in engineering is essential to justify wide use of modeling in the 1980's.

These remarks have been essentially confined to numerical modeling. The thermal problem is tractable by such methods for real geometry. However, the fluid flow methods do not exist for casting geometries at a cost one can afford, if at all. The analog model is often forgotten these days. The dramatic visual understanding of the Glicksman and Schaefer presentation of dendritic growth reminds us of the power of the analog model. The revival of the analog method of fluid flow simulation in complex geometries should be considered as another tool for production problem solving. The work in die casting flow completed in the 1960's by Wallace and others again is a reminder that the technology is available. Developments in castable, transparent polymers can provide a cost effective model of the die or mold.

The conference clearly demonstrated that the modeling method must be tailored to the goals of the modeler and the economic restraints imposed upon him or her. The research and academic institutions clearly have different goals in introducing modeling to their staff and students, as compared to a production plant attempting to solve problems in a timely manner. There is today a distinct danger of a widening gulf in capability between the universities and the industries that will employ their graduates or use their products. The cost of modern computer facilities is very high for the low budgets of the universities and is further aggrevated by the rapid pace of change. In contrast, the market for graphics systems in industry is one of the fastest growing segments of the computer industry. This is a nationwide problem of our times that must be solved. The contrast to the technological '60's is vivid: where the universities were in the forefront of technological development.

ROBOTS IN THE FOUNDRY

THOMAS A. CHURCH

VICE PRESIDENT
CONVEYERSMITH, INC.
LAGRANGE, ILLINOIS

The application of Robots in the foundry industry outside of die casting and investment casting plants has, to date, been limited. Only recently, have robots been introduced to the environment of the ferrous foundry. Three of the American manufacturers of robots have some sixty five applications currently "on stream".

The use of robots in foundries has, so far, been directed toward jobs that are either hazardous or boring. It now appears that jobs requiring accurate repeatability or precision are being considered for robot installation.

Typical applications are core dipping, mold spraying, flame drying, core setting, vent drilling, mold clamping and gating removal.

This paper will cover robot and manipulator applications at Caterpillar's new foundry at Mapleton, Illinois and seeing eye robot control research at General Motors.

To begin with, it would be proper to define the two types of equipment and point out the basic differences between manipulators and robots.

The subject of manipulators is a relatively simple one. In general, they fill a gap between common material handling equipment, such as hoists, and more sophisticated robots. This type of equipment assists the operator with his job, providing help in lifting, balancing and positioning or isolation from hazardous duties.

Manipulators are available as very simple lifting devices with capacities from a few pounds up to a range of approximately 1,000 lbs. or, as more complex pieces of equipment with capacities up to several thousand pounds with the operator sitting in an air conditioned enclosure operating the equipment with joy sticks. Foundry applications include casting handling, core dipping, core setting and riser knock-off.

Robots, on the other hand are, for the most part, self-sufficient pieces of equipment. The definition of a robot, according to the Robot Institute of America, is "a robot is a programmable, multi-function manipulator designed to move material, parts, tools or specialized devices through variable programmed motions for the performance of a variety of tasks". Currently, there are some twenty manufacturers of robots here and in foreign countries. Their products range from small to very large machines with controls that dedicate the operation to one job or allow the robot to be taught to perform many different or varying tasks. These machines are being purchased to work at specific jobs or, in some cases, moved from job to job. It should be pointed out that many manufacturers of machine tools and foundry equipment have built machines that approach the function of a robot. But, only recently have true robots, with their own identity, made their way into the industry.

The applications of robots in Europe seem to be on a level similar to that here in the United States. One of the more popular units is made in Sweden and is completely electric powered through motors and ball screws. This device is fully programmable and has the unique feature to be able to anticipate tool wear and compensate for it during operations.

One example is the grinding of fins on castings. (Figure 1) The robot knows, by electric feedback from the grinder held in it's hand, how much fin has been removed and how much more to grind.

Figure 1- Robot grinding fins on castings

This same type unit is used to cut castings from the gating system (Figure 2). Here, the operator loads castings on one side of a turntable while the robot cuts off on the other side.

Perhaps, the hardest lesson most foundry operators have had to learn in the last ten to fifteen years is that as equipment becomes more complex, the demands on the maintenance department become more acute. The modern foundry can no longer rely on unskilled mechanics to keep it running.

Figure 2- Operator loads castings onto one side of turntable while the robot cuts on the other

The introduction of robots has furthered this call for highly trained maintenance personel. Experience in hydraulics, servo motors, programmable controllers, air logic and computers is necessary to keep the robots healthy.

The robot manufacturers have had to learn their lesson, also. The foundry environment is perhaps the hardest on equipment of any industry. Early designs proved too weak and unreliable. Engineers were sent "back to the board" to provide designs that would stand up and provide reasonable up times. Robots require a scheduled preventative maintenance program. The foundrymen have learned this was necessary on their other equipment and have had to apply the same programs to robots.

Approximately one and a half years ago, Caterpillar Tractor Co. started operations at it's second gray iron foundry in Mapleton, Illinois. This plant was designed to produce "V" blocks and about a dozen types of cylinder heads. Because of this limited number of castings, the opportunity to automate was great. To this end, two types of manipulators and three types of robots were installed.

Caterpillar's decision to use robots or manipulators was based on the following advantages:

1. Improved production
2. More consistant performance
3. Application to jobs that were tedious, boring or would place the employees in a less than desirable environment. These jobs include handling hot castings, dipping heavy cores, spraying, drying and venting of molds.

Because manipulators have some common motions and similarities to robots, we will take a look at some manipulator applications before covering the robots.

Manipulators, called balancers, were installed to assist in the lifting of heavy cores. A balancer is used to lift a crankcase core assembly from the end of a group of conveyers and load it onto a carrier of a power and free conveyer. (Figure 3) Valves located near the handle of the unit allow the operator to float the core from one position to another.

Figure 3- Operator floats a crankcase core with the use of a balancer

In a similar application, the operator uses a balancer to lift a barrel core assembly by inflating rubber bladders in the hollow bores of the core. (Figure 4) Another arrangement using parallel arms keeps the core level during movement from conveyer to the rack. (Figure 5)

Figure 4- Balancer lifts barrel core assembly

Figure 5-
Moving crankcase core from conveyor to rack

About ten years ago, a hydraulically powered manipulator was developed that allowed the operator to be remote from the actual unit. In some cases, in an air conditioned enclosure. Caterpillar installed its first unit of this type in 1974 to remove cylinder heads from a shakeout conveyer. This installation was necessary due to the hot, smokey and dirty conditions. Prior to the installation, many cracked castings resulted from being dropped from the conveyer into the power and free conveyer cooling buckets. The excellent result of this installation paved the way for the application of several more units in the new facility.

The cylinder block line has two units. (Figure 6) The first loads castings on a power and free cooling conveyer hook and the second loads the runners and down sprues in a carrier on the same conveyer.

Figure 6-
Two unit cylinder-block line

Unit One has a rotating head with a hydraulic clamp on one side and a blade on the other. The operator uses the clamp to remove a hook that is hung on the conveyer carrier. (Figure 7)

Figure 7-
Removing hook
hung on the con-
veyor carrier

Castings and sand pass from the shakeout to the apron conveyer. The #1 operator, by rotating the head, can use the blade to position the block and knock away any excess sand. He then picks up the block with the hook and hangs it on the conveyer carrier. (Figure 8) Castings range from 450 lbs. to 1200 lbs. Production rate is 50 to 100 blocks per hour.

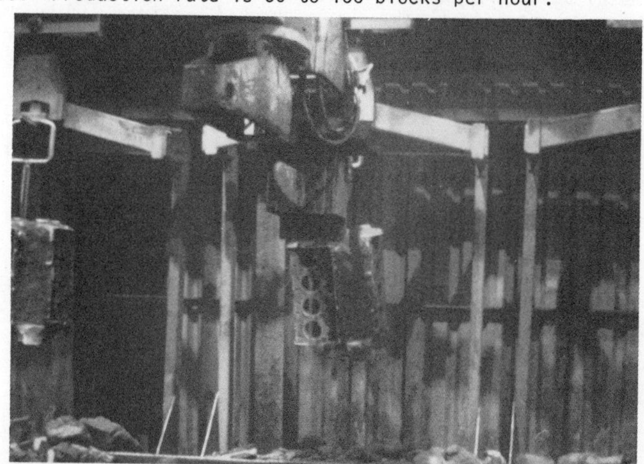

Figure 8-
Picking up block
with hook

The second unit downstream from the first, picks up the gates and sprues and deposits them in a bucket on the conveyer. (Figure 9) This unit can also remove castings if the first unit is down.

Another mold line produces heads. This line has one manipulator to remove the heads from the shakeout deck and place them in a cooling bucket on a power and free conveyer. (Figure 10) Gates and sprues are swept off the shakeout into the same bucket. (Figure 11) Castings weigh up to 530 lbs. and are produced at a rate of 70 plus per hour. This unit has a remotely located operator enclosure. (Figure 12) This arrangement was necessary so as to give the operator the best view of the work area.

Figure 9- Second unit picks up gate

Figure 10- Manipulator removes heads from shakeout deck

The third manipulator is used to stack head castings that have been cooled and are ready to be transported to the cleaning room. (Figure 13)

In this operation, the heads are picked up by an end, allowing the sand to drain and stacked on a pallet (Figures 14 & 15)

Figure 11- Sweeping gates and sprues off shakeout

Figure 12- Remotely located operator enclosure

Figure 13- Stacking head castings

The operators, after a short training period, utilize the manipulators as extensions of their bodies. After watching them work for awhile, one begins to think they are human, because the motions are so direct and humanlike.

Figure 14- Picking up head and draining sand

Robot installations at Caterpillar are many and varied. There are three different type units, each one selected on the basis of application, lowest initial investment and return on investment.

Consistant results from core dipping are always desireable. Manual operations can produce cores with too much wash, not enough or other variables that affect casting quality. Caterpillar used this reasoning to justify the installation of a robot to handle and dip 63 lb. cylinder block barrel cores.

Figure 15- Stacking heads on a pallet

The robot has seven possible motions and is equipped with an inflatable grab to pick up the cores. It is hydraulically powered and, through the use of a digital computer, can be programmed to provide point to point movement, resulting in the shortest path and time cycle.

The robot is hung from an overhead traversing carrier with its rotating unit and arm hanging below. The program calls for five motions, horizontal, vertical, extend or retract and rotate.

Figure 16- Robot lowering inflatable bladders into hollow bores of core

As a core is ejected from the core box, on strip pins, (Figure 16) the robot moves in over the core and lowers the inflatable bladders into the hollow bores of the core. (Figure 17) Two core assemblies are removed from the core machine, at one time. The robot then retracts and traverses to the dip tank where it lowers the cores to the proper depth. (Figure 18) The cores are raised from the tank and oscillated to remove excess dip. (Figure 19)

Figure 17- Robot removing two core assemblies

Figure 18- Lowering cores to proper depth in dip tank

Figure 19-
Oscillating cores to remove excess dip

Rotating 180°, (Figure 20) the robot moves the cores toward a power and free trolley conveyer. The conveyer is equipped with a feedback sending unit that syncronizes the speed of the robot with the conveyer. The two core assemblies are loaded on the lower shelf of the carrier and the robot returns to pick out the next set of cores. (Figure 21).

Figure 20-
Robot moves cores to trolley conveyor

Mold spray and drying can be a tedious, as well as hot, job. Manual operations can produce varying results. Caterpillar has eleven robots drying molds and two for spraying. At present, some manual spraying is still being performed.

The spray units have a standard spray gun attached to the hand at the end of the arm. (Figure 22) The unit is programmed to follow an exact path to evenly apply the wash. (Figure 23) Experience has shown that robot spraying has produced a more uniform coating with less material used. The robot has been taught that after completing a mold, it puts its spray tip in a container of water to keep it from plugging.

Figure 21-
Loading cores
onto carrier

Figure 22-
Spray unit

Drying mold wash is accomplised with a similar program to the spray operations. (Figure 24) Here, the robot provides uniform drying without either hot or wet spots.

In another building, Caterpillar has two robots in an investment casting operation. (Figure 25) In two parallel systems, wax patterns are dipped in a slurry, spun to remove excess dip and coated with a ceramic material either in a fluid bed or rainfall sander. (Figures 26, 27, 28)

Another type of robot is used to vent molds. Prior to this installation a man using a pneumatic hammer punched the vents as the mold moved past him. This sometimes resulted in missed vents and was a very tedious job.

Figure 23-
Sprayer applying
even coat of wash

Figure 24-
Robot drying
mold wash

Figure 25-
Robots in invest-
ment casting
operation

Figure 26-
Wax pattern
being dipped in
a slurry

Figure 27-
Spinning to
remove excess
dip

Initial installation of these four robots utilized air hammers in the robots hand. After awhile, it was found that the robot was strong enough to punch the vent with only a rod.

As each mold moves through the vent area, the robot moves through its programmed path to vent the mold. (Figure 29) At the end of the cycle, the robot uses the rod to depress a paddle attached to a limit switch. This ends the cycle and shows the controls that the robot has not lost or damaged its rod (Figure 30).

While each robot has its own control center, Caterpillar is looking into connecting all of the units into a central computer. This would allow programming of tooling or changes for new jobs, scheduling maintenance and trouble shooting.

Figure 28- Coating wax pattern with ceramic from rainfall sander

Figure 29- Venting molds

Up until very recently, the application of robots required that the part or product arrive at the work station at a specific location or orientation. This requirement was based on the fact that the robot had no intelligence to recognize randomly placed parts.

Figure 30- Robot using rod to depress paddle-limit switch

In many foundry applications, parts arrive at work stations in random order and sometimes as a mix of several parts. Mechanical sorters have been tried, but, generally have proven unsuccessful or expensive.

If a robot could see and recognize a part as it entered its work area, a vast improvement in operations would result.

Several years ago, General Motors research labs demonstrated a vision controlled robot system for transferring mixed parts from belt conveyers to a predetermined location. The concept for this system was based on earlier work performed by S.R.I. International to develop intelligent machines.

The system demonstrated by General Motors was named Consight and showed the capability to determine position and orientation of parts, was easily reprogrammed by insertion of new data, and was adaptable to the foundry manufacturing environment. The vision system developed for this application did not require light tables, colored parts, or other impractical means of enhancing contrast. These factors make the systems suitalbe for production plant use.

Following the demonstration of feasibility, a decision was made to build a prototype production unit. The systems and hardware for this unit were developed at General Motors Manufacturing Development with assistance from General Motors Research Laboratories and Central Foundry Division. The objectives of this project were to:

1. Build a system that would be capable of operating in a foundry environment.
2. Reduce system cost.
3. Improve system reliability.
4. Reduce robot cycle time.

A vision based robot guidance system has the capability of controlling a robot to perform non-repetitive operations. This type of system could be utilized to perform the following types of operations:

1. Transfer of parts from moving belts to process operations.
2. Sorting of castings.
3. Packing of castings in trays or on belts for anealing.
4. Placing of castings on pallets or in containers for shipping.

Central Foundry is currently evaluating these and other potential applications. Their objective is to develop a suitable application for this sytsem in one of their manufacturing operations. In addition to demonstrating technical feasibility in manufacturing, the goal will be to increase process productivity and to reduce cost.

Other companies are working on their own form of sighted robots. It's the first step toward truly intelligent units. Computers have already been taught to speak. Who knows, the next generation of robots could completely run a facility and call for help when something goes wrong.

The application of robots and manipulators in the foundry has been limited to jobs that were hazardous, dirty or repetitive. In the future, they will no doubt replace workers in other jobs. This could lead to union complaints. Managers will have to cope with yet another problem in today's complex workplace.

CREDITS

AMERICAN FOUNDRYMEN'S SOCIETY
TC-1 PLANT ENGINEERING COMMITTEE

ROBOT PROGRAM MEMBERS

JOHN SCHOEN - ASSISTANT TO PRESIDENT
KENZLER ENGINEERING CO.
MILWAUKEE, WISCONSIN

BILL SCHULZE - SUPERVISIONG PLANNING ENG.
CATERPILLAR TRACTOR CO.
MAPLETON, ILLINOIS

NORRIS LUTHER
LESTER B. KNIGHT
CHICAGO, ILLINOIS

ROBOERT ROWLAND
SYSTEMS & CONTROL DESIGN
LITTLETON, COLORADO

WILLIAM CHEEK
WORKS ENGINEER
CENTRAL FOUNDRY DIV. GMC
SAGINAW, MICHIGAN

REFERENCES

ROBOT INSTITUTE OF AMERICA
ONE S.M.E. DRIVE
P.O. BOX 930
DEARBORN, MICHIGAN 48128

PROCESS MODELING

M. C. Flemings

Ford Professor of Engineering
Director, Materials Processing Center
M.I.T.

Abstract

The understanding and control of complex processes such as industrial solidification processes comes about through a complex interplay of analysis and synthesis. Mathematical modeling is a powerful new tool to aid in this task, especially when supplemented by experimental modeling.

This paper presents a general discussion of mathematical and experimental modeling of solidification processes using ingot solidification as example and then gives related specific examples.

Introduction

Only a few decades ago, academic researchers and industrial technologists in the solidification field were worlds apart, and the types of problems they chose to work on often seemed unrelated. For example, it was widely believed then that if one wanted to learn something about the manufacture of a casting or ingot of a given size, it was necessary to study the full size object. This, of course, eliminated serious study in universities on solidification of castings above a few hundred pounds and eliminated work on ingot solidification altogether.

There was, after all, some justification for the prevailing view that these problems could not be simply scaled down to laboratory size. They cannot. The problem is intrinsically different from that of scaling, say, fluid flow problems. Chemical engineers have taught us the scaling laws for fluid flow problems and we can, if we wish, simply model our full size flow problem on a laboratory scale and then use the scaling laws to predict behavior in the real system. In solidification of castings or ingots, the problem is more complicated. The scaling laws are different for different aspects of the problem. For example, bulk liquid convection may vary as the cube of the linear dimension of our casting, solidification time may vary as the square, and thermal stresses may vary linearly with the dimension.

So what is to be done? The modern approach is to analyze and then synthesize. First, we break the problem into its component parts ... into parts sufficiently fine that we can understand qualitatively the important details of the part of the process on which we are focusing. We may hope to understand these quantitatively as well. Following this step of analysis of individual building blocks of the process, we synthesize by incorporating these blocks into a model of the entire process. Now, if we have done our job properly, and have incorporated sufficiently fine detail, we understand the whole and can predict behavior of the process in practice. We can then modify and improve it, scale it to other sizes or shapes, or use our understanding to develop a wholly new process.

Analysis

Until recently, in fact, working on large-scale synthesis has not only been beyond the reach of academic researchers, but beyond that of most industrial researchers as well. Witness, for example, the fact that until recently ingot solidification problems were not a part of the purview of the technical centers of large steel companies. Such problems were left exclusively to the shop floor technologists who actually worked with the ingots themselves. There has, however, been a strong continuing underlying belief of technologists that the fundamental building blocks of industrial solidification problems are fair game for researchers and that development of a fundamental understanding of these building blocks would lead to important industrial advances.

So, many of us in the research and academic communities have spent much time working on these fundamental problems while trying to keep the studies relevant to the industrial problems as well. I will turn now to some examples using, for the most part, ingot solidification as the framework.

Atomic Scale Processes

The solidification of an ingot must begin with at least one nucleation event and, in the 40's and 50's, Hollomon and Turnbull (1) began to study this process analytically and experimentally on a small scale, with startling success. This work has served as the foundation of studies that continue on to this day, including the particularly interesting current ones of Perepezko (2).

Hollomon and Turnbull gave us a clear physical picture of the nucleation process. Atoms congregate into "embryos" in the liquid or on a foreign substrate and when, at some temperature below the thermodynamic equilibrium melting temperature they reach a critical radius, r^*, they become a stable nucleus and grow rapidly to form a solid grain.

The nuclei, and later full size crystals themselves, grow by attachment of atoms onto sites on the growing lattice. This process has been much studied analytically, experimentally, and now with the aid of computer analysis. Much discussion continues to center on the character of the interface (rough, smooth, etc.) during growth, and on the amount of undercooling necessary to drive the interface forward (3,4).

The Dendrite Tip

On a much larger scale, that of the dendrite tip, a suitable topic for academic research for over 20 years has been the mathematical problems of the shape, undercooling, and growth rate of the dendrite tip growing into a liquid melt. The heat and mass transfer problems are solved mathematically or numerically using a differential element or finite element smaller than the dendrite tip. Some remarkable success has been obtained with these analyses in predicting, for example, dendrite growth rate in supercooled pure liquids (5).

More recently, the problem of the "dendrite array" ... treating the interactive effects of a group of dendrite tips ... has become of academic interest. One noteworthy outcome of this work has been an understanding of factors affecting primary dendrite arm spacing, and development of analyses to predict this spacing. An interesting physical result is that primary dendrite arm spacing is not only determined by what happens at the tip, as above, but by the processes also going on well behind the tip (6).

The Liquid-Solid Zone

If one were to picture the spacing between the secondary dendrite arms of a growing dendrite array as being, say, a foot wide, then the zone where liquid and solid coexist might extend back from the dendrite tip by a half-mile or more. The liquid-solid "mushy" or pasty region is very long indeed. When we quantitatively study this zone, we usually use, not a differential element, but a finite "volume element" that is large enough to enclose at least several dendrite arms, but small enough that it can be treated (for some purposes) mathematically as a differential element or as a small finite element for numerical analysis. Purposes for which we do this include heat flow, interdendritic fluid flow, and microsegregation formation.

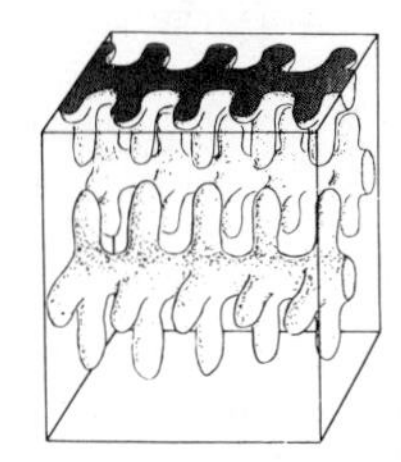

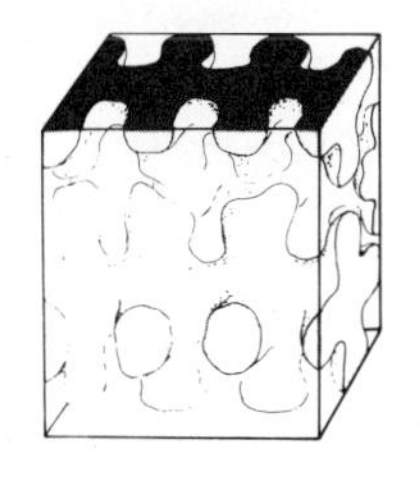

Figure 1. A "volume element" from a liquid-solid zone showing material (a) about 20% solid, and (b) about 50% solid.

Figure 1 is a sketch of a suitable "volume element" at two different stages of solidification. A close comparison of these sketches shows some rather considerable changes. The fraction solid is, of course, large at the later stage of solidification but so is the spacing of the structure (the secondary dendrite arm spacing). This spacing increases by a "ripening" or "coarsening" process whereby smaller arms melt (or dissolve) during solidification while larger ones get more than their share of the growth. In addition, there is the progressive movement(not shown in these figures) of arms in the direction of the temperature gradient, and the tendency for spaces between arms to fill in, in the latter stages of solidification, to form plates (especially in the heat flow direction) (3,7,8).

The simplest mathematical relation to describe the solute redistribution that takes place in this volume element during solidification is the Scheil equation which, expressed in differential form, is:

$$\frac{df_L}{dC_L} = - \left(\frac{1}{1-k}\right) \frac{f_L}{C_L} \tag{1}$$

where f_L is fraction liquid, C_L is liquid composition and k is the equilibrium partition ratio. For constant, k, the expression is:

$$C_L = C_0 f_L^{k-1} \tag{2}$$

This simple equation has been modified by various workers to account for solid diffusion, temperature gradient migration of the dendrite, tip undercooling, and other factors, but it remains a widely used and useful equation to predict the microsegregation in dendritically solidified structures.

A second important equation relating to this structure is the well known empirical equation which can be at least semi-quantitatively understood on the basis of coarsening theory:

$$d_s = a \, \varepsilon^{-n} \tag{3}$$

where d_s is secondary dendrite arm spacing, a is a constant, ε is local cooling rate and n is an exponent which experimentally is usually found to lie between about 1/4 and 2/5, and theoretically is expected to be 1/3. Figure 2 shows a plot based on a wide range of experimental data for aluminum 4.5% copper alloy of both primary dendrite arm spacing and secondary dendrite arm spacing.

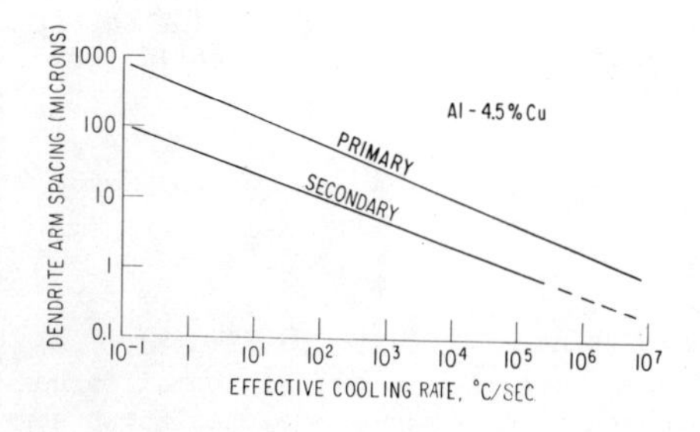

Figure 2. Primary and secondary dendrite arm spacings in aluminum-copper alloy based on experimental data (mostly at the lower cooling rates) of a number of investigations.

The same "ripening" process that controls secondary dendrite arm spacing also aids in the breakup of dendrites under the action of fluid flow, as shown in upper left of Figure 3. It is now understood that in many cases, the final grain size of a casting or ingot is determined not by nucleation frequency but by extent of dendrite breakup. In the limit, very vigorous agitation can eliminate the dendrite structure altogether, producing a semi-solid material that has the flow characteristics of a slurry. This is also shown in Figure 3. The process for doing this has been termed "Rheocasting" (9,10).

In the real world, we may not think only of binary alloys. Inclusions, porosity, and alloy second phases may form, and Figure 4 shows an example of inclusions ... e.g., oxide inclusions in steel. The inclusions sketched

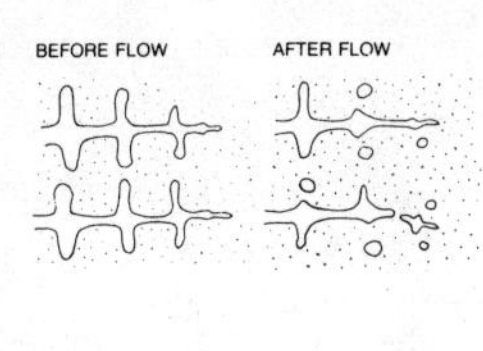

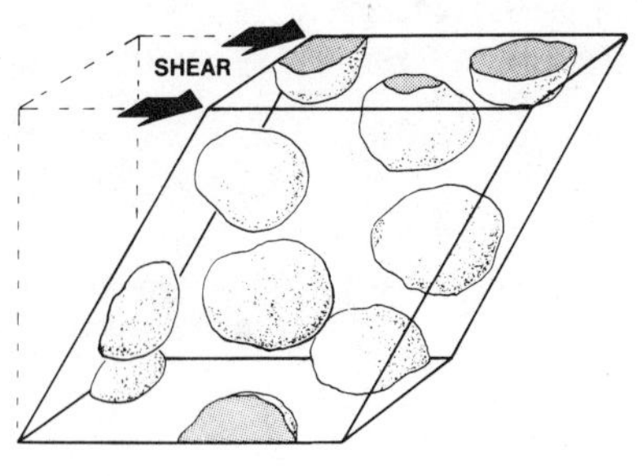

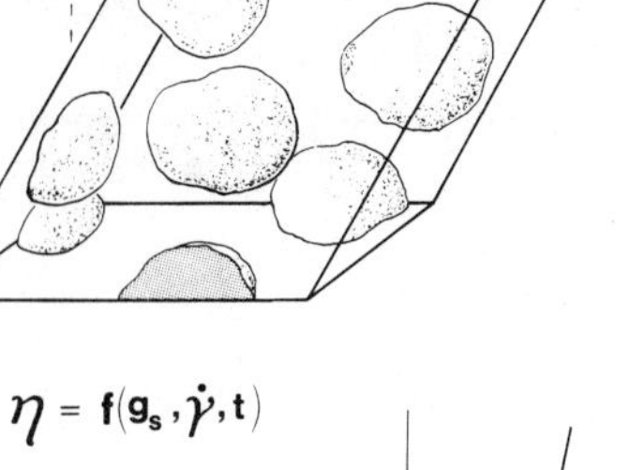

Figure 3. Breakup of a dendritic structure from convection (top left). Rheocast structure resulting from complete breakup (center). Typical behavior of viscosity versus fraction solid for the Rheocast material (lower right).

are ones such as SiO_2 inclusions in steel; they are "pushed" ahead by the growing dendrites. They may float, collide, and coalesce into larger inclusions. Other types of inclusions, such as Al_2O_3 in steel are entrapped by the growing dendrites. In either case, we can describe the solidification behavior of inclusions with a modified "Scheil" equation which, for the stage of solidification shown, can be written in differential form as (3):

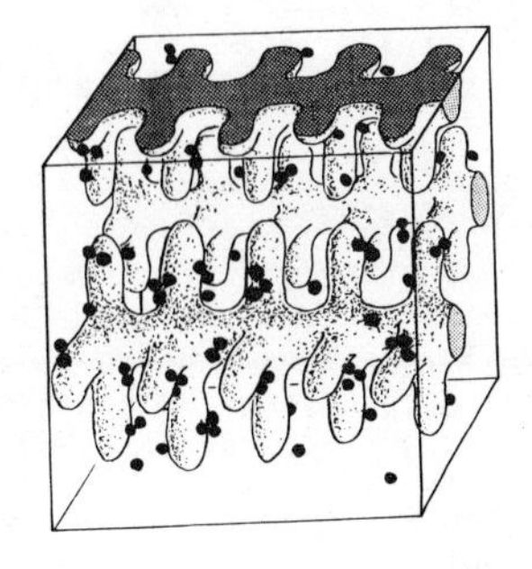

Figure 4. Secondary inclusions forming during dendritic solidification.

$$\frac{df_L}{dC_{Lm}} = -\left(\frac{1}{1-k_{\alpha m}}\right)\frac{f_L}{C_{Lm}} - \left(\frac{k_{\beta m}-k_{\alpha m}}{1-k_{\alpha m}}\right)\frac{df_\beta}{dC_{Lm}} \tag{4}$$

where C_{Lm} is liquid composition of element m, f_β is weight fraction of second phase β, $k_{\alpha m}$ is partition ratio between β and liquid with respect to m and $k_{\alpha m}$ is partition ratio between α and m with respect to element m. As we have learned in recent years, we must not neglect the flow of interdendritic liquid through the growing dendrite array. This flow, Figure 5, results from solidification shrinkage, gravity driven convection, and

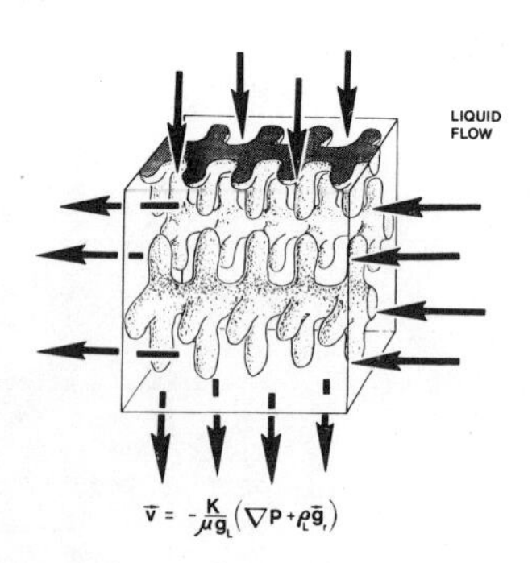

Figure 5. Fluid flow through a solidifying volume element. The component of flow perpendicular the isotherm results in local composition change.

solid deformation. It modifies the Scheil equation for binary alloys, and therefore the final local solid composition (11,12), as follows:

$$\frac{\partial g_L}{\partial C_L} = -\left(\frac{1-\beta}{1-k}\right)\left(1 + \frac{\vec{v}\cdot\nabla\vec{T}}{\varepsilon}\right)\frac{g_L}{C_L} \tag{5}$$

where g_L is volume fraction liquid. This flow is the primary cause of macrosegregation in castings and ingots.

Finally, we must describe heat flow in the solidifying liquid-solid zone. Heat is transported primarily by conduction in this zone. Its source is specific heat and heat of fusion. Generally, heat is not added during solidification, although this is possible, as from a high burning temperature exothermic.

The Fully Solid and Fully Liquid Regions

In the fully solid region of a solidifying casting or ingot, we are also interested in heat flow behavior. And, we are interested as well in the thermal stresses and strains which form during cooling. Because of the industrial importance of problems such as ingot bulging and cracking, much attention is being paid to detailed analysis of the materials and theoretical aspects of this problem. In the fully liquid zone, heat is transported both by conduction and convection, mostly by the latter. We concern ourselves with other aspects of the liquid zone as well ... for example, the inclusions which form, coalesce, coarsen, and then float to the top of the ladle, or perhaps plug a bottom pouring ladle, or end up in the ingot we are producing.

Studies on heat flow and thermal stress behavior in solidification are an important part of solidification science and technology today, and a number of other sections in this book deal with that subject in detail.

Interfaces

Interfaces pose some special problems in solidification. These interfaces include mold-metal, mold-atmosphere and metal-air. From the heat flow standpoint, we have relatively simple analytic expressions with which to work, but determining correct numerical values for heat transfer coefficients in these simple expressions is not easy. In the case of metal-mold heat transfer, for example, the coefficient may vary significantly and sometime erratically during solidification as the metal pulls away from the mold. Considerable laboratory experimental effort is being devoted to measuring these coefficients. Mold-metal reactions are another problem area, and one suitable for study. One of the more interesting interfacial problems in solidification is the entire question of formation of surface finish ... for example, how laps or other forms of surface roughness form in ingot solidification. This involves considerations of heat flow, fluid flow, solidification, and surface energy.

Synthesis

Beyond the question of academic curiosity, the value of the foregoing analyses lies in our ultimate ability to integrate them into a coherent, useful understanding of a complex process. Historically, we have done this, when we have done it at all, by qualitatively synthesizing fundamental knowledge to understand a process, or a part of a process. With this understanding, we then aim to improve the process or its control. This is shown schematically in Figure 6. Alternatively, we may qualita-

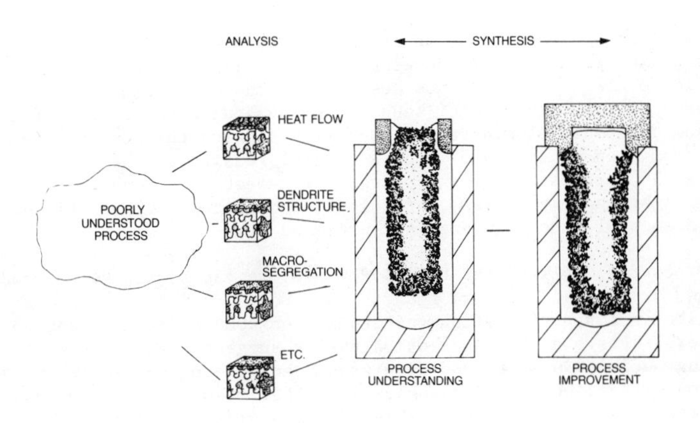

Figure 6. Process understanding and improvement (without modeling).

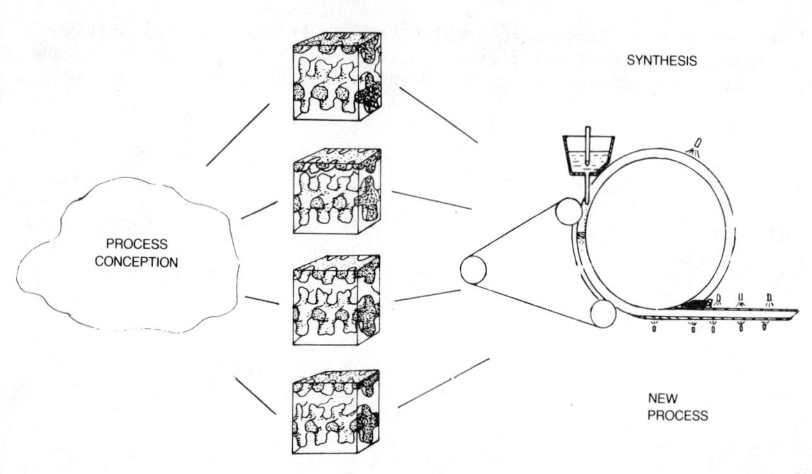

Figure 7. Process invention and development (without modeling).

tively synthesize our fundamental understanding (with, we may hope, some imagination as well) to develop an entirely new process, Figure 7.

An excellent example of synthesizing of fundamental studies to achieve practical gains has been the development of nucleation theory, and its combination with other aspects of fundamental understanding of liquid metals, and the solidification process to achieve practical grain refiners.

Much work has been done in attempting to predict characteristics of a good grain refiner and to develop specific, improved grain refiners for specific metals. Often, it has seemed that the commercially useful grain refiners that were developed (such as improved aluminum-boron refiners for aluminum or cobalt-aluminate for superalloys) were the result more of empiricism than of theory ... but a careful reading of the literature over the last 30 years strongly supports the argument that much of this work was "enlightened empiricism", stimulated and guided, however blindly, by the fundamental understanding that preceded it.

Of course, as with the Delphic oracle, we can sometimes misread the message, or take it for more than it is. So, perhaps, we have spent too much time in our grain refinement studies on some approaches and not enough on others. For example, nucleation theory suggests that a method of increasing nucleation is to add significant energy to the system in the form of strong vibrations by ultrasonics, or other means. Much time and effort have been spent trying to make this idea useful in practice without, to my knowledge, any significant "pay off". Perhaps, our too narrow focus on classical nucleation theory blinded us to the fact that gentle convection results in significant dendrite breakup and is therefore a powerful grain refiner. This is a method we now see being commercially exploited.

The intimate, and sole, dependency of secondary dendrite arm spacing on local cooling rate is another fundamental aspect that is qualitatively (as well as quantitatively) used in design of castings and ingots today. The improved properties resulting from fine dendrite arm spacing are widely recognized, and in processes where high properties are desired, we take special pains to increase this cooling rate. Other aspects of fundamental understanding that are qualitatively used in practice today relate to macrosegregation, dendrite alignment, inclusion coalescence, heat treatment of casting structure, etc.

A limitation of this technique of synthesizing qualitatively is that, as processes become increasingly complex, it becomes increasingly difficult to qualitatively combine even a part of the building blocks to understand the real process. Whatever understanding is attained must be verified by experiment, and an important additional limitation is that experimentation on large scale processes is enormously expensive. To cast and analyze a single large ingot, for example, may cost well in excess of $100,000. Yet, it usually is futile to try to directly model the process experimentally on a small scale, since there is no assurance whatsoever that the laboratory results can be extrapolated to the production system.

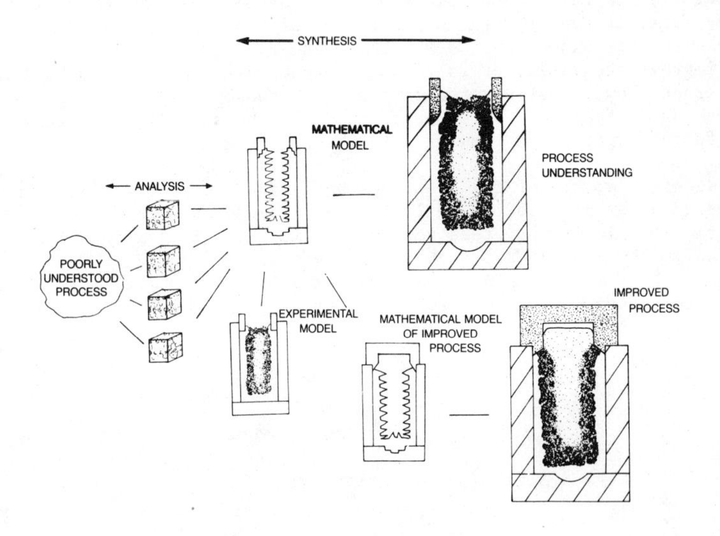

Figure 8. Process understanding and improvement with the aid of modeling.

It is here that mathematical modeling will play an increasingly important role in the years to come ... aided by increasingly powerful computers and increasingly sophisticated methods of experimental modeling to test the mathematical models. Figure 8 shows this schematically, for process understanding and improvement. The component parts are first analyzed. This is done in much more detail than was formerly possible, with the aid of numerical techniques for small-scale modeling. An example is describing in detail dendrite growth behavior by mathematical modeling.

Next, the component parts of the problem are quantitatively combined with a mathematical model for the process as a whole. For example, it is well within our capabilities today to formulate a mathematical model for ingot solidification that will describe isotherm movement, dendrite arm spacing, microsegregation and macrosegregation behavior, and thermal stresses. Many examples are already in the literature where at least some of these variables are modeled.

Before proceeding to tests on, say, a 400-ton ingot, we will be well advised to do some experimental modeling on a much smaller ingot ... perhaps even with a different metal than we are using in production. Here we are interested not in fully duplicating on a small scale what happens on the large scale, but simply confirming the correctness of the mathematical model. Once this is done, as sketched in Figure 8, we are ready to test the model against the process. In our ingot example, we must first, of course, be sure we have good input data (for thermal properties of the metal and mold materials, etc.). Now we test the model against the real ingot by selective nondestructive testing on the production floor or perhaps by destructive examination of an ingot or two. If our mathematical model is inaccurate, we must move back a step or two and improve the model or the input data, before we are ready to move on to improve the process.

But the centrally important goal is to develop a model of sufficient accuracy that it can be used to modify the process in any way we desire. We may wish, for example, to scale the process up to a still larger size, or to improve the yield, or to reduce macrosegregation, or to substitute an alternate lower cost hot top material for the one we are using. All of these things we can test by computer on our mathematical model before going to the expense and risk of a full scale production trial.

Essentially, all of the arguments above apply when modeling is used to develop a new process, as well as to improve an existing process. A schematic example of the steps to be taken is shown in Figure 9. In this case, as in the one discussed above, the process is an iterative one. The mathematical model is improved based on experimental modeling; additional basic information may be required to complete the model, etc.

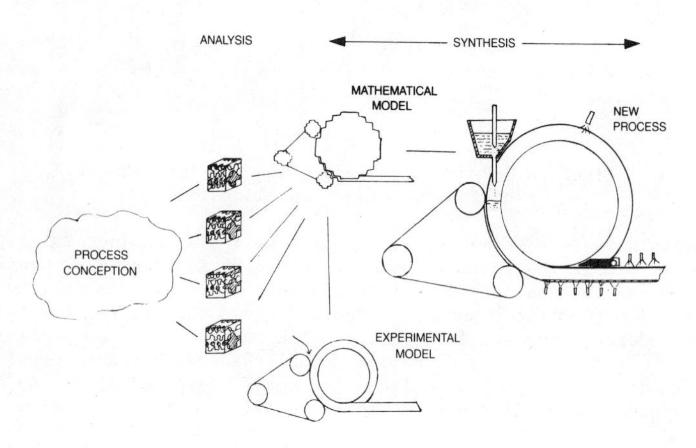

Figure 9. Process invention and development with the aid of modeling.

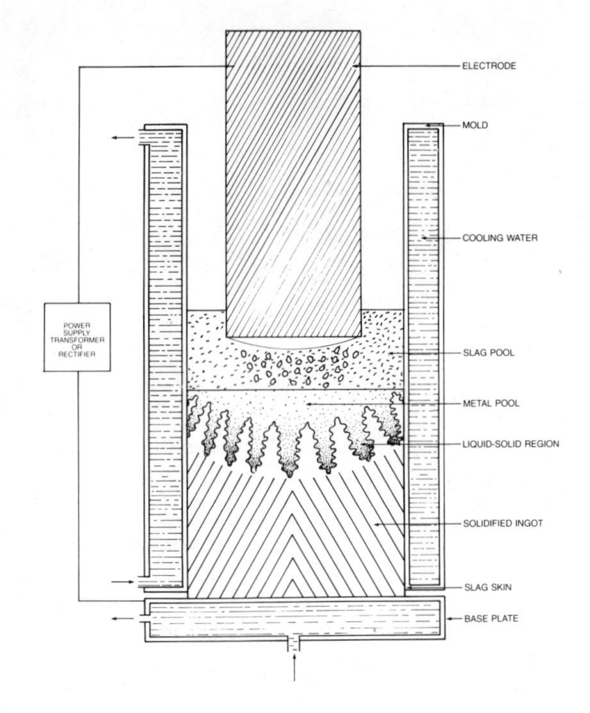

Figure 10. Schematic illustration of the electroslag remelting process (ESR).

Some Process Models

One example of mathematical modeling to illustrate the foregoing is work on modeling the electroslag remelting process (ESR). Figure 10 is a schematic illustration of the process. Modeling of this process is complicated by the facts that melting occurs in situ and that there are strong electric currents with their resultant joule heating and electromagnetic stirring.

Mathematical models have been developed for this process for the electromagnetically driven fluid flow in the liquid pool, thermal fields, and macrosegregation (13-18). In a number of cases, the mathematical models have been checked using small scale experimental models, before going to the full scale production units. In important instances, these models have proven their worth industrially.

Figure 11 shows some experimental data on macrosegregation in one such laboratory unit and compares it to results of a mathematical model for the macrosegregation. Results are closely similar, and comparable results (at least for some of the elements of this complex alloy) have been achieved at the full scale plant trials of Jeanfils, Chen, and Klein (17) of Figure 12.

Many other examples of modeling of large scale solidification systems could be cited. These include studies on continuous casting, large ingot

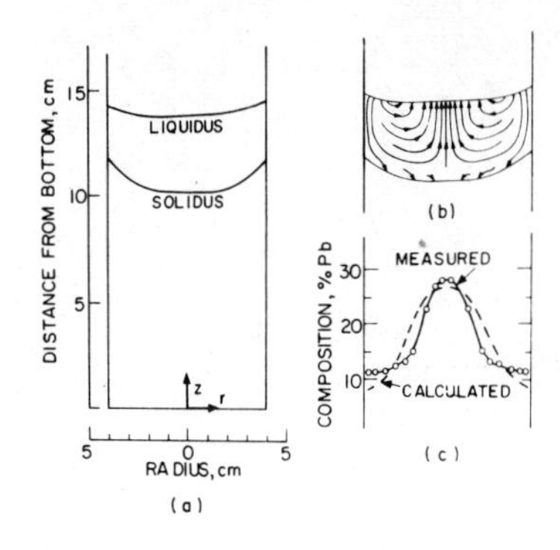

Figure 11. Macrosegregation in experimentally modeled ESR process (16).

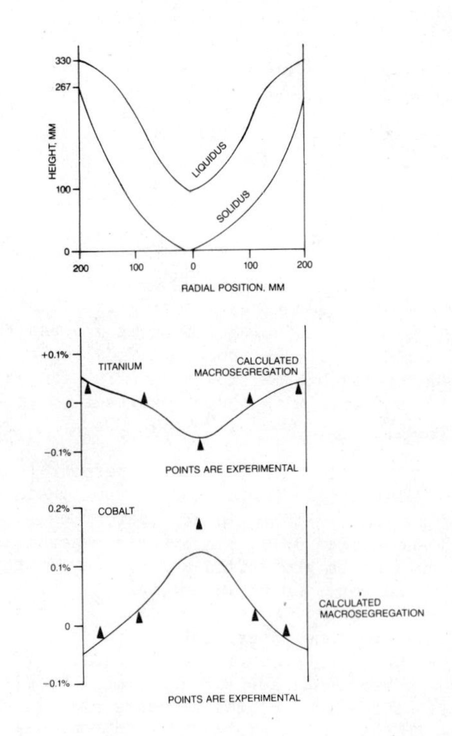

Figure 12. Macrosegregation in full scale ESR process (from Jeanfils et al (17).

solidification and hot-topping, shaped sand casting, and die and permanent mold casting. We will see much more of this type of work, and much of value coming from it.

In closing, following are some examples of studies from my own laboratory waiting to be developed into useful processes. Small scale mathematical and experimental modeling have been done as part of each of these studies and large scale modeling could play an important role if the techniques, or ones like them, are eventually scaled up to commercial usage.

We have known for many years that a steady magnetic field can greatly reduce convection in liquid metals because of the added $\vec{J}x\vec{B}$ term that appears in the fluid flow equation. The metal behaves as if the magnetic field greatly increased its viscosity. Application of this process to solidification problems was pioneered by H. Utech in his thesis (19,20) and later studied by Jackson et al (21) and others. Years ago, we showed the process results in improved homogeneity of single crystals, and coarser grain structure in small ingots (by eliminating the convection that produces new grains). Now, Sony has announced successful application of the technique to production of silicon crystals for manufacture of large scale integrated circuits. The magnetic field produces significantly improved homogeneity of the crystals, with a resultant 20% improvement in "yield" of the finished silicon "chips" (22).

It seems likely that this process may have other useful applications in solidification processing. One example could be in production of columnar and "single crystal" turbine blades to avoid the growing of "stray" grains. Another could be to alter the flow behavior of metal in a mold during filling to improve surface finish.

Another laboratory technique my students and I have been studying recently is the "squeezing" of liquid out of semi-solid material either when the metal is held isothermally, Figure 13, or is heated during squeezing, Figure 14. These techniques can be used to purify bulk metals

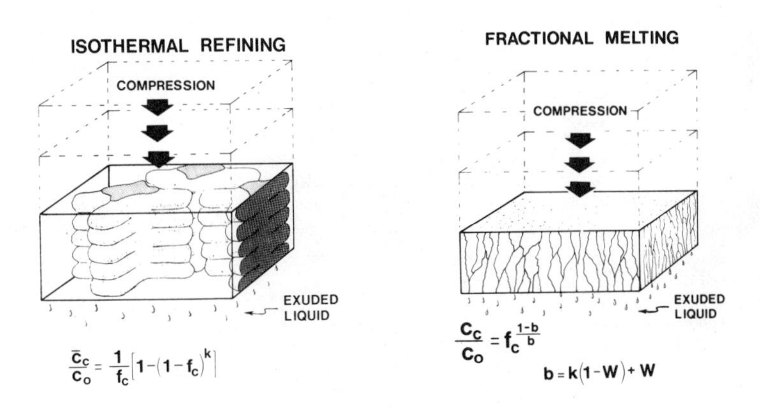

Figure 13. Fractional Solidification. Figure 14. Fractional Melting.

(e.g., scrap recycling or upgrading of high purity metals) (23,24). They can also be used to strengthen metals, by melting and "squeezing out" just the last little bit of interdendritic material ... which otherwise would solidify to form to embrittling nonmetallics or intermetallics (25). Perhaps, this technique will one day be scaled to a commercial process.

As a last example, there is the model of Matsumiya who built an experimental and mathematical model of a strip casting process employing "Rheocast" material (26). The experimental model is shown in Figure 15a. Rheocast material, with the metallurgical structures shown in Figure 3, is cast onto a moving belt which then travels horizontally into a space between the belt and a rotating drum. The semi-solid metal simultaneously flows and flattens under the wheel, and solidifies. Matsumiya modeled the thermal flow and solidification behavior of the semi-solid alloy and successfully compared it with experiment. Some results of his model are shown in Figure 15b. Perhaps, this process will prove too complicated to scale up to production of strip of metals such as steel, or perhaps it will some day prove a practical method to assure casting thin steel strip with good surface finish and freedom from cracks.

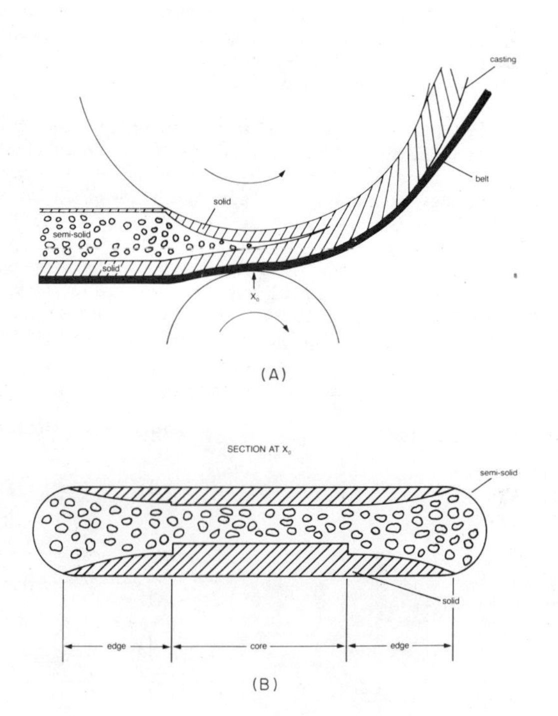

Figure 15. (a) Continuous strip production by Rheocasting (after Matsumiya and Flemings) (26).
(b) Results of process model for continuous trip production(after Matsumiya and Flemings)(26).

Closure

The central aim of this paper has been to demonstrate by examples that understanding and control of complex processes can only come about by a complex interplay of analysis and synthesis. Mathematical modeling is a powerful new tool to aid in this task, especially when supplemented by experimental modeling.

Acknowledgements

The research at Massachusetts Institute of Technology referred to in this paper was carried out by a large number of graduate students and staff members, many of whom are noted in the references. The research at MIT was supported by a number of government and private agencies including the National Science Foundation, National Aeronautics and Space Administration, Army Research Office, Army Materials and Mechanics Research Center, and Department of Energy.

References

1. J. H. Hollomon and D. Turnbull, Progr. Met. Phys., 4 (1953) 333.

2. J. H. Perepezko, C. Galaup, D. H. Rasmussen, Proceedings, 3rd European Symposium on Materials Sciences in Space, Grenoble (1979) 375.

3. M. C. Flemings, Solidification Processing, McGraw-Hill, New York (1974).

4. G. H. Gilmer, Proceedings, Modeling of Casting and Welding Processes, Engineering Foundation, August, 1980.

5. M. E. Glicksman, R. J. Schaefer, J. D. Ayers, Met. Trans., 7A (1976) 1747.

6. M. H. Burden, J. D. Hunt, J. Crystal Growth, 22 (1974) 109.

7. D. J. Allen, J. D. Hunt, Met. Trans., 7A (1976) 767.

8. L. A. Lalli, Proceedings, Modeling of Casting and Welding Processes, Engineering Foundation, August, 1980.

9. D. Spencer, R. Mehrabian, M. C. Flemings, Met. Trans., 3 (1972) 1925.

10. M. C. Flemings, K. P. Young, Yearbook of Science and Technology, McGraw-Hill, New York (1978) 49.

11. M. C. Flemings, G. E. Nereo, Trans. Met. Soc. AIME, 239 (1967) 1449.

12. R. Mehrabian, M. A. Keane, M. C. Flemings, Met. Trans., 1 (1970) 1209.

13. A. S. Gallantyne, A. Mitchell, Ironmaking and Steelmaking, 4 (1977)222.

14. J. Kreyenberg, K. Schwerdtfegger, Arch. Eisenk., 50 (1979) 1.

15. J. Szekely, M. Choudhary, Proceedings, 6th International Vacuum Metallurgy Conference, April (1979) San Diego, California.

16. S. Kou, D. R. Poirier, M. C. Flemings, Met. Trans., 9B (1978) 711.

17. C. J. Jeanfils, J. H. Chen, H. J. Klein, "Modeling of Macrosegregation in Electroslag Remelting of Superalloys" (to be published).

18. S. D. Ridder, F. C. Reyes, S. Chakravorty, R. Mehrabian, J. D. Nauman, J. H. Chen, H. J. Klein, Met. Trans., 9B (1978) 415.

19. H. P. Utech, M. C. Flemings, J. Appl. Phys., 37 (1966) 2021.

20. M. C. Flemings, H. P. Utech, "Process for Making Solids and Products Thereof", U.S. Patent No. 3,464,812 (September 2, 1969).

21. K. A. Jackson, J. D. Hunt, D. R. Uhlmann, T. P. Seward III, Trans.Met. Soc. AIME, 236 (1966) 149.

22. Business Week, August 25 (1980).

23. A. L. Lux, M. C. Flemings, Met. Trans., 10B (1979) 71.

24. A. L. Lux, M. C. Flemings, Met. Trans., 10B (1979) 79.

25. F. E. Goodwin, P. Davami, M. C. Flemings, Met. Trans., 11A (1980) 1777.

26. T. Matsumiya, M. C. Flemings, Met. Trans., 12B (1981) 17.

Subject Index

Author Index